Introduction to Kinesiology

Studying Physical Activity

FOURTH EDITION

Shirl J. Hoffman, EdD

EDITOR

Human Kinetics

Library of Congress Cataloging-in-Publication Data

Introduction to kinesiology / Shirl J. Hoffman, editor. -- 4th ed.
    p. ; cm.
Includes bibliographical references and index.
I. Hoffman, Shirl J., 1939-
[DNLM: 1. Kinesiology, Applied. 2. Motor Activity. 3. Physical Education and Training. WE 103]
612.7'6--dc23
               2012030828

ISBN-10: 1-4504-3432-0 (print)
ISBN-13: 978-1-4504-3432-4 (print)

Permission notices for photographs reprinted in this book can be found on pages xv-xvi.

The web addresses cited in this text were current as of September 2012, unless otherwise noted.

Acquisitions Editor: Myles Schrag; **Developmental Editor:** Amanda S. Ewing; **Assistant Editors:** Kali Cox and Casey A. Gentis; **Copyeditor:** Julie Anderson; **Indexer:** Andrea J. Hepner; **Permissions Manager:** Dalene Reeder; **Graphic Designer:** Nancy Rasmus; **Graphic Artist:** Denise Lowry; **Cover Designer:** Keith Blomberg; **Photograph (cover):** © Greg Epperson/fotolia.com; **Photo Asset Manager:** Laura Fitch; **Photo Production Manager:** Jason Allen; **Art Manager:** Kelly Hendren; **Associate Art Manager:** Alan L. Wilborn; **Art Style Development:** Joanne Brummett; **Illustrations:** © Human Kinetics, unless otherwise noted; **Printer:** Courier Companies, Inc.

Printed in the United States of America   10 9 8 7 6 5 4

The paper in this book was manufactured using responsible forestry methods.

Human Kinetics
Website: www.HumanKinetics.com

United States: Human Kinetics, P.O. Box 5076, Champaign, IL 61825-5076
800-747-4457
e-mail: humank@hkusa.com

Canada: Human Kinetics, 475 Devonshire Road Unit 100, Windsor, ON N8Y 2L5
800-465-7301 (in Canada only)
e-mail: info@hkcanada.com

Europe: Human Kinetics, 107 Bradford Road, Stanningley, Leeds LS28 6AT, United Kingdom
+44 (0) 113 255 5665
e-mail: hk@hkeurope.com

Australia: Human Kinetics, 57A Price Avenue, Lower Mitcham, South Australia 5062
08 8372 0999
e-mail: info@hkaustralia.com

New Zealand: Human Kinetics, P.O. Box 80, Torrens Park, South Australia 5062
0800 222 062
e-mail: info@hknewzealand.com

E5768

For Sharol and Marcel, Dale and April,
Brigitte, Philippe, and Barbara

contents

For Instructors

Writing an introductory text for any discipline is a daunting task, but it is especially so in a young discipline undergoing rapid evolution. Substantial questions about the discipline's nature, future, organization, purpose, name, and relationship to the professions have yet to be answered to the satisfaction of all kinesiologists. In one way or another, all these issues converge in the writing of an introductory text. This text takes bold positions on many of these issues, such as the name for the discipline, the central phenomenon of study, and definitions of critical terms. Issues that seemed less critical for understanding the big picture of kinesiology were purposely kept in the background so as not to distract students from the primary theme of the text.

Even more problematic for authors of introductory texts is the changing demographic land-scape of students enrolling in departments of kinesiology. Gone are the days when physical education professors could confidently assume that all students entering as majors in their departments were planning to be physical education teachers and coaches. Now it is just as likely that students have their sights set on careers in sport management, athletic training, cardiac rehabilitation, fitness counseling and leadership, aquatic leadership, rehabilitative exercise, sport information, adapted physical education, or related professions. Faculty are encountering many students who elect the kinesiology major as preparation for graduate study in physical therapy, chiropractic, podiatry, or medicine. Still others choose kinesiology as a course in liberal studies, without specific professional goals in mind. And of course, a substantial percentage of students come to the introductory course lacking any clear vision of what career they will pursue. The likelihood that students with a diverse range of needs and interests would use this book was very much in mind at every stage of the writing process.

The authors of this text believe that a shared interest in and curiosity about physical activity are what unite students with these different professional aspirations and make kinesiology the logical discipline for them to study. This book emphasizes physical activity as the core element of kinesiology and devotes a great deal of space to analyzing the importance of physical activity in our lives and in the professions of kinesiology. Students who lack insight into the pervasiveness and indispensability of physical activity, or who doubt its worth in their personal or social lives, will not easily be convinced that a field of kinesiology is needed; thus, much of the text is designed to convince them that physical activity is central to our lives. Most students come to an introductory class in kinesiology with a wealth of knowledge and experience about physical activity, but that is no guarantee they will appreciate its variety, complexity, and elegance. Like art educators, who often must teach students to view the world from a fresh and novel perspective, so the teacher of this introductory course must assist students in viewing physical activity in new and exciting ways if they are to appreciate fully the importance of the discipline of kinesiology.

In the fourth edition we have continued to shape the text to students' needs, incorporating helpful comments from instructors' feedback. As in previous editions, the overarching purpose of this text is to foster in students a deep appreciation for the breadth and importance of physical activity, orienting them to the discipline of kinesiology, helping them understand its relationship to physical activity, and introducing them to the physical activity professions. The perspectives on physical activity are both broad and narrow. In part I, the discussion ranges far beyond exercise, sport, or health-related physical activity. Taking such an expansive view of physical activity and kinesiology is not to suggest that the dominant emphasis traditionally given to sport and exercise in kinesiology should be diminished. These topics have been and will continue to be the dominant forms of physical activity studied in the discipline. At the same time, the book will have missed its mark if it fails to expand students' conceptions of physical activity or fails to develop an appreciation for the potential contributions that kinesiology can make to fields far removed from sport and exercise.

Plan of the Text

The text is organized around a new model of kinesiology. In this ambitious undertaking, the discipline is envisioned as dependent on three sources of knowledge: participating in and observing physical activity, studying and conducting research, and delivering professional services. By defining the discipline as an integration of knowledge derived from experience, formal study, and professional practice, the authors hope that students will recognize the critical importance of all three of these dimensions in kinesiology.

Chapter 1 provides an overview of the book, offers key definitions, and presents the model around which the text is organized. The model, devised jointly by co-author Janet Harris and me, is intended to explain to those unfamiliar with the traditions of our field how the field is organized and to help those who already are members of the profession understand how their work fits into the larger whole. Chapters 2 through 4 use a wide lens to view the many ways in which our physical activity experiences contribute to our understanding of kinesiology. Chapter 2 introduces "the spheres of physical activity experience" as a way of helping students appreciate the pervasiveness and importance of physical activity, not only in sport and exercise but also in work, rehabilitation, daily living, and other spheres of existence. Chapter 3 directs attention to the importance of physical activity experiences, concentrating on the way kinesiologists (physical activity experience specialists) design and manipulate physical activity experiences to bring about predetermined ends. Discussions of the benefits of physical activity experiences can, often quite unintentionally, lead students to appreciate only the physical aspects of physical activity. The holistic approach taken in chapter 4 is intended to pique students' interest in the subjective aspects of physical activity. The chapter invites students to consider the deep emotional and spiritual meanings, as well as the many different kinds of knowledge, that they can glean from physical activity.

Part II (chapters 5-11) introduces students to the academic subdisciplines of kinesiology, portraying them as elements in three spheres of scholarly study of physical activity: the biophysical, sociocultural, and behavioral spheres. This section of the text is ambitious in its effort to give adequate coverage to subdisciplines in single chapters written for undergraduate, first-year students. We have benefitted from feedback furnished by instructors who have used the text in its prior edition, and we hope that this edition makes these chapters more "digestible" and accessible to undergraduate students, who may not have in-depth background in the various subdisciplines. Each chapter gives a brief overview of major historical events in the development of the subdiscipline; the research methods used in the subdiscipline; what professionals such as biomechanists, exercise physiologists, and others do in the course of their professional work; and how students' present knowledge can form a foundation for more advanced study. Each chapter presents practical, real-world applications from the subdiscipline and is organized to help readers understand why the subdiscipline is important and how it may relate to a variety of professional endeavors.

The contributors to this text are experienced scholars and professionals who have mastered their fields of study and are keenly interested in teaching students. The roster of authors for part II remains the same except for Kathy Jamieson, who assumed the role of senior author in chapter 7, Sociology of Physical Activity, co-authored with Margaret Duncan. Other returning authors featured in part II include Scott Kretchmar, Richard Swanson, Jerry Thomas, Kathi Thomas, Robin Vealey, Kathy Simpson, and Jennifer Caputo, all of whom have reworked their chapters in an effort to reach the critical audience of this text. When teaching the material in these chapters, I have found it helpful to incorporate lectures from colleagues in my department who are specialists in the relevant subdiscipline. Students usually enthusiastically welcome these visits.

The goal of part III (chapters 12-17) is to stimulate student interest in the professions in which physical activity plays a central role. Throughout this section, authors have kept in mind that many students entering kinesiology departments may have tentative professional goals at best. To this end, chapter 12 takes a general approach to the physical activity professions, introducing students to what it means to be a professional and what will be required of them if they are to become professionals. This chapter also leads students through a series of steps intended to assist them in determining whether their interests, talents, and motivations suit them to a career in kinesiology.

Chapters 13 through 17 include a wealth of information about clusters of careers drawn from the spheres of professional practice. Each chapter focuses on a collection of professional opportunities within a sphere, including health and fitness, therapeutic exercise, teaching physical education, coaching and sport instruction, and sport management. The authors of all these chapters explain the need for professions in the given area and offer a brief historical background of the sphere. In addition, each chapter includes an analysis of work settings in the sphere; up-close views of specific professions; an overview of educational qualifications needed to practice in the sphere; resources, case studies, and profiles of working professionals; and a section offering inside advice to those planning careers in that cluster of professions. All of these chapters have been updated and expanded for the fourth edition.

Returning authors include Jeremy Howell, Sandra Minor Bulmer, Chad Starkey, Kim Graber, Tom Templin, Joe Luxbacher, and Clay Stoldt. A new name appears in the roster of authors in part III: chapter 17 has been co-authored by Mark Vermillion. Brief biographies of all of the authors can be found in the back pages of the text. Again, in teaching the course, I have found that outside visitors representing various career areas of the physical activity industry are valuable adjuncts to the introductory course.

Features of the Text

Schematics guide the organization of the book and, although they may appear redundant, help students keep the big picture in mind as they read individual chapters. Key points and interesting sidebars are helpful supplements to the main text. As in the third edition, interactive exercises throughout the text involve students actively in their learning. These exercises engage students more deeply with the content, helping them put themselves into the picture described in a chapter. The study questions at the end of the chapter also have proven helpful in focusing students' attention on key points in the text. In the final analysis, however, the effectiveness of the text will depend on the creative uses to which it is put by the teacher. Rather than relying on the text as a single resource, teachers should integrate the various topics that it covers with challenging in-class and out-of-class projects that will encourage students to view themselves as professionals-in-training and take the initial steps toward identifying and attaining their career goals.

An addition to this edition of the book is the web study guide. This rich, interactive tool offers students a variety of learning features:

- **Interactive activities.** Using a variety of multimedia experiences, these activities engage students in the material. Scenario-based audio and video activities demonstrate an aspect of kinesiology and then require students to complete an activity based on the demonstration and explanation. Other activities involve drag-and-drop, selecting from a menu, or responding with short answers to photos and scenarios presented.

- **Review of key points.** The online review allows students to test themselves on the key points by using interactive activities. Students can print out the key points as a study aid after completing the activity.

- **Activity feedback.** Students receive targeted information to correct or reinforce their answers to activities.

- **Web searches.** Students can seek out specific information on the Internet and then provide their findings on forms that can be printed out and turned in to you.

This text is also accompanied by a full array of instructor resources:

- **Presentation package plus image bank:** More than 370 PowerPoint slides present the key concepts from the chapters, including select illustrations and tables. The image bank contains all of the figures, tables, and content photos. You can use these images to supplement lecture slides, create handouts, or develop other teaching materials for your classes.

- **Instructor guide:** The instructor guide includes information on how to use the various ancillaries, a syllabus, and chapter-by-chapter files that include a chapter outline with key points, additional student assignments, ideas for additional teaching topics and guest speakers, resources, and answers to the in-text review questions.

- **Test package:** The test package contains 470 questions in true-or-false, multiple-choice, fill-in-the-blank, and essay question formats.

The instructor ancillaries can be accessed at www.HumanKinetics.com/Introduction ToKinesiology.

For Students

If you are like most students, you probably entered college having decided on a major before you knew a great deal about it. You may not have thought much about several questions: Why is the major important? What types of careers do people in the major enter after graduating? What subjects do students study in this major? How is the course of study organized? Even though you have decided on or are considering a major in kinesiology, you may not know much about the field of study and may have only vague ideas about the career opportunities available. Not knowing the answers to these questions at the beginning of your undergraduate experience makes it difficult to put course work in perspective and could leave you questioning the value of many of the courses that you are required to take.

eBook
available at
your campus bookstore
or HumanKinetics.com

We wrote this textbook to address those questions and concerns. The book is intended to help you understand what the discipline of kinesiology is, why it is important, what kinesiology majors study, what types of knowledge you will acquire over the course of 4 years, and what types of careers are available to you. If you read and study it carefully, this book will help you reach an early decision not only about whether you want to major in kinesiology but also about what type of career most interests you. You may see no pressing need to decide on a career at this early stage of your college experience, but as you will see in this book, there are advantages to identifying early the career path that you want to follow.

Kinesiology may be a term that you have never heard before. In fact, you may be enrolled in a major called exercise and sport science, or physical education and exercise science, or something else. One of the confusing things about this major is that the discipline is known by many titles in various colleges and universities. This circumstance isn't all that uncommon for a young and evolving area of study such as kinesiology; the discipline simply hasn't been around long enough for everybody to agree on a single term! Over the past decade, however, more and more scholars and college and university departments have decided to use the term *kinesiology*. For this reason we have selected it as the title of this text.

Because kinesiology integrates knowledge from a variety of sources and draws on theories, concepts, and principles from many other disciplines, students often find it difficult to describe or explain the field to their friends who aren't familiar with it. This text presents the discipline in a new and exciting way. It will help you understand not only the enormous breadth of the discipline but also how the many parts of the discipline are tied together. In the pages that follow, you will learn that physical activity is the centerpiece of the kinesiology discipline. Indeed, this idea may be the most important point in the book. Whether you plan to be a sport administrator, athletic trainer, physical education teacher, or physical or occupational therapist, you should have an interest in and curiosity about physical activity.

Just as the discipline of psychology is the study of human behavior and the discipline of biology is the study of life forms, the discipline of kinesiology is the study of physical activity. But don't interpret *study* to mean merely reading books, writing papers, and taking exams. Kinesiology integrates knowledge from three different yet related sources: your experiences performing and observing physical activity, the formal study of physical activity, and professional practice centered in physical activity. This new approach will help you understand the interrelationships of the parts of the discipline.

How the Text Is Organized

Part I of the text will spark your thinking about physical activity, why it is important, and how it intersects our lives at a variety of different junctures that we call *spheres of physical activity experience*. If you think that sport and exercise are what kinesiologists study, you are correct, but these are not the only forms of physical activity that kinesiologists study. After reading this book you should appreciate the fact that kinesiology also contributes to our understanding and improvement of physical activity performances in venues far beyond the athletic field or exercise gym. It also will become clear that human experience in physical activity is a unique source of knowledge; that humans have a unique ability to plan and engage in physical activity experiences to improve the quality of their lives; that the interaction of our thoughts and emotions (subjective experiences) with physical activity may be as critical as the activity itself; and that the unique contribution of teachers, coaches, fitness leaders, sport managers, athletic trainers, cardiac rehabilitation therapists, and many other related professionals consists of their skill in manipulating physical activity experiences to achieve predetermined objectives.

Whereas part I invites you to reflect on knowledge you have gained through the experiences of participating in and watching physical activity, part II presents broad coverage of the knowledge that we can derive from the scholarly study of physical activity. The chapter titles in this part of the book outline the core body of academic (theoretical) knowledge that all kinesiology students should master before graduation. At your institution a single course may cover each of these areas, or one course may cover two or more areas. To help you better organize this vast body of knowledge in your mind, we have divided it into larger categories called *spheres of scholarly study*. Nationally recognized scholars who are particularly skilled at introducing you to their areas of specialized study have written these chapters. Each chapter is intended only to introduce you to the area of study, not to provide you with all the scholarly knowledge that you will need to master in that area before you graduate.

Part III introduces you to the world of professional practice. Students enrolled in kinesiology departments often have widely disparate career goals, ranging from careers as athletic trainers or physical educators to careers as corporate fitness leaders or cardiac rehabilitation specialists, among many other possibilities. Each of these is a form of professional practice. This part of the book groups the many possible careers into *spheres of professional practice* according to similarities in the types of work and general educational requirements. Outstanding scholars in various areas of professional practice wrote the chapters. These writers are eminently qualified to describe the different types of work that physical activity professionals do, as well as to offer details concerning what you can do to become an outstanding professional.

Features of the Text

The book contains a great deal of information; absorbing it all in a single course can be a challenge. We have designed the text in a way that should help you master the material presented: Key points, interesting sidebars, interactive items, special elements, and some truly exciting photos should make this an enjoyable read. Don't skip over the objectives listed at the beginning of each chapter; they are your map for the pages that follow. Take time to read the many interactive items sprinkled throughout the text. These features will stimulate your thinking and reinforce what you have learned by reading the previous paragraphs. Don't miss the unique opportunity that they offer for learning.

Focus on the key points in every chapter. They are brief summaries of what your authors believe to be the most important points made in the previous pages. Read each key point carefully and think about what you have read in the chapter. You will find also that many of the captions for photographs in this text are information packed. In many cases they constitute key points in their own right. The study questions at the end of each chapter will direct you to what the authors believe are some of the most important points they have covered.

Another don't-miss feature is the web study guide. Boxed cross-references throughout the text trigger you to go to the web study guide, where you'll find a variety of multimedia experiences that'll help you learn, understand, and apply the information in the text.

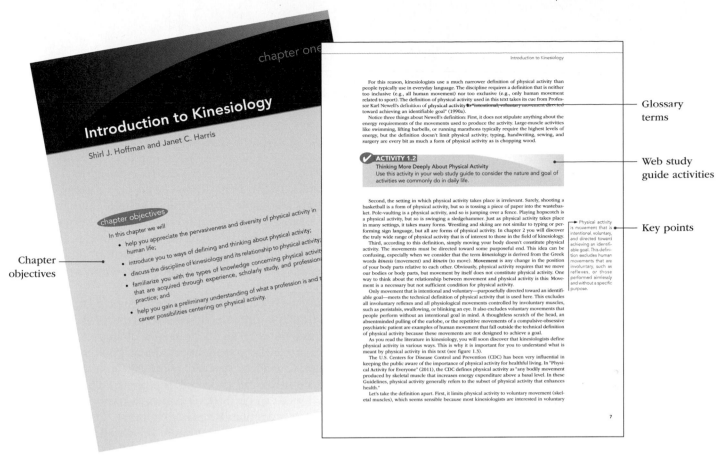

Chapter objectives

Glossary terms

Web study guide activities

Key points

The web study guide material is essential to your complete understanding of the content in your course. Here are the types of activities you'll find in the study guide:

These activities use a variety of multimedia to demonstrate various aspects of kinesiology. Audio, video, drag-and-drop activities, self-ratings, interviews, and more make the content come alive.

Sometimes you'll be asked to search for specific information on the Internet. Then you'll complete an associated activity in a downloadable document that you can submit to your instructor if requested to.

The study guide provides an online review of key points in each chapter and lets you test yourself on how well you understand the material. You can print out the key points as a study aid.

At the end of each of the three main parts of the book, you'll find an activity that allows you to engage with and reflect on the material contained in the chapters in that part. For instance, the part I portfolio activity asks you to reflect on your subjective experience of the Olympic Games.

In the practical plug-ins, you'll be presented with a problem or challenge that a practicing professional faces. You'll consider how the professional might tackle the problem or challenge, and then you'll find out how the scenario was actually resolved. Remember to use these "lessons" as you continue through the course and when you face similar situations or challenges in the field.

The activities called "Day in the Life" are designed so that you can compare your ideas about how people in various professions typically spend their days with reports from actual professionals doing those jobs.

Good luck in your reading, and good luck in your career in kinesiology!

acknowledgments

A project of this sort is the work of many hands. I have been fortunate to play a central role in shaping its direction, a rare and thoroughly enjoyable opportunity. I would compare being editor of a work such as this to having been given the opportunity to drive a new automobile; the pleasure of steering it and guarding its fenders comes only if those at the factory have assembled a roadworthy product. My knowledgeable co-contributors and a team of highly competent people at Human Kinetics have indeed made this text—now in its fourth edition—a roadworthy product.

Although this book features the work of many authors, it isn't an anthology or a compilation: It is a textbook in every sense of the word. It is organized around a central theme; structured to accomplish specific purposes; and integrated in terms, concepts, objectives, and graphics. Each author brings to the project a unique professional and academic background and a track record of achievement in his or her respective specializations, adding a richness and breadth to the text that are largely responsible for the book's success. Achieving consistency and flow and explaining what sometimes can be difficult concepts haven't been easy for the authors, the editor, or the editorial staff at Human Kinetics. We think the fourth edition is much improved; we will wait for your feedback to let us know how well we did. Ultimately the responsibility for pulling together the diverse contributions of authors, providing a structure for the text, ironing out inconsistencies in terms, and interpreting material fell to the editor. Any errors that have been made, whether of commission or omission, are attributable to me, not to others who have given so generously of their time.

At the risk of inadvertently omitting a name I would like to give special acknowledgment to those who made key contributions in the production of *Introduction to Kinesiology.* My co-contributors caught the spirit of the project, sensed its importance, and produced highly readable and informative pages. In this fourth edition they have worked hard to tailor the content to the capabilities of first-year students. The finished product is as much theirs as it is mine. Special credit goes to my colleague Janet Harris, who shared editorial duties for the first edition and whose initial influence continues to help shape the book. The time I found to carry out the editor's role was extracted from the hours I might have spent with my wife, Claude Mourot. Her understanding and steadying influence were more important than she knows.

Only those who have been fortunate enough to author books published by Human Kinetics can appreciate the talents of its publication teams. Those specifically involved with this book are listed on the copyright page, and each did his or her work superbly. Pam Kocher Brown, a faculty member at the University of North Carolina at Greensboro, created the activities in the student study guide. She did an outstanding job of making the material in the book interactive and easier to learn.

photo credits

Page 296	John Albright/Icon SMI
Page 298	Reprinted from W.L. Kenney, J.H. Wilmore, and D.L. Costill, 2012, *Foundations of sport and exercise psychology*, 5th ed. (Champaign, IL: Human Kinetics), 7.
Page 307	© Human Kinetics
Page 316	Monkey Business/fotolia.com
Page 324	© Stanislav Stepanov/Dreamstime.com
Page 338	© Enrique Gomez/Dreamstime.com
Page 343	© Human Kinetics
Page 348	© Human Kinetics
Page 358	Bananastock
Page 362	New York Times
Page 365	Andres Rodriguez/fotolia.com
Page 370	© Human Kinetics
Page 374	© Human Kinetics
Page 388	© Human Kinetics
Page 393	© Human Kinetics
Page 396	Bananastock
Page 398	© Human Kinetics
Page 401	© Human Kinetics
Page 404	© Human Kinetics
Page 409	© Human Kinetics
Page 418	© Human Kinetics
Page 420	© Human Kinetics
Page 429	© Gregory Perkins/Dreamstime.com
Page 436	© Human Kinetics
Page 448	© Human Kinetics
Page 461	© Human Kinetics
Page 478	© Human Kinetics
Page 486	© PA Photos
Page 495	© PA Photos

chapter one

Introduction to Kinesiology

Shirl J. Hoffman and Janet C. Harris

chapter objectives

In this chapter we will

- help you appreciate the pervasiveness and diversity of physical activity in human life;

- introduce you to ways of defining and thinking about physical activity;

- discuss the discipline of kinesiology and its relationship to physical activity;

- familiarize you with the types of knowledge concerning physical activity that are acquired through experience, scholarly study, and professional practice; and

- help you gain a preliminary understanding of what a profession is and the career possibilities centering on physical activity.

You may not have realized it, but when your alarm clock went off this morning and you bounded out of bed, you embarked on a daylong physical activity expedition. Even if you didn't go to the gym or athletic field or engage in hard labor, you did get out of bed, walk to the bathroom, brush your teeth, get dressed, eat breakfast, and make your way to class. After your morning classes you probably ate lunch, visited the library, went back to your room, and cruised the Internet. All of these are forms of physical activity. They are remarkable not only for their complexity but also for their variety. In fact, physical activity is so interwoven into your life that it's probably far easier to count the number of ways in which you weren't physically active today than the ways in which you were.

 ACTIVITY 1.1

Who's More Active?
Use this activity in your web study guide to consider the daily routines of two people and decide which one is more physically active.

If you take a moment to reflect on how physically active you have been, you will see that your life is an endless universe of physical activity. You walk, reach, run, lift, leap, throw, grasp, wave, push, pull, move your fingers and toes, adjust your head for a better line of vision, adjust your posture, and perform thousands of other movements as part of living a normal human existence. Physical activity is essential in your work, whether you perform hard physical labor or low-energy tasks. We use physical activity to express ourselves in gesture, art, and dance.

Even more than this, physical activity is part of human nature. It is an important means by which we explore and discover our world. It helps us define ourselves as human beings. A significant part of our lifetime is spent learning to master a broad range of physical activities, from the earliest skills of reaching, grasping, and walking to enormously complex skills such as hitting a baseball, performing a somersault, or playing the piano. Most of us master a broad range of physical activities at a moderate level of competence. Others concentrate on a limited number of skills, a focus that can lead to extraordinary performances ranging from the incredible consistency with which NBA players can accurately sink baskets from more than 30 feet (9 meters) away to the amazing finger dexterity of a concert pianist.

In this chapter we talk about physical activity in very general terms and explain its relation to the field of **kinesiology**. Taking time to read the chapter carefully will help you appreciate the complexity, diversity, and importance of physical activity to human life. It also will help you understand how the discipline of kinesiology is organized. If you've been physically active throughout your life, you already have some knowledge of physical activity. This background will be of enormous benefit to you as you roll up your sleeves and begin to probe the depths of knowledge of kinesiology. But prior experiences can also hinder your understanding, especially when you are required to think about those experiences in new ways. At times you will have to set your assumptions aside so that you can examine physical activity from a fresh and exciting point of view. This may be more difficult than you might imagine.

Ways of Developing an Understanding of Kinesiology

Because people are now more aware than ever of the importance of physical activity, enrollment in college and university curriculums devoted to the study of physical activity is on the rise. In some universities, the department of kinesiology is one of the largest academic units on campus. There has been an explosion of career opportunities for college-trained professionals with in-depth knowledge of the scientific and humanistic bases of physical activity coupled with training in professional practice. Forty years ago departments of kinesiology (then called "physical education") were designed exclusively for preparing physical education teachers and coaches. Now kinesiology serves as the academic base for a diverse assortment of careers such as physical education teaching and coaching, physical therapy, cardiac rehabilitation, sport management, athletic training, and fitness leadership and management.

With the growing recognition of the importance of physical activity in our daily lives has come the realization that it deserves to be studied just as seriously and systematically as other disciplines in higher education such as biology, psychology, and sociology. No doubt you've heard the word *discipline,* but you may not really understand what it means. A **discipline** is a body of knowledge organized around a theme or focus (see figure 1.1). Disciplines embody knowledge that learned people consider worthy of study. The focus of a discipline identifies what those who work in the discipline study. The central focus of biology, for example, is life forms; the focus of psychology is the mind and mental and emotional processes; and the focus of anthropology is cultures. Although debates are still being waged about the focus of kinesiology, it is now generally regarded as the discipline that focuses on physical activity.

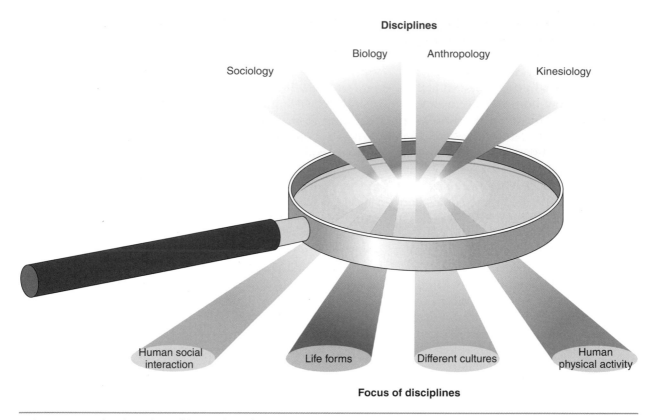

Disciplines

Sociology   Biology   Anthropology   Kinesiology

Human social interaction   Life forms   Different cultures   Human physical activity

Focus of disciplines

FIGURE 1.1   The discipline of kinesiology focuses on human physical activity.

Physical Activity: The Focus of Kinesiology

You may have noticed in the college courses you've taken so far that all disciplines are not learned or studied in the same way. Art, for example, may be studied through reading, writing, and experimentation with artistic projects in the studio. People learn history, literature, and philosophy largely through reading, writing, memorization, and discussion. Reading, writing, memorization, and discussion are also important in learning chemistry and biology, but these disciplines involve active participation in laboratory exercises as well.

People learn kinesiology in three different but related ways. One is by watching or performing, just as students in the disciplines of art or music learn to appreciate art or music by watching and listening or performing. Watching or performing physical activity may not seem as important as learning through scholarly study, but don't underestimate its value in developing your understanding of physical activity. This direct experience with physical activity is an important source of knowledge about physical activity and contributes to your understanding of kinesiology.

A second way of developing an understanding of kinesiology is through systematic scholarly study. This way of learning involves reading about, studying, and discussing with colleagues theoretical and practical matters related to physical activity. It also involves experiences in laboratories. Mastering subjects in the kinesiology curriculum requires these forms of study. Where does the knowledge contained in such subjects come from? Mostly from the work of those in the field of kinesiology who have developed and added to the knowledge base through systematic research and scholarship. Scholars who conduct research in biomechanics, motor learning and control, psychology and sociology of physical activity, exercise physiology, and history and philosophy of physical activity develop important foundational knowledge for the scholarly study of kinesiology.

A third way of learning about physical activity is through professional practice, although here the focus is not so much on learning to *perform* physical activity or on learning *about* it as it is on learning to use physical activity to help those whom you serve in your professional role. Professionals such as physical education teachers, personal trainers, and cardiac rehabilitation specialists, for example, systematically manipulate the physical activity experiences of students, clients, patients, and others whom they serve as a way of helping them achieve personal goals. Ensuring that these goals are achieved requires a specific type of knowledge, but how is this knowledge gained? Some of it comes from applied research that compares the effectiveness of different styles of teaching, different methods of treatment, different types of exercise routines, and other professional regimens. Some of it comes from hands-on experience—a knowledge base developed through helping clients, students, or patients to achieve their goals. And, as we shall see, knowledge gained from experiencing physical activity and the scholarly study of physical activity can be applied to professional practice as well.

> Experiencing physical activity, engaging in scholarly study of physical activity, and conducting professional practice centered on physical activity are the three sources of physical activity knowledge that constitute the discipline of kinesiology.

Exercise therapists, teachers and coaches, athletic trainers, and other physical activity professionals develop a rich knowledge base of physical activity through their personal experience engaging in and watching physical activity and through scholarly study about physical activity. This knowledge can serve as a foundation for professional practice knowledge. If this knowledge is grounded in careful, systematic observations of the effects of manipulations of physical activity on others such as students, patients, and clients, it often becomes incorporated into kinesiology curriculums offered by colleges and universities and is taught to students. As such it also is part of the discipline of kinesiology.

You will notice that we have carefully defined kinesiology as knowledge derived from (1) experiencing physical activity, (2) engaging in scholarly study of physical activity, and (3) conducting professional practice centered on physical activity. Such knowledge becomes part of the discipline of kinesiology only when that knowledge is embedded in a college or university curriculum in kinesiology or is used by kinesiologists in their physical activity research. The reason for this is to clarify precisely what is part of the "official" discipline and what is not. People perform, study, and use physical activity in many venues outside the college or university setting. These activities may be important and valuable in their own right, but they do not constitute practicing kinesiology any more than the use of elementary psychological principles by a businessperson to motivate her sales force constitutes practicing psychology. The discipline of psychology remains tied to the college and university curriculum and to the research of psychologists. Similarly, people may use the principles of kinesiology outside the discipline, but kinesiology per se remains a function of curriculums and research in colleges and universities.

> Only knowledge about physical activity that is included in a college or university curriculum or used in research is part of the body of knowledge of kinesiology.

There is another reason for limiting our definition of kinesiology to knowledge contained in college or university curriculums or used in research. The knowledge that you will acquire in your major curriculum is more highly organized and more scientifically verifiable than the knowledge of physical activity that laypersons use. Universities use rigorous methods to monitor the authenticity of the knowledge included in their curriculums and in the research of their faculty. Think about it: Would you have more confidence in the recommendations of a university kinesiologist who specializes in fencing than those of a police officer who fences as a hobby? Would you have more confidence in the scientific accuracy of recommendations for exercise programs offered by an exercise physiologist than in those offered by a television exercise guru who lacks formal training in kinesiology? Would you be more likely to trust the recommendations of a university specialist in pedagogy concerning how to organize a large group of young children for physical activity instruction than those of a volunteer coach who has no formal training in kinesiology? Given your decision to invest several years of hard work in preparing for a career in the specialized field of kinesiology, we might presume that your answer to all three questions is yes.

Many people experience physical activity (e.g., walk through a supermarket collecting groceries, play pickup basketball), study physical activity (e.g., read popular trade books on fitness or sports), or engage in some form of "professional" practice (e.g., volunteer as a youth league coach) outside the confines of the university curriculum, but these may not constitute practicing kinesiology and these people do not become "kinesiologists" under the strict definition being used in this text.

Later in this chapter you will learn how all three components of physical activity fit together to make up the discipline of kinesiology. People gain the most complete knowledge of kinesiology from engaging in all three components—experience, scholarly study, and professional practice—although knowledge from each component is valuable in its own right. These three components are depicted in figure 1.2. This figure appears throughout the book to remind you of the component of physical activity being discussed in the text at that point.

FIGURE 1.2 Sources of knowledge in kinesiology.

What Is Physical Activity?

You may think it is silly to ask what is meant by *physical activity*. Everybody knows what it is, so why waste time defining it? But definitions are important, especially in scientific and professional fields where terms often are defined somewhat differently than they are in everyday language. These **technical definitions** ensure that people working within a science or profession have a common understanding. You may be surprised to learn that kinesiologists do not all agree on the technical definition of *physical activity*. Thus, before we go much further, let's ensure that we all have the same understanding of what physical activity is.

In everyday language, almost any muscular action is considered physical activity. Throwing a javelin, driving a car, walking, performing a cartwheel, swimming, digging a ditch, hammering a nail, scratching your head—all are examples of physical activity, as are the kick you exhibit when the doctor taps your patellar tendon to test your reflexes, and the blinking of your eye; some would even describe the peristaltic action of your small intestine brought on by muscular contractions or the contraction of your diaphragm when you sneeze as physical activity. But are all these muscular actions of equal concern to kinesiologists? Not really. Although all are examples of human movement, they're too diverse in form and purpose for any single discipline to study. Indeed, if kinesiology focused on all forms of human movement, then kinesiologists would study everything that humans do, because living is moving (see the sidebar Physical Activity and Doing)!

Defining Physical Activity

Although we are all familiar with physical activity, we rarely think deeply about it. Before you continue reading this chapter, think carefully about the following questions. Take time to compare your answers with those of your classmates.

1. How would you define physical activity?
2. What do you consider the most important characteristics of physical activity?
3. What methods can you think of that kinesiologists might use to study physical activity?
4. Name five professions that use physical activity.

After you have thought about or discussed these questions, rewrite your definition of physical activity. Keep your answers handy so you can see how they compare with the discussion that follows.

For this reason, kinesiologists use a much narrower definition of physical activity than people typically use in everyday language. The discipline requires a definition that is neither too inclusive (e.g., all human movement) nor too exclusive (e.g., only human movement related to sport). The definition of physical activity used in this text takes its cue from Professor Karl Newell's definition of **physical activity** as "intentional, voluntary movement directed toward achieving an identifiable goal" (1990a).

Notice three things about Newell's definition: First, it does not stipulate anything about the energy requirements of the movements used to produce the activity. Large-muscle activities like swimming, lifting barbells, or running marathons typically require the highest levels of energy, but the definition doesn't limit physical activity; typing, handwriting, sewing, and surgery are every bit as much a form of physical activity as is chopping wood.

✔ ACTIVITY 1.2

Thinking More Deeply About Physical Activity
Use this activity in your web study guide to consider the nature and goal of activities we commonly do in daily life.

Second, the setting in which physical activity takes place is irrelevant. Surely, shooting a basketball is a form of physical activity, but so is tossing a piece of paper into the wastebasket. Pole-vaulting is a physical activity, and so is jumping over a fence. Playing hopscotch is a physical activity, but so is swinging a sledgehammer. Just as physical activity takes place in many settings, it takes many forms. Wrestling and skiing are not similar to typing or performing sign language, but all are forms of physical activity. In chapter 2 you will discover the truly wide range of physical activity that is of interest to those in the field of kinesiology.

Third, according to this definition, simply moving your body doesn't constitute physical activity. The movements must be directed toward some purposeful end. This idea can be confusing, especially when we consider that the term *kinesiology* is derived from the Greek words *kinesis* (movement) and *kinein* (to move). **Movement** is any change in the position of your body parts relative to each other. Obviously, physical activity requires that we move our bodies or body parts, but movement by itself does not constitute physical activity. One way to think about the relationship between movement and physical activity is this: Movement is a necessary but not sufficient condition for physical activity.

Only movement that is intentional and voluntary—purposefully directed toward an identifiable goal—meets the technical definition of physical activity that is used here. This excludes all involuntary reflexes and all physiological movements controlled by involuntary muscles, such as peristalsis, swallowing, or blinking an eye. It also excludes voluntary movements that people perform without an intentional goal in mind. A thoughtless scratch of the head, an absentminded pulling of the earlobe, or the repetitive movements of a compulsive-obsessive psychiatric patient are examples of human movement that fall outside the technical definition of physical activity because these movements are not designed to achieve a goal.

As you read the literature in kinesiology, you will soon discover that kinesiologists define physical activity in various ways. This is why it is important for you to understand what is meant by physical activity in this text (see figure 1.3).

The U.S. Centers for Disease Control and Prevention (CDC) has been very influential in keeping the public aware of the importance of physical activity for healthful living. In "Physical Activity for Everyone" (2011), the CDC defines physical activity as "any bodily movement produced by skeletal muscle that increases energy expenditure above a basal level. In these Guidelines, physical activity generally refers to the subset of physical activity that enhances health."

Let's take the definition apart. First, it limits physical activity to voluntary movement (skeletal muscles), which seems sensible because most kinesiologists are interested in voluntary

> ➤ Physical activity is movement that is intentional, voluntary, and directed toward achieving an identifiable goal. This definition excludes human movements that are involuntary, such as reflexes, or those performed aimlessly and without a specific purpose.

Too inclusive
"All movements, voluntary or involuntary, performed by humans."

Too exclusive
"Voluntary movement intentionally performed for specific purposes; requires a substantial amount of energy."

Just right!
"Voluntary movement intentionally performed to achieve a goal in sport, exercise, or any other sphere of life experience."

FIGURE 1.3 Definitions of physical activity. Only movement that is voluntary and goal oriented meets the technical definition of physical activity.

rather than involuntary movements of the body. At the same time the definition seems too narrow, because it limits the definition of physical activity to only movements that increase energy expenditure above a base level.

Obviously, kinesiologists are interested in high-energy forms of physical activity such as running, lifting weights, and exercising, but kinesiologists also study skilled movement that may require little energy expenditure. For example, physical education teachers and coaches often teach bowling or archery; occupational and physical therapists may treat stroke patients using gentle movements or teach elderly residents of a nursing home how to use a knife and fork, to bathe, or to comb their hair. Motor control researchers may be interested in measuring fine motor skills or reaction times or the time required to move one's hand a short distance under different experimental conditions. Obviously, none of these requires expenditure of large amounts of energy. Thus, the CDC's definition seems too restrictive because it eliminates many types of physical activity that interest kinesiology scholars and professionals.

Physical Activity and "Doing"

The distinguished neurophysiologist Sir Charles Sherrington wrote the following in his classic book, *Man on His Nature:* "All [humans] can do is to move things, and [their] muscle contraction is [their] sole means thereto" (Sherrington 1940, p. 107). What do you think Sherrington meant when he said that all we can do is "move things"? Can you think of anything you can do without moving? What does your answer imply about the importance of physical activity in our lives? (Hint: What is the dictionary definition of *doing*?)

Of course we could limit our technical definition of physical activity to include only those physical activities that relate directly to exercise and sport. Because kinesiology has its historical roots in exercise and sport and because most students entering the field have their sights set on a career in exercise or sport, this definition may be appealing.

But even though sport and exercise receive the most attention in the field of kinesiology—and they certainly take center stage in this text—kinesiology embraces a far wider range of physical activity than sport and exercise. For example, kinesiologists study basic postural mechanisms, the physiology and body mechanics of work, the development of reaching and grasping behaviors in infants, and the daily life support activities of elderly people. Physical education teachers teach children how to perform fundamental movement patterns such as hopping, running, and skipping or expressive physical activities such as dance; and therapists working in rehabilitation programs teach patients to recover lost capacities to walk, sit, rise from a chair, or drive a car. None of these strictly relate to sport or exercise. As the range of occupations pursued by graduates of kinesiology departments continues to expand, clearly a technical definition of physical activity that reaches beyond exercise and sport is in order.

Context-Based Definitions

Sometimes kinesiologists focus so closely on the specific area of physical activity in which they work that they define physical activity only in terms that make sense to them. Physiologists concerned about the health ramifications of not engaging in daily vigorous exercise wrote the definition promoted by the CDC and therefore included the qualification about energy expenditure. (Health benefits of physical activity are roughly proportional to energy levels.) How might coaches using only their focused experiences of physical activity define physical activity? Physical therapists? Dance teachers?

What Is Kinesiology?

Kinesiology is a discipline or body of knowledge that focuses on physical activity. To reiterate, the discipline derives and incorporates knowledge from three different yet related sources:

- Experiencing (or doing) physical activity (experiential knowledge)
- Studying the theoretical and conceptual bases of physical activity (theoretical knowledge)
- Engaging in **professional practice centered on physical activity** (professional practice knowledge)

We have already mentioned that this book is organized around the unique ways in which we learn about physical activity. Not surprisingly, these also are the ways in which we approach the study of kinesiology. Typically, we tend to associate learning with reading, writing, discussion, and memorization. We have seen, however, that experience with physical activity, theories and concepts about physical activity, and professional practice all are important sources of knowledge for kinesiologists. They are also important sources of knowledge for you as a student studying kinesiology. For this reason, this text has been divided into three parts, each of which examines a distinct source of knowledge that is incorporated into the discipline of kinesiology.

Figure 1.4 depicts the three major dimensions of kinesiology. They will help you envision in a new way what you have learned. Each dimension corresponds to a different source of knowledge of physical activity as shown in figure 1.2. The part of the figure marked A1 represents knowledge acquired from our hands-on experiences with physical activity. Usually, this involves performing physical activity, but we can also acquire knowledge by observing others perform. Students often acquire this **experiential knowledge**, described in part I of this book, through physical activity classes (e.g., classes in soccer, weight training, swimming) offered in kinesiology departments. The experiential knowledge included in a formal college or university kinesiology curriculum is part of the discipline of kinesiology. This experiential knowledge hasn't yet been divided into formal subdisciplines within the field of kinesiology, although each form of activity represents a specialized form of experience.

Of course, experiential knowledge can be acquired outside a formal college or university curriculum; this kind of experiential knowledge is represented by part A2 of figure 1.4. This

A1 Knowledge gained through experiencing physical activity that is systematically incorporated into the discipline of kinesiology (e.g., college tennis class offered for academic credit, weight-training class offered for academic credit).

A2 Knowledge gained through experiencing physical activity that is not incorporated into the discipline of kinesiology (e.g., learning a dramatic stage movement, taking tennis lessons at a country club, playing Little League baseball, firefighting).

B1 Knowledge gained through scholarly study of physical activity that is systematically incorporated into the discipline of kinesiology (e.g., sport history, exercise physiology, motor development).

B2 Knowledge gained through scholarly study of physical activity that is not systematically incorporated into the discipline of kinesiology (e.g., research about playing a musical instrument, reading a popular book on fitness).

C1 Knowledge gained through professional practice centered in physical activity that is systematically incorporated into the discipline of kinesiology (e.g., knowledge gained in roles such as certified athletic trainer or elementary physical education teacher that is included in university kinesiology classes).

C2 Knowledge gained through professional practice centered in physical activity that is not systematically incorporated into the discipline of kinesiology (e.g., knowledge gained in roles such as certified athletic trainer or physical education teacher that is not included in university kinesiology classes).

FIGURE 1.4 The three major dimensions of kinesiology.

knowledge may be on the fringes of the discipline of kinesiology or not part of the discipline at all. This knowledge becomes a central part of the kinesiology discipline when it is formally incorporated into college and university kinesiology classes. These aspects of experiencing physical activity are also discussed in part I of this text.

Part B1 of figure 1.4 represents disciplinary knowledge acquired from the scholarly study of theories and concepts about physical activity. This **theoretical knowledge** usually is divided into categories or **subdisciplines** in university curriculums. Taken together, the subdisciplines constitute the spheres of scholarly study of physical activity examined in part II of this book. This disciplinary knowledge is derived from research and scholarly study and is taught in its most systematic and comprehensive form in college and university curriculums, usually in departments of kinesiology.

Because physical activity is such a broad category of human behavior, however, other university departments sometimes engage in the scholarly study of physical activity as well, as represented by part B2 in figure 1.4. For example, departments of drama, dance, and music teach about the scholarly aspects of physical activity, as do departments of engineering and medicine. This knowledge, however, is considered to be on the fringes of the discipline of kinesiology. Departments of kinesiology are the only academic units within colleges and universities that identify the study of physical activity as their sole mission and offer an organized curriculum to teach it. You surely have noticed that bookstores stock many books and magazines that deal with sports and fitness. People who lack in-depth education in kinesiology wrote many of these books, which therefore may have no basis in science or systematic analysis. Thus, we cannot consider them part of the discipline of kinesiology.

Part C1 of the model in figure 1.4 represents disciplinary knowledge acquired from professional practice centered on physical activity—such as managing a fitness center, teaching physical education, engaging in personal training, or working in cardiac rehabilitation. This **professional practice knowledge** is part of the discipline when the knowledge is discovered or tested in preprofessional or professional settings and is incorporated by faculty into college and university kinesiology classes. Professional practice knowledge usually deals with appropriate ways of manipulating physical activity experiences to bring about specific results. For example, personal trainers may manipulate the exercise experiences of their clients in one way to increase muscular strength and in another way to increase flexibility. Part III of this book describes this component of physical activity knowledge in detail.

Of course, not all knowledge about how to coach or teach or perform other types of practice is part of professional practice knowledge. For example, a volunteer coach who is planning drills for her team may rely less on the latest research on pedagogy, skill learning, and fitness and more on knowledge gained from people who coached her in the past. Such knowledge is often flawed. For this reason, some of the knowledge acquired by practitioners is not incorporated into the discipline. To indicate this, such knowledge is located on the perimeter of C2 in figure 1.4.

Professional practice knowledge is most valuable when combined with knowledge from the other dimensions of kinesiology (experience and scholarly study). Together, these can provide an important framework for conceiving of and using knowledge about professional practice. Sometimes people who possess only a fraction of the disciplinary knowledge (as represented in parts A1, B1, and C1) are hired to perform professional roles anyway. This can occur because the demand for physical activity professionals is so great that companies and institutions often hire people who lack adequate qualifications. School districts, for example, sometimes hire coaches who have little or no background in kinesiology. Likewise, fitness centers may hire personal trainers who lack adequate qualifications. Obviously, one can develop a modest level of competency in almost any profession without mastering the total body of knowledge. Through trial and much error, such a person may muddle through. This approach, however, can be dangerous. For example, you may have known or heard about

> The discipline of kinesiology consists of experiential knowledge, theoretical knowledge, and professional practice knowledge. Experiential knowledge derives from experiencing physical activity, theoretical knowledge derives from systematic research about physical activity, and professional practice knowledge derives from and contributes to the process of delivering physical activity services.

laypersons who managed to learn enough about the law to represent themselves in court. But such a person runs a high risk of making a mistake. (This is the basis for the old saying "He who chooses to represent himself in court has a fool for a client.") Similarly, an unqualified individual who assumes the role of a physical activity professional lacks the informed judgment of one who has studied kinesiology.

Let's quickly review our description of kinesiology. It is a discipline or body of knowledge that concentrates on physical activity. The knowledge of kinesiology is acquired through experience (watching or performing physical activity), scholarly study of theories and concepts about physical activity, and professional practice centered in physical activity. Because kinesiology is a formal discipline, it typically is taught in a college or university by people highly trained in certain areas of specialization, whether this specialization is in performance, scholarly study and research, or professional practice. Thus, faculty who are able to instruct you about the broad knowledge base of the discipline teach in departments of kinesiology. The curriculums of these departments represent organized sequences of classes that are structured in ways to ensure maximal opportunities for integrating all three types of knowledge.

American Kinesiology Association's Definition of Kinesiology

The American Kinesiology Association (AKA) defines kinesiology as "the academic discipline which involves the study of physical activity and its impact on health, society, and quality of life." As a discipline, kinesiology draws on several sources of knowledge including personal and corporate physical activity experiences, professional practices centered on physical activity, and scholarly study and research of physical activity itself. Although the discipline is most often associated with the third source, scholarly study and research, the AKA recognizes that the body of knowledge of kinesiology is informed by and defined by the other two sources as well. Ultimately, the uniqueness of kinesiology as a discipline is its embrace and integration of a multidimensional study and application of physical activity—biological, medical, and health-related aspects but also psychological, social-humanistic, and a variety of professional perspectives as well. Although individual departments may choose to shape their curriculums and research

Reprinted, by permission, from American Kinesiology Association.

agendas around select aspects of the discipline, such institutional preferences should not be interpreted as a complete and comprehensive definition of the discipline (www.AmericanKinesiology.org).

The Focus of Kinesiology: Exercise and Skilled Movement

Kinesiology focuses on two general categories or forms of physical activity: exercise and skilled movement (see figure 1.5). You know much about both forms of physical activity. In fact, your interest in one or both of these is probably responsible for your decision to seek a degree in kinesiology.

▶ Kinesiology is a discipline that focuses on physical activity, especially two categories of physical activity: exercise and skilled movement.

Exercise

Exercise is one principal form of physical activity. People engage in **exercise** to (1) improve athletic, military, or some other kind of physical performance, (2) improve their health, or (3) regain performance that has been reduced as a result of injury or disease. Because exercise involves many different types of physical activity, breaking the term down into three major categories is helpful:

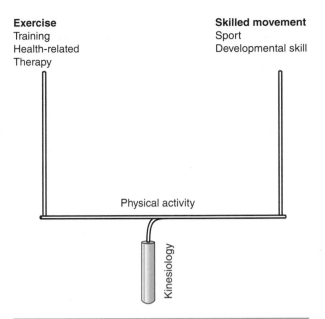

FIGURE 1.5 The two principal categories of physical activity studied by kinesiologists.

- **Training** is exercise performed for the express purpose of improving athletic, military, work-related, or recreation-related performance (see chapter 3). Kinesiology graduates who embark on careers as conditioning specialists for university or professional sport teams are specifically concerned with physical activity as training.

- **Health-related exercise** is undertaken specifically to develop or maintain a sound working body, free of disease and able to perform daily tasks and deal with emergencies. Kinesiology graduates who work as fitness leaders and personal trainers focus almost exclusively on physical activity as health-related exercise.

- **Therapeutic exercise** (or therapy) is performed to restore capacities previously acquired or developed that have been lost due to injury, disease, or behavioral patterns. For example, patients who have had a heart attack usually require physical activity regimens to help them regain cardiovascular health. Kinesiology graduates who embark on careers to work in these cardiac rehabilitation programs, or who work as athletic trainers or physical therapists, focus on physical activity as therapy.

Skilled Movement

Skilled movement is the second area of focus of kinesiology. It is seen in performances in which accuracy of direction, force, and rhythm or timing are essential to accomplishing predetermined goals. Normally, people learn these qualities of physical activity through systematic practice rather than by exercising. Although strength, cardiorespiratory endurance, and flexibility may be important foundational elements upon which skilled movement can be developed, they themselves are not elements of skilled movement. (The differences between the development of these qualities and skilled movement are covered in greater detail in chapter 3.) Two categories of skilled movement are of primary interest to kinesiologists.

Sport involves skilled movement that is organized in game contexts. In sports, players try to achieve a goal in a manner specified by rules, usually in competitive contexts. Note three things about this definition. First, the physical activity in sport is "skilled," which means that it is performed efficiently and effectively. Not all forms of physical activity require a great deal of skill, but in every type of sport the advantage belongs to competitors who have learned to move their bodies in skillful ways. The soccer player who passes the ball deftly to her teammate, the golfer who strikes the ball squarely, and the gymnast who successfully completes a double rotation on dismount all are expressing skill in their performances. Second, rules are essential in sport. Rules provide a level playing field for competitors, and they are essential

> Exercise is one of two major forms of physical activity; skilled movement is the other. Exercise usually is one of three types: training for the purpose of improving performance in sport or other types of physical performances, health-related activity to maintain or improve health, or therapy to rehabilitate individuals from debilitating disease or injury. Skilled movement, the second major form of physical activity, includes physical activities in which accuracy of direction, force, and rhythm or timing are essential to attaining predetermined goals. In kinesiology, skilled movement is studied in sport and in developmental skills.

for organizing games. Without rules, players could do whatever they felt like doing, and the game would become chaotic. Third, note that the physical activities performed in sport tend to be framed in competition, against other teams, against individuals, against established records, or against "personal bests."

Developmental skills involve skilled movements that are not usually performed in sport settings. For example, elementary physical education teachers teach 1st graders how to perform such fundamental movement patterns such as skipping, throwing, or hopping. Acquiring these developmental skills at an early age may lead to high levels of proficiency in sport in later years, but when they are taught they have no direct correlation with a specific sport. Likewise, graduates of kinesiology programs who pursue careers in occupational or physical therapy or who work in nursing home facilities as physical activity specialists may spend a great deal of time teaching poststroke, disabled, or elderly patients the developmental skills required to eat, dress, and groom. The range of developmental skills is enormous. Improving walking and running, improving grasping or reaching, and improving the efficiency of movements used in industrial or military settings all fall within the purview of kinesiology. The airplane pilot, NASCAR driver, seamstress, orthopedic surgeon, and dentist all rely on developmental skills. In chapter 2 we explore this broad range of physical activity and discover how important it is in our daily lives.

Obviously, these categories of physical activity are not mutually exclusive. Some people engage in exercise and sport simultaneously. For example, you might compete in racquetball

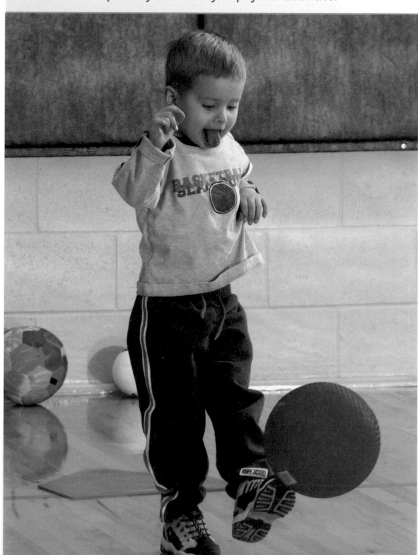

| Learning developmental skills such as kicking helps prepare this child for competency in a variety of physical activities.

with the hope of becoming good enough to win your city's championship but also to improve your body's functioning or appearance. Individuals might participate in judo competition because they enjoy it but also because of the health benefits it brings. Use the categories as guides to understanding and appreciating the types of physical activity that concern kinesiologists, not as hard and fast distinctions.

Kinesiology and Your Career

One thing that makes kinesiology an exciting field is that it opens the door to a wide range of careers. Because undergraduate programs expose you to broad knowledge about physical activity, they are excellent preparation for careers as fitness leaders and consultants, teachers and coaches, cardiac and neuromuscular rehabilitation specialists, sport management specialists, athletic trainers, strength training specialists, and numerous other physical activity professions. Undergraduate programs in kinesiology also serve as critical preparation for students who wish to further their studies in physical therapy, occupational therapy, podiatry, chiropractic, and medicine. Some students study kinesiology as a liberal studies subject. These students may have no immediate plans to enter a physical activity profession but simply are interested in learning about physical activity.

Many undergraduates pursue master's degrees in kinesiology after graduation (see figure 1.6). Some, like those preparing to be physical education teachers, do so to become more knowledgeable about their profession and to meet special certification requirements. Others pursue graduate work to meet the educational requirements of a profession that is related to but not normally included in kinesiology. Increasingly, master's degrees are becoming the minimal requirement for entering some professions. Physical therapy and athletic training,

FIGURE 1.6 An undergraduate degree in kinesiology can be the starting point for many different physical activity careers.

for example, value the knowledge that students acquire in undergraduate kinesiology programs as critical **preprofessional** preparation for entering these fields. Sometimes students continue their studies beyond the master's level to obtain the doctoral degree in kinesiology so that they can become college or university faculty members or researchers.

Professionals trained in kinesiology share a common interest in and curiosity about physical activity in its broadest dimensions. Physical education teachers may entertain the broadest perspective of physical activity because of the enormous breadth of their teaching assignments, which can range from fitness and sport skills to dance and developmental skills. In addition, these practitioners may have a special interest in using physical activity to develop social responsibility and other desirable personal traits in children. Other practitioners may have much more narrow interests.

For example, fitness leaders are most interested in cardiorespiratory fitness, strength, and exercise that involves large muscles of the body. These practitioners are especially interested in how such activities can alter the physiological functioning of the body. Professionals working in athletic training or rehabilitation exercise are particularly interested in physical activity as a medium of rehabilitation. Sport marketers have a different orientation to physical activity altogether: how physical activity performances by elite athletes can be made more appealing to paying audiences. Thus, in addition to developing an understanding of and appreciation for physical activity, you should seek to understand physical activity in the context of whatever specialized professional practice you choose to enter.

Why *Kinesiology*?

Scholars have debated at length the label that best characterizes an academic discipline focused broadly on physical activity. For many years, the label *physical education* was considered appropriate for the rather limited mission of kinesiology departments—preparing school physical education teachers and coaches. Today, physical education is generally understood to refer to broad-based *instructional programs* in sport and exercise offered by public schools, colleges, and universities. Most colleges and universities continue to sponsor an activity program (sometimes called "the basic instruction program") for the general student body. *Physical education* is also the name for the academic specialization within many kinesiology departments that prepares teachers of these programs. (Often this is known as PETE—physical education teacher education.)

Because most departments originally referred to as "physical education departments" now prepare students for a panorama of careers in physical activity, the term *physical education* fails to capture the broad essence of the field. Many names have been proposed for the discipline; but the emerging label, and the one used in this text, is *kinesiology*. Your department might be called exercise and sport science, kinesiology, human performance, health and human performance, human movement science, sport studies, exercise science, physical education, or any of the following departmental names:

- Department of physical education
- Department of health, physical education, recreation, and dance
- Department of physical education, health, and leisure sciences
- Department of physical education and fitness
- Department of exercise and health science
- Department of sport science and physical education
- Department of kinesiology
- Department of exercise and sport science
- Department of physical education and movement science
- Department of movement sciences and leisure studies
- Department of food, nutrition, and exercise science

- Department of human movement studies
- Department of sport studies

This diverse range of names can be confusing. Older fields such as anthropology, sociology, psychology, and history don't have this problem. Because kinesiology is a young, evolving discipline and because disciplines often need a long time to define themselves, it isn't surprising that a variety of names still are used to identify the field. Variation in departmental names also stems from the fact that in some departments—usually at smaller colleges—kinesiology is but one of several different majors offered to students. Thus, a department that also prepares health educators or recreation specialists may call itself "kinesiology and leisure studies" or "health, exercise science, and recreation" to reflect the multiple degree programs that are offered. At one large southern U.S. university, kinesiology degrees are offered in the Department of Nutrition, Food, and Movement Sciences, along with two additional degrees. Thus, you should be careful not to confuse the title of departments in particular universities with the title of the field as it is coming to be known across the United States.

Although none of the various names used for our field are wrong, most scholars emphasize the need for a single term broad enough to describe the discipline. Although not all scholars believe that *kinesiology* is the best name for the discipline, we believe that it is. So, regardless of the name of the department in which you are enrolled, we encourage you to refer to the discipline as kinesiology.

Surging Support for *Kinesiology*

At last count, there were more than 170 departments of kinesiology in colleges and universities in the United States. Most of the departments at prestigious doctoral universities have adopted the term *kinesiology* because it best characterizes a discipline that deals with many different forms of physical activity in diverse professional settings. Support for the label *kinesiology* has come from the National Academy of Kinesiology, an honorary society of approximately 140 scholars. The organization, founded in 1930 as the American Academy of Physical Education, changed its name in 1993 to American Academy of Kinesiology and Physical Education. In October 2010, the name was again changed to National Academy of Kinesiology in recognition of the increasing popularity of the term *kinesiology*. An organization of college and university faculty known as the National Association for Kinesiology and Physical Education in Higher Education changed its name recently to National Association of Kinesiology in Higher Education to reflect the growing acceptance of *kinesiology* as a name for the discipline. Also of note is the establishment of the American Kinesiology Association in 2008, a consortium of kinesiology departments in North America. According to its mission statement, AKA "promotes and enhances kinesiology as a unified field of study and advances its many applications . . . by advocating for kinesiology at national and international levels as well as by supporting its member departments by providing resource materials and leadership and educational opportunities for university administrators in kinesiology" (www.americankinesiology.org/About-us).

Sources of Confusion About the Label *Kinesiology*

One could argue that the best term for our field is *academic kinesiology,* to distinguish the field from some forms of alternative medicine that use *kinesiology* in their titles. For example, "applied kinesiology" is an alternative medicine that uses muscle testing to determine the course of therapy. Practitioners of chiropractic often use applied kinesiology. "Dental kinesiology" applies virtually the same techniques in analyzing the motions of the jaw and tongue. "Behavioral kinesiologists" treat patients using biofeedback, and "spiritual kinesiologists" claim to promote healing by uniting body and soul. To further complicate matters, courses in functional anatomy that teach students how muscles and bones produce specific movements are sometimes called "clinical kinesiology." So far these alternative meanings of kinesiology have not proved to be problematic. On its website (www.americankinesiology. org), the American Kinesiology Association has clarified the differences between academic kinesiology and the other forms of kinesiology in this statement:

> "The academic discipline of kinesiology is taught and researched in colleges and universities and is to be differentiated from Applied Kinesiology and other fields that use the term 'kinesiology' (dental, spiritual, holistic, bio-spiritual), some of which lack grounding in the scientific study of physical activity. The AKA supports the academic discipline of kinesiology."

Allied Fields

Kinesiology has historical bonds with disciplines such as health, dance, and recreation. Two of the largest professional organizations concerned with physical activity are the American Alliance for Health, Physical Education, Recreation and Dance (AAHPERD) and the Canadian Association for Health, Physical Education, Recreation and Dance. The membership of these organizations includes not only people with a central interest in exercise and sport but also those working in the fields of health, recreation, and dance.

Health problems arising from inactivity are of interest to physical activity professionals as well as health professionals, but health professionals are also interested in problems that have little to do with physical activity, including issues related to sexually transmitted diseases, HIV (human immunodeficiency virus), smoking cessation, and drug abuse.

Recreation specialists are interested in physical activity as a leisure pursuit, but leisure pursuits stretch far beyond those in which physical activity is a primary concern. Travel, crafts, and nature study are examples.

Dance professionals are members of professional organizations, and although dancers are interested in physical activity, they are interested in a much narrower slice of the physical activity pie. Expressive and artistic forms of movement form the centerpiece of the field of dance.

Combining this assortment of fields into a single professional organization may seem strange, but they have a long, historic relationship. For most of the 20th century these areas were represented in a single college or university department across the land; in some cases they are still housed within the same department. Today, each field has become more specialized and isolated; each has its own professional organizations and journals. In many cases, each is taught in a separate department. But because of their shared histories and because they are often interested in the same problems, faculty from these areas frequently work together as colleagues.

Holistic Nature of Kinesiology

Although the bodily aspects of kinesiology receive most of the attention, in a kinesiology program it is important to remember that human beings are multidimensional creatures with interrelated cognitions, emotions, body, and soul.

When you complete this introductory study of kinesiology, we hope that you will be convinced of the holistic nature of physical activity. **Holism** refers to the interdependence of mind, emotions, body, and spirit. Although some people think kinesiology deals exclusively with the body and bodily movement, in reality it has a much broader reach. Physical activity involves our minds, emotions, and souls as much as it does our bodies. It is convenient to speak of *physical* activity because the physical aspects are so easily observed, but physical activity also is *cognitive* activity, *emotional* activity, and even *soul* activity.

Thus, studying kinesiology will take you far beyond the study of the biological aspects of physical activity. Kinesiologists study the psychological antecedents and outcomes of physical activity; the sociological, philosophical, and historical foundations of sport and physical activity; the dynamics of skill development, performance, and learning; and the human processes involved in the teaching and learning of physical activity.

▌Experiencing the beauty of the world around us speaks to our holistic nature, which is an important focus of the kinesiology discipline.

Think about the last time you went for a run or a brisk walk. This physical activity may have seemed so easy as you let your mind wander, enjoying the scenery of a lake or the beauty of a mountain trail, that you underestimated precisely how complex its underlying physiology and psychology were. Physical activity is so much a part of our everyday lives that we easily forget how wonderfully complex the human body is. Most students who enroll in kinesiology programs are curious about the functioning of the human body. In your kinesiology program you'll study anatomy and physiology along with taking advanced courses in exercise physiology, biomechanics, sport and exercise psychology, and motor behavior. These classes will help you understand and appreciate the many mechanisms and systems that enable our bodies to perform physical activity.

You may also be required to take a course in philosophy of physical activity. If so, you are fortunate, since this course will help you understand how truly holistic physical activity is. By thinking philosophically you will come to understand more fully the personal meaning you discover in physical activity. This thinking also will sensitize you to the myriad ethical issues associated with sport, exercise, and other types of physical activity. And through your study of psychology and sociology of physical activity you will learn to appreciate how attitudes and social settings influence our interpretation of the significance of our physical activity experiences. Although it is easier to think about our bodies or body parts such as the heart, muscles, or bones as though they were machines or instruments that our minds or souls "use" to achieve our purposes, our bodies are an indivisible part of our humanity. It may not be an exaggeration to say that no other discipline is so diverse in its aims, so interdisciplinary in its subject matter, or so complex in its organization as kinesiology.

The three-dimensional analysis of physical activity offered in this book—experience, study, professional practice—is designed to help you organize your thinking about the discipline of kinesiology and the broader field of physical activity. This approach not only will help you develop a framework for understanding physical activity and the physical activity professions but also will help you understand the basis for your course work in kinesiology and assist you in planning and implementing career goals. Let's look briefly at each of the three dimensions.

Sources of Kinesiology Knowledge

We have already pointed out that kinesiology derives its knowledge from three distinct sources; this text is organized around these sources. We begin with physical activity experience, the least technical of the sources. Gaining experience moving in a variety of contexts for a variety of purposes is an important contributor to foundational knowledge of the discipline. Scholarly study of physical activity is a much more technical source of knowledge because it depends on advanced knowledge about exercise and skilled movement. The third source of knowledge comes from professional practice centered on physical activity, the application of experience and scholarly study to serve various client and patient populations. All of these forms of knowledge interact in complex ways; each is dependent on and builds on knowledge from other sources.

Experiencing Physical Activity

Figure 1.7 highlights physical activity experience as a way of underscoring its importance to your knowledge of kinesiology. By this point in your life, you have had millions of physical activity experiences. You've learned to master thousands of complex motor skills including tying your shoelaces, driving your car, playing Frisbee, and maybe even performing cartwheels. Many of these forms of physical activity are essential in everyday life. Other experiences such as sport and leisure-time pursuits are discretionary—you choose to do them because they are fun. All of these experiences offer the potential to teach you something about physical activity, yourself, and the world around you.

Experiencing physical activity—performing or watching it—generates highly personal and often unique sensations, emotions, thoughts, and feelings. For example, watching an Olympic gymnastics championship is likely to produce different feelings than watching a dance performance or a football game; digging a ditch is likely to induce different feelings than swinging a golf club. Precisely how we experience physical activity depends on the

FIGURE 1.7 Experience and the other two sources of kinesiology knowledge.

Inventory of Personal Experiences With Physical Activity

To think more about your personal experiences with physical activity, answer the following questions.

1. In what form of physical activity are you most competent? The least competent?
2. What sport or form of exercise do you enjoy the most? Why?
3. What is the hardest physical work you have ever done? What made it difficult?
4. What sport do you most enjoy watching? Which aspects do you find most enjoyable?
5. What is the most dangerous physical activity you have ever performed?
6. What is the longest distance you have run?
7. What physical activity would you most like to learn to perform well?
8. What physical activity have you performed that you never want to perform again?
9. How do you usually feel before exercising? After exercising?
10. What have you learned about yourself because of performing your favorite physical activity?

nature of the activity itself, the context in which it is performed, and our own peculiar sets of attitudes and experiences. Watching an instructor demonstrate a golf swing when we are learning it is likely to be a completely different experience than watching Tiger Woods hit the ball.

Our experiences have a significant influence on the meanings we attach to particular physical activities. They also influence our future decisions about whether to engage in a particular activity again. Obviously, we have little choice about some types of physical activity; we must engage in physical activities associated with our work, in those required to eat, and in those required to maintain our personal hygiene, whether we enjoy them or not. Beyond this, however, we have a great deal of discretion about which activities to perform, with whom, under what conditions, and for how long. Ultimately, our decisions about playing basketball or playing hockey, running 5 miles (8 km) or swimming 50 laps, taking the elevator or climbing the stairs, mowing the yard or hiring someone else to do it, and seeking out physical activity or living the life of a "couch potato" will depend, to a large degree, on our personal experiences with those activities.

Thus, it is important to understand that even though classrooms, laboratories, and libraries offer kinesiological knowledge, direct participation in physical activity is an important source of knowledge as well.

When we participate in physical activities as part of the course work in kinesiology, we not only learn about the activity, we also learn about ourselves as moving human beings. Climbing a rock face, skating on a frozen pond, or parachuting from an airplane can be the unique means by which we gain not only a greater understanding of the activity itself but also knowledge of our own capacities and limitations and knowledge about those around us. Also, certain types of physical activities such as surfing, hiking, or long-distance running offer unique opportunities for reflective thought not normally available in other aspects of our lives.

As a student of kinesiology, you must appreciate the place of physical activity in each sphere or category of your life experiences. In this book, we refer to these various aspects of our everyday lives in which physical activity plays a distinct role as the **spheres of physical activity experience** (see figure 1.8). You will discover how physical activity penetrates all spheres of our personal lives.

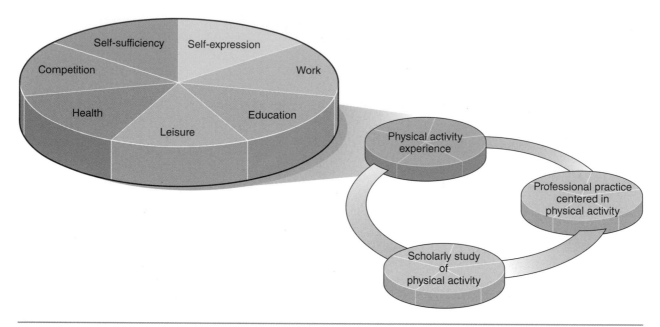

FIGURE 1.8 The spheres of physical activity experience. Physical activity plays an important role in every major sphere of our lives.

Scholarly Study of Physical Activity

Interest in studying and learning about physical activity appears to be at an all-time peak. More than 800 colleges and universities in the United States, and many more in other countries, have academic programs devoted to the study of physical activity (figure 1.9). In addition, unknown numbers of institutes and centers outside academe—medical complexes, military and space research programs, and industrial engineering centers—systematically investigate various dimensions of physical activity. Judging from the numbers of books on sport, exercise, and fitness flooding popular bookstores, laypeople have a keen interest in studying physical activity as well.

Faculty who design formal programs of study in colleges and universities are specialists in organizing knowledge to promote the fullest understanding of the subject. But because physical activity is such a pervasive and diverse phenomenon, it can be difficult to organize that knowledge around a single framework that will enable it to be studied systematically. In fact, debates continue about how to organize the discipline of kinesiology (Newell 1990b).

For the present, scholars have been content to divide the scholarly study of kinesiology into a number of subdisciplines or areas of study. Each of these subdisciplines is related to some larger, broader, and more established "parent" discipline, such as psychology, physiology, sociology, biology, history, or philosophy. For instance, exercise physiology draws on basic concepts and theories from physiology, motor behavior draws on psychology, and philosophy of physical activity draws on philosophy. This means that kinesiology students must develop a working knowledge of the language, theories, and

FIGURE 1.9 Scholarly study and the other two sources of kinesiology knowledge.

conceptual frameworks of a number of major disciplines and learn to apply them to physical activity. You will find this to be a formidable challenge!

You will also discover that each separate area of study addresses a different set of questions about sport and exercise. Exercise physiologists typically ask and answer questions related to short-term changes that occur because of exercise or long-term changes that occur because of training. Questions having to do with forces acting on and generated by the body and questions about the mechanical basis of performance techniques or the interplay of muscles and bones are the province of biomechanics. Specialists in motor behavior study the neural mechanisms underlying the performance of motor tasks, the effects of practice on learning motor skills, and how human development affects our capacity to move. Sport and exercise psychologists are most concerned about the mental aspects of performance, psychological techniques that help elite athletes attain optimal performances, and motivational factors that cause people to drop out of exercise programs. Sport philosophers study questions of values and ethics as they relate to sport and exercise, such as "What values should guide our design of sport and exercise programs?" or "Can sport be a form of art?" Sport historians study the factors that have influenced the development of sport and kinesiology in different eras and cultures, and sport sociologists analyze the influence of such factors as gender, race, class, media, politics, and religion on participation in sport and exercise.

Over the past four decades, the subdisciplines have developed into specialized areas of study, each with its own place in undergraduate and graduate curriculums in kinesiology, its own academic and professional societies, and its own scholarly journals. For example, members of the Association for the Advancement of Sport Psychology meet annually to hear the latest research reports about the psychological aspects of sport and exercise. They keep abreast of developments in this subdiscipline by reading *The Sport Psychologist,* a journal devoted to applied psychological research in sport psychology, or *Journal of Sport and Exercise Psychology,* a journal about more theoretical aspects of the subdiscipline. Members of the North American Society for Sport History, the North American Society for the Sociology of Sport, the North American Society for the Psychology of Sport and Physical Activity, the Canadian Society for Psychomotor Learning and Sport Psychology, the American College of Sports Medicine, and the Canadian Society for Exercise Physiology, along with several other specialized societies, also meet annually to discuss the latest findings in their fields. Many kinesiology departments now offer master's degrees and doctoral degrees in these specializations.

 ACTIVITY 1.3

Specialized Areas of Scholarly Study
Go to your web study guide to see if you can identify the focus of seven subdisciplines in the field of kinesiology.

Most kinesiology professors have received in-depth education in one or two subdisciplines rather than in all of them. Your professors, for example, may identify themselves as biomechanists, exercise and sport psychologists, or exercise physiologists or by some other title that denotes their special expertise. This process of specialization has advanced our knowledge of kinesiology by permitting researcher-scholars to narrow their scholarly focus on specific aspects of physical activity. The disadvantage is that communication among kinesiologists in the various subdisciplines is not always easy. Sometimes these divisions hinder efforts to integrate the different kinds of knowledge and apply them to professional practice.

Figure 1.10 shows the **spheres of scholarly study of physical activity**. The subdisciplines presented in this book are those that are dominant right now. Because kinesiology is a dynamic, rapidly expanding discipline, the subdisciplines that compose it will likely expand also. In fact, some have already moved toward dividing into smaller, more specialized subfields. Biomechanics, for example, could be further separated into anatomical biomechanics

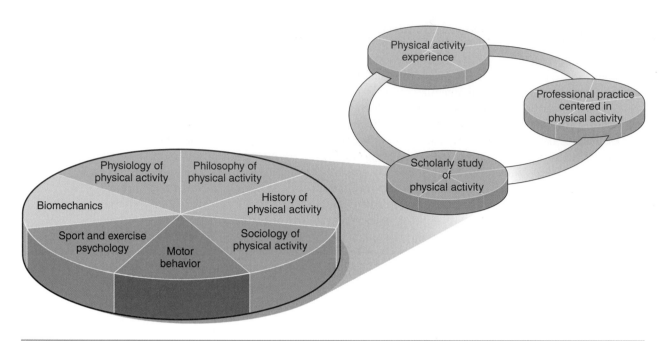

FIGURE 1.10 Spheres of scholarly study of physical activity.

and sport biomechanics. Sport psychology and exercise psychology are developing into separate subdisciplines. Exercise physiology, the oldest of the subdisciplines, has divided into a number of specializations, including sport physiology, work physiology, exercise epidemiology, and exercise and nutrition. This process of expansion of the boundaries of scholarly knowledge and the increased specialization that follows likely will continue for many years. This development, of course, underscores the need for kinesiologists and physical activity professionals to stay abreast of knowledge contained in the spheres of scholarly study.

University and college departments differ in how they cover each subdiscipline. Sometimes each is represented in the curriculum by a separate course (e.g., sport history, sport psychology, or motor learning). In some cases two or more subdisciplines may be combined in one course. A course in the sociocultural aspects of physical activity, for example, may include the study of history of physical activity, sociology of physical activity, and philosophy of physical activity. As kinesiology continues to flex its muscles and the knowledge base continues to expand, the number of courses devoted to the scholarly study of physical activity will probably increase.

As you dig into the material in part II of this book, remember that the subdisciplines and spheres of scholarly knowledge are merely frameworks to help us study and organize our thinking about physical activity. Mastery of the knowledge in each subdiscipline will require you to think in slightly different ways, to master different theories, and to use different terminology. The danger is that as you move from a course in exercise physiology to a course on psychology of sport and exercise and then to a course in history and philosophy of sport, you will compartmentalize your thinking about the field. Unfortunately, our most popular curriculum models encourage such thinking. But you can resist this kind of fragmented thinking by keeping in mind that kinesiology is a unified body of knowledge and that when you draw on this knowledge in professional practice you draw on the entire body of knowledge simultaneously, not in the tidy boxes in which you learned it in your undergraduate program. For the present this method of organizing the scholarly study of kinesiology seems to be the best way to approach it. As the discipline evolves, however, students and professors who are best able to make connections between facts, concepts, and principles drawn from different subdisciplines and different spheres of scholarly study will be in the best position to make significant contributions to the overall discipline.

> Research and systematic analyses form the backbone of the scholarly study of physical activity. Our scholarly knowledge about physical activity has been organized into subdisciplines, each providing a unique perspective from which to view the dynamics and processes of physical activity.

Practicing a Physical Activity Profession

Shelly McBeal lives to dance and perform gymnastics; eventually she wants to own a dance and gymnastics studio for youngsters. Carmen Riaz wasn't a football star, but he so much enjoyed the game that his professional goal is to become a coach at the college level. Yolinda Matthews was on the varsity basketball team in high school but spent most of her junior year undergoing rehabilitation for a knee injury. Now she's interested in a career as an athletic trainer or physical therapist. Toni Bullington was overweight and unfit until a friend got her interested in weight training and bodybuilding. Now she's a fit and confident young woman who would like to become a personal trainer to help others discover what she has learned. Gabe Chauvigne completed an undergraduate degree in engineering and has returned to school to complete a baccalaureate degree in kinesiology. He hopes to study sport biomechanics and become a college professor and researcher.

Like most college students, you probably didn't decide to enroll in college simply to learn about physical activity. You (and surely your parents!) hope that your studies and your degree will lead eventually to a job, preferably a job in the physical activity field. But you probably have many questions. What types of jobs are available to kinesiology graduates? What do these jobs entail? What are their requirements? Is special certification or a graduate degree required? What types of certifications are appropriate? What social environments surround the workplace? Can you be happy in this profession for the rest of your life? These and many more questions face students exploring careers in physical activity.

Part III of this book introduces you to the third dimension of kinesiology—professional practice (see figure 1.11). Professional practice may be envisioned as a process of putting knowledge to work.

When we enter professional practice, we learn new things about providing physical activity services to the public. For example, if you decide to become a physical education teacher, your course work will include a stint as a student teacher where, under the tutelage of an experienced physical education teacher, you will learn to apply what you learned in your teacher education classes: how to structure lessons, how to provide practice experiences, how to organize a curriculum, and how to relate to students with special problems. As you pursue your career you will probably learn a lot that you were never taught

FIGURE 1.11 Professional practice and the other two sources of kinesiology knowledge.

in college classes about how to work productively and collaboratively with other teachers in your school. This knowledge, gained through hands-on experience under the supervision of a specialist as well as on-the-job experience, is different from knowledge acquired through research, and it is different from knowledge acquired through your own direct experiences performing or watching physical activity.

Becoming a Professional

Now let's turn briefly to your career plans. One of the first decisions you will have to make is whether you really want to serve in a professional capacity. But what does it mean to be a professional? In chapter 12 you will learn that becoming a professional entails special responsibilities such as devoting yourself to serving others, giving priority to your clients' needs and interests, performing your duties in an ethical manner, and keeping abreast of developments in your field to ensure that your decisions are based on the best available evidence. Whereas those who specialize in scholarly study focus on discovery of knowledge

---

Preliminary Thinking About Your Career

It is not too early to begin preparing for your professional career. As a way of taking stock of how carefully you have explored a career in the physical activity professions, place a check in the appropriate box next to each question.

- ☐ I have investigated the types of careers available to graduates of a kinesiology program.
- ☐ I have investigated whether a master's degree in kinesiology or a related field is required to practice my chosen profession.
- ☐ I have a timetable for completing my undergraduate degree and a plan for preparing myself for a career in the field of physical activity.
- ☐ I have consulted at least two physical activity professionals concerning how I can best achieve my career aspirations.
- ☐ I have visited a professional setting similar to the one in which I hope to work when I graduate.

---

and how to communicate it effectively to others, those who specialize in professional practice focus on enhancing the lives of those they serve.

Not all types of work qualify as professions. Usually professions require a formal college or university education in which one learns how to master a complex body of knowledge and to perform critical professional skills. In chapter 12 you will see how various types of work can be located on a continuum from strictly nonprofessional (e.g., day laborer) to strictly professional (e.g., surgeon). Most types of work are located between these two extremes. More important, you will learn about the expectations society holds for professionals, and you will have an opportunity to decide whether you really want to pursue a professional career.

Careers in the Physical Activity Field

Once you have decided which physical activity career you will pursue, you must take primary responsibility for developing that career. A career is a lifelong pursuit that may move through various stages as you progress from one type of employment to another. For example, a career in the teaching profession may begin at the public school level and wind up at the community college or university level. A career in the fitness industry may take you from a position as a personal trainer at a local commercial center to a position in a corporate fitness program and ultimately to the position of director of a hospital-linked wellness center. Career counselors now predict that young people graduating from college, on average, will change careers up to four times throughout their life spans. Of course, where you finish your career is not a pressing issue at this time, but getting a good start on your first career is.

One of the exciting aspects of kinesiology is the diverse range of careers that you may develop from a solid undergraduate education. Each of these careers may have specialized requirements within a kinesiology curriculum, but all can be viewed as professional applications of kinesiology. Each of these occupations can be assigned to one of the **spheres of professional practice in physical activity** displayed in figure 1.12.

Part III of this book covers essential information about the various careers pursued by people who begin their higher education with an undergraduate degree in kinesiology. This information will help you compete in the market for a professional position. From this point forward in your college experience, you need to realize that you are involved in a healthy competition, first with yourself as you work to achieve good grades on exams, papers, and projects. When you graduate, you will continue to compete (now against others), vying with other kinesiology graduates when you interview for professional positions. In many ways, this competition is the most important you will enter. How should you prepare? If you were

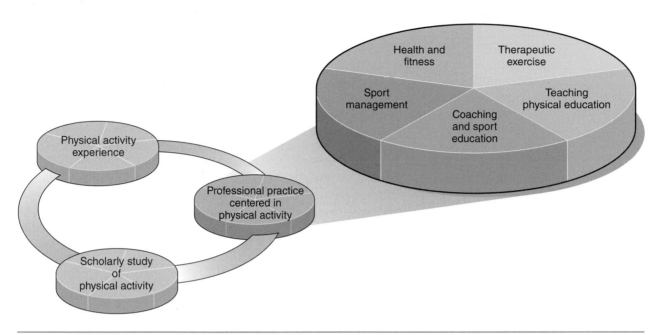

FIGURE 1.12 Spheres of professional practice centered on physical activity.

preparing to compete for a world championship in the marathon, you probably would commit yourself totally to planning and implementing a strategic training program that offers you the best chances for achieving victory. If preparation is important for an athletic competition, how much more important is it for your lifelong career?

Throughout this book you will be reminded repeatedly of this important point: You and only you are responsible for preparing yourself for your chosen career. Your academic department is responsible for offering you a well-sequenced, relevant curriculum with learning experiences designed to achieve your career goals. Your professor is responsible for offering competent advisement and excellent teaching. But you are responsible for taking advantage of these resources to develop yourself into the best professional possible.

To become a marketable, sought-after professional, you must take advantage of experiences both on and off campus. Obviously, committing yourself to mastering the material of your course work is the first step, but it is only the first step. Hours spent outside the classroom in professionally relevant voluntary service work or summer employment are also critical. As a preprofessional you should show an interest in your profession that goes beyond excelling in course work. Becoming a student member of professional physical activity organizations is one way to demonstrate early commitment. Independently exploring developments in your chosen field is another. The chapters that follow will encourage you to think on your own and act on behalf of your educational and professional development. It is never too early to take these vital steps.

> For most students, the goal of pursuing a degree in kinesiology is to obtain a job in a physical activity profession. Those who identify with a profession early, perform well in their course work, and seek out professionally related experiences outside the classroom will be the top competitors for jobs in physical activity professions.

Wrap-Up

With this bit of introductory knowledge under your belt, you are now ready to dig into the topic of physical activity and kinesiology more deeply. You have learned, for example, that kinesiology is the study of physical activity, and you know that physical activity has a technical definition that is quite different from our use of the term in everyday life. You also have learned that kinesiology centers around two general forms of physical activity: exercise and skilled movement.

You've also learned something about how the field of kinesiology is organized. One source of knowledge in kinesiology is experience that comes from participating in and watching physical activity; for purposes of study this realm of experience can be broken down into seven spheres of physical activity experience. Another source of knowledge comes from the scholarly study of physical activity, and this realm also can be broken down into seven spheres or subdisciplines. Finally, knowledge comes from professional practice centered on physical activity, and this realm, too, can be broken down into seven spheres of professional practice.

This text is organized into three parts, each part related to each of these forms of kinesiology knowledge. Part I will help you appreciate the importance of experiencing physical activity as a source of enjoyment; as a means to health, education, and rehabilitation; and as a necessary ingredient of our lives at play and work. Part II contains an overview of the major facts, concepts, and theories in the subdisciplines that together make up knowledge gained from the scholarly study of physical activity. Each chapter in this part offers a brief overview of a separate subdiscipline that has been authored by an expert in the area. These chapters will give you a solid orientation for the in-depth study of those subdisciplines that you will undertake later as courses in your kinesiology major. Finally, part III focuses on professional practice centered on physical activity. The goals of that section are to introduce you to the most prominent physical activity professions and to point out ways in which undergraduate course work in kinesiology connects to them.

 ACTIVITY 1.4

Review this chapter's key points in your web study guide.

Review Questions

1. What is the difference between movement and physical activity? Give an example of an instance in which human movement does not meet the technical definition of physical activity.

2. What is meant when kinesiology is described as a holistic discipline?

3. What two general categories of physical activity receive the most attention in kinesiology? What are the subclassifications of these categories?

4. What are the three sources of knowledge of kinesiology?

5. List the spheres of physical activity experience.

6. List the spheres of scholarly study.

7. List the spheres of professional practice centered in physical activity, and give an example of a career in the physical activity professions for each sphere.

Experiencing Physical Activity

One of the everlasting mysteries of sports is why people watch and perform them. Several decades ago writer Joseph Epstein penned his popular essay "Obsessed With Sport" in an effort to get to the bottom of his own attraction to sports (1976, p. 67). "What is the fascination? Why is it that, with the prospect of a game to watch in the evening or weekend, the day seems lighter and brighter?" Epstein's efforts were brave, but hardly convincing. Despite decades of musing by philosophers, journalists, and fans themselves, there remains no single, agreed upon explanation of why some of us sit for hours in front of our televisions watching baseball, or don silly hats and paint our faces and brave the snow and cold of stadiums to watch our favorite football teams, or put off doing important things in life in order to watch young men and women play basketball.

Nor have we, despite years of thinking and research, been able to put our finger on precisely why some of us become so hooked on playing sports or exercising that we grant them—mere pastimes—priorities in our lives. The avid golfer who, shunning work at the office, sneaks off to the course for a "quick 18"; the dedicated tennis player who fills up every weekend with matches and tournaments; and the runner who—seemingly unable to stop herself—runs in the rain, cold, snow, and unbearable heat are propelled by some invisible force that isn't easy to explain. The former editor of *Runner's World*, trying to understand his attraction to running, put it this way: "I write because the thoughts inside have to be put in more visible form. I run because it's inside pushing to get out" (Sheehan 1978, p. 74).

Clearly, "doing" physical activity and observing it intently are unique experiences, both can attract us in inexplicable ways and, more importantly, both focus our attention on the critical importance of experience in the realm of physical activity. *Experience* is a broad and pervasive term. Consider how each of the following questions focuses on a distinct aspect of the term *experience*.

- Why are physical activity experiences important, and with what aspects of our lives do they intersect?

- What types of physical activity experiences have been most important in helping you master skill, training, and physical fitness?

- How have your thoughts, emotions, understandings of yourself and others, and the meaning you attach to physical activity been influenced by your physical activity experiences?

The chapters in this part of the book will help you answer these three related, but separate, questions. The first question deals with the part physical activity plays in the living of your life. Obviously, if physical activity didn't contribute to our lives, there would be little justification for a field of study centered on it. To help you answer this question, chapter 2 describes the different ways in which physical activity penetrates our daily lives. The chapter will introduce you to the seven spheres of physical activity experience—distinct aspects of our lives in which physical activity plays an important role.

Even though you have been immersed in physical activity from the day you were born, this chapter should help you develop a new respect and appreciation for the role that activity plays in your daily life. If you are like most kinesiology majors, your decision to major in kinesiology can be traced to some particular, in-depth experiences you had in a particular sport or form of exercise. These experiences will help provide you with a frame of reference as you study this field, but you should remember that as a student of kinesiology—a field that embraces all forms of physical activity—you should be interested in physical activity experience in all its various forms. Your program of study in kinesiology is designed to help you develop a refined curiosity about what physical activity is.

The second question refers to the particular effects experience has had on your physical capabilities. Much of your course of study in kinesiology will be aimed at showing you how physical experience can be manipulated to bring about specific changes in performance capacity or health or increases in skill or development. Chapter 3 introduces you, in a general way, to the dynamics of this process. Skill in manipulating physical activity experiences intelligently and systematically is the exclusive domain of those who work in the field of kinesiology, just as skill in diagnosing various forms of disease is the exclusive domain of physicians. Kinesiologists are physical activity experience experts. They understand the connections between types and amounts of physical activity experiences and specific outcomes, whether the goal is the development of skillful movement (skill); improved health (physical fitness); improved strength, **endurance**, and flexibility (conditioning); recovery from injury or disease (rehabilitation); or merely enjoyment. Knowing the most efficient, most effective, and safest way to achieve these goals is the kinesiologist's stock-in-trade.

The third question is directed toward your interior experiences in physical activity. Knowing how physical activity experiences affect our skill, endurance, strength, flexibility, health, and fitness is one thing; knowing how they affect our inner lives and why we enjoy them is another. Those who don't enjoy physical activity are not likely to do it voluntarily. Chapter 4 introduces you to this subjective dimension of experience. After reading and studying this chapter, you will understand that physical activity is not only a physical experience but an emotional, cognitive, and spiritual experience as well. How physical activity affects our thoughts and feelings, how these feelings may affect our future physical activity choices, and how we derive knowledge and meaning through physical activity are all part of what we call **subjective physical activity experience**. All of our involvement in physical activity ultimately has a bearing on this subjective domain.

By nature, our subjective experiences are difficult to describe, and sometimes they affect us in highly personal ways. Because of this we sometimes ignore this profound interior world of human experience, choosing instead to focus on observable dimensions of experience such as our time in a 5K race, how much weight we lost during an exercise program, how many points we scored, or how quickly we recovered from an injury. After reading chapter 4, you might wonder why we even refer to it as *physical* activity. Clearly, it is more than physical. (Perhaps *human activity* would be a better term.)

Physical activity experience is so prevalent in our lives that we tend to take it for granted. Yet it is the foundation of the study of kinesiology. In the following chapters we invite you to view physical activity experience from an entirely new perspective, beginning with the seven spheres of physical activity.

 PART I PORTFOLIO

Responding to the Olympics: Subjective Experience

When you've worked your way through the first four chapters, go to the Part I Portfolio activity in your web study guide. In the activity, you will consider scenarios related to the Olympic Games and respond based on both your own experience and the knowledge you gained from part I.

chapter two

Spheres of Physical Activity Experience

Shirl J. Hoffman

chapter objectives

In this chapter we will

- broaden your understanding of the universe of physical activities,

- familiarize you with the different types of physical activity in different social compartments called spheres, and

- introduce you to some of the potential benefits and limitations of physical activity.

Take a moment to examine the photograph of the jazz dancers. What aspects of the performance capture your attention? Is it the position of the male dancer's legs, the height of his jump, or his outstretched arms? Some will be attracted to the way he seems suspended in space; others to the orientation of his head, neck, and trunk; and still others to the graceful position of his hands. Some will be drawn to the arched back of the female dancer, the lack of symmetry in the positions of her arms and legs, and the grace she manages to achieve without even moving. Do the body positions and facial expressions express a particular attitude or emotion? Do you see the movements as examples of grace and beauty or of sheer athletic skill?

The photo can help you understand something important about physical activity and kinesiology. Although sport and exercise are the primary focuses of kinesiology, dance is an important form of physical activity that is of interest to kinesiologists as well. So are hundreds of other forms of physical activity. The dancers in the photo help to underscore the main point of this chapter: Kinesiologists are interested in a broad range of physical activity, including physical activity that takes place in the arts. You probably are exploring kinesiology as a career because you have a special interest in sport or exercise. Because of this you may assume that these are the only forms of physical activity that interest kinesiologists. If so, prepare to broaden your horizons. For only as you come to understand and appreciate the beauty, elegance, drama, force, precision, and variety of all forms of physical activity can you truly appreciate physical activity that occurs in sport and exercise settings.

If you're like most people, you probably haven't given much thought to how important physical activity is in your daily life and you only think seriously about the question when your capacity for moving is limited due to disease or injury. For example, nothing makes us appreciate the importance of the ankle joint in walking quite like having a sprained ankle or the importance of the thumb in grasping quite like having a broken thumb. Physical activity affects our lives in thousands of ways and in countless forms and levels of intensity. Although kinesiologists traditionally have concerned themselves primarily with sport and exercise, they now recognize that the boundaries of kinesiology extend to all forms of exercise and skilled movement. Physical activity that occurs on sport fields and in fitness centers is important to physical activity professionals; but so is physical activity that occurs in the workplace, the rehabilitation center, the dance studio, the nursing home, and many other venues.

This chapter is intended to stretch your conception of physical activity. It leads you on an expedition through a vast expanse of physical activities done for different reasons and in different settings. The purpose, as we have said, is to expand your frame of reference for thinking about physical activity, but it also is to increase your appreciation of how critically important kinesiology is to our daily lives.

One way to begin to appreciate the enormous variety of physical activities and the way they intersect with your daily life is to ask yourself how often and in what ways you were physically active during the past week. In answering this question, you might first think about the various social situations you were in and then the role physical activity played in each situation. For example, you might think, "I worked at the restaurant 20 hours last week, and using this conceptual framework I estimate that I waited on 22 tables, poured more than 50 cups of coffee, walked to the kitchen nearly 100 times, and prepared 52 salads. Or you might think, "I played golf on Saturday," and with the framework of playing golf to guide your thinking you might recall that you walked about 4 miles and executed approximately 18 drives, 25 shots with midirons, 13 approach shots, and 30 putts.

But it isn't simply the amount of physical activity that is important; the variety of physical activity is important too. The types of physical activity and movements performed in working at a restaurant are much different from those involved in playing golf. So, as you relate this chapter to your everyday experiences, don't think only about how much physical activity you performed; think about the various kinds of physical activity as well.

As noted in chapter 1, physical activity is movement intended to accomplish specific purposes. As you act out your daily life, the purposes and goals you wish to achieve—whether the activity is something as simple as moving to the other side of the room or as complicated as juggling three balls—usually determine both the types and amount of physical activity that will be required. But when you think about the enormous variety of goals and purposes that require some sort of physical activity, you can easily get overwhelmed. To get a handle on the importance and pervasiveness of physical activity in our lives, we need some system of organizing our personal experiences with physical activity—a conceptual framework.

In chapter 1 we used the term *spheres of physical activity experience* as a conceptual framework for classifying the different life experiences in which physical activity plays an important role (see figure 2.1).

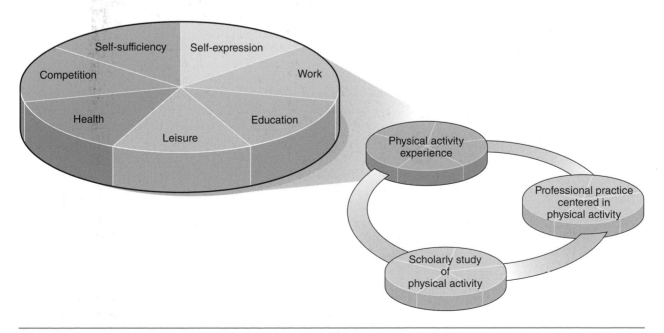

FIGURE 2.1 Spheres of physical activity experience.

We defined spheres of physical activity experience as dimensions of everyday life in which physical activity plays an important and distinctive role. Generally, the spheres identify different purposes of physical activity, usually but not always in different social contexts. For example, the work sphere includes all physical activity done in the workplace to meet the demands of employment. The education sphere includes physical activity carried out in educational settings as a way of learning.

You will note that some spheres are less definitive in terms of the social context in which the activity is usually performed. The self-sufficiency sphere, for example, includes physical activity carried out to survive and live an independent life. Although most of these activities are carried out in the home, they also can be performed at work, during leisure time, or in educational settings. Likewise, physical activity in the self-expressive sphere may occur in a variety of social contexts but always serves the purpose of allowing us to express our emotions. Keep these three things in mind as you study the spheres:

- The spheres are not intended to classify specific types of physical activities. They simply highlight the different compartments of our life experiences in which physical activity plays an important part.

- Some activities may be common to more than one sphere of experience. For example, if you run 3 miles (4.8 km) each day, this activity may be assigned to the sphere of leisure but also might be related to your health life or your self-expression.

- The purpose of this chapter is to provide you with a general framework for thinking about the importance and pervasiveness of physical activity in your life, not to compartmentalize the activities themselves.

In the sections that follow, we explore each of the seven spheres of physical activity experience—self-sufficiency, self-expression, work, education, leisure, health, and competition—by examining some activities that they encompass. We discuss both positive and negative issues related to the spheres and learn about professionals who specialize in some of those issues. Keep in mind that these spheres serve merely to help you look at the many ways and contexts in which you experience physical activity. We hope that this discussion will challenge you to think more deeply about the many levels of this field, levels that go far beyond the merely physical.

The Sphere of Self-Sufficiency

You devote a significant part of your life to taking care of yourself, and physical activity is an important means by which you accomplish this (see figure 2.2). Although tasks such as walking from your bed to the bathroom, eating breakfast, or brushing your teeth are perfunctory, they are critical to your ability to live independently or even survive. If you couldn't perform them you would have to rely on the help of others. In other words, you would have lost your self-sufficiency. Other activities that are not essential for your survival or independent living still contribute to your self-sufficiency and personal comfort. These activities might include cleaning your apartment, washing the dishes, ironing clothes, or driving your car.

> To live functional, independent lives, we must perform ADLs and IADLs, the latter of which tend to be more physically demanding. We also become self-sufficient by performing physical activities intended to maintain or improve the home. These activities are called "home maintenance activities."

Physical and occupational therapists work with people who, through injury or disease, have lost the ability to perform the important tasks of daily life. Typically, therapists divide self-sufficient physical activity into two major categories: **activities of daily living (ADLs)** and **instrumental activities of daily living (IADLs)**. The former include more personal behaviors related to grooming, using the toilet, dressing, eating, or walking. People who require daily assistance in these tasks are usually disabled due to injury or disease or frail because of age. Activities classified as IADLs include the less personal activities of telephoning, shopping, cooking, or doing laundry (Katz et al. 1963). Generally, IADLs are more physically demanding than ADLs. Health insurance companies often use these two classifications of physical activity in daily living to gauge the level of disability of patients and to determine the type of health support they require. For example, an older person may not qualify to collect

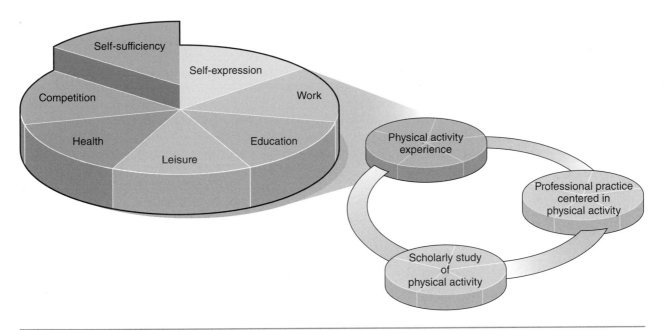

FIGURE 2.2 Self-sufficiency as a sphere of physical activity experience.

long-term health care insurance unless a therapist documents that the individual is unable to perform two or more ADLs.

The physical activities we perform around our homes are often more complicated and demand more energy than typical ADLs and IADLs. Shoveling snow, fixing our automobiles, painting our apartments, repairing electrical fixtures, and similar tasks are examples of **home maintenance activities** that also reflect our level of self-sufficiency. All require relatively high levels of energy or skill or both. Although many people hire others to perform home maintenance activities for them (e.g., gardeners, painters, automobile mechanics, plumbers), the explosion of sales of do-it-yourself manuals and home improvement television shows suggests that more people are performing these household activities themselves.

Limitations in Physical Activity for Self-Sufficiency

A variety of movements are required to carry out daily tasks such as housecleaning, doing laundry, bathing, cooking, opening jars, writing checks, and shopping for groceries. For example, using a vacuum cleaner includes the fundamental movements of walking, standing, grasping, reaching, pushing, and maintaining an upright posture. Remaining self-sufficient requires us to transport objects of different weights, ascend or descend stairways, and perform other fairly complex actions. When disease or injury compromises our movement capabilities, occupational and physical therapists can help us learn or relearn self-maintenance activities.

ACTIVITY 2.1

Self-Sufficiency
Use this activity in your web study guide to practice distinguishing between ADLs and IADLs.

Before devising treatment plans, or "interventions," for rehabilitating people with physical disabilities, physical and occupational therapists must completely understand the physical activity requirements of each task. Obtaining this understanding usually involves a thorough analysis of the movements that a person must perform to carry out the task. Only then can the therapist decide which muscles the patient must strengthen or which movement patterns the patient must refine to regain self-sufficiency. These activity analyses can also help us understand the amazing complexity of what at first may seem to be simple tasks.

An activity analysis of the seemingly simple self-care task of standing up from a seated position, a major challenge to the victim of a cerebral stroke, reveals at least four critical phases: (1) the feet must be well placed on the floor in a position to receive weight evenly divided on the two legs; (2) the trunk must be flexed forward at the waist while remaining extended; (3) the knees must move forward of the ankles; and (4) the hips and knees must extend for final alignment (Carr & Shepherd 1987). By skillfully comparing the patient's movements to a model of correct performance, the therapist can design a training program to speed recovery in this fundamental task. Such activity analyses require an in-depth knowledge of body mechanics and anatomy, something the therapist learns from studying biomechanics, an important topic in kinesiology.

Self-Sufficiency and Aging

You may have given little thought to the importance of ADLs and IADLs in your life because you may have experienced little trouble performing them. But a surprisingly large portion of the population require assistance in even these basic self-care tasks, a condition that deprives them of independence in their daily living. Limitations may result from accident, injury, congenital disorders, or aging. Aging is a major factor limiting performance of ADLs and IADLs. The population of those who are elderly is growing rapidly. Worldwide, the number of people aged 65 and older will increase from an estimated 524 million in 2010 to 1.5 billion in 2050. Most of the increases are expected to take place in developing countries (National Institute on Aging n.d.). As of 2010, more than 40 million Americans were 65 years of age or older (median age of 74), and this demographic segment is expected to reach 72 million by 2030 (U.S. Department of Health and Human Services [USDHHS] 2010). In 2011 there were an estimated 5 million Canadians over the age of 65, a number that is expected to double by 2036 (Human Resources and Skills Development in Canada 2011).

In 2010, 7.3% of the noninstitutionalized U.S. population (16.7 million) found it difficult or impossible to walk a quarter of a mile. The number of adults with general physical functioning difficulty was 35.8 million (15.6%). Nearly 5 million adults (2%) required the help of another person to perform ADLs, and more than 9 million (4%) required help with IADLs. An estimated 11% of people over the age of 75 require help performing at least one ADL; approximately 19% must have help performing at least one IADL. Socioeconomic factors greatly affect these numbers. For example, poor adults were about four times as likely to require help with ADLs or IADLs as those who were not poor. Obviously, had the institutionalized population (people in nursing homes and assisted living residences) been included in this database, the numbers in each category would be significantly higher (U.S. Centers for Disease Control and Prevention [CDC] 2010a). Figure 2.3 shows how disabilities are associated with increases in age. In 2026, approximately 54% of people with disabilities in Canada will be seniors versus 42% in 2001 (Human Resources and Social Development Canada [HRSDC] 2005).

Now the good news! A decline in the rate of disabilities among elderly people in the United States first observed between the years 1982 and 1999 (Manton & Gu 2001) appears to be continuing. According to data from the National Long Term Care Survey, the prevalence of chronic disability among 65-year-olds decreased from 26.5% in 1982 to 19% between 1999 and 2004. In addition to the general decline in disability, the average annual rate of decline accelerated from 0.6% in 1984 to 2.2% between 2004 and 2005. Researchers believe that environmental modifications, assistive technologies, and biomedical advances may be largely responsible. Other encouraging changes between 1982 and 2004 through 2005 include these (Manton 2008):

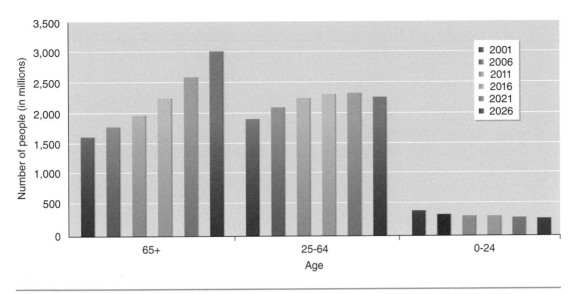

FIGURE 2.3 Disabled and projected disabled population in Canada, 2001-2026.

Adapted from Human Resources and Social Development Canada (HRSDC) 2005.

- The proportion of people without any disabilities increased the most in the oldest age group, increasing by 32.6% among those 85 years and older.
- The percentage of Medicare enrollees age 65 and older who lived in long-term care institutions such as nursing homes decreased dramatically—from 7.5% to 4%—perhaps a result of the availability of more assisted living options and improved Medicare reimbursement policies.
- Chronic disability among elderly people continues to decline.

Older people are often injured in their attempts to perform ADLs and IADLs. Common accidents include falling on stairways or in bathtubs and suffering burns or scalds. Movement limitations cause most of these accidents, but environmental factors such as poorly lit stairways, frayed rugs, and poorly maintained homes play a part. Because older people realize that the risk of accidents around the home is high, many simply stop performing certain ADLs and IADLs. This situation results in a severely diminished quality of life (Czaja 1997). When persons who are elderly lose either the ability to perform self-care activities or confidence in their ability to perform them, they can suffer a devastating setback that deprives them of the satisfaction of being able to live up to their physical, mental, and emotional potential. The prospect of a growing population of individuals who are dependent on others to carry out their daily tasks is also undesirable from an economic standpoint. When major segments of the population lose their ability to perform ADLs and IADLs, the national health care system takes on a heavy burden. As a taxpayer, you have a real stake in the costs of Medicare and Social Security programs. An elderly population that requires assistance in the performance of ADLs has a significant effect on these programs.

These trends represent an awesome challenge to those in the physical activity profession who work with older populations. Although an increase in the number of elderly people over the next few decades is inevitable, an increase in the proportion of this population who are hampered by limitations in physical activity is not. Many of the people who suffer the most limitations in later years are those who failed to make physical activity a daily part of their normal lives throughout their lives. Recreational activities and exercise programs designed and administered by physical activity professionals will play an increasingly larger role in preventing and rehabilitating these age-related disabilities.

> Limitations in the performance of ADLs and IADLs among elderly people require them to depend on others or institutions to perform the tasks of daily living. This problem is of great personal and economic importance.

Spotlight on Research: Physical Activity Is the Frontline of Defense Against Limitations in Self-Sufficiency

Research on elderly populations is booming as scientists begin to focus on the many problems accompanying aging. Many of these problems can be ameliorated by physical activity. Here are two studies that examined the benefits of exercise in older people.

Exercise May Reduce Risk of Dementia

The limitations imposed on self-sufficiency by dementia are receiving more and more attention by medical experts and social commentators. After an extensive review of the literature, researchers at the Mayo Clinic reported finding "a significantly reduced risk of dementia associated with midlife exercise; similarly, midlife exercise significantly reduced later risks of mild cognitive impairment." They concluded that "exercise should not be overlooked as an important therapeutic strategy" in the treatment of Alzheimer's disease (Ahlskog et al. 2011, p. 874).

Fitness Decline With Age After 45

Researchers from the Kinesiology Department at the University of Houston studied 3,429 women and 16,889 men aged 20 to 96. The researchers found that fitness levels declined continuously although the decrease wasn't linear or steady. Specifically, cardio-respiratory fitness declined more rapidly after age 45, and the decline was greater for men than women. The results also showed that being active and maintaining a healthy body weight were associated with higher levels of cardiorespiratory fitness in the life span studied. The authors noted that given the high levels of obesity and low levels of physical activity in the general population, the results suggest that in the long term, more men and women will qualify for the designation of "disabled" by the Social Security Administration at a younger age (Jackson et al. 2009).

The Sphere of Self-Expression

The urge to express our inner feelings is one of the most basic human instincts (see figure 2.4). All of us would like to demonstrate, in one way or another, what is unique about us, what makes us special. Obviously, we are limited in the ways we can do this. People often hesitate to express themselves in speech, and few possess the talents of poets, songwriters, or artists. One way we can give outlet to our inner feelings is by moving our bodies or readjusting our postures.

Have you ever considered how often we express our feelings and convey other messages in our body movements and postures? We walk rapidly and with a spring in our step when we feel happy and in a slow, plodding manner when we feel sad. We jump up and down for joy, stretch our arms overhead in elation, hang our heads in sorrow, and clap our hands in appreciation. We use body movements, often unconsciously, to express our emotional states. Our body movements seem to be wired directly to our emotions so that changes in our movements occur without voluntary effort on our part. Although pointing to interesting aspects of our human identities, these examples do not meet our technical definition of physical activity because they do not involve deliberate or voluntary movement directed toward a goal.

Many times, however, we deliberately use movements to express feelings and moods and to convey other messages. Have you ever thought about how much you depend on physical activity to help you communicate when you are talking? By using gestures, hand signals, and changes in body posture, we can deliberately and intentionally emphasize such verbal messages as describing shapes and showing directions. We also use physical activity intentionally to express our feelings and emotions in dance, religious liturgy, and various ceremonies. One

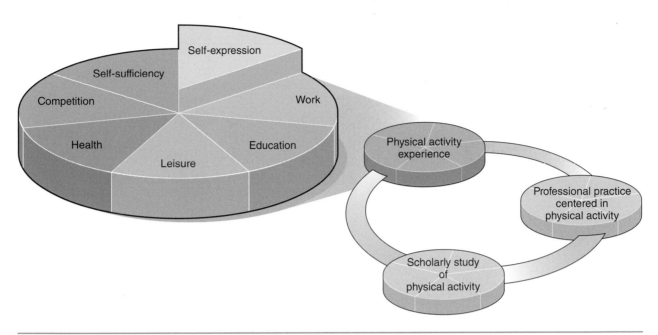

FIGURE 2.4 Self-expression as a sphere of physical activity experience.

could even argue that we express something about ourselves every time we move, whether in shooting a jump shot, running a race, or diving from a platform into a pool. Let's examine some ways in which we use physical activity to express ourselves.

Gestures

Movements of our hands, fingers, or other body parts used to communicate our intentions to others are called **gestures**. We use them in place of or in conjunction with talking. Beckoning someone to approach by flexing and extending the index finger with the palm up, signaling someone to stop by holding up a hand with the palm away, or nodding instead of saying yes are examples of gestures. Scientists who study nonverbal communication distinguish between three types of intentional gestures: emblems, illustrators, and regulators.

Emblems

Emblems are body movements, usually hand movements, that can be directly translated into words and are easily understood by those in the culture or subculture in which they are used (Morris 1994). People may use emblems with or without accompanying words, but even when used without words, emblems convey a great deal of information. For example, they can be used to communicate at a distance or in environments in which verbal communication is difficult or impossible, as in construction settings where crane operators receive messages from ground workers or in noisy athletic arenas where coaches must send signals to players on the court (see figure 2.5). Unable to talk to each other under water, scuba divers often use an elaborate set of hand signals to communicate. In noisy arenas or playing fields, referees and umpires use emblems to indicate to players, coaches, and audiences that a player has fouled, whether a base runner was out or safe, whether the tennis ball was in or out, and when time of the game has expired.

Illustrators

Illustrators are gestures that we use to illustrate or complement what we are saying. When you talk about yourself, you may point to yourself, and when you talk about someone

FIGURE 2.5 Emblems used for sending messages in sport: *(a)* personal foul (football); *(b)* 30-second time out (basketball); *(c)* offside (soccer); *(d)* out (cricket); *(e)* slider (baseball); *(f)* replay serve (volleyball); *(g)* good (tennis).

else in the room, you may point to him or her. People use illustrators to describe the motion of objects. If you were to tell someone about the path of a foul ball that narrowly missed your head when you were sitting along the third base line at the baseball park, you might use an illustrator gesture to describe it. Illustrators can also convey a particular tone in a verbal message. When a coach pounds the fist of his right hand into the palm of his left while talking, he is adding a sense of determination and seriousness to his verbal message.

Regulators

Regulators are body movements used to guide the flow of conversation. Hand and body movements used in greetings (shaking hands, waving, nodding) and in partings (waving, shaking hands, hugging) are examples of regulators. A person may also use regulators to signal to other parties that he or she is finished with the conversation. These signals include shifting weight from one foot to the other or turning toward the door. A person may use hand gestures, shifts in gaze, and head movements in conversations to signal that he or she has not yet finished talking or, conversely, that it is the other person's turn to talk.

Cultural Differences in Gestures

The meaning of a gesture often is specific to a particular culture, and not knowing its meaning can lead to embarrassing situations. For example, beckoning someone with your forefinger and an upraised palm is an obscene gesture in some cultures, as is making an "O" with your thumb and index finger (as if to signal "okay"). Some of the most culturally specific gestures are regulators, especially those used for greetings. If you were to meet an acquaintance in America and he folded his arms across his chest when you extended your hand to shake his, you might think him rude, but this movement indicates a very respectful greeting in Malaysia. In Eskimo country, a hit on the shoulder doesn't mean that you are being challenged to a fight. It means "hello." Nodding your head up and down in most countries means yes, but in Greece and Bulgaria it means no. Even within cultures, the meanings attached to gestures can change over time. Fifty years ago, if you saw two Americans bumping the knuckles of their clenched fists together after one of them made a spectacular golf shot, you might have thought it was some form of secret greeting. Today, the "fist bump" has become a popular form of greeting or a gesture of congratulations. Can you think of other relatively recent innovations in American gestures?

▶ We use physical activity as a form of communication and expression in combination with or in place of words. Gestures can supplement or substitute for spoken words.

▌The fist bump is a modern gesture that has replaced the high-five as a way to say "good job."

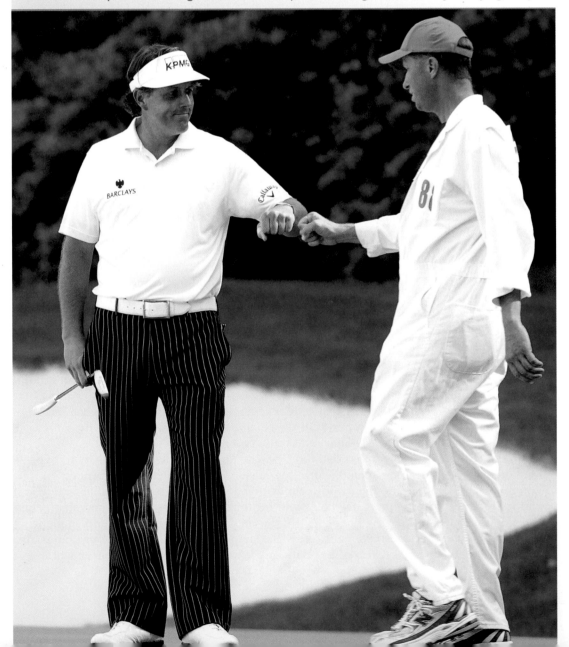

Walking Patterns Can Signal Potential Muggers

Walking is an enormously vivid means of expressing ourselves, and street criminals often take notice. Obviously, many attacks on pedestrians are random, but in some cases, would-be muggers take into account the movement patterns of potential victims before deciding to assault them. In a classic study, psychologists asked a group of career criminals who had been sent to prison for violent crimes such as rape, robbery, and murder to view a video of people walking down a New York City street and identify those they would most likely target. Surprisingly, size, sex, age, and gender were not major factors in their decisions, but the walking patterns of the people were. The criminals were assessing the ease with which they could overpower the targets by keying in on such nonverbal signals as posture, pace of walking, length of stride, and awareness of environment. Interestingly, neither the criminals nor victims were consciously aware of these cues. One of the main determinants of whether a victim would be assaulted was a walking style that lacked "interactional synchrony" and "wholeness"—walking patterns that lack organized movement and flowing motion. Such people were viewed by the criminals as less self-confident—perhaps because their walk suggested they were less athletic and fit—and therefore easier targets (Hutsmyre & Dixit 2009).

Writing and Dance

Often, we express our feelings in the way we execute physical activity. We may combine expressive elements with movement used to accomplish specific tasks. For example, we may not run, swim, or lift weights specifically to express something about ourselves, but often we cannot help it. The manner in which a basketball player bounces the ball before shooting free throws and the movements she selects to propel the ball toward the basket involve instrumental and expressive movements. **Instrumental movements** are the critical movements required to attain the goal of the activity; **expressive movements** are idiosyncratic movements that are not required for goal attainment but that express something about the individual. No movements are more goal oriented or more reflective of our emotions, thoughts, and feelings than those we use for handwriting. Handwriting provides a permanent tracing of the movements one uses to produce it. Graphologists—analysts who study handwriting—believe that how we write is affected by our personalities and our emotional state at the time of the writing. Therefore, they use handwriting samples to predict aspects of human behavior and emotions that may not be amenable to measurement using other means. Extreme right slant in writing, for example, is thought to reflect a more intense personality. Larger writing is associated with a more expansive, outgoing personality; small writing, with a more controlled personality. Although graphology has come under criticism by scientists on a number of counts, it continues to be used by marriage counselors to identify compatibilities, by career guidance counselors, and as a part of preemployment screenings to select executives. Separating instrumental movements from expressive movements may be impossible in such athletic performances as a twisting layup, a run around the bases, or a marathon run.

Sometimes, however, we use physical activity deliberately to express sentiment and emotion with no instrumental goal in mind. In dance and in ritual, physical activity serves the express purpose of conveying feelings and symbolic meaning.

From the earliest periods of recorded time, people have used physical activity to express human emotion and meaning symbolically. One reason is that human movement enables us to communicate complex thoughts and feelings that are difficult or impossible to transmit verbally. Much the way painters use color and form, sculptors use texture and shape, and musicians use tones and rhythm, dancers use the mediums of force, time, and sequencing of movements to express aesthetic messages. The shape and dimensions of our bodies also affect the aesthetic characteristics of our dance. A leap by a tall, slender dancer will evoke

What Does Your Signature Say About You?

Writing your signature is an instrumental act that is loaded with expressive information. Graphologists—experts who study handwriting—claim that personality and mood are reflected in our writing styles. Psychologists have discovered that signature styles vary with such factors as self-esteem and status. When self-esteem was manipulated experimentally by informing subjects that they had done well or poorly on a prior experimental task, those told they had done poorly used less space (smaller movements) to sign their name than those told they had done well. Status also seems to affect the size of movements used to sign your name. Professors' (higher status) signatures have been found to be significantly larger than those of their students. Figure 2.6 shows the result of an informal record made by Karl Scheibe of the size of his signature on the textbooks he purchased as a freshman in college, then during each year of undergraduate school, again during graduate school, and then once more after he became a member of a college faculty (Zweigenhaft 1977). It is readily apparent that the space used to sign his name systematically increased over an 11-year period. Scheibe and others believe that the changes reflect the elevated status he enjoyed moving from undergraduate school to graduate school and eventually to a professorship. (Notice the large increase that occurred just as he was about to enter the professoriate.) Obviously, this wasn't a controlled study. Can you think of other factors that might have accounted for the change in Scheibe's signature? Could you design an experiment to test his hypothesis?

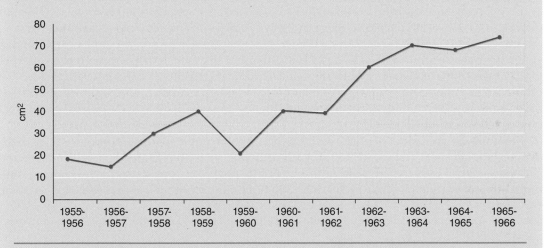

FIGURE 2.6   How changes in status can affect the size of a signature.

Adapted from Zweigenhaft 1977.

much different emotions in observers than the same leap executed by a short, overweight dancer. In either case, the moving body can tell a story that is inexpressible in word or song.

The physical requirements of dance can be as exacting as—or even more exacting than—the most strenuous sport or exercise routines. The energy requirements of ballet, disco, jazz, Latin, modern, and tap dancing are approximately the same as the energy requirements for hunting, kayaking, gymnastics, climbing, and team sports (Ainsworth et al. 2000). Muscular and cardiorespiratory endurance, flexibility, strength, balance, agility, and coordination all are critical to dancing. Long and arduous practice and conditioning regimens are required to achieve the standard body positions and movements required in ballet, for example. Likewise, the free-flowing, creative routines of modern dance, in which the dancer may have more freedom in interpreting the notations of the choreographer, require intense training and conditioning. Regardless of the form of dance performed, the result is the same: "the presentation of a significant emotional concept through formal movement materials" (Phenix 1964, p. 197).

Dance is an art form that uses physical activity to express attitudes and feelings that may be difficult or impossible to express in normal verbal communication. Rituals often use physical activity to express symbolically sacred values or beliefs.

Sometimes we dance not to communicate our feelings to others but simply as a way of giving outlet to our feelings, even when we are alone. Certainly, an audience, a partner, or a band is not required to enjoy dancing! Did you ever dance to music on the radio when no one else was in the room? Sometimes we dance spontaneously, merely to express joy and jubilation, as, for example, football players sometimes dance in the end zone after scoring a touchdown. You may like to dance because you enjoy social dances such as swing, tango, electric slide, or country line, or you may like to demonstrate your skill in moving your body in relation to a partner or to the beat of the music. Perhaps through dance you communicate something about yourself that you can't communicate in any other way.

The Sphere of Work

Work constitutes a significant portion of the total life experiences of most of us, and work always involves physical activity (see figure 2.7) even if it is minimal. What different kinds of work have you done in your life? Have you worked as a construction laborer, a truck loader for a shipping company, a lawn maintenance technician, a pizza delivery person, a house-painter, a server at a restaurant, a kitchen worker for a fast food chain, an office assistant, or a checkout attendant at a grocery store? If so, what kinds of physical activity have these different jobs required?

In thinking about the answer to this question, you probably recognized that the type and intensity of physical activity can vary enormously from job to job. The checkout attendant at the grocery store and the construction laborer both are engaged in physical activity, but they are not physically active in the same ways. Although you may have easily recognized the importance of physical activity to the work of the pizza delivery person or truck loader, you may not have appreciated the equal importance of physical activity (though of a different kind) in the work of secretaries, accountants, computer consultants, and graphic artists. The physical actions of typing, filing, and operating an adding machine, although not strenuous, require precise positioning of the limbs and fingers, movements that are essential to satisfactory execution of these jobs.

Throughout recorded history, societies have associated manual labor, with its high demand for physical activity, with lower status, while according higher status to managerial and supervisory jobs that require little physical activity. Yet low-status jobs requiring high levels of physical activity may offer a health bonus for workers. Increased levels of some

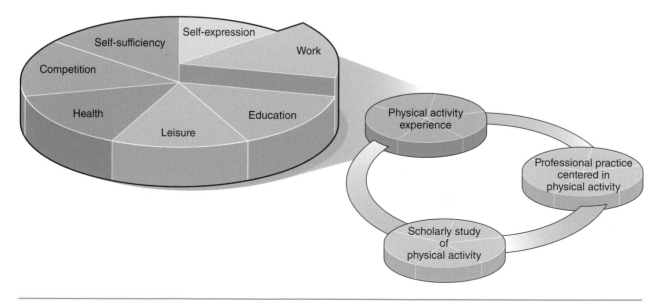

FIGURE 2.7 Work as a sphere of physical activity experience.

forms of physical activity may ward off certain diseases associated with physical inactivity (sometimes called "hypokinetic diseases," or "low-physical-activity diseases"). CareerCast.com's list of the 10 worst jobs in 2010 included the job of mail carrier, citing the physical demands as part of the reason. But it discounted the health bonus carriers receive from their daily walking (Strieber 2010). An early, classic study comparing incidence of heart attacks among drivers and conductors of English double-decker buses revealed that drivers, who were mostly sedentary (and had the easier, higher-status job), had higher incidence of heart disease than did conductors (lower-status workers), who were constantly in motion walking up and down stairs (Morris et al. 1953). This finding suggests that the types of work that society places a high value on (those with a low amount and intensity of physical activity) may not be conducive to good health.

As the workplace in the United States increasingly adopts technological innovation and the nature of work shifts from activity-intensive manufacturing, agriculture, and construction to service and professional occupations, the amount of physical activity required to perform work will continue to decrease. Current projections on the workforce show that the jobs of the future are much more likely to be found in the service and managerial sectors, in which physical activity requirements tend to be more modest than they are in manufacturing, construction, or agriculture. Furthermore, workers in the United States continue to log

As technology continues to shape the character of work, the amount of physical activity required on the job is likely to decrease, placing workers at higher risk for diseases brought on by physical inactivity.

As the workplace becomes more sedentary, workers must find ways to incorporate activity into their everyday lives.

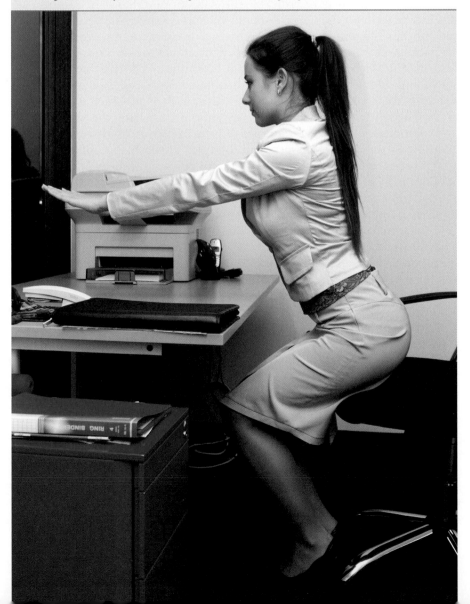

more working hours than in other countries. As recently as 2007 Americans worked nearly 5 weeks longer each year (>180 hours) than their European counterparts (Focus 2011), and given that workers' average commute to work has increased to an average of 46 minutes per day in the United States, workers may not have the time required to exercise daily (Gallup 2007). This has led some health scientists to recommend that people put more emphasis on integrating physical activity into their working lives, for example, by parking their cars farther from their offices or walking up stairs rather than taking the elevator (Dong et al. 2004).

The bottom line is that workers are expected to be at increased risk for diseases associated with reduced physical activity unless they supplement their work lives with physical activity during leisure (nonwork) time.

Physical Activity, Efficiency, and Injury in the Workplace

Work can be viewed as a process in which workers trade physical activity for compensation. Employers, in turn, translate workers' physical activity into production of goods or services. The more efficient this production is, the greater the profit the employer will enjoy. Thus, workers' physical activity—whether vigorous and expansive or mild and limited in scope—is the foundation on which the commercial enterprise rests.

Businesses can increase the efficiency of production in two ways. The first is to replace inefficient physical activity with lower-cost technology. Machines and robots now allow car manufacturers to produce many more cars per day at lower cost than they did when manual labor did all the assembly. But technology cannot replace all physical activity; in many jobs, the movement of workers remains critical to the efficient production of goods or services. In these cases, increases in production are sought via improvements in the efficiency and safety of workers' movements and working conditions. For example, the work of assembly line employees packaging television sets in boxes, of those picking fruit from trees or vegetables from fields, or of those operating machines in furniture and textile factories can be made more efficient and safer through changes in their movements, redesign of the machines with which they do work, or reconfiguration of the surroundings in which work is performed.

The physical activity professionals who specialize in improving the efficiency, safety, and well-being of workers are called **ergonomists** or **human factors engineers**. Although experts continue to discuss the differences between the roles of these specialists (some claim that the study of human factors is more theoretical and relies more heavily on psychology and that ergonomics is more practical and relies more on anatomy and physiology), the trend is for these terms to be used synonymously (Kroemer et al. 1994).

Ergonomists or human factors engineers apply their knowledge of exercise physiology, psychology, anatomy, and biomechanics in studying the demands of various jobs. (For this reason, a background in kinesiology is excellent preparation for advanced study in human factors or ergonomics.) These engineers then make recommendations regarding changes in movements used to perform the task, changes in the work environment, or changes in programs for training workers. Often, simply rearranging the location of objects to be assembled, adjusting the height of workers' chairs or tables, redesigning the shape of workstations, or modifying the order in which workers grasp components in the assembly process can markedly improve worker efficiency. Ergonomists usually are expert observers of the movements used in various kinds of work, and they make recommendations for change based on this critical analysis. Often ergonomists look for risks inherent in the work. A typical risk analysis of a work site is presented on page 48. Obviously, a thorough knowledge of the structure of the body as well as a complete understanding of how the body moves is essential to the work of ergonomists.

Productivity in the workplace suffers its greatest decline when workers are injured and miss work. This circumstance is especially troublesome when the injuries occur directly because of performing the physical activity required to carry out the job. Of course, workplace accidents remain a serious problem, but ergonomists mostly focus on injuries caused by the physical activity that a job requires. Many of these are musculoskeletal disorders that arise either from overexertion or from repetitive motion. According to the National Institute for Occupational

Safety and Health (NIOSH), women are particularly susceptible to such injuries, suffering an estimated 63% of all work-related repetitive motion injuries (NIOSH 2000). From 2009 to 2010, the incidence rate for musculoskeletal injuries due to overexertion that resulted in days away from work increased 4% among all workers; the rate among nurses and orderlies who often lift and transport patients increased approximately 10% during this time (U.S. Bureau of Labor Statistics 2011). Overexertion injuries result from application of maximal forces by the body, usually when the worker is not physically capable of such exertions (e.g., weak back muscles) or is poorly prepared from a positional standpoint (e.g., lifting without bending the knees). Most of these injuries occur when a worker is lifting, pushing, or pulling objects or holding, carrying, or turning objects (USDHHS 1997).

Repetition injuries are those that result when workers repeat the same series of movements over long periods. These injuries are also known as **cumulative trauma disorders**. Usually they result from repeated physical stress to joints and muscles, which in turn damages tendons, nerves, and skeletal structures. Assembly line workers who package items perform the same arm, hand, finger, and trunk movements several hundred, or even several thousand, times each day. Even the small range of motion involved in typing can produce cumulative trauma disorder. Although our bodies are remarkably durable and adaptable, sometimes work asks too much of them, and they suffer deterioration or collapse.

One of the most common types of cumulative trauma disorders is **carpal tunnel syndrome**, a painful injury to the wrist caused by repetitive movements. The many tasks that can cause this injury include manual assembling, packing, typing, driving, lifting, hammering, or simply sitting at a desk for long periods performing clerical work. One way in which ergonomic specialists attempt to reduce the likelihood of this disorder is to redesign the tools or workspace to facilitate safe and efficient operational movements. For example, the photo ahead shows a modified garden trowel that attaches to the forearm and has a right angle grip that is designed to reduce pressure on wrist joint and arm muscles and help ward off carpal tunnel syndrome.

Some forms of work can lead to overexertion, which can produce severe musculoskeletal injury, most often in the back. Other tasks require repetitive physical activity, which over long periods can damage muscles, tendons, and nerves. Such injuries usually require rehabilitation by physical and occupational therapists. Even sitting for long periods at your desk can lead to musculoskeletal injury. In an effort to prevent such disabilities, companies have designed a number of ergonomic chairs that place the body in the most anatomically correct position for doing desk work.

❚ Ergonomic analyses often lead to the design of special equipment to make work safer and more productive, such as this Easi-Grip Arm Support Cuff.

Physical and Psychological Demands of Work

What types of work require the most physically demanding forms of physical activity? This question is important not only to employers but also to labor unions, insurance companies, and employees, as well as to ergonomists. To determine the physical demands of work, the U.S. Bureau of Labor Statistics applies a formula that takes into account the amount of weight that a person is normally required to lift on the job along with other factors such as the amount of stooping, climbing, or balancing.

Using these criteria and taking into account how many hours beyond 40 hours per week a job requires, *Jobs Rated Almanac* (Kranz 2002) presents a calculation for a total physical demand score for each of 250 occupations. A low score signals relatively low amounts of physical activity and exertion. Although the ratings suggest that high-status jobs such as managerial and professional jobs require relatively low levels of physical activity, this is not always the case. Some types of blue-collar work are fairly sedentary, whereas many white-collar jobs have high physical activity demands. For example, the jobs of musician, photojournalist, physical therapist, undertaker, and fashion model have an average ranking of 175, whereas less prestigious jobs of office machine repairer, guard, electrical equipment repairer, automobile assembler, and railroad conductor have a mean ranking of 155.

From a total health perspective, we are also interested in the psychological stresses that various jobs place on workers. We know that the physical activity experienced in sport,

Typical Ergonomic Analysis of Worksite

Ergonomists often evaluate workplaces in order to improve safety conditions for workers. This is a type of worksheet that an ergonomist might complete when doing a very general review of a worksite.

Job: General Maintenance Worker in a Manufacturing Company

Work includes cleaning and mopping (wet/dry) floors and stairways; wiping, vacuuming, scrubbing, and polishing surfaces; receiving and storing shipments; shoveling snow; making minor repairs of blinds; cleaning windows; and maintaining equipment.

Musculoskeletal Demands: Indicate Yes/No

Does the job require the following?

Awkward posture

Bending or twisting at the waist to handle tools, loads, or parts

Lifting or carrying loads of more than 60 pounds (27 kilograms)

Pinch, pull, or power grip

Climbing a ladder

Frequent, repetitive motions

Flexing of the neck, shoulder, elbow, wrist, or finger joints

Forceful or sudden movements

Sustained contraction of muscles

Excessive flexing or extending of joints

Working Environment: Indicate Yes/No

Can the workplace be described as follows?

Too hot or too cold

Poorly lit so that it poses a danger

Noisy such that it distracts or produces hearing loss

Crowded so as to prevent comfortable work positions or movements

Dirty, slippery, cluttered, or obstructed

Lacking railings on stairways

Manipulation of Tools: Indicate Yes/No

Can the tools required for the job be described as follows?

Weigh more than 9 pounds (4 kilograms)

Expose worker to risk of cuts, punctures, or strains

Difficult to access and store

Require severe flexing of the wrist, neck or back

Vibrate excessively

games, and exercise during nonworking hours can help to reduce or eliminate stress. Our experience in the workplace suggests that high-prestige jobs that tend to require less physical activity may also be the most stressful. Are high-prestige jobs always the most psychologically stressful? *Jobs Rated Almanac* (Kranz 2002) also includes a ranking of jobs based on the stress levels they induce. These rankings are based on such considerations as whether the worker faces deadlines, is in competition with others, encounters physical hazards or uncomfortable environmental conditions, is expected to perform precise movements, or is responsible for the lives of others through his or her actions.

Of the jobs rated, the five least stressful (low score indicates low stress) were musical instrument repairer, florist, medical records technician, actuary, and forklift operator. The five jobs rated highest in stress were president of the United States, firefighter, senior corporate executive, racecar driver, and taxi driver. Although managers, supervisors, and professional workers do not always experience the highest levels of stress, the types of employment most likely to raise stress levels generally tend to be the jobs requiring low levels of physical activity. Thus, certain classes of workers—usually those responsible for making the major decisions for institutions and corporations—find themselves in a potentially health-threatening situation, subjected to this double whammy of high stress and low physical activity. These people may be especially vulnerable to heart attacks, hypertension, and strokes unless they pursue exercise outside of work.

The loss of key personnel because of death or disease is potentially devastating to businesses. To combat the health-eroding effects of high-prestige work, to ward off some forms of work-related injuries, and to ensure a vigorous and energetic workforce, many companies offer physical activity and fitness programs or provide discounts for employees to join commercial fitness centers. Employers that recognize the practical importance of such programs have maintained their vitality even during economic downturns. Between 2009 and 2010, for example, the number of companies (20,000 or more employees) that offer fitness programs or provide discounts grew 11% (Bernstein 2011).

With the advent of industrialization and technology, the physical activity requirements of work have diminished. The result has been the creation of jobs that threaten workers by imposing higher levels of stress and lower levels of physical activity. Businesses have attempted to solve the problem by sponsoring exercise and sport programs for employees.

The Sphere of Education

Physical activity plays an important role in the sphere of education (see figure 2.8). Education is essential both to preserve cultural traditions and to provide the knowledge and skills that enable society to progress. Physical activity is involved in all phases of education—from the eye movements required in reading and the wrist, finger, and arm movements required in writing to the more expansive forms of physical activity required in learning how to play a musical instrument or operate a power saw in vocational arts class. In almost every form of educational program, the aim is to change the behavior of those being instructed, in other words, to change physical activity patterns of those being instructed.

Although we will be most interested in educational programs in sport, exercise, and recreation, these are only the tip of the instructional iceberg. Look around and you will see physical activity instruction occurring everywhere! Physical therapists teach patients how to walk; dental and medical school faculty teach students how to perform the intricate movements required in surgery; fathers and mothers teach their children how to dribble a soccer ball; ergonomists train employees in new assembly techniques; American Red Cross staff teach would-be lifeguards how to rescue swimmers; senior automobile mechanics train new mechanics how to use tools at the local garage; and business school instructors teach secretaries-in-training how to operate word processors. Anywhere physical activity is important, you will find some form of instruction, whether formal or informal. Instruction in physical activity may be as universal as physical activity itself.

> The education sphere includes that aspect of our lives in which we set out to learn new skills or knowledge. Usually, physical activity plays an important role in this sphere, whether in connection with learning cognitive material or learning to perform physical skills.

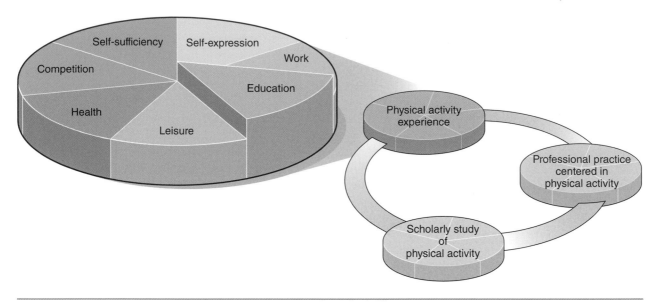

FIGURE 2.8 Education as a sphere of physical activity experience.

Instruction in Sport and Exercise

Not long ago, instruction in sport and exercise was limited largely to public school and college physical education programs, municipal recreation programs, and the military. Today, sport and fitness centers, resorts, corporations, hospitals, and clubs (tennis, golf, swimming) are also in the sport and fitness instruction business. Millions of children are taught how to play sports by their fathers or mothers or older siblings as well as by youth sport team coaches. Companies devoted to specialized instruction of young athletes, sport camps offered each summer by famous coaches, and age-group swimming and private gymnastics schools all are involved in physical activity instruction. Adults used to shy away from instruction in sport activities but now they seek it out, often paying as much as $75 to $100 per hour for lessons from top skiing, scuba, golf, or fitness training specialists.

Private clubs and commercial spas, gyms, ice rinks, riding stables, dance studios, and martial arts centers now compete with school-based programs in the instruction business. Many large corporations offer recreation programs for their employees, and many of these include basic instruction in sports and exercise. Instruction through the media has also become a huge business. Television's Golf Academy Channel offers instruction in golf 24 hours a day. How-to books on every conceivable sport line the shelves of the mega-bookstores, and sport specialty magazines include regular instructional features.

Coupled with this interest in learning to perform sports and exercise is a growing demand for knowledge about how to live a healthful lifestyle. Corporate managers, keenly aware of the economic advantages of a healthy workforce, are making instructional programs available at the worksite, often in collaboration with their health maintenance organizations or health insurers. Typically, these classes use a variety of methods to integrate theoretical information about exercise, nutrition, and body mechanics with practical experiences in the exercise room. The explosion of interest in learning about sport and exercise has led to an expansion of instruction beyond the walls of schools and colleges.

Physical Education

Although instruction in sport, fitness, and exercise occurs in myriad formal and informal settings, it is most visible and accessible in physical education classes offered as part of the school curriculum (see chapter 15). Our educational system is based on the belief that it is in the best interest of a democracy to make available to its citizens a free and accessible program of education. Instruction in sport, exercise, and fundamental movements has long been considered an integral component of the educational enterprise.

In this section we give cursory attention to the life experience you probably knew as "phys ed" or "gym class," which physical activity professionals more appropriately call "physical education." Because these programs have enormous potential for influencing the physical activity patterns of large segments of the population, they have attracted the attention of all physical activity professionals as well as public health officials. For this reason, you—as a future physical activity professional—should appreciate the potential as well as some of the problems associated with school-based physical activity programs.

Chances are that you spent a significant part of your public school education experience in physical education class. But although most of us have attended physical education class, we have not all had the same experience. School districts do not all teach the same program of physical activities. Sometimes teachers within the same school district or even within the same school teach different subject matter and use different methods. Perhaps you were fortunate to have had a competent and enthusiastic teacher who offered carefully structured and sequenced classes. If so, you probably learned much about physical activity. However, if you were unlucky you may have had a bored or unimaginative teacher who "threw out the ball" and offered little in the way of instruction. If that was your physical education experience, you probably learned less than your colleagues did who attended classes taught by more competent and energetic teachers.

Elements of a High-Quality Physical Education Program

- Adequate instructional time is allowed (at least 150 minutes per week for elementary school students and 225 minutes per week for middle school and high school students).
- All classes are taught by a qualified physical education teacher.
- Class sizes are reasonable.
- Proper equipment and facilities are available.
- Classes are inclusive of all students.
- The program provides for students with disabilities.
- Students are active for most of class time.
- Lessons are well designed.
- Students are given out-of-class assignments.
- Physical activity is NOT used as a punishment
- Appropriate use of physical activity and assessment tools.

Adapted from *Physical education analysis tool.* Centers for Disease Control and Prevention 2006.

Inevitably, your personal experiences in physical education will influence your conception of the subject. In any case, you should recognize that your experience—whether favorable or unfavorable—may not accurately reflect what physical education is like in most schools.

Effective educational programs are founded on clearly stated objectives. Objectives are vital maps for educators, pointing them toward the ends to be achieved with their instruction. Objectives determine what and how teachers will teach. Students should be aware of the objectives that are guiding the teachers' actions so that they can be partners in the learning experience. In your experiences in physical education classes, your teachers' objectives may not have been clear to you. Sometimes it may have seemed that the objective was developing fitness. At other times learning sport skills seemed to be the goal, and sometimes the teacher may have seemed most interested in developing students' sense of fair play. What, then, should be *the objectives* of physical education?

Physical education is the only near-universal program of sport and exercise instruction available to young people. For this reason, it should be of the highest quality possible.

School superintendents, principals, classroom teachers, parents, physical education teachers, and students all may have slightly different views concerning the objectives of physical education. Physical educators even disagree among themselves concerning the most important objectives of school physical education programs. As a result, experts have been unable to come to a universal agreement about what constitutes a physically educated person. In an effort to bring the factions closer together, the National Association for Sport and Physical Education (NASPE), working hand in hand with public school teachers and university professors, published a list of content standards for physical education programs. They are presented in the form of behaviors that should be demonstrated by a physically educated person (NASPE 2004).

Objectives of Physical Education

A physically educated person does the following:

1. Demonstrates competency in motor skills and movement patterns needed to perform a variety of physical activities.
2. Demonstrates understanding of movement concepts, principles, strategies, and tactics as they apply to the learning and performance of physical activities.
3. Participates regularly in physical activity.
4. Achieves and maintains a health-enhancing level of physical fitness.
5. Exhibits responsible personal and social behavior that respects self and others in physical activity settings.
6. Values physical activity for health, enjoyment, challenge, self-expression, or social interaction.

Reprinted from *Moving into the future: National physical education standards: A guide to content and assessment*, 2nd ed. (2004) with permission from the National Association for Sport and Physical Education (NASPE), 1900 Association Drive, Reston, VA 20191-1599.

Although developing physical fitness probably is the top objective of most physical education teachers, only two of the six NASPE standards focus on fitness and achieving a physically active lifestyle. Others emphasize the need to attain competency in a diverse number of activities, to experience social and psychological growth, to learn concepts about motor skills, and to have an opportunity for fun and self-expression. A growing trend is to view physical activity programs in schools as an opportunity to teach children social responsibility. You can begin to see the problem. Even this narrow list of standards leaves a great deal of room for diversity in objectives, instructional approaches, and content. For the time being, school physical education programs will likely continue to aim at a cluster of different objectives, each requiring slightly different approaches and content. Individual teachers will also likely continue to emphasize the objectives that most closely align with their personal philosophies of physical education.

Trends and forces in society influence what physical education teachers emphasize in their classes. This makes sense because public education must be responsive to the needs of society. For example, the growing popularity of sport during the early 20th century and the burgeoning interest among educators in the phenomenon of play as an educational experience were among the reasons that schools began to substitute sports in the physical education curriculum for the dull and often painful formal systems of exercises and gymnastics that were typical at the close of the 19th century (see chapter 5).

Physical educators have many objectives but traditionally have given priority in the curriculum to exercise and sports. Why is it important that public education pursue these objectives? Let's briefly examine each one.

Teaching Physical Education for Physical Fitness

The promotion of physical health through exercise has been an objective of physical education from its inception. The extent to which this objective is emphasized in public schools has varied according to pressure exerted by society. Today, fitness has been moved to the top of the agenda, primarily because public health officials have become alarmed at the overall poor fitness and sedentary lifestyles of children.

ACTIVITY 2.2

Education
Evaluate three case studies in this activity in your web study guide.

The population of "couch potatoes" (or "cyber potatoes," those who spend excessive leisure time at the computer) is growing, and many of them are children who haven't yet reached middle school. The Neilson Company estimates that kids spend on average 7 hours and 38 minutes each day watching television or videos or playing video games. Preschoolers spend 32.5 hours per week in front of the television. Older preteens (ages 6-11) spend nearly as much time (28 hours) (Neilson Wire 2009). Estimates are that 71% of 8- to 18-year-olds have TVs in their bedrooms. Adding to the problem is that in 53% of households that include 7th to 12th graders, there are no rules governing TV watching (Rideout et al. 2010).

Are You a Physically Educated Person?

On the basis of the NASPE standards and objectives for physical education listed previously, do you consider yourself a physically educated person? Think about your elementary and high school physical education experiences. Was your teacher enthusiastic? Did he or she model a physically active lifestyle? How often did you have physical education class? Was it a required subject in your school? Were you taught mostly lifetime sport activities (e.g., tennis, golf, archery) or team sports? What factors made the experience enjoyable or unenjoyable? What would you have done to increase its effectiveness?

One effect of these sedentary lifestyles has been an increase in health problems associated with physical inactivity. You will become painfully aware by reading this text that overweight and obesity have been declared by many health professionals as America's number one health problem. How bad is it? The percentage of children aged 6 to 11 years in the United States classified as obese increased from 7% in 1980 to nearly 20% in 2008. The percentage of adolescents aged 1 to 19 years who were obese increased from 5% to 18% over the same period (CDC 2012a).

What physical educators find especially alarming are the growing numbers of overweight and obese children and adolescents in their classes. Figure 2.9, which shows rates of obesity from ages 2 to 19, depicts an alarming jump in obesity from ages 2-5 to ages 6-11, the period when elementary school physical education programs should be playing a role in keeping the epidemic under control. Poor diet is a contributing factor, of course, but so is the lack of participation in regular vigorous physical activity. In Canada, it is estimated that 25% of children between 5 and 17 are obese. Unfortunately, only around 4% of Canadian children and youth accumulate 20 minutes of vigorous physical activity at least 3 days a week; 6% accumulate 10 minutes; and 11% accumulate 5 minutes (Statistics Canada 2011).

One might think that the public schools would have recognized the dangers posed by this state of affairs and increased the number of hours of physical education required each week.

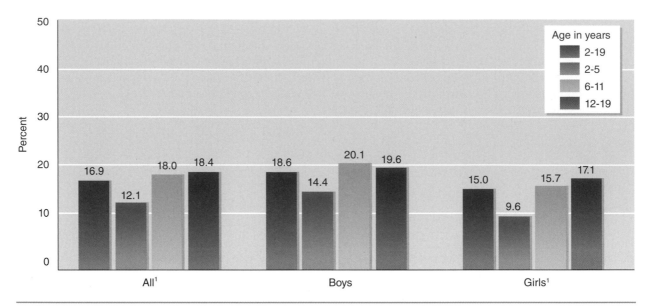

FIGURE 2.9 Prevalence of obesity among children and adolescents aged 2 to 19 in the United States from 2009 to 2010. [1]Significant increasing linear trend by age ($p < .005$).

Center for Disease Control and National Center for Health Statistics.

After all, as Lawrence Locke, professor emeritus at the University of Massachusetts, has pointed out, "School physical education is an experience through which almost all of our population passes, at least in some form, and it is the single place where most of them encounter formal sport and exercise fitness activities before adulthood. Those encounters also take place at precisely the time when lifestyles are being examined and adopted, and when possibilities for the self are being accepted or rejected" (Locke 1996, p. 429). Although the logic of Professor Locke's argument is impeccable, the trend has been to cut rather than expand physical education programs because of pressures on schools to devote more time to math and science. For example, data reported by the CDC (2012c) show that in 2011, 52% of high school students (68% of 9th graders but only 38% of 12th graders) attended physical education class at least once in an average week. The percentage of high school students attending physical education class daily in 2011 (31%) was only a fraction higher than it was in 1995 despite increasing public concerns about obesity and lack of physical activity among young people.

Moreover, access to physical education class is related to socioeconomic status (SES); as of 2007 only 49% of low-SES students went to schools where physical education was required compared with 59% of high-SES students, this despite the fact that obesity affects students of lower SES disproportionately. Contrary to popular opinion, varsity sports do not make up for the relatively low participation rates in physical education. Only 33% of girls and 37% of boys participate in varsity programs (Johnson et al. 2007).

Recognizing the future effect that childhood obesity could have on the national health budget, some state legislatures have mandated daily physical education for all students, reversing the trend of the past few decades. Still, changes will have to be made in the way some physical education classes are conducted if they are to yield their intended health benefits. Sometimes physical education classes include very little moderate to vigorous physical activity. For example, a study of California public schools found that students were sedentary most of the time during physical education classes; only 4 minutes every half hour were spent in what scientists would classify as vigorous activity. Students enrolled in larger classes were appreciably less active than students in classes with fewer than 45 students. Thus, it appears that mandating physical education in the schools does not always ensure that such programs will have their intended effects. Clearly, more attention needs to be given to increasing the amount and level of physical activity children engage in while in class (California School Boards Association 2009).

Television: A Formidable Challenge to Physical Educators

A study that tracked the physical activity level, body mass index, and television watching of nearly 12,000 Canadian young people aged 5 to 19 underscores the uphill battle that physical education teachers face. The study found that 17% of subjects were overweight and 11% were obese. Adjusting for each child's age, sex, and parental education, the investigators found that the odds of a child's being obese decreased by 20% for every 3,000 steps walked or run per day and increased by 21% for every 30 minutes of television watched per day. The authors concluded, "Television viewing is the more prominent factor in terms of predicting overweight, and it contributes to obesity, but steps/day attenuates the association between television viewing and obesity and therefore can be considered protective against obesity" (Tudor-Locke et al. 2011, p. 66).

But there are some encouraging signs to indicate that the nation is beginning to take the obesity problem in children seriously. The CDC, a national agency responsible for overseeing the health of the nation, called attention to the sedentary ways of the American population in 1996 with the publication of *Physical Activity and Health: A Report of the Surgeon General* (USDHHS 1996). This was preceded by the USDHHS publication *Healthy People 2000* (USDHHS 1990), a statement of health goals for the year 2000. After assessing the limited success at attaining these goals, the USDHHS modified and restated them in *Healthy People 2010* (USDHHS 2000) and has since published *Healthy People 2020* (USDHHS 2012), which sets even higher expectations. All these reports give a great deal of attention to the problem of physical inactivity among young people and recommend an increase in the percentage of schools requiring physical education.

Teaching Physical Education for Sport Skill Development

Developing physically fit students is an important objective, but most teachers reach for objectives far beyond fitness. Sport education can help achieve some of these objectives, such as developing responsible personal and social behavior and providing opportunities for enjoyment, challenge, and self-expression. Sport also can be an important means for attaining and maintaining physical fitness. If students do not learn to incorporate moderate to intense forms of physical activity into their lifestyles for the remainder of their lives, making students fit while they are in school will achieve little. What good will we have done if we create a population of fit young people who eventually slip into sedentary lifestyles after they graduate?

Coupled with this issue is the need to develop competency in these lifetime sport skills. People who have confidence in their ability to swim, play tennis, golf, in-line skate, or play squash are more likely to engage in these sports throughout their lives. Unless physical education classes are conducted daily, teachers will have little chance to develop skill competencies in their students.

Because of this emphasis on lifelong patterns of physical activity, high school physical education programs should include a heavy concentration of lifetime activities such as tennis, golf, handball, or racquetball rather than team sports, which people are less likely to play following graduation. Yet, despite health and education leaders' emphasis on the importance of featuring lifetime sports in the curriculum, available data suggest that team sports, not lifetime sports, continue to dominate the American public school physical education curriculum. Basketball, baseball, volleyball, and soccer head the list of sports taught in high school curriculums. Swimming, jogging, tennis, racquetball, and hiking—physical activities that can be carried over into adulthood—are much less likely to be taught (USDHHS 1996). Clearly, much work needs to be done if physical education classes are ever to realize their potential for increasing the levels of physical activity of the general population.

Equally discouraging is a finding by researchers at Cornell University showing that increasing the physical education requirement—without changing the way physical education

Surveys suggest that America and Canada are experiencing an epidemic of adult and childhood obesity. Increasing time allotted to physical education programs in the schools would seem to be one way to counteract this trend, but simply requiring physical education may not lead to a reduction of overweight and an increase in physical activity in youths.

programs are organized or taught or increasing the general physical activity patterns of children when out of school—is unlikely to remedy the childhood obesity epidemic. Students who were required to attend physical education class were active only 31 more minutes per week than students whose schools had no requirement. Although requiring physical activity increased the number of days per week that girls reported exercising vigorously or engaging in strength-building exercises, it was not associated with a lower incidence of overweight in students (Cawley 2007).

The Sphere of Leisure

Another area of our lives in which physical activity takes place is leisure (see figure 2.10). Although leisure is not the same as free time (discussed subsequently), most laypeople think of it that way. What do you do in your free time—play sports, exercise, read, attend concerts, watch television, hike? The term *leisure pursuits* is often used synonymously with *play* or *recreation.* But leisure is more a state of mind rather than simply free time. Sometimes we choose to work in our free time. Have you ever given up an opportunity to spend a summer weekend (free time) camping or at the beach, instead volunteering to work an extra shift at your job? If so, was your work time "free time"?

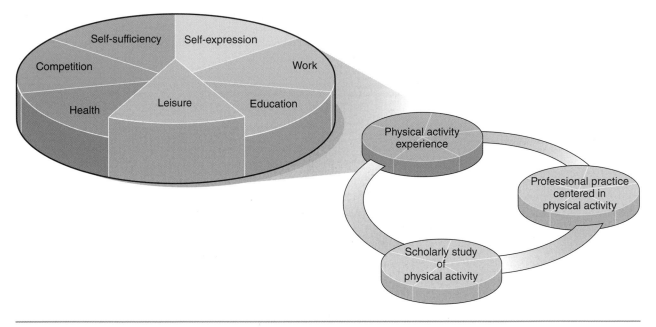

FIGURE 2.10   Leisure as a sphere of physical activity experience.

As you can see, the terms *leisure* and *free time* are a bit more ambiguous than they first appear. Can you really work in your leisure time? You might have noticed that some people play softball or soccer with an intensity that seems to be anything but leisurely. Conversely, can people play (be at leisure) while they work? Perhaps. Artists, novelists, and professors sometimes seem to approach their work almost as if it were a leisure experience. Thus, a good place to start in our examination of the leisure sphere of physical activity experience is to clarify what we mean by free time and leisure.

Distinguishing Leisure From Free Time

When we are able to disengage from our everyday lives and carve out minutes and hours to do what we want to do, we often refer to this period as **free time**. We are free from our everyday routines and free to do what we want to do, when, where, and with whom we choose. Although we often use the terms *free time* and *leisure* synonymously, experts distinguish between the two.

Leisure theorists describe **leisure** not as free time but as "a state of being" (De Grazia 1962). Philosophers, beginning with the ancient Greeks, have struggled to describe this state, but none have done so with complete success. In its purest state, leisure is a feeling that all is well with the world, a feeling of supreme contentment. Leisure can provoke feelings of celebration and wonder, creativity and discovery, excitement and reflection. Pieper (1952) described leisure as the "basis of culture," the psychological and spiritual disposition that is the fundamental prerequisite for the great works of art, music, theater, and philosophy.

How do we achieve this state of being that we call leisure? Sometimes we achieve it while engaged in sedentary activities such as sitting beside a gurgling stream, reading a novel, or listening to music. At other times, we can achieve the state of leisure through large-muscle activities such as running, hiking, surfing, or skiing. Precisely what this state of being is or how it comes about may never be explained fully. One thing seems clear. Leisure has the effect of clarifying to ourselves and others who we truly are when we are free from the sense of obligation, anxiety, and pressure that can confront us in our everyday lives. And when we experience leisure, we look forward to experiencing it again and again.

To use our free time to achieve a state of leisure, we must be able to divorce ourselves psychologically from other aspects of our lives, a challenging task for many people. In addition, we must choose an activity that is conducive to a state of leisure. Although many of us choose to watch television during our free time, a study of teenagers showed that they do not find it a satisfying experience (Csikszentmihalyi 1990b). Hollow free-time experiences are not leisure experiences. They don't challenge us, stimulate our imaginations, or help us reveal our identities the way that true leisure experiences do. Thus, if we are to convert free time into leisure time, we must use it for participating in activities that tend to nourish and maintain the state of being known as leisure.

> Leisure is a state of being, and free-time activities can help us attain this state. Large-muscle physical activities such as sport and exercise have the potential for nourishing and maintaining a leisure disposition.

❚ Any activity that allows us to feel at peace with the world and brings us a sense of contentment can be labeled a leisure activity.

Physical Activity as Leisure Activity

The entire range of leisure pursuits—from sedentary activities such as chess or reading to large-muscle activities such as water skiing or softball—is the focus of study in the discipline known as **leisure studies or recreation**. Kinesiology concerns itself primarily with the large-muscle forms of leisure pursuits that require substantial physical activity. If this leads you to wonder whether the interests of kinesiologists overlap with those of leisure professionals, the answer is yes. Faculty and students in departments of leisure studies and departments of kinesiology both view sport and exercise as a legitimate part of their fields of study. Although it might seem that this blurring of the boundaries of the two disciplines would be problematic, it has existed for several decades and has not deterred the continuing expansion and refinement of each of these exciting areas of study.

Activities such as golf, folk dancing, softball, hiking, long runs in the park, and boating have great potential for putting us at leisure. But we can approach these in ways that make such a leisurely disposition impossible. For example, becoming upset over an official's call, cheating, intentionally injuring an opponent, and exercising under the compulsive stimulus of anorexia nervosa are not compatible with a leisure state. The context in which the activity takes place is also important. Is it possible for athletes to attain a state of leisure when 80,000 spectators are sending up a deafening roar from the stadium? In the final analysis, whether or not a sport or exercise is truly a leisure pursuit depends on its nature, the context in which it is pursued, and the motivations and attitudes that participants bring to the activity. Professionals in the leisure industry try to help people make choices about their free time that increase the state of being that we know as leisure.

At best, free-time activities of sport and exercise offer us only the potential for achieving the state of leisure. Whether or not a physical activity is truly a leisure pursuit depends on the nature of the activity and the context in which we pursue it. The challenge of the physical activity professions is to teach people to participate in free-time physical activity pursuits in ways that nourish the disposition known as leisure.

Because your future as a physical activity professional will likely be affected by participation trends in large-muscle **leisure activities**, it is worth taking the time to examine trends of participation in this important sphere. Remember, kinesiologists are particularly interested in leisure-time pursuits that involve relatively large amounts of physical activity. Here the data are somewhat encouraging. The proportion of adults reporting no leisure-time physical activity decreased from 31% in 1989 to 25% in 2002 and was still at 25% in 2008 (CDC 2011). The rate of participation for men is higher than that for women across all age groups, and participation declines as a function of age. You may not be as active in your leisure time today as you should be, but the data suggest that your activity levels are likely to decrease even more as you take on responsibilities of job and family.

What do people do in their leisure time? (Although competitive sports clearly fit under the leisure activity umbrella, we will look at their popularity under the competition sphere and focus here on large-muscle, noncompetitive forms of leisure activities.) In 2011, the most popular form of leisure-time physical activity in the United States was exercise walking (>97 million people over 7 years of age reported walking for exercise at least once during the past year). The next highest participation rates were in exercising with equipment (56 million), swimming (46 million), camping (43 million), bicycle riding (39 million), and bowling (35 million) (National Sporting Goods Association 2011).

Because surveys of physical activity tend to vary widely, in terms of both the questions asked and the populations polled, research on leisure-time physical activities is, at best, an inexact science. With this caveat in mind, those tempted to think that Americans are fairly active in their leisure time should note that the same National Sporting Goods Association survey cited previously showed a marked decline in participation in many high-active leisure activities over the preceding 10 years. For example, participation in wilderness backpacking and camping declined 24% between 2001 and 2010, golf by 18%, wrestling by 17%, and

Although participation in some recreational leisure-time activities involving moderate to intense physical activity remains high, the rate of growth appears to have slowed and in some cases declined, whereas participation in more sedentary activities appears to be on the rise.

National Physical Activity Plan

In 2006, a National Physical Activity Plan for the United States was developed by participating agencies and nongovernmental organizations. Its purpose is to issue policies, programs, and initiatives to increase physical activity in all segments of the population by coordinating efforts from business and industry; health care; education; mass media; parks, recreation, fitness, and sports; public health; transportation; land use and community design; and volunteer and nonprofit agencies. The coordinating committee along with more than 300 experts appointed to working groups from each of the sectors listed here drafted an initial plan, which then went through several revisions. The final plan has been published and can be downloaded at www.physicalactivityplan.org/.

Reprinted, by permission, from *National physical activity plan.*

Spotlight on Research: The Hidden Benefits of Leisure

For most people, the enjoyment that leisure brings is enough justification for seeking it out. Now researchers have discovered that taking time for leisure activities has hidden benefits: It can improve both physical and mental health. In a study of 1,400 adults, researchers found that those with high levels of engagement in a variety of leisure activities, even activities requiring low levels of energy such as visiting friends or family, engaging in religious activities, going to clubs, or playing sports, had lower blood pressure, waist circumference, and body mass index than those with lower levels of engagement. The participants who engaged in wide ranges of leisure activities also had more diverse social networks, higher feelings of satisfaction, and lower levels of depression; they slept better and exercised more. The authors point out that "engaging in multiple types of leisure activities plays a role in buffering the negative psychological impact of stress" (Pressman et al. 2009, p. 730).

baseball by 16%. Can you think of reasons why participation in these activities may have declined?

Watching Sports

One form of sedentary leisure activity clearly on the rise is watching sports. People spend more than $1 billion on tickets to professional and amateur sporting events each year, approximately the same amount that they spend on all movies and 60% of the total that they spend on all theater, opera, and nonathletic entertainment sponsored by nonprofit organizations. More than 188 million spectators attended professional and college baseball, basketball, football, and professional hockey games in 2010, up from approximately 150 million in 1985. Annual spectator attendance more than tripled in women's college basketball between 1985 and the turn of the 21st century (U.S. Census Bureau 2002).

Such data, however, don't give the complete picture. For example, they don't take into account the 11.5 million fans who watched high school sports or the millions of others who watched youth sport teams and other amateur sports. Nor do they include the even greater numbers who watch sports on television. An estimated 111.3 million people watched the New

York Giants defeat the New England Patriots in the Super Bowl in 2012, narrowly bettering the record of 111 million who watched the event the year before (Bauder 2012).

The growth in sport spectatorship represents something of a paradox for the physical activity profession. The jobs of many physical activity professionals—coaches, athletes, and athletic trainers—depend on the continued viability of sport as mass entertainment. Sport management, an increasingly popular career option for physical activity professionals, prepares specialists who plan, market, stage, publicize, and supervise mass spectator sport events. Their job is to make watching sports a profitable and enjoyable enterprise by increasing the number of people watching sport events. At the same time, the rise in mass spectatorship, a largely sedentary activity, works at cross-purposes with the profession's efforts to increase physical activity in the general population. When one considers the markedly upward trend in spectators alongside troublesome data presented in this chapter regarding declining or stable participation rates in physical activity, it is difficult not to gain the impression that we are becoming a nation of spectators rather than active participants.

Of additional concern to physical activity professionals is whether watching sports constitutes a viable leisure pursuit as that term is understood by leisure theorists. Does it nourish the psychological or spiritual state called leisure? Some believe that it can. Michael Novak's popular book *The Joy of Sports* (1976) is an eloquent testimony to the fact that watching sports may add to the quality of our leisure lives. Yet an obsessive preoccupation with the fortunes of one's favorite sport team can be socially unhealthy, especially when it leads to neglect of more important issues such as human relationships. As enthusiasm for watching sports continues to increase, the profession will have the challenge of differentiating between healthy, constructive spectatorship and addictive, destructive preoccupation.

Although participation in many sports appears to have stabilized and in several cases declined, interest in watching sports continues to grow, abetted by a thriving commercial and television market. This trend toward mass watching or mass vicarious participation through fantasy sport leagues as opposed to actually playing sports may not be compatible with national health efforts to increase the physical activity participation levels of the population at large.

Aging and Leisure Activities

Not long ago, most people believed that life essentially ended at 65. Persons who were elderly faced the problem of finding sedentary ways to fill their free time until they died. Thanks to advances in health maintenance and the growth of the field of gerontology, we now know that people over 65 have enormous potential to learn new skills, explore new vistas, and engage in forms of physical activity that most elderly people wouldn't have dreamed of engaging in 20 years ago. As one example, more than 4,800 athletes from 92 countries participated in the 2011 World Masters Athletic Championships in Sacramento, California. Sixty-four-year-old Phil Raschker of the United States won the high jump (1.35 meters) and the long jump (4.27 meters). Ninety-five-year-old New Yorker Ida Keeling, at 4 feet 6 inches (137 centimeters) and 83 pounds (38 kilograms), set a record for her age group in 2011 by running 60 meters in less than 30 seconds in a local race. Seventy-three-year-old Sandy Pailis of Tempe, Arizona, had started resistance training 10 years previously when she was diagnosed with osteoporosis, a deteriorating disease of the bones. She competed in the local senior Olympics, where she squatted 135 pounds (61 kilograms), bench pressed 80 pounds (36 kilograms), and deadlifted 185 pounds (84 kilograms).

Clearly, getting old does not mean that people have to curtail their leisure physical activities. One study found, for example, that in 2009 more than 2.5 million people older than 65 participated in aerobic exercise programs, nearly 2 million bicycled, more than 6 million exercised with equipment, and more than 12 million walked regularly for exercise (U.S. Census Bureau 2012). As older populations continue to recognize the benefits of leading a healthy lifestyle—throughout the span of their lives—we can expect the numbers of physically active older adults seeking physically active leisure activities to increase over the next decade. Working with this population provides an exciting professional opportunity for today's students who are majoring in kinesiology.

The Sphere of Health

Attending to our personal and community health needs consumes a large part of our normal daily experiences. We bathe, dispose of our garbage, ensure that we drink clean water, eat proper foods, and visit physicians when we are ill to maintain our health, which in turn enables us to be productive citizens and enjoy life. All these health-related activities involve physical activity in varying degrees. We can't, for example, brush our teeth, bathe ourselves, or carry out the garbage without performing physical activity. But physical activity also intersects with this sphere in a much more direct way (see figure 2.11). We now possess scientific evidence that physical activity performed in the right amounts and with sufficient frequency contributes to our health in many important ways.

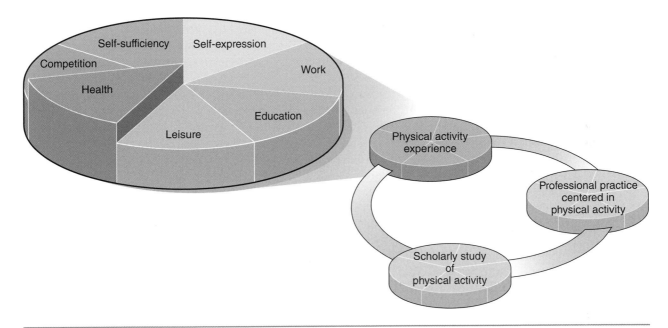

FIGURE 2.11   Health as a sphere of physical activity experience.

We now know that moderate to vigorous physical activity performed regularly and at safe levels almost always results in health benefits. We also know that the payoffs do not result only from workouts or painful exercise routines performed specifically to get in shape or become fit. Physical activity performed as part of work, sport participation, or another leisure pursuit also provides benefits. At the same time, it seems equally clear that the safest, most effective, and most efficient routes to attaining health benefits from physical activity are carefully designed programs supervised by exercise professionals who are well versed in the science of physical activity. As the public comes to understand this more, the demand for highly trained fitness leaders and consultants will continue to increase.

Physical Activity, Health, and the National Interest

Why is it important for Americans to maintain and improve their health through physical activity? For one thing, healthy people are vital to the national economy. A sick population not only incurs medical expenses but also drains the economy through lost productivity. Consider this: The United States spends a greater proportion of its gross domestic product on health care than any other industrialized nation. In 2011, total health care expenditures in the United States were $2.7 trillion, reflecting a growth rate of 3.9%, the same observed in 2010. The figure is expected to reach $4.6 trillion by 2020 (U.S. Centers for Medicaid and Medicare Services 2012). By offering consultation and leadership in increasing the physical activity of the general population as a way of combating these figures, kinesiology offers great promise to our society.

In 2009, approximately 600,000 people in the United States died from heart disease—25% of all deaths. An additional 129,000 died from stroke, nearly 69,000 from diabetes, and 25,000 from ailments related to high blood pressure (CDC 2012b). As a future kinesiologist you should know that the effects of all of these diseases can be mitigated by regular exercise and healthy eating. As a taxpayer, you have a large stake in this issue. As health care costs increase in the private sector, the amount spent by state and federal governments on health care costs increases also. Obviously, these expenditures represent substantial drains on state and federal budgets, limiting funding for education and other social programs. The massive spending also poses a potentially significant problem for future generations who will have to foot the bill. None of this, of course, measures the personal effect of poor health. Each incidence of illness brought on by lack of physical activity represents one person who has lost his or her potential for living an active, productive, enjoyable life.

Many of these diseases are related to obesity, which is a problem not only of childhood but also of adulthood. Sedentary lifestyles are a major contributing factor. The U.S. Bureau of Labor Statistics (2012) estimated that in 2011, people older than 15 watched television 2.8 hours per day, which represented about half of their total leisure time. This is more than three times the amount of time spent on the next most popular leisure activity—socializing with friends and family. According to the CDC, 34.2% of the American population older than 20 can be classified as overweight, another 33.8% are obese, and 5.5% are extremely obese (CDC 2010b). There is a strong association between obesity and socioeconomic status. People who are poor and uneducated are more likely to be obese than those with higher incomes and levels of education. Likewise, people of lower levels of income and education are less likely to engage in physical activity and less likely to eat a well-balanced, nutritious diet. Regular physical activity performed at moderate to high levels of intensity is a front line of defense not only against obesity but also against heart disease, stroke, and diabetes.

Physical activity figures so prominently in U.S. health objectives because it has been shown to prevent and help rehabilitate the costly diseases described earlier. In addition, physical activity has been associated with a reduced risk of colon cancer and appears to help lower the risk of developing non-insulin-dependent diabetes mellitus. Regular physical activity may enable those who are elderly to live independently for a longer time and avoid falls and other injuries. Regular exercise also lowers the risk of mortality for both younger and older adults. The response of U.S. health agencies to the alarming increase in obesity rates was to draft a list of national health objectives for the year 2010, which has been updated with a target date for 2020. *Healthy People 2020* includes recommendations for increasing the level of physical activity among Americans. Among the goals are to reduce the proportion of the population who engage in no leisure-time physical activity to 32.6% (from the current 36.2%), increase to 47.9% (from 43.5%) adult participation in physical activity of at least moderate intensity for at least 150 minutes per week, and increase to 24.1% (from the current 21.9%) the proportion of adults who engage in muscle strengthening exercises 2 or more days per week (USDHHS 2002). Clearly, the profession of kinesiology is well positioned to make a significant contribution to the attainment of national health goals.

The failure to make physical activity a part of our daily lives has led to a high incidence of deaths because of stroke, heart disease, and diabetes. Regular moderate to vigorous physical activity could prevent many of these deaths, yet large segments of the population remain sedentary.

The Darker Side of Physical Activity

At this point, a reality check might be helpful. Obviously, physical activity is a valuable adjunct to healthy living, but it is not an unqualified benefit. Just because some physical activity is good, it doesn't necessarily follow that more is better. What physical activities you select and how you go about performing them also can have detrimental effects. The modern physical activity professional must be acutely aware of the negative as well as positive aspects of physical activity.

Not all injuries have their basis in acute trauma. Often, overexercise, whether in the form of weight training, running, aerobics, or repetitive motions in work settings, can result in stress fractures, strained muscles, inflamed tendons, and psychological staleness. Sometimes the damage brought on by long-term wear and tear, although more insidious, can be more traumatizing. Repeated injuries to the knees, fingers, wrists, hips, and backs take their toll on athletes not only in the weeks and months following the injury but in later years as well. Years of pitching a baseball can lead to rotator cuff injuries; stress fractures are common to gymnasts.

We have already noted that carpal tunnel syndrome, a painful and debilitating injury to the hand and wrist brought on by repeating the same hand motions day after day as in typing and assembly tasks, plagues thousands of workers.

As noted in the sidebar, football produces a stunning number of injuries, many of them serious. But sports injuries are not limited to football. Most sports involve some degree of risk. According to the American Association of Neurological Surgeons, 446,788 sports-related

Injuries Plague Football Players

Few sports are more popular in North America or more dangerous than football. Football captures the imagination of North Americans in almost inexplicable ways from the time we are very young until we are old. Sometimes it is very difficult to objectively examine those aspects of our culture with which we are most familiar. It is like this with football, a sport associated with the highest number of direct catastrophic injuries for any sport as reported by the National Center for Catastrophic Injury Research (Boden et al. 2007). A person from another culture who wasn't familiar with the game might view with horror the violence and injuries that occur on the football field. But for those of us who grew up with the game, the injuries suffered in association with the game seem somehow "normal." Although the rate of catastrophic injuries suffered in football is relatively low in terms of the 100,000-player exposure rate, other less dramatic but no less serious injuries have been traced to the game.

For example, recent media accounts have highlighted the terrible, lifelong damage incurred by professional football players. First-person accounts of men in their late 30s unable to walk unassisted due to repeated joint trauma, of elbows that don't work, or of fingers broken so many times they can no longer hold a coffee cup have flooded the sport pages. Perhaps of greatest concern is the scourge of brain concussions suffered by football players. The disorienting effects of a hard collision, often joked about by sportscasters as "having your bell rung," occur when the brain collides against the interior of the skull, injuring vital nerve tissue. The effects of repeated head trauma (often glorified by players as "head banging") are cumulative, and those who have spent many years playing the game sometimes suffer permanent disability including *dementia pugilistica*, a brain disorder more commonly associated with boxers. Newspapers report heart-rending stories of professional football players in their 40s who are unable to keep a job or function as effective members of society as a result of mental disorders traced to repeated trauma they suffered playing a brutal game. Recently attention has shifted to high school football, where, because the athletes' brains are still developing, catastrophic head injuries can have even more serious and long-lasting effects. More difficult to identify are the long-term effects of repeated "minor" head injuries, because the symptoms may not appear for years. Sports are wonderful experiences, but our enthusiasm should never blind us to the real risk of injury they pose.

1. Do you believe football is a body-threatening sport? Is it a violent sport?
2. If so, what factors do you believe contribute to the violence?
3. What recommendations would you make to lower the injury rate in football?
4. What are the implications for kinesiology majors who plan careers in athletic training and sports medicine? Coaching? Sport management? Sport psychology?

head injuries were reported in 2009: 47,000 were incurred while playing football, approximately 85,000 while bicycle riding, and 39,000 while playing softball and baseball (American Association of Neurological Surgeons 2011).

Physical activity can also become problematic when the urge to do so is compulsive and habitual, and when people develop an unhealthy emotional dependence on exercise. Those with a physical activity addiction focus so single-mindedly on an activity that they neglect daily obligations important for a healthy and well-balanced life. When does devotion to an activity such as running or golf become addictive and harmful? No clear boundaries separate healthy from unhealthy exercise, but concern might arise when physical activity interferes with work or harms important personal relationships. Another signal is the onset of emotional disorders when the opportunity to engage in the activity is taken away. Many avid runners, for example, experience acute bouts of depression when they are injured and unable to run. High school wrestlers who must "make weight" to compete within a specific weight category often engage in harmful practices. Young women interested in gymnastics or figure skating can become preoccupied with weight loss and continue to exercise when body mass is already minimal. As with most other activities in our daily lives, the best prescription for a happy, healthful, and balanced life is to pursue physical activity with a sense of dedication and commitment but in moderation.

The response of kinesiology professionals to the problem of injury during physical activity has been vigorous. Kinesiology departments have cooperated in establishing excellent programs of athletic training and sports medicine and overseeing the tremendous growth of the National Athletic Trainers' Association. As you will see in chapter 14, preventing and treating athletic and exercise injuries make up one of the fastest-growing specializations in the field of kinesiology.

> When pursued in moderation with an eye toward a balanced life, physical activity is desirable. When performed under circumstances that put the integrity of the body at risk or induce questionable behavior patterns and psychological states, physical activity is undesirable.

The Sphere of Competition

Do you like to compete against others? Does the opportunity to compete with classmates for the highest grade, to compete with coworkers for the worker of the month award, or to compete against others for the league championship heighten your interest in the activity? Chances are that competition plays an important role in some part of your life. In fact, it is difficult for anyone to avoid competition completely, and chances are that physical activity figures prominently in most forms of competition that you pursue (see figure 2.12).

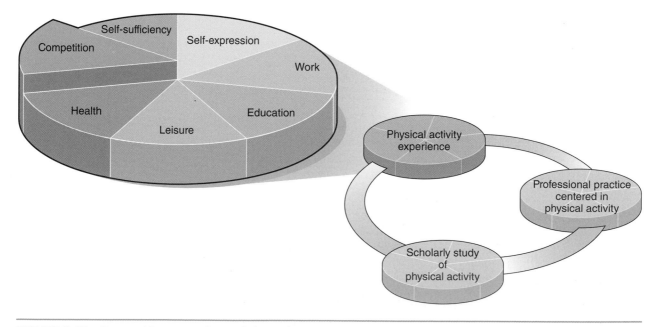

FIGURE 2.12 Competition as a sphere of physical activity experience.

Competition is not itself an activity but rather an organizing principle for activity. It can add to or detract from the enjoyment of activities, but it also usually increases one's level of performance. Skipping stones on a pond can be an exciting experience for children; when they decide to see who can skip a stone the farthest, the activity becomes even more exciting. As they become more intent in their efforts to outskip their playmates, their stone-skipping performances may also improve. Some people thoroughly enjoy shooting baskets alone on the playground or playing golf by themselves or in leisurely social contexts with friends, whereas others seek the peculiar kind of excitement that can come only from competition.

Nature of Competition

Competition is derived from a Latin word meaning "to strive together," but most of us think of it as "striving against." We compete *against* opponents. But if you think about it, competition between teams or individuals requires an element of cooperation or striving *together*. Cooperation is necessary to hold the game together. Sport philosopher Robert Simon defines competition at its best as "a mutual quest for excellence in the intelligent and directed use of athletic skills in the face of challenge" (Simon 2004, p. 24). Opponents who play their best help us to play our best. When players become spoilsports or give up, or when they cheat, the game falls apart.

When most people think of competitive physical activity, they think immediately of sports—physical activities that place a premium on skill, strength, endurance, or other physical qualities and in which one's performance is assessed in relation to a standard or the performance of others. Unlike physical activity in the other spheres of experience, most sports depend on the organizing principle of competition. Extract competition from most sports and you are left with goalless, aimless physical activity. For this reason we have designated a separate sphere for competition.

Competition can be added to physical activity in different ways. In **side-by-side competitive activities** such as golf, swimming, running, bowling, shot put, or cycling, no direct interaction takes place between competitors. Competitors do not interfere with each other's chances of succeeding; in fact, doing so is against the rules. In **face-to-face noncontact activities** such as volleyball, baseball, tennis, or racquetball, competitors interact by trying to maximize their own chances and decrease their opponents' chances of winning. In this form of competition, players deliberately attempt to thwart the efforts of their opponents and take advantage of their competitive weaknesses, although the rules permit no direct contact between opposing players. In **face-to-face contact activities** such as football, rugby, hockey, or wrestling (and increasingly soccer and basketball), players thwart the efforts of opponents by direct physical

Identifying Good and Bad Competition

When competition goes well, sport is a beautiful thing. When competition goes astray, sport can be ugly. Briefly explain how you would determine whether competition was contributing to or detracting from the beauty of sport in each of the following situations.

- A player's conduct after losing a game
- A player's conduct after winning a game
- How players conduct themselves before a game
- A coach's conduct before, during, and after a game
- The coach's behavior during the week before a game
- How a coach goes about recruiting a star athlete
- The crowd's reaction to an obviously incorrect call by a game official
- The crowd's reaction when their team wins the World Series or Super Bowl
- A parent's reaction to his or her child's failures in Little League baseball
- A university's policy regarding awarding scholarships to students

Zaballa's Choice: Agree or Disagree?

On November 2, 1969, in the enormous Bernabéu Stadium in Madrid, the Real Madrid football (soccer) team was pitted against Sabadell in the Spanish football league match before 80,000 spectators. At the 50-minute mark with the score tied, Pedro Zaballa received a pass, and just as he was about to shoot, the Madrid defensive back and goalkeeper collided, both falling to the ground unconscious. Later, it was learned that one of them had broken his jaw. Zaballa, now facing a defenseless goal, had only to nudge the ball into the net and, very likely, secure the championship for his team. Yet in that instant when the crowd, his teammates, and the egocentric impulses of the game all conspired to urge him to blast the ball into the net, he did something that shocked the fans. Without hesitating, he kicked the ball away from the mouth of the goal and over the end line. He was roundly criticized by his teammates and supporters, although ultimately he was given the Fair Play Award by the International Fair Play Association. Asked to explain, Zaballa said, "I followed a natural impulse not to score under the circumstances."

What would you had done had you been Zaballa in each of the following situations.

1. If you faced the same situation that confronted Zaballa?
2. If your team had been ahead by 3 goals at the time of Zaballa's action?
3. If your team had been losing by 1 goal at the time of Zaballa's action?
4. If the game had not been for the league championship?
5. If the game had been on a sandlot between two "pick-up" teams?

manipulation (blocks, tackles, checks). Finally, **impersonal competition** occurs when participants compete against records set by themselves or others. Trying to set a new record for the time taken to swim the English Channel and trying to climb a mountain by a route never before taken are activities in which one's competitors (holders of the record) are not present. In fact, they may not even be alive!

Problems arise with competition when it becomes more than an organizing principle to increase our enjoyment or push us to better performances. Often competition becomes an end in itself. When coaches and players elevate the goal of winning a contest above the sheer enjoyment of the activity or the basic values of friendship, caring, and cooperation, they convert competition into an end as opposed to a means to an end. Mihalyi Csikszentmihalyi, a psychologist who has taught us much about sport, says, "The challenges of competition can be stimulating and enjoyable. But when beating the opponents takes precedence in the mind over performing as well as possible, enjoyment tends to disappear. Competition is enjoyable only when it is a means to perfect one's skills: when it becomes an end in itself, it ceases to be fun" (Csikszentmihalyi 1990a, p. 46).

 ACTIVITY 2.3

Competition

Check your understanding of the four types of competitive activity and consider the effects of competition itself on sport in your web study guide.

Effects of Competition

Competition is an ingredient we add to play to make it more enjoyable, just as poetry is added to words or music added to sound to add to our enjoyment of words and sounds. Shooting a ball at a hoop in the gym is fun, but shooting in the context of a game makes it even more fun. It enables you to test your skills against others. Likewise, competition creates the drama that is played out in our major sports arenas. If winning or losing wasn't at stake, few people would be interested in watching. Another reason often heard for competition is that it improves performances. In trying to prove yourself better than your competitors, you reach higher levels of performance than you would otherwise. But is this always the case?

Some critics of competition, dating to anthropologist Margaret Mead in the 1930s, would say that competition is problematic. For example, Kohn (1992) claims that competition unnecessarily increases levels of anxiety and lowers motivation among those who think they have no chance of winning. Other critics would say that competition fosters aggression toward others. They would be especially critical of zero-sum competitions in which winning by one individual or team comes at the expense of losing by another individual or team. As to assertions that competition improves performance, these critics would point to an exhaustive analysis of 75 studies conducted at the University of Minnesota on the effect of competition on the performance of motor skills. This investigation found that cooperating on motor tasks more than competing improved performance, although the investigators noted that the effects depend on the type of competition used (Stanne et al. 2009).

Perhaps the best approach for coaches and others whose work centers on competition is to realize that competition can be fun and can help players develop both physically and emotionally. At the same time, coaches need to recognize that anytime two teams or individuals vie for superiority in their talents, it is a delicate interpersonal interaction. Care must be taken to keep games in their proper perspectives.

Imagine that you are the coach of a Little League baseball team. What rules will you enforce to ensure that competition improves performance, is socially rewarding, and promotes positive character development? What rules will you enforce for parents? How will you ensure that competition does not tempt players to cheat? For example, will you tell players to inform the umpire if, when rounding second base, they fail to touch the bag?

Wrap-Up

We have covered a lot of ground in this chapter. We began by recognizing that although kinesiology focuses primarily on exercise and sport, kinesiologists often study a much broader spectrum of physical activity. By now you should realize that physical activity is important not only to our recreational lives but also to virtually everything we do. You should know about the various dimensions or spheres of daily living in which physical activity plays a vital role. We saw, for example, that physical activity is central to carrying out the daily chores through which we remain self-sufficient. It also is the means by which we express ourselves and do our work. Physical activity plays a significant role in education, leisure, health, and our competitive pursuits.

In examining the spheres of physical activity experience, it becomes apparent that we rely on physical activity every day, not only to survive but also to lead full and rich lives. Surely, physical activity has many practical benefits, although we do not seek these benefits through physical activity as much as we should. Having said this, it would be misleading to leave you with the impression that physical activity is important only because it is useful to achieving some end. Physical activity may be as important to us for the special types of personal experiences it offers as for any benefits we derive from participating. Put simply, life becomes richer as we immerse ourselves regularly in a variety of interesting physical activity experiences. We will explore this personal or subjective side of physical activity in chapter 4. But first let's take a detour to help you appreciate the vital importance of physical activity and learn how physical activity professionals use it to serve their students, clients, and patients.

ACTIVITY 2.4

Use the key points review in your web study guide as a study aid for this chapter.

Review Questions

1. Why are ADLs and IADLs important to kinesiologists who work with elderly and disabled people?

2. What type of physical activity professional is likely to be involved in treatment of an individual with carpal tunnel syndrome? What professional is likely to be involved in redesigning the workplace to reduce the risk of carpal tunnel syndrome?

3. What are gestures, and what purposes do they serve in our daily living?

4. Why is public school physical education important? What objectives do physical education teachers pursue?

5. List three health benefits of regular physical activity.

6. What does the element of competition add to physical activity? When is this helpful, and when might it be harmful?

7. Describe a situation in which physical activity may help nourish and maintain a state of leisure. Describe a situation in which physical activity may diminish the possibility of attaining a state of leisure.

chapter three

The Importance of Physical Activity Experiences

Shirl J. Hoffman

chapter objectives

In this chapter we will

- help you gain an appreciation for physical activity as it contributes to and expresses the signature of humanity;

- encourage your appreciation for the various ways that experience can expand the human potential for physical activity;

- help you develop insights into factors that should be taken into account in prescribing physical activity experiences;

- explain some principles governing the relationship among skill, practice, and learning and among physical capacity, training, and conditioning;

- discuss the relationship between physical activity experience and physical fitness; and

- discuss how heredity and other factors can modify the effects of experience.

If you followed the competition at the 2012 Summer Olympics, no doubt you marveled at the performances of 16-year-old gymnast Gabrielle ("Gabby") Douglas. Gabby, or "The Flying Squirrel" as she is affectionately called by friends and teammates, astounded judges, audiences, teammates, and competitors with near perfect floor and balance beam routines. Her light bearing on the uneven bars reminded some sports writers of a hummingbird. She won the all-around gold medal for women's gymnastics and, along with the four other members of the "Fierce Five" United States' team, won the team gold medal, the first U.S. team to do so in 16 years. No other American gymnast has ever won a gold medal in both the individual and team competition. No other African American gymnast has ever won the all-around.

How did Gabby become so proficient in gymnastics at such a relatively young age? Giving the question some thought you might conclude that her rise to prominence in gymnastics was the result of inherited abilities. There is little question that her 4'11" size is an advantage in a sport that involves rotating around multiple body axes. Perhaps she also was blessed with an extraordinary sense of balance. Regardless, these traits are secondary to the wealth of gymnastics experience she had almost from the day she was born. Although she didn't begin formal gymnastics training until the ripe old age of 6 and didn't become a Virginia State Champion until she was 8 years old, she had been performing gymnastics since the age of three. At the age of four she taught herself to perform a one-armed cartwheel. Her experience in the sport wasn't random; it was guided by expert coaches from the beginning. In 2010 when she was 14 her mother allowed her to move from her home on the east coast to live with a family in Iowa so that she could be taught by Liang Chow, one of the best coaches in America. When she stepped onto the floor of the North Greenwich Arena in London, she literally was drawing on her lifelong experiences in performing gymnastics. This experience was the most important factor in her development as an international athletic star.

The word *experience* can have many different definitions, so we should first clarify precisely how we will use the term in this chapter. One definition is "individual reaction to events, feelings, etc." (Neufeldt 1988, p. 478). In this sense we may "experience" sadness, joy, beauty, and a host of other emotions. This version of experience, which we will call **subjective experience**, is always a part of physical activity. We explore the topic in the next chapter. Here we focus on **activity experience**, which is defined as any "activity that includes training, observation of practice, and personal participation" (Neufeldt 1988, p. 478). Activity experience includes performing physical activity as well as observing it. If you are *experienced* as opposed to *inexperienced* in skiing, you will have skied on many occasions in many contexts and probably with many different people. Gabby Douglas, for example, has extensive experience with the physical activity of gymnastics. Liang Chow, her coach, has extensive experience watching and analyzing gymnasts.

> ▶ Physical activity experience refers to our history of participation, training, practice, or observation of any particular physical activity. It differs from subjective physical activity experiences, which are our reactions, feelings, and thoughts about physical activity.

As you move through this chapter, you will be reminded repeatedly of the critical importance of physical activity experience as essential to developing our capacities to perform physical activity and to improve health.

Although physical activity is fundamental to human life, we do not come into the world readily equipped to perform it. Birth may innately endow us with an ability to move, but developing control over our movements requires experience. Newborns spend a lot of time kicking their legs and waving their arms; these are aimless and spontaneous movements that occur without prompting or encouragement from their parents. Infants also move automatically in response to external stimuli. For example, holding a newborn in the prone position over water may elicit swimming-like movements of the arms and legs; applying pressure to the sole of the foot elicits a crawling pattern; and turning the head of a supine infant to one side causes the arm and leg on the same side to extend. Much experience and maturation are required to convert these spontaneous and automatic movements into the conscious, goal-directed movements that kinesiologists call *physical activity*.

Think back to a time when you were first learning a skill. It is likely that your movements were jerky and inefficient. With practice, your movements became smoother, allowing you to be not only more accurate but more forceful. Usually this involves an important change in the sequence of movements we use. If you carefully examine the two photos that follow, you can see this principle in action. In the first photo (on left), the youngster locks shoulders, elbows, and wrists in an effort to control the movement but in doing so is able to generate little power. The more advanced player on the right rotates the hips, trunk, shoulders, elbows, and wrists independently, allowing her to generate much more force.

Almost everybody engaged in physical activity wants to improve his or her performance, whether it means performing activities faster, more forcefully, for a longer time, or more skillfully. The only way to do this is to expose yourself systematically to appropriate physical activity experiences. Such experiences enable people enrolled in fitness programs to become stronger or increase their cardiorespiratory endurance. It helps athletes become more skillful, dancers to become more graceful, and factory workers and military personnel to become more efficient.

Physical activity experiences also are important because they are the means by which we achieve important health and psychological benefits of physical activity. These benefits usually go hand in hand with increased capacity for performing physical activity, but not always. As we will see, physical activity often leads to a reduction in body weight, a decline in percentage of body fat, and decreased stress levels, but it is possible to achieve these

■ Physical activity experience with skills leads to more powerful movements.

The central concern of all kinesiologists and physical activity professionals is the systematic manipulation of physical activity experiences based on sound scientific principles to improve skill, performance, and health and well-being.

changes without significantly improving physical activity performance or, for that matter, without engaging in physical activity at all.

In part II of this text you will study the scientific bases for the effects of experience by learning how physiological systems respond to repeated performances of activities. You will appreciate at a deeper level why training experiences that stress the cardiorespiratory system will improve your capacity for running a mile in a shorter time or why your capacity for lifting a heavier weight will improve if you progressively increase the resistance against which your muscles work. You will also learn why we become better tango dancers, typists, or golfers by immersing ourselves in well-organized practice routines.

Obviously, not all types of physical activity experience induce positive changes. Poorly planned training or practice programs, for example, may not help you achieve your goals. They may waste your time and energy and, even worse, retard progress toward your goals by causing injury or leading you down unproductive paths of experience. Trained physical activity specialists are important safeguards against such dangers. As you progress through your undergraduate program you will come to view physical activity professionals such as physical therapists, coaches, physical education teachers, and personal trainers as experts in using scientific principles to systematically manipulate physical activity experiences in order to improve the skill, performance level, and health and well-being of those they serve. (*Manipulation* as used here refers to skillfully handling physical activity experiences rather than exerting control over others.) Scholars and researchers in kinesiology also are in the experience business; through their work they identify the types of activity experiences that are most likely to produce the intended results. Many of the courses required for kinesiology majors center on ways to safely and effectively manipulate physical activity experiences to bring about positive changes in clients' capacities for physical activity. Other experts such as sport historians or sport sociologists do not help students learn how to manipulate physical activity as much as help them understand the significance of physical activity in our culture.

Physical Activity as a Signature of Humanity

Although it is tempting to think of physical activity experience as inferior to mental activity, such a limited view wouldn't do justice to this far-reaching and important concept. Physical activity draws on aspects of humanity that help define us as the highest order of living creatures. Were it not for this connection, physical activity experience could not work its marvelous effects. Let's step back from the topic of experience for a moment to explore how physical activity helps define our humanity and how our innately human capacities enable us to develop our physical activity capacity through experience.

What characteristics set humans apart from animals or machines? Your first thought might be that we possess souls or a moral sense, that we are able to engage in complex forms of social communication, or that we possess a capacity for high-level reasoning that animals and machines can never replicate. Good cases can be made for each of these arguments. What you probably didn't think about, however, is that our extraordinary capacity to translate complex mental concepts into precise and creative physical actions is also a defining human characteristic.

But what about animals? Don't animals have a capacity to link movements with intentions and plans for moving, and if so can't we say that they perform physical activity? Certainly they engage in goal-oriented, voluntary movements. For example, chimpanzees use a variety of tools to probe termite mounds and even use weapons to hunt small mammals. They also use small stones to crack palm nuts using the same techniques that they have seen used by local villagers. Dolphins have been known to mimic human postures. Archerfish, which stun insects with a sudden blast of water, learn how to aim their squirts by watching an experienced fish perform the task. A small ground finch can uncover food using its feet to push aside stones weighing 14 times its body weight—equivalent to a 200-pound (90-kilogram) human moving a 1.5-ton (1,360-kilogram) rock—by bracing its head against a large rock for leverage (Gill 1989). Yet, although animals can perform goal-oriented movements, the human capacity for physical activity is of an entirely different order than the capacity of these and other lower animals. This is a rather bold assertion; let's look at it more carefully.

Recall that physical activity is movement that is intentional, voluntary, and directed toward achieving an identifiable goal. Understanding the technical definition of physical activity is essential to understanding the discussion that follows. Having reviewed the definition, you are now ready to consider four ways in which physical activity may be considered unique to the human species.

> A unique capacity for performing physical activity is one of the major features that contribute to the distinctive character of the human race.

Intelligence-Based Physical Activity

First, because humans are big-brained, highly intelligent creatures, their physical activity tends to be rooted in more intricate plans and directed toward more sophisticated goals than is the case with lower animals. Motor plans or mental images serve as dynamic maps to guide our movements in skills as simple as performing a standing long jump or as complex as piloting an airplane, performing brain surgery, or catching a Frisbee. Animals, by comparison, entertain relatively simple plans. Kangaroos have a capacity for cardiovascular endurance that humans can only dream of having, hopping in 15- to 20-foot (4.5- to 6-meter) strides at all-out pace for up to 20 miles (32 kilometers), but its endurance capacity cannot be used to play the rule-bound game of soccer. A cheetah can easily outrun a human in a contest in which the goal is simply to run as fast as possible. But, unable to formulate a clear understanding of goals and constraints imposed by rules, a cheetah will not be able to adapt its extraordinary running skill to the complexities of an Olympic relay race.

| Kangaroos are excellent jumpers and have great endurance but would not make very good basketball players.

Ethically and Aesthetically Based Physical Activity

Second, because humans are essentially spiritual creatures possessing unique moral and aesthetic senses, we can use our movements to express our imagination and moral reasoning. We can use our movements to express beauty, joy, wonder, and other deep and complex moods. Physical activity is integral to rituals that form an important part of our religious and civic lives. This statement isn't to deny that other animals have emotional lives, too. Chimps, for example, can become so depressed when their mothers die that they sometimes die too (Heltne 1989). Although some chimps can use sign language to express emotions and can daub paint on paper, animals cannot translate emotions through muscular actions into symbolic works of art or other elaborate expressions of sorrow, joy, or wonder. If thousands of chimpanzees were each given a hammer and chisel, perhaps, by mere chance, one of them might be able to create a recognizable work of art but never something as profound as Michelangelo's *David* or Rodin's *The Thinker*. Wading birds may engage in elaborate mating dances, but the choreography is the product of instincts, not intentional expression of mood in movement as occurs, for example, in *Swan Lake*.

Flexibility and Adaptability of Physical Activity

Third, human physical activity is further distinguished from that of other animals by virtue of the unique combinations of movements permitted by our anatomy. At first, this may seem an exaggeration. After all, elephants are equipped for performing far more forceful movements than humans are. Greyhounds can surpass us in speed, dolphins in swimming, and monkeys in agility. In what ways, then, can we say that humans hold a movement advantage?

Two properties of our anatomies give us great advantage in moving. Animals may be endowed with considerable specific talents for moving; humans are the best equipped for performing a wide range of activities. First, with our upright posture and bipedal gait, we are the only animals whose forelimbs have been totally freed from assisting with walking, flying, brachiating, or swimming. The human body has a foot specially constructed to bear weight and give leverage to the leg, a pelvis specifically designed for attaching strong muscles needed to help maintain bipedal balance, and thighbones to permit long strides in walking. Our ability to walk on two feet has been described as "the most spectacular physical trait of human beings" (LaBarre 1963, p. 73).

Second, humans have available a dexterity of movement made possible by a unique complex of hands, arms, shoulders, and stereoscopic vision. The human hand possesses a true opposable thumb, a much more beneficial construction than that of the typical primate hand, whose five fingers all operate more or less on the same plane. This arrangement allows us to perform delicate grasping, manipulating, and adjusting movements not possible for others in the animal kingdom. But this is not the only advantage. Our hands are positioned at the end of a series of long arm bones, joined to an amazingly movable shoulder girdle that a dog or cat can only dream of having. (Can you imagine a dog, cat, or horse being able to scratch its back with its forelimb?) This movable shoulder girdle enables us to position our hands through an enormously large path in space. This upper-arm advantage is complemented by a facility for stereoscopic vision (lacking in many animals) that not only gives us advantages in depth perception but also allows us to perform most of our movements in our field of vision.

Humans and Animals at Play

One form of physical activity engaged in by both humans and animals is play. Cats, dogs, dolphins, monkeys, and otters play. Penguins play a sort of "king of the hill," rats wrestle and emit a chirping noise that scientists have likened to laughing, and ravens can throw sticks. Although the purpose that play serves is still being debated, it likely is essential for human survival. Which of the following types of play are likely to be a part of the lifestyles of both humans and animals?

- Exercise play, involving wide locomotor movements such as running, jumping, and climbing
- Object play, involving use of fine movements to move objects in some way
- Construction play, in which materials are combined to create a new product, as in clay modeling
- Social contingency play, motivated by the pleasure found in producing contingent responses in others (peek-a-boo, tickling)
- Rough and tumble play, which involves vigorous physical contact accompanied by playful signals that distinguish it from hostile aggression (wrestling, pursuit games)
- Fantasy play, in which real or imaginary objects are given properties different from those they actually possess
- Games with rules, in which behavior is guided by explicit rules

Adapted from S.T. Parker, 1984, Playing for keeps: An evolutionary perspective on human games. In *Play in animals and humans*, edited by P.K. Smith (New York: Blackwell Science), 271-293.

Ability to Improve Performance Through Planned Experience

The fourth—and perhaps the most significant—distinguishing characteristic of humans is our ability to improve our capacity for physical activity through planned, systematic practice and training. Only humans possess the intelligence that allows them to use physical activity in planned, systematic, and scientifically verifiable ways as a means of improving their health, performance, or skill or as a means of physical rehabilitation. Obviously, the cardiorespiratory efficiency of the young lion or eagle improves as its hunting range expands, but the driving force of its activity is hunger and survival, not a systematic conditioning plan to improve the physiological functioning of its body. Although some species of animals teach their young how to hunt and fish, the methods are relatively primitive, lacking the sophistication of human plans.

In summary, the capacity to link physical activity to complex plans and deep moral and aesthetic sentiments, an anatomy that permits these sophisticated thoughts and feelings to be played out in complex sequences of skeletal movements, and a highly developed ability to plan and implement activity experiences to improve physical performance have contributed immensely to the advance of human civilization. In this sense, physical activity enables us to explore, test, and manifest our humanity.

Our rational capacity for systematically manipulating physical activity experience has given rise to the need for kinesiology, a discipline devoted to the intelligent application of experience to the solution of physical activity problems.

Human physical activity is distinguished from that of lower animals by four characteristics:

- Ability to match movements to sophisticated plans to achieve more complex physical activity goals
- Ability to apply physical activity to artistic creations
- A high degree of flexibility and adaptation in physical activity made possible by unique anatomical structures
- Ability to plan experiences that improve performance

Factors Influencing the Kinds and Amounts of Performance Experiences

▶ Although we possess a unique facility for performing, planning, and implementing physical activity experiences to improve our performance of physical activity, most of us are not inclined to explore this potential to the fullest.

Because our potential for physical activity helps us to express what is unique about our humanity, it might seem that each of us naturally would maximize opportunities to engage in physical activity and incorporate it into our lives. But that isn't the case. Although it is natural for us to engage in physical activity, as we saw in the last chapter, relatively few of us are inclined to do so, at least regularly and in its more vigorous forms. There is an enormous disconnect between what we know about the benefits of vigorous physical activity and our behavior patterns. Likewise, few of us are inclined to tap our full potential for developing skill in a variety of physical activities. We know that when people are competent in a sport, they enjoy engaging in it during their leisure time, yet not many people devote themselves to the considerable practice necessary to reach even a moderate level of competence. It's probably fair to say that average adults perform most types of motor skills at levels far below their capabilities.

Why this paradox? Why do most of us fail to explore to its fullest potential something so fundamental to our human nature and so beneficial to us? In contrast, why do some people engage in physical activity over their entire lives? This is one of the most pressing questions confronting the physical activity professions today, and as you will see when you study chapter 9, it is an area of study central to exercise psychology. Unfortunately, there are no simple answers. Many factors in our lives encourage us to explore physical activity and also to discourage us from doing so. Among the barriers that keep us from living physically active lives are lack of time, lack of access to an exercise or sport facility, and lack of a safe environment in which to be physically active. But these aren't the only factors to consider. Let's look briefly at how two major factors might affect the type and extent of our physical activity experiences: our social environments and our personal circumstances (see figure 3.1).

FIGURE 3.1   Many factors influence our decision to engage in physical activity.

Social Environment

The people with whom you regularly interact can significantly affect the types and amounts of physical activity experiences you pursue, both as a child and as an adult. As a child you may have received substantial social support to engage in certain kinds of physical activity, causing you to be physically active in your younger years. Friendships and social alliances have been shown to have a strong influence on the amount of vigorous exercise undertaken by both young and middle-aged men and women (King et al. 1992; Sallis & Hovell 1990). If your girlfriend or boyfriend has a physically active lifestyle, you probably feel encouraged to do the same. If your boss plays tennis or racquetball, you might also be attracted to the game. Although each of us is ultimately responsible for our decisions regarding physical activity, those with whom we associate can affect our dispositions toward it.

Parents

When you were a child, did your parents lead a physically active life or were they largely sedentary? If they modeled an active lifestyle, took you with them to their workouts at an exercise facility, took a daily run in your neighborhood, or were avid tennis or squash players, they can take some credit for the fact that you are an active person too. If, on the other hand, they were "couch potatoes" who avoided sport, exercise, and other physical activity, you may have had to resist their adverse influence. From a research standpoint, the extent to which parental levels of physical activity affect their children's levels of physical activity remains somewhat unclear. Researchers have found, for example, that the activity levels of children who were participating in an obesity treatment program bore a remarkable correspondence to the activity levels of their parents (Kalakanis et al. 2001). Yet other researchers have found no correlation between the levels of engagement of parents and children in physical activity (Jago et al. 2010)

Even if parents' participation in physical activity doesn't have a strong influence on the physical activity levels of their children, encouraging them—especially preschoolers and adolescents—to be physically active is likely to have an effect. This is a good point for future parents to keep in mind. Merely showing support for their children's physical activity by transporting them to physical activity settings or assisting with the organization of the activity (e.g., serving as a Little League coach or sponsor of the cheerleading squad) may increase the probability that their children will be physically active (U.S. Department of Health and Human Services [USDHHS] 1996).

Interestingly, researchers haven't found a strong link between levels of physical activity experiences in younger years with levels of physical activity later in life. For example, there does not appear to be a strong relationship between the extent of sport participation in youth and the extent of physical activity participation in adulthood. Participating in school or college sports does not appear to influence the amount of physical activity that one engages in as an adult (Brill et al. 1989; Cauley et al. 1991; Dishman & Sallis 1994). You are probably familiar with people who were star athletes in college and became physically inactive following graduation, just as you can probably think of individuals who did not lead physically active lives as children but became avid adherents of exercise and sport in later years. Obviously, we have much to learn about this connection. Until a more complete picture emerges, however, it seems best to ensure that constructive physical activity experiences are integrated into the early years of development.

Spotlight on Research: Do Inactive Parents Produce Physically Inactive Children?

Researchers in the Department of Exercise, Nutrition, and Health Sciences at the University of Bristol in England wanted to know whether physically active parents produce physically active children. Sixth graders and their parents from 40 primary schools participated in the study to examine parents' and children's physical activity patterns. Levels of parental and child physical activity and inactive time were assessed using accelerometers, small devices that are worn by subjects while they go about their daily activities. The devices provide reasonably reliable measures of physical activity. The investigators found no relationship between parents' and children's physical activity levels but did discover something very significant. The more parents watched TV, the more their children watched as well. For girls, the relative probability of watching more than 4 hours of TV per day was 3.67 times higher if their parent watched 2 to 4 hours of TV per day compared with girls who watched fewer than 2 hours of TV per day. The chances of watching more than 4 hours of TV per day by boys was 10.47 times higher if their parent watched more than 4 hours of TV per day compared with boys who watched fewer than 2 hours of TV per day. These findings show that although parents may not have a direct influence on their children's physical activity patterns, they can have a direct influence on their sedentary behavior. This is something of considerable concern to kinesiologists given that research has shown that men who reported more than 23 hours a week of sedentary activity have a 64% greater risk of dying from heart disease than those who were sedentary fewer than 11 hours per week, even though many of these men exercised regularly (Jago et al. 2010; Warren. et al. 2010).

Peers

The extent of your friends' involvement in physical activity also may have influenced your physical activity decisions. Do you hang around people who like to be physically active? If so, you are likely to be physically active too. Some researchers believe that the physical activity patterns of your peers may be as important as those of your family in determining whether you become involved in sport (Kirby et al. 2011).

As you grew older, did you come into contact with more physically active peers, or did those around you become more inactive? Unfortunately, the statistical probability is that your peer group became less rather than more active as you grew older. For example, in 2011 only 29% of high school students surveyed had participated in 60 minutes of physical activity on all 7 days before the survey (U.S. Centers for Disease Control and Prevention [CDC] 2012).

If your peer group is immersed in online social networking, playing video games, or other sedentary pursuits, you will likely lead an inactive lifestyle, too, rather than allow various types of vigorous physical activity to form the nucleus of your social life. As you grow older, peer groups may continue to affect your physical activity decisions. If you are married, your spouse and your spouse's friends are likely to influence your decisions to participate or not

---

What Factors Have Influenced Your Physical Activity Patterns?

What leisure-time or competitive physical activities did you participate in as a child? Did you play Little League or soccer or football? Did you take private dance or gymnastics lessons? Did you ride bikes, climb trees, or roller skate with your friends in the neighborhood? Did physical education motivate you to engage in physical activity outside of class? Did you participate in the activities of the YWCA or the YMCA? Do you think that these early physical activities influenced your engagement in physical activity as an adult? Do you see similarities in the general types of physical activities that you pursued as a young child and those that you pursue now? What factors have contributed to your continuing (or discontinuing) engagement in these activities?

---

participate in physical activity (Loy et al. 1978). Your relatives', friends', and social groups' predispositions toward physical activity, indirectly but predictably, will probably continue to influence your decisions throughout life.

Teachers and Coaches

If you were an athlete in school, your coaches—of youth sport teams, junior high school teams, high school varsity teams—probably had a significant effect on your physical activity decisions. Athletes, for example, tend to associate their participation in sport with the earlier influence of coaches, especially during adolescence (Ebihara et al. 1983). Physical education teachers and coaches are influential because they are in a position to confirm or disconfirm a young person's competency in an activity. A remark by your father or mother that you performed well in a soccer game will probably encourage you to continue to develop your skills in that sport, but a similar remark from the coach is likely to be more influential.

Obviously, physical education teachers and coaches also can create social environments that discourage young people from seeking out physical activity experiences. In this sense, teachers and coaches may act as gatekeepers to the physically active life for thousands of children each year. Thus, it is not an exaggeration to say that the behavior of a teacher or coach may have profound and lifelong effects on a student's physical activity experience.

> The people closest to you, including your parents, peers, coaches, and teachers, are major influences on the kind and amount of experience you have with a particular physical activity, especially when you are young.

Personal Circumstances

In addition to personal circumstances, the social environment, such as the availability and accessibility of facilities and play spaces, can determine the amount and kinds of people's physical activity experiences (Garcia et al. 1995; Zakarian et al. 1994). Geography is another important factor (see the Spotlight on Research sidebar). Those living in northern climates are more likely to develop competency in skiing, skating, and outdoor activities such as hunting, hiking, and fishing than they are in swimming or golf, simply because the weather dictates the availability of appropriate sport environments. Temperate climates with warm, sunny days tend to encourage jogging, in-line skating, and walking. Sometimes local cultural traditions emphasize particular sports. High school football is uniquely important in western Pennsylvania, Florida, and Texas; boys' wrestling and girls' basketball are popular in Iowa. The game of the inner city is basketball, a cultural tradition owing at least partly to the fact that space isn't available for sports such as baseball and football.

Economic considerations can also dictate activity preferences. Generally, highly educated people in higher income brackets with high-status jobs tend to be more active than their poorer, less educated counterparts (see figure 3.2). Part of the reason for this disparity is that economic considerations often limit opportunities for participation in physical activities to those with higher income levels. The difference is especially noticeable in sports requiring expensive equipment, high admission fees, or, in cases such as swimming, transportation or admission to fee-based private clubs. The effect of income on participation in skiing and golf

Spotlight on Research: Geographical Hot Spots for Producing Professional Athletes

Where you were born and raised can play a part in your decisions to play sports. Researcher Christopher Storm (2005), using a database of the hometowns of every American-born major league professional athlete to play in one of the four recognized "major" leagues (Major League Baseball, National Basketball Association, National Football League, National Hockey League) during the league's last completed season (2005), was able to identify the production of states and counties to the professional leagues:

- Overall: The areas that produced the highest per-capita (age 20-44) production were the Southeast (Texas to South Carolina), the Northern Plains (Utah to the Dakotas and Iowa), California, and Ohio. The least productive region was New England.

- Baseball: The most productive regions tended to be rural; the Pacific Coast and Southeast, along with Ohio, the Dakotas, and Alaska.

- Basketball: The Great Lakes states, the Pacific Coast, and the Southeast produced the most players per capita. Basketball players come mostly from urban counties.

- Football: Counties producing the most football players were in the Southeast. Counties in New York and New England produce the fewest.

- Hockey: Predictably, the upper Midwest and the Northeast, along with Utah and Colorado, produced the most professional hockey players.

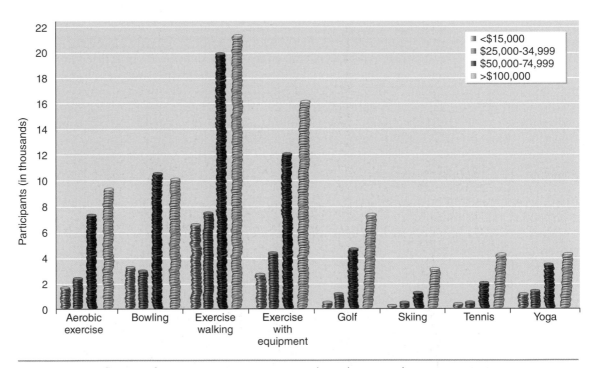

FIGURE 3.2 Influence of income on participation in selected sport and exercise activities.

Adapted from US Department of Commerce, Census Bureau 2012.

is particularly noticeable, with more than 50 times more skiers and 12 times more golfers from the upper income bracket compared with the lowest. Those who exercise with equipment (probably at clubs that require dues) are more than 5 times as likely to be from the highest than the lowest income bracket. Given that exercise walking is essentially free and readily available to all, economic differences may not account for the fact that three times more people in the highest income bracket walk for exercise than in the lowest income bracket.

Our own attitudes and personal attributes also affect our physical activity patterns. As we have noted, decisions to become involved in physical activities are not always outside your control. Your own perceptions, feelings, and decisions play a major part in the process. For example, a decision to prepare yourself for a marathon or train for bodybuilding competition may have to do with your personal attraction to certain unique characteristics of these activities. If you participate in hang gliding, it may have something to do with your willingness to take risks, the exhilaration you feel from heights, and your perceived ability to keep the glider in the air. As we will see in chapter 4, we tend to involve ourselves in activities that we enjoy, and only with difficulty can we pin down precisely those factors that cause us to enjoy some activities and shun others. Evidence suggests that decisions to become involved in or avoid physical activities may have to do with our feelings of self-esteem and our perceptions of our competence in the activities. Nevertheless, an almost endless number of personal factors can affect these perceptions.

> Often, our physical activity experiences—and consequently, those in which we develop proficiency—are determined by factors that lie outside of our control, such as climate, regional culture, and economic considerations.

▌ Would you consider doing this physical activity? What individual characteristics lead you to make that decision?

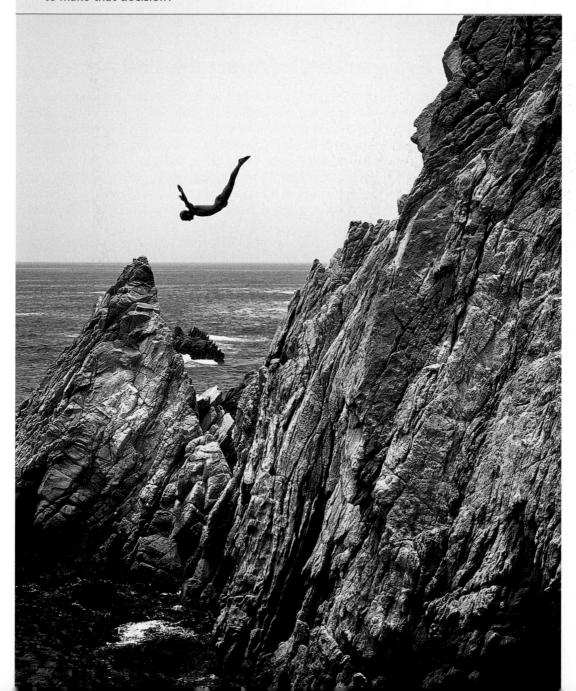

In the end, it may be as difficult to determine precisely why one person chooses to exercise regularly whereas another chooses to avoid exercise, or why one chooses to play tennis whereas another prefers water skiing, as it is to determine why some people like broccoli and some can't tolerate it or why some people like heavy metal music and others prefer classical. We have seen that physical activity reflects something of our humanity. It also seems that our physical activity preferences may reflect something about our uniqueness as individuals. Thus, as a future physical activity professional, you will need to pay careful attention not only to the developing research in this area but also to the particular needs, desires, and attributes of the individuals who will seek your services.

Let's stop and retrace the path we've taken to arrive at this juncture. We've seen that the capacity to elaborate physical activity into complex plans and actions is a unique characteristic of humans. Coincidental to this is our remarkable capacity to use experience to our advantage to learn motor skills; to increase our strength, endurance, and flexibility; and to improve our health. Yet evidence suggests that most of us do not come close to exploring the limits of our physical activity potential. This idea led us to consider in a general way some factors that affect our decisions to incorporate physical activity into our lives. These influences included the effects of parents, peers, and teachers and coaches, as well as individual circumstances such as geographical region, cultural traditions, and economic considerations, not to mention our own attitudes and dispositions.

> After we have accounted for all the factors in our social and ecological environments that affect our decisions about physical activity, we are left to consider the indeterminable factors, such as our perceptions of ourselves, our competency in the activity, and the activity itself, that also may affect our decisions about physical activity.

Two Ways Experience Can Affect Physical Activity

As a student who is considering majoring in kinesiology, you should realize how vital it is that you become familiar with the processes and mechanisms by which performance experience improves our capacity to perform physical activity. You will learn about theories and research that consider such information in detail when you enroll in courses such as biomechanics; exercise physiology; motor learning, development, and control; and sport psychology. Before you begin this advanced study, however, you will find it helpful to step back and look at the effects of physical activity experience in much broader terms. This perspective will provide you with a conceptual framework upon which you can build a more specialized knowledge of the scientific bases of physical activity.

Think back to a particular form of physical activity in which you had a special interest. How did your experience with this activity affect your capacity to perform it on successive occasions? If it was a skill such as volleyball, the effects of your accumulated experiences probably were most evident in changes in your ability to perform the sequences of movements associated with serving, setting, or spiking. If the activity was cardiorespiratory training, the effects of your accumulated experiences may have been evident not so much in changes in the nature of your movements but in your improved capacity to perform the activity over longer periods at greater intensity. These represent the two fundamental effects of experience: development of skill through practice, which is called *learning*, and improvement in physical performance capacity through a type of exercise called training, which is termed *conditioning* (see figure 3.3).

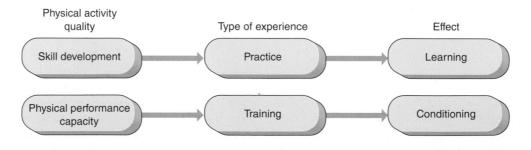

FIGURE 3.3 The nature of improvement in physical activity depends on the types of physical activity experience.

Practice, Learning, and Skill

Earlier we saw that infants face the challenge of recombining and refining their primitive motor responses into goal-directed voluntary actions to meet the demands of daily life. Throughout life we continue to confront the challenges of assembling sequences of movements and body positions to accomplish specific tasks. Activities that place a premium on efficient, coordinated motor responses are called **motor skills**. Although threading a needle, performing a cartwheel, pitching a baseball, and typing a letter are all motor skills, each places distinctly different demands on our perceptual and motor processing systems. The general quality that underlies the performance of motor skills is called **skill**. Skill is reflected in the efficiency and accuracy with which an individual is able to attain the goal of a motor skill.

Becoming a skilled mover is a never-ending process of gaining more and more control over your motor system by gradually refining the nerve and muscle systems through performance experience. The type of performance experience required to do this is known as **practice**. Laypersons tend to think of practice as merely an endless repetition of the same movements by rote, but that isn't how practice works. Practice of motor skills, especially during early learning, involves a lot of cognition. It involves higher brain function and integration of spinal neural activity as well as muscular action. (In fact, muscular action is merely the result of these higher cognitive processes.) Practice is a deliberate effort to "get it right" by modifying the erroneous movements we made on earlier trials and revising our cognitive plans for moving on the next trial. On each practice trial we vary the level of force or direction of our movements, adopt new strategies, and analyze feedback from our movements so that we may, for example, throw a ball farther or more accurately. This is why one scientist has referred to practice as "a particular type of repetition without repetition" (Bernstein 1967, p. 134).

> The type of physical activity experience that brings about changes in skill is called *practice*. The relatively permanent effect of practice is called *learning*.

Sometimes we refer to practicing as a process of acquiring experience, implying that we can collect and store our practice experiences. This metaphor is helpful. If we were unable to store memories of past physical activity experiences, each trial would be like our first. We would be in the same situation as people who have lost their short-term memory because of damage to an area of the brain known as the basal ganglia. Because such people can't store information about their previous response in memory, successive repetitions of a skill usually do not bring about learning. They may have played the piano 100 times, but each attempt is an entirely new experience. With a normally functioning brain, however, we can benefit from memories of our experiences and gradually reshape our motor responses in ways that allow us to attain skill goals more accurately and with greater efficiency.

Keeping in mind the importance of cognitive processes during early skill learning shouldn't blind us from the fact that learning can also proceed on an unconscious or "tacit" level. For example, individuals are able to improve their performance in rapidly responding to stimuli (reaction time) that appear in a complex temporal pattern but may not be able to explain what the temporal pattern is. Catching a fly ball in baseball is an extraordinarily complex skill, and most components of this action are achieved at an implicit (unconscious) level. Calculating the thousands of trajectories and movements that can result in a successful catch is virtually impossible for a computer, but a young child can learn, implicitly, the movements that allow her to catch the ball. People who have had strokes that limit or eliminate their short-term memory may not remember having practiced a skill previously but may still show improvement from practice to practice (Meehan et al. 2011).

The refinements in the nervous system that result from practice are referred to as **learning**. Although much has yet to be discovered about learning, scientists agree that learning is a central rather than a peripheral phenomenon. This is another way of saying that learning is a result of a rewiring of neural circuits in the brain and spinal cord rather than changes in peripheral organs such as bone or muscle. (Even though we often hear the term *muscle memory*, it is the brain, not the muscles, that remembers.) We also know that these changes tend to be relatively permanent. If, after having played many years of tennis you do not return to the court for several years, the level of your play will deteriorate, but with a little practice you will regain it rapidly. Amazing, isn't it?

When you learn a motor skill, you actually remap the neural routes in your central nervous system. You know from your own experience that this remapping doesn't happen overnight. Depending on the skill, it may take weeks, years, or even decades to master. Obviously, the extent to which changes in skill become permanent depends a great deal on how much you practice the skill.

Training, Conditioning, and Physical Performance Capacity

The second type of change brought about by experience is much different from learning. Improvements in some physical activities hinge more on improving strength, developing the capacity to exert greater force, or increasing the range of motion at joints rather than improving the accuracy or timing of your movements. These factors are not elements of skill as much as elements of **physical performance capacity**. To put it in a slightly different way, strength and flexibility aren't learned—they are developed.

Consider the case of two tennis players, Sam and Jeff, both of whom had devoted long hours of practice to the sport and had played in competitive leagues. Each had developed approximately the same level of skill in tennis, as evident by their equal proficiency in serving, backhand and forehand strokes, and net play. Then Sam—a perceptive kinesiology major—began to supplement his tennis practice with a rigorous training program designed

▌ Olympic silver medalist Wu Jingbiao of China had to undergo rigorous training and conditioning to be able to lift such massive amounts of weight.

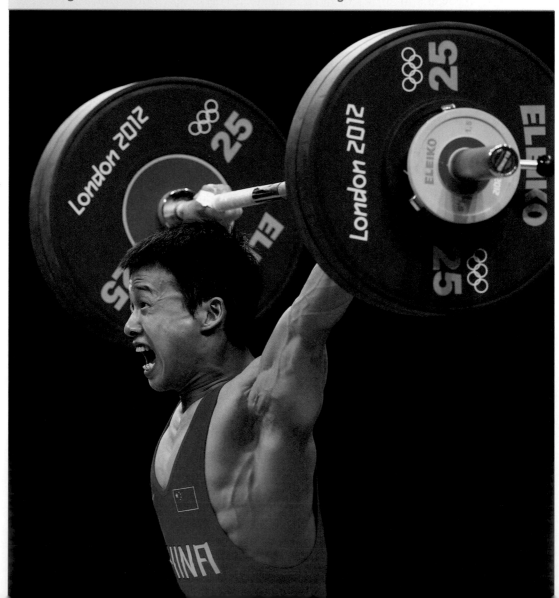

to increase his physical capacity through development of muscular endurance, cardiorespiratory endurance, and flexibility, all of which he believed to be almost as important as skill in the sport of tennis. As a result, he became a more proficient tennis player than Jeff. Although both players had learned the *skills* of tennis to the same level, Sam ended up the better player because he supplemented his skill development with improvement of other performance qualities that were important in the sport.

The supplemental types of experiences that Sam engaged in were weightlifting, along with "cardio-drills," stretching, and running, kinds of physical activity experiences known as training. The changes brought about by training are called **conditioning**. Training is physical activity experience designed to improve muscular strength and endurance, cardiorespiratory endurance, flexibility, and other aspects of performance that condition the performer. Training increases the capacity to move our bodies more forcefully (strength), over longer durations (endurance), and in more flexible ways (flexibility). Generally, kinesiology majors study the experiences of practice that lead to learning in courses on motor behavior and study the experiences of training that lead to conditioning in courses on physiology of physical activity.

Practice Versus Training

The differences between practice and training can be summed up this way: Practice affects memory, cognition, perception, and other central nervous system processes associated with problem solving. Training, in contrast, produces effects that are largely peripheral to the central nervous system, usually on muscle, bone, soft tissue, and the cardiorespiratory system. Training is most effective when done systematically under the direction of a professional exercise instructor, but often training effects occur as a default or unintended by-product of physical activity. For example, one chops wood or shovels snow to accomplish ends that have nothing to do with training per se, but if you perform these activities enough, you will become conditioned. Quite simply, if you repeat movements that appropriately stress muscles, tendons, and bones; elevate the heart rate to certain predetermined levels; or move your joints through large ranges of motion, some level of conditioning will occur. On the other hand, mindlessly performing a skill, no matter how many times, is not likely to lead to learning.

Is practice or training the most appropriate physical activity experience for improving your performance in an activity? Most large-muscle activities require some degree of both practice and conditioning, but usually one or the other is most important. The key to determining which should be given dominant emphasis is to ask whether success in the skill depends primarily on skill or on physical performance capacity. If skill is the primary quality, practice is the most important experience; if physical performance capacity is the most important quality, training is the most important. If both skill and performance capacity are important, both practice and training are important. Even at this early stage in your education, you should be able to determine the extent to which an activity depends mostly on skill or physical performance capacity. Activities in which success depends primarily on learning precise and coordinated sequences of movements are located at the right end of the continuum in figure 3.4. Improvement in these activities is largely a product of practice. At the extreme left end of the continuum are activities that require little in the way of skill but depend heavily on factors such as strength, endurance, and flexibility. We improve our performance in these activities primarily by training experiences. Notice how activities might be distributed along the training-to-practice continuum.

> Physical activity experiences known as *training* are used to develop such performance qualities as muscle strength and endurance, cardiorespiratory endurance, and flexibility. The state of having developed these qualities is known as *conditioning*.

> In American football, wide receivers who catch passes, quarterbacks who throw passes, and halfbacks who run the ball are often said to play the "skill positions." Other positions such as lineman (tackles, guards, and ends), whose primary responsibilities are blocking and tackling, are not given this title. Why do you think this is so?

✔ **ACTIVITY 3.1**

Practice or Training?
Distinguish between practice and training in this activity in your web study guide.

Primary quality of activity

Physical performance capacity

Skill development

Jumping jack | Sawing a board | Blocking an opponent | Handspring | Basketball dribble | Tennis serve | Golf putt | Video game

Training

Practice

Type of experience needed to improve performance

FIGURE 3.4 Emphasis given to practice or training depends on the relative importance of skill or physical capacity factors in the activity.

> Generally, practice without training, or training without practice, is an incomplete formula for developing excellence in sport.

Because most physical activities incorporate both physical capacity and skill elements, improvement often requires both practice and training experiences. Rehabilitation programs designed to help individuals learn to walk following a stroke focus on practice because much of the problem relates to neurological impairment, but given that loss of muscular strength also may impinge on efforts to support the body, training may also be required. But, as shown in figure 3.4 some activities, such as playing video games, rely almost entirely on practice for improvement whereas others, such as performing jumping jacks, rely almost exclusively on training.

If your goal were to become a member of the varsity soccer team, much of your preparation would involve practicing such skills as dribbling, tackling, and passing the ball. At the same time, because the game requires you to run continuously, you would need to run sprints and long distances to develop cardiorespiratory endurance. You also might train with weights, especially with your legs, to increase kicking force. Like many sports, soccer requires attention to both practice and training experiences.

Performance Experience and Physical Fitness

Training experiences are important not only for developing proficiency in a particular sport or rehabilitation activity but also for increasing overall capacity to perform activities of daily living. The term that we use to describe this capacity is **physical fitness**. As we learned in chapter 1, the type of exercise specifically intended to improve fitness, health, and well-being is called *health-related exercise*.

A physically fit person is one who has used exercise to develop the capacity to perform the essential activities of daily living at a high level, has sufficient energy remaining to pursue an active leisure life, and is still able to meet unexpected physical demands that emergencies may impose. For example, our conception of a physically fit person is one who is physically able to carry out the demands of his or her job; play tennis, racquetball, or other sports on the weekend; and still be able to run downstairs in case of a fire, walk 5 miles (8 kilometers) to get help when his or her automobile breaks down, or shovel snow from the driveway after a big storm.

Because this type of fitness is reflected in a capacity to *perform* physical activities, it is called **motor performance fitness**, which has been defined as "the ability to perform daily activities with vigor" (Pate 1988, p. 177). Motor performance fitness may be reflected in how many sit-ups or push-ups you can do; how long it takes you to run or walk a given distance; or your ability to perform other activities that rely on strength, endurance, or flexibility. Because balance, agility, coordination, quickness of response, and similar factors are also important in performing daily activities, some kinesiologists include these as additional indicators of motor performance fitness. Unlike strength, endurance, or flexibility, which are developed through training, these latter qualities are developed more by practice.

Definition of Physical Fitness

Although we all have a general idea as to what is meant by *physical fitness*, the term is defined in many different ways. Here are a few examples:

- U.S. Centers for Disease Control and Prevention: Physical fitness is a set of attributes a person has with regard to his or her ability to perform physical activities that require aerobic fitness, endurance, strength, or flexibility and is determined by a combination of regular activity and genetically inherited ability (CDC 2011).

- Public Health Agency of Canada: Physical fitness is a set of attributes that are either health related or performance (or skill) related. Health-related fitness comprises those components of fitness that exhibit a relationship with health status. Performance- or skill-related fitness involves those components of fitness that enable optimal work or sport performance (Public Health Agency of Canada 2012).

- *Journal of the American Medical Association:* Five components of fitness include **cardiorespiratory** (heart and lungs) endurance, muscular strength, muscular endurance, body composition, and flexibility. Cardiorespiratory endurance is the ability to perform sustained physical activity, such as walking, swimming, or running. Muscular strength and endurance are linked and are improved through use of weight-bearing exercise, such as lifting weights or using resistance bands. The proportions of muscle, fat, and body water make up body composition. Flexibility is related to range of motion and is improved by gently and consistently stretching muscles and the connective tissues surrounding them (Torpy et al. 2005).

What are the similarities and differences among these definitions of fitness? Which definition do you think comes closest to that discussed in the preceding paragraphs?

A second type of physical fitness, generally regarded by those in the health-related professions as more significant than motor performance fitness, is **health-related fitness**. This type of physical fitness refers to having developed, through physical activity experience, the traits and capacities normally associated with a healthy body, specifically in relation to diseases known to result from a physically inactive lifestyle (sometimes called **hypokinetic diseases**). Heart disease is one type of hypokinetic disease; high cholesterol, high blood pressure, and obesity also are associated directly with an inactive lifestyle. Health-related fitness may be assessed directly using various technologies without actually assessing motor performance. Using a skinfold caliper to measure percentage of body fat is one example. In other cases, health-related fitness may be assessed while the subject engages in motor performances. For example, recording heart activity (via electrocardiography), blood pressure, and heart and respiration rates while a person exercises can give kinesiologists a fairly clear picture of a person's health in relation to cardiovascular-related hypokinetic diseases.

Measuring Physical Fitness

Most physical activity professionals, whether working in gymnasiums, clinics, fitness testing laboratories, fitness centers, or athletic conditioning programs, face the task of assessing the fitness of their clients. Because of this, knowing how to assess physical fitness is as important as knowing how to define it; in fact, how you decide to measure physical fitness is a clear reflection of your definition. How would you measure fitness in a population of college students? If you approach the task by recording the number of sit-ups these students can perform in a minute, timing them on a 100-meter (109-yard) dash and in an agility test, or measuring range of motion of their spines or at their hip and shoulder joints, your conception of fitness is probably closely related to motor performance fitness.

If you decide to measure the students' fitness by having them run on a treadmill for a specified period during which you monitor heart rate, blood pressure, oxygen consumption, and heart activity (via electrocardiography) and follow this with skinfold measurements to determine body fat and a test of blood cholesterol level, your definition is much more associated with health-related fitness.

Table 3.1 lists the various components assessed by motor- and health-related fitness tests. As you can see, some of the motor fitness measures bear little relation to the health of the individual. Could an individual with a diseased cardiovascular system score high on a balance, strength, or agility test? Such a result is quite likely. As shown in the table, some components of fitness are assessed by both motor- and health-related tests. Flexibility tests, for example, are a measure of motor performance capacity but also reflect something about the health of the joints. (Regular physical activity guards against the onset of diseases that limit motion, just as regular vigorous activity guards against diseases that destroy the circulatory system.)

> Training experiences that improve our general capacity for performing daily activities and preventing disease processes associated with low levels of physical activity are known as fitness activities.

TABLE 3.1 Components of Physical Fitness

Component	Motor performance	Health-related fitness
Anaerobic power	*	
Speed	*	
Muscular strength	*	*
Muscular endurance	*	*
Cardiorespiratory endurance	*	*
Flexibility	*	*
Body composition		*
Agility	*	

*Indicates that the test measures this particular component of physical fitness.

Adapted, by permission, from R.R. Pate, 1988, "The evolving definition of physical fitness," *Quest* 40: 178.

Type of Physical Fitness

Which type of physical fitness should physical activity professionals be most concerned about? Generally, health-related fitness is of most interest to those in the medical community and, increasingly, to those in kinesiology. Given the increasing interest among public health officials in the role of physical activity in disease prevention, the emphasis on health-related fitness is likely to continue. At the same time, it is reasonable to continue to conceive of physically fit persons as those who can effectively carry out their daily activities. This will become increasingly important as our population grows older. In light of this, assessment of both motor performance and health-related fitness would seem to be in order for most populations.

Depth or Breadth of Experience: Generalists Versus Specialists

Do you have in-depth experience with a particular physical activity (a specialist) or a broad range of experiences with a number of different activities? If you are like most people, you are more of a generalist than a specialist. Limiting practice and training to one particular activity allows you to develop **depth of capacity** in that activity. In contrast, exposing yourself to a broad range of different kinds of training and practice experiences increases your **breadth of capacity** for a number of different types of skills and activities.

Physical Activity Generalist

Physical activity generalists are individuals with experience in a broad range of skills and activities. A person who has developed low-average to above-average competency in, say, rock climbing, wrestling, ice skating, football, and baseball is a generalist. The advantage that accrues to generalists is the enjoyment and satisfaction that come from being able to engage in a variety of activities. The disadvantage is that competence is not highly developed in any single activity, thus depriving the individual of the experiences that come from demonstrating excellence.

Physical Activity Specialist

Physical activity specialists are those who devote themselves to developing depth of capacity in a single activity or a narrow range of activities. The young girl who wants to become an Olympic gymnast must commit herself to training, practicing, and competitive schedules; she is unlikely to be found on the tennis court, golf course, or in the swimming pool on a regular basis. Tiger Woods is extraordinarily skilled in golf but apparently is not a great dancer. The advantages that accrue to specialists are the pride and satisfaction that come from being able to do one activity or a small number of activities at an above-average level. The disadvantage is that the individual misses out on opportunities to engage in a number of different activities. Concentrating efforts on a single physical activity can result in remarkable proficiency, as anyone who watches national- or international-class athletes can appreciate. These in-depth experiences can result in some amazing capacities.

- As a result of long hours of training, Bob Natoli was able to lift a total of 51,640 pounds (23,423 kilograms) in 1 hour of upright barbell rowing.
- Nancy Schubring trained herself to run a half marathon in 1 hour, 30 minutes, and 51 seconds while pushing a youngster in a baby carriage.
- With practice and training, Ashrita Furman developed the ability to run a mile (1.6 kilometers) in 7 minutes and 5 seconds while balancing a baseball bat on his finger.
- Intensive training enabled Paddy Doyle to perform 1,940 pushups in 1 hour *using the backs of his hands.*
- With extensive practice, Michael Kettman learned to spin 28 basketballs simultaneously.
- Chris Gibson developed the capacity to perform 3,025 consecutive somersaults on a trampoline.

These examples are striking testimony to the way in which in-depth training and practice experiences can improve the sophisticated nerve, muscle, and cardiorespiratory systems that allow us to perform.

How Quality and Quantity Affect Physical Activity Experience

By now it should be clear that the route to improving your skill or physical capacity is by selecting appropriate practice and training experiences. The type of physical activity experience must be matched appropriately with the desired performance goal to bring about the intended effects. Knowing how to match physical activity experiences with physical activity goals is a competency that all practicing kinesiologists must master.

By studying exercise science, you will develop a clear understanding of the types of practice and training experiences and their underlying rationales. Two principles can guide your thinking about the relationship between experience and physical activity: the principle of quality and the principle of quantity.

The **principle of quality** states that experiences that engage us in the most critical components of an activity are most likely to increase our capacity to perform that activity. Critical components are the elements of an activity that are most important for performing it at a high level. As a first order of business, you must decide whether skill or performance capacity or both are critical to performance of the skill (see figure 3.3). For example, ask yourself how the principle of quality applies to lifting a heavy weight or serving a tennis ball. In the first case, your answer was probably "strong arm muscles" or, more appropriately, "strength in the large muscles of the entire body." In the second case, your answer may have been "good form," "the correct movements," or, more appropriately, "a coordinated pattern of muscle action that brings the hip, trunk, shoulder, and wrist joints into play at precisely the correct times to propel the ball forcefully into the opposite service court."

This principle suggests that if you want to be a weightlifter, your physical activity experiences must engage you in training exercises that focus on the critical component (strength). If you want to learn to serve a tennis ball, your physical activity experiences must engage you in practice that focuses on the movement pattern that you must learn. Running 4 miles (6.4 kilometers) a day will increase your cardiorespiratory endurance but will have minimal effect on arm strength. Thus, running isn't an appropriate training experience for weightlifters because it fails to take into account the critical component of strength, just as pitching horseshoes is an inappropriate practice experience for learning the tennis serve because it involves a different movement pattern and skill goal.

According to the **principle of quantity**, when all other factors are equal, increasing the frequency of our engagement with the critical components of an activity usually results in the largest performance improvement in that activity. Generally, the performer whose experiences have engaged him or her *most often* in the critical components of an activity becomes the most competent in that activity. Remember that the principle of quantity incorporates the principle of quality. Extensive experience—whether practice or training—that does not engage us in the most critical aspects of the activity is unlikely to result in substantial improvement in performance.

Whether your client is a postcardiac patient, an athlete recovering from injury, a teenager embarking on an exercise program to lose weight, a student in a physical education class, or an older adult hoping to improve her capacity for performing activities of daily living, you must first direct your attention to the critical components underlying the performance of the activity. The process of systematically identifying the critical components of an activity is called **task analysis**.

Let's take these concepts into the workplace. If a marathoner embarking on a training program to decrease her time in the 26.2-mile (42.2 kilometer) event asked you to plan training experiences that would help her achieve her goal, your first challenge would be to conduct a task analysis of the activity to identify its critical components. Rather quickly, you would conclude that the critical components in this activity relate more to physical performance capacity than to skill. This judgment, coupled with your knowledge of exercise physiology and clinical knowledge you have acquired as a practicing professional, would lead you to conclude that she will need muscular endurance if she is to run for more than 2 hours without fatiguing her leg muscles and that she will need cardiorespiratory endurance if her circulatory system is to develop the capacity for delivering large amounts of oxygenated blood to the working muscles. Thus, training activities that emphasize these components would assume a priority in the training plan that you prescribe.

Centering Practice on the Correct Critical Components

Based on your understanding of the principles of quality and quantity, which of the following coaching practices would you recommend? Think about whether the activities seem to be appropriate training experiences for the players. Specifically, what critical components of performance do the activities seem designed to develop? What critical components might a coach neglect to account for in preparing players for competition? What practice activities might have been more effective in developing skill?

- A trainer for an industry that ships its products overseas trains workers how to pack components into 12-inch by 12-inch mailing boxes. He devotes much of each practice session to developing the workers' arm and shoulder muscles.

- A basketball coach has his players spend much of practice time jumping sideways over benches, jumping rope, and tossing medicine balls as a way of conditioning them for competition.

When you begin to study kinesiology in depth, you will discover that the principles of quality and quantity are subject to many qualifications. You also will discover that identifying the critical aspects of various physical activities and determining the appropriate types of practice or training experiences are not easy tasks. For example, what are the critical elements in batting a pitched ball? Is skill or physical capacity the primary consideration, or are both important? A player must develop the skill to hit the ball, but he or she must also develop the strength to swing rapidly and hit the ball a long distance. Is watching the ball the most important element? Developing force with the arms? Developing limb–target accuracy? Developing force with the hips and trunk? How much and what kinds of practice and training experience does each case require? Your competence as a physical activity specialist will depend a great deal on your skill at carrying out such analyses and using the results to plan appropriate practice or training experiences.

To understand this concept more deeply, consider the four possible choices a practitioner can make (see figure 3.5). The decision regarding which critical components (quality of activity) to emphasize can be correct or incorrect, as can the decision regarding the quantity of physical activity experiences to provide. When the practitioner selects the appropriate critical component or components and provides the maximal amount of engagement with these components, chances for success will be good (decision A). But if the practitioner provides maximal amounts of engagement in inappropriate critical components (decision B), the client will not likely make progress. In fact, large amounts of practice in unimportant aspects of an activity may even lead to a decrease in performance. In decision C, the practitioner has identified the correct critical components but has provided less than adequate amounts of practice, training, or therapy. Although the client will make some progress, it won't be as substantial as it would have been if the practitioner had provided more experience with these critical components. In decision D, the practitioner fails on both counts, providing infrequent engagement in inappropriate (irrelevant) critical components. Although we would not expect anything to come of these physical activity experiences, this decision may not be as harmful as decision B. Although with decision D practice or training was focused on inappropriate critical elements, at least only minimal amounts of experience were provided, making damage less likely than might be the case with decision B.

Experience will improve physical activity capacity only if it engages us in the essential aspects or components of the activity (principle of quality). Generally, the more often our training or practice focuses on these components, the better will be our performance of the activity (principle of quantity).

Kinesiologists, physical education teachers, and coaches constantly conduct task analyses to identify the critical components of physical activities; they use this information to design appropriate practice and training experiences.

ACTIVITY 3.2

Improving Performance
Determine the best approach to helping two athletes improve their performance in this activity in your web study guide.

Principle of quality
Appropriateness of critical component selected

	Appropriate	Inappropriate
High level of engagement	Decision A Optimal progress	Decision B No progress or possibly regress
Low level of engagement	Decision C Little progress	Decision D No progress

Principle of quantity Frequency of engagement

FIGURE 3.5 Results of decisions about the quality and quantity of physical activity experiences.

Identifying Critical Components of a Physical Activity

How do physical activity specialists zero in on the critical components of an activity? In many instances, perhaps most, professionals use informal, intuitive task analyses developed over many years of practice. Like veteran physicians diagnosing patient ailments, experienced physical activity professionals can identify the critical elements of a task quickly without resorting to a checklist or other formal system. For beginners, however, a general framework is usually recommended. Some examples of frameworks for conducting task analyses of physical activities follow.

The first task is to identify the quality of experience that will bring about improvement. If the activity is located more toward the skill development (right) end of the upper continuum displayed in figure 3.4, the experiences should be those that are likely to improve skill. Thus the goal of your analysis will be to identify critical aspects of the performance that can be learned through carefully constructed practices. If the activity is located near the physical performance (left) capacity end of the upper continuum, your focus will be on training experiences that will promote conditioning appropriate for the activity.

Determining Skill Components Critical for Learning

If the physical activity in question falls near the skill development end of the continuum in figure 3.4, the analyst's first task is to identify components of the activity that must be practiced to maximize learning.

In making such important decisions, physical activity professionals often consult **motor skill taxonomies**, classification systems that categorize skills according to their common critical elements. Obviously, we can classify motor skills in many ways. We could, for example, classify them according to whether they require large movements or small movements, whether they are performed in the water or on land, whether they involve fast movements or very slow movements, or whether they require equipment. But it is doubtful that any of these factors could be considered critical because they do not appear to have direct implications for how to plan practice experiences. Remember, if a taxonomy is to be useful, it must at least supply us with hints about how to design practice experiences.

One of the most illuminating classification systems designed for analyzing motor skills in the past 40 years is a simple scheme that locates skills on a continuum from **closed skills** to **open skills**, as shown in figure 3.6 (Gentile 1972; Poulton 1957). The classification is based on the presumption that a critical component of all skills is the predictability of the environmental events to which performers must adapt their movements. In this view, performing a motor skill is essentially a matter of adapting one's movements to relevant objects, persons, or environments. For example, to execute the simple task of picking up a ball, performers must move toward the ball, reach their hands in the direction of the ball, and position their hands and fingers in such a way as to grasp the ball. Moving their arms or hands away from the ball or keeping their fingers in a stiff and extended position will not allow the performers to accomplish the goal. In this sense, then, the location of the ball in space at any given time, along with the shape of the ball, controls or regulates the movements of the performer.

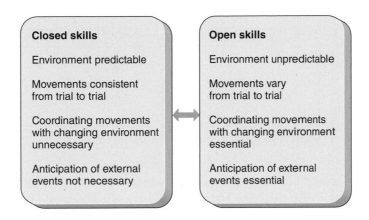

FIGURE 3.6 The open skill–closed skill continuum.

If the ball to be picked up is resting on a table, its position is fully predictable before and during the movement. In this circumstance, the movements that the performer must execute to achieve the goal are also predictable and may be planned before the actual trial. Motor skills performed in such highly predictable environments and requiring highly predictable movements are called *closed skills*. The appropriate practice experience, therefore, would be one that focuses on stereotypical and consistent arm, hand, and finger movements ("grooving in" the movement). Also, because the ball is stationary, performers need not be concerned about coordinating the timing of their movements to coincide with changes in the environment, as a baseball batter has to, for example. In this sense, performers are free to execute the movements in their own time.

If one were to change the goal of the task by requiring the performer to pick up the ball as it is rolled at various speeds across the table, a variable rather than a stereotypical motor response would be needed. In addition, performers could no longer plan and execute their responses in their own time. The start and duration of their movements would be controlled by the speed at which the ball is rolled. Now what was originally a closed skill has become an open skill. The skill is open in the sense that it requires performers to monitor a constantly changing environment so that they can adjust their movements accordingly.

Thus, when the skill is "opened up," the ball not only controls the direction and location of performers' movements, as was the case when it was stationary, but also determines when performers must initiate movements. (Thus, performers can make errors not only by being inaccurate in the direction of movements but also by being early or late.) If the ball's velocity and direction change on each trial, performers face an even more complex challenge. Because the ball might roll at high speed or low speed and in any direction, it could well topple over the edge of the table before the performer has time to decide on a response and execute it.

Thus, open skills present performers with unpredictable environments. This simple change in the structure of the task—creating a moving rather than a stationary environment to which performers must adapt their movements—makes for fundamental changes in the critical components of the task. Whether the skill is to be performed in an open or closed environment must be taken into account in the planning of practice experiences.

The determining factor for locating any skill on the open skill–closed skill continuum is the predictability of the movements required to attain the goal. Figure 3.7 shows how some skills might be located on the continuum. Remember, skills near the closed end are performed in highly predictable environments. Thus, the movements required for successful execution are also highly predictable. Hitting a golf ball, throwing darts, and hitting a softball from a batting tee are examples of closed skills.

FIGURE 3.7 Possible locations of some skills on the open skill–closed skill continuum.

Skills may be located on a continuum anchored at one end by skills in which performers must adapt their movements to fixed and therefore highly predictable environments (closed skills) and at the other end by skills in which performers must adapt their movements to changing and highly unpredictable environments (open skills).

At the extreme open end of the continuum are activities in which the movements required are highly unpredictable because the locations of the environmental objects to which performers must adapt change in an irregular way. Returning table tennis serves, which requires reacting to balls directed to multiples angles on the table and hit with enormous spin, would be placed near the open end of the continuum. In contrast, a golf drive is performed in a stable environment because no reaction is required, and thus it is considered a closed skill. Skiing meets the general definition of a closed skill, but because the body moves swiftly against a stationary terrain, skiing takes on some of the characteristics of an open skill as the performer must constantly adjust to a changing landscape. Dancing with a partner requires constant adjustment in position and steps, but as these are learned, changes in the position of one's partner become more predictable.

Determining Practice Experiences Critical for Improving Skills

The types of practice experiences essential for improving skills vary depending on the location of the skill on the continuum. For example, skills near the closed end of the continuum may be best practiced in situations in which the environment is structured the same way on each trial. In learning to hit a golf ball, for example, a player would focus on grooving a theoretically correct technique that could be repeated trial after trial. Practice regimens for skills near the open end of the continuum are more likely to involve structuring the environment in various ways on each trial so that the direction and velocity of movement of the relevant objects, persons, or environment change on each trial. Thus, the goal of practice with open skills is to develop a flexible technique that allows one to adapt to a variety of environmental stimuli.

Referring to figure 3.5, you can see that if a coach has her softball team spend a lot of time hitting balls from a tee (closed skill) rather than hitting pitched balls (open skill), her prescriptions fall within the upper right cell of the matrix (large amounts of engagement with inappropriate critical components). Such practice would not focus on the critical aspect of the task: coordinating the swing with the direction and speed of the ball. As a result, players would waste their time and effort and might learn inappropriate techniques. Practice for open skills typically concentrates on developing strategies for anticipating likely changes in the environment. Baseball batters often do this when facing pitchers whom they have studied or watched on video. This approach provides performers with more time to plan and execute their responses. Obviously, skills located near the center of the continuum in figure 3.7 will require appropriate application of all these practice strategies.

Remember that the open skill–closed skill continuum focuses on only one critical component of motor skills. Other critical components may be present as well. Chapter 8 will introduce you to many other elements that you must take into account in planning practice experiences for skills.

Determining Experiences Appropriate for Training

Having considered how to determine the types of *practice* experiences essential for *learning* skills, how can we determine the types of *training* experiences likely to lead to the *conditioning* essential for improving physical capacity? Remember, unlike skill, physical performance capacity depends on training experiences involving such critical components as muscular endurance, cardiorespiratory endurance, strength, and flexibility. Knowing the critical components that underlie a particular activity permits you to plan appropriate training strategies for developing them.

How would you determine which types of physical activity experiences are essential for improving competitive performance in bicycling and performance in golf? To start, you must determine the critical components of golf and bicycling. You might approach the problem by referring to the continuum depicted in figure 3.4. You should recognize immediately that the key to successful performance in golf is learning to perform a coordinated sequence of movements and that these movements must be acquired through practice experiences. This is not to say that strength, endurance, and flexibility are not important in golf, only that they are less important than skill in this case. You will also realize that bicycling, although depending on the development of an efficient posture and a coordinated sequence of efficient movements, is primarily determined by the contribution of such performance components as cardiorespiratory endurance, muscular endurance, strength, flexibility, and balance. Given the dominant importance of physical capacity elements in bicycling and their secondary importance in golf, how might we identify the specific critical physical capacity elements around which training programs should be developed?

> ► One way to identify the critical performance components underlying any physical activity is to ask experts to rate how important selected components are in the successful performance of the activity.

This problem can be approached in two ways. One is to consult experts. If we asked a panel of exercise physiologists or conditioning experts to use a scale of 0 to 3 to rate the relative contributions of five critical performance components in golf and bicycling, we could draw up a chart like table 3.2. (This table depicts the total ratings assigned by seven experts to five physical capacity elements for each of 10 physical activities.) Examining the table, we can see that four of the five physical capacity components listed were judged critical to bicycling, suggesting that any training regimen should focus on each of these. In the opinion of the experts, none of the components figured prominently in golf performance, something that won't surprise you. Although strength and endurance training may improve golf performance to some degree, golf is predominately a game of skill and hence is more effectively improved through practice. Softball also received relatively low ratings. Swimming and jogging, which depend primarily on cardiorespiratory endurance, were rated high in this critical component.

A second approach to determining critical performance components is to answer a series of questions about the activity based on your knowledge of it and your understanding of kinesiology. Binary decision trees are often used by physicians when diagnosing diseases or by electricians when they troubleshoot a circuit defect. Binary decision trees focus your attention on the most important questions and can help you approach task analyses in a clearheaded, systematic way.

Obviously, the decision tree in the web study guide is not intended to be an exhaustive analysis of all the critical performance factors that might be important in an activity. Although

TABLE 3.2 Ratings by Experts of Physical Capacity Elements

Physical fitness	Jogging	Bicycling	Swimming	Handball or squash	Skiing (alpine)	Basketball	Tennis	Calisthenics	Golf	Softball
Cardiorespiratory endurance (stamina)	21	19	21	19	16	19	16	10	8	6
Muscular endurance	20	18	20	18	18	17	16	13	8	8
Muscular strength	17	16	14	15	15	15	14	16	9	7
Flexibility	9	9	15	15	14	13	14	19	9	9
Balance	17	18	17	17	21	16	16	15	8	7

Combined ratings of seven experts as to the contribution of cardiorespiratory endurance, muscular endurance, muscular strength, flexibility, and balance on a scale of 0 to 3. A maximum score is 21.

Adapted, by permission, from D.J. Anspaugh, M.H. Hamrick, and R.D. Rosato, 1991, *Wellness* (St. Louis: Mosby Year Book), 165. © McGraw-Hill Companies, Inc.

► Binary decision trees that focus your attention on the most important questions to ask about a performance often can help identify the critical performance components underlying the task.

it is a general taxonomic model, it will help you appreciate the many critical components underlying physical activities. The model will also lay the groundwork for learning in a more detailed fashion all that goes into determining how to match performance experiences with critical performance components.

 ACTIVITY 3.3

Binary Decision Tree
In this exercise in your web study guide, you'll experience a systematic approach to determining critical physical performance capacity components.

As important as physical experiences are in determining how well we are able to perform skills, they are not the only factor that we have to take into consideration. Think back to the case of Gabby Douglas. Can we say that the only thing that separates her from most other gymnasts is her extensive experience on the balance beam and performing floor exercise? Probably not. For one thing, she has a small, compact body that lends itself to generating explosive power and rotating in the air, a strong lower body that helps her maintain balance, and, quite possibly, a genetically endowed nervous system that gives her unusual control over her movements and postures. She also has had, since her earliest years, an unflagging commitment to become a world-class gymnast. All of this raises the question of how much heredity contributes to athletic performance. As we have seen, athletes cannot move to elite status unless they have spent years gaining experience with an activity. However, heredity contributes as well.

Heredity and Experience

You and a friend may have had about the same amount of competitive basketball practice, and both of you may have adhered to rigorous conditioning programs to improve your jumping ability and overall strength. Yet you may be a much better player than your friend. Why? Perhaps you were more dedicated and more highly motivated to succeed. You may have had the benefit of superior coaching. But part of the answer may also be that you were lucky enough to have parents whose stature and abilities were passed on to you. For example, you may be taller, faster, and more agile, or you may have what seems to be a natural facility for changing direction, jumping, or coordinating the movements of your arms and legs, all of which are important to success in basketball. Practice can modify most of these characteristics somewhat, but the genetic contribution of your parents may have played at least as important a role. Before we close the cover on this study of the role of experience, then, let's briefly consider ways in which the effects of heredity might modify the effects of physical activity experience.

Abilities as Building Blocks for Experience

In research on motor learning and control, genetic predispositions that offer advantages or disadvantages for particular activities are usually termed **abilities**. Although physical activity scientists still do not completely understand abilities, researchers believe that abilities are genetically endowed perceptual, cognitive, motor, physiological, and personality traits that underlie human performance. For example, batting a baseball requires special visual abilities, surfing requires certain balance abilities, and orienteering requires spatial ability. People who possess great amounts of the unique abilities required for a particular activity have greater potential for success in that activity than those who did not inherit those abilities. Notice that we said that these individuals have greater potential. Usually, potential by itself is not sufficient for achieving high levels of skill or performance. The key to exploiting and eventually realizing potential is engagement in appropriate physical activity experiences.

> Ultimately, how well we are able to perform an activity is determined by both our physical activity experiences and our abilities.

Think of abilities as the foundational building blocks on which we construct our experiences. People with greater amounts of the abilities required in an activity have potential for higher achievement, but a person will not realize that potential unless he or she also capitalizes on opportunities for improving performance through practice and training. Thus, the sum of your abilities and your practice and training experiences will determine the highest level of competency that you are able to achieve in a particular activity. Sometimes, people inherit the abilities required by an activity but fail to exploit those abilities through practice and training (underachievers). Others appear to have little natural ability for an activity, but they compensate for this by unusually ambitious practicing and training schedules (overachievers).

Genetic factors also influence our body proportions, skeletal size, bone mass, limb length, limb circumference, and distribution of muscle and fat, and these features clearly have a bearing on our capacity to perform various physical activities. If you inherited a small frame, you probably won't excel in American football, rugby, or sumo wrestling, no matter how much you practice and train. If you inherited a very large frame, don't expect to make the Olympic team in gymnastics, ice skating, or diving—sports in which a compact body configuration is essential to rapid twisting and turning. Figure 3.8 shows some theoretically ideal body proportions for selected sport activities. Keep in mind that although inherited body characteristics may be a limiting factor for those aspiring to elite status as highly specialized performers, for most people this consideration is not important. Those hoping to achieve moderate levels of performance that will enable them to participate at a recreational or social level will find training and practice sufficient stimuli to achieve their goals.

Interaction of Experience and Abilities

The contribution of inherited abilities to the performance of physical activity is a much more complicated issue than most people imagine. Geneticists are just beginning to probe the way heredity contributes to performance, and their explorations so far have led them to be cautious about generalizing about the topic. One method scientists have used has been to measure performance on a battery of tests by people with common genetic backgrounds. For example, correlating the performances of children with those of their parents on a reaction time test presumably would shed some light on the contribution made by hereditary factors. (A correlation of 1.0 is a perfect correlation [rarely found] and would suggest that almost all of the performance can be explained on the basis of heredity. In contrast, a correlation of 0.50 suggests that approximately 25% of the performance [0.50^2] is accounted for

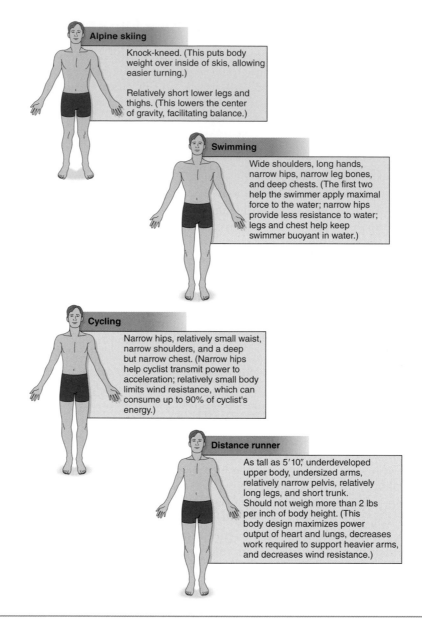

FIGURE 3.8   Ideal body proportions for selected sports.

Adapted, by permission, from R.B. Arnot and C.L. Gaines, 1984, *Sportselection* (New York: Viking Press).

by common hereditary factors.) Table 3.3 shows that correlations between various relatives and motor performance are not generally large.

One thing that complicates the issue is that genetics not only may supply us with the abilities required to perform certain activities (movement speed, hand–eye coordination, explosiveness) but also may determine how our bodies respond to experience. (We might say that heredity "interacts" with experience.) As a case in point, consider long-distance running, in which a runner's ability to deliver oxygen to the muscles (and use it) and to eliminate carbon dioxide, a phenomenon known as *maximal oxygen uptake*, largely determines performance. Endurance training generally increases runners' maximal oxygen uptake, but it doesn't affect everyone in the same way. A standard bout of conditioning may improve endurance performance by 16% for some individuals and 97% for others (Lortie et al. 1984). Why this wide variability of physiological response to training? One reason may be genetic differences in the capacity for oxygen uptake. In this case, experience (training)

TABLE 3.3   Correlations for Selected Aspects of Motor Performance and Selected Relatives

Relatives	Reaction time	Movement time	Eye–hand coordination	Strength (push-up)	Strength (grip)	Endurance (sit-ups)
Parent–child	0.13	0.09	0.04	0.25	0.20	0.24
Siblings	0.37	0.33	0.32	0.38	0.29	0.34
Uncle or aunt–nephew or niece	0.20	0.18	0.23	0.51	–0.12	0.54

Adapted from Bouchard, Malina, and Perusse 1997.

appears to interact with various genetic traits. Thus, those lucky enough to have inherited this training-response ability will probably show greater improvement as a result of a standard bout of training than those who lack it. It also appears that heredity may play a role in heart rate response to training, stroke volume, and cardiac output during submaximal exercise. Genetic differences also have been able to explain some performance variables such as adaptation to endurance exercise, muscle strength, and vertical jump height as a measure of explosive power.

Research regarding genetic contributions to athletic performance are in their infancy, and we are far from having answers that would clarify this important research area. One group of researchers who exhaustively examined the possible contribution of genetic variants to elite athletic performance summarized the state of the art this way: "There is still no evidence that any of these (genetic) variants have any substantial predictive value for prospectively identifying potential elite athletes" (MacArthur & North 2005, p. 336).

Like the effects of training on conditioning, the effects of practice on learning may be modified by inherited abilities, although the process appears to operate somewhat differently. Theoretically, each motor skill is supported by certain motor abilities, which are distributed unevenly among the population. If you are fortunate enough to have inherited the speed-of-response ability, for example, you will likely perform well in tasks that require this ability (e.g., sprinting or batting). If you inherited an ability for mechanical reasoning, you may do well in motor tasks that require this ability (e.g., the operation of complex equipment).

The issue becomes more complicated when we look at the contribution of abilities to skills when they are practiced for long periods. Practicing skills usually increases levels of performance, but scientists have discovered that the abilities required to be successful at low levels of performance may be different from those required to perform at very high levels. In some classes of skills, cognitive abilities such as mechanical reasoning and spatial ability may be important during the initial stages of learning but not during later stages. In contrast, motor abilities such as speed of movement or reaction time might not be important in early stages but might be critical at higher stages of learning (Fleishman & Hempel 1955). Thus, the effects of practice, like the effects of training, interact with the influence of inherited traits.

The average person would be wise not to base her decision to engage in a particular activity solely on whether she seems to possess large amounts of requisite abilities for that activity. Genetically endowed ability might be a legitimate concern for the young person (and her parents) who envisions competing in the Olympics someday, but for most of us, our collection of inherited abilities—whatever they may be—coupled with practice and instruction will enable us to play most sports at a level adequate to bring us enjoyment. Besides, none of us can ever know for sure what our abilities are. The best advice is to engage in the physical activities that you enjoy, realizing that regardless of your abilities, the only avenue to achievement will be through experience. In most cases, the greater the quantity of high-quality experience you have with an activity, the better your performance will be.

The upper limits of learning and conditioning are determined by both our physical activity experiences and inherited factors called *abilities*. Because experience is the only variable over which we have control, the best strategy for achieving excellence is to avail ourselves of the highest-quality physical activity experiences possible.

Human Desire: The Incalculable Ingredient of Elite Performance

Summing up their exhaustive analysis of the role of genetics in athletic performance, researchers Elaine Ostrander and her colleagues wondered whether perhaps the really important variable in attaining athletic success is something called "heart":

"In this review we have considered several classes of performance enhancing genetic variants that lead to improved athletic performance. It is worth stating, as has Helen Pearson, that we may have missed the most important class of all, those that control the brain and nervous system. All of the body's critical organs including lungs, muscles, and heart will, when taxed beyond comfort, send signals to the brain ordering the body to cease. Yet in elite athletes we find an extraordinary ability to disobey those signals. . . . When watching the extraordinary performances of some of the athletes mentioned in this article, we perceive an intangible quality that, for lack of a better term, is often called "heart." In every elite athlete there seems to be the ability to "dig down deep" and find the extra energy that is needed to train the extra lap, finish first, break the record, or capture the medal. . . . With the right early coaching, training, and nutritional guidance, together with the innate desire to excel, most of us could have achieved far more as athletes than we ever dreamed. As we muse every four years how we would look with an Olympic gold medal dangling from our necks, it is truly an intriguing thought." (Ostrander et al. 2009, p. 429)

Wrap-Up

We began this chapter by making the obvious observation that the key to improving our physical activity performances is to accumulate physical activity experiences. Thus, we should engage ourselves and those we serve in experiences of the right quality and quantity. Knowing the specific types of experiences essential for learning skills or improving physical capacity in any activity or for developing physical fitness is a key element in the expertise of physical activity professionals. Although you will not be able to appreciate fully all the variables that come into play in prescribing physical activity experiences for clients until after you have completed your study of kinesiology, this chapter introduced you to a number of general factors that you must take into account. These include considering whether the activity consists of skills that must be learned through practice or whether the activity consists of physical capacities that must be conditioned through training. We also saw that any practice or training regimen must focus on the critical elements of the activity and provide in-depth experiences if performance is to improve. Finally, we looked briefly at the traits that we inherit as factors that interact with experience to bring about performance changes across individuals.

 ACTIVITY 3.4

Use the key points review in your web study guide as a study aid for this chapter.

Review Questions

1. List four unique characteristics of human physical activity.

2. What factors influence our decisions regarding what physical activities we shall engage in and how physically active we shall be?

3. What type of activity can we improve by practicing? What do we call the improvement brought about through practice?

4. What is the principle of quality? What is the principle of quantity? Give an example in which a physical activity professional would use each principle to design an appropriate physical activity experience.

5. What is meant by an ability? In what way might abilities limit the level of proficiency that we attain in a physical activity?

chapter four

The Importance of Subjective Experiences in Physical Activity

Shirl J. Hoffman

chapter objectives

In this chapter we will

- discuss four commonly overlooked truths about physical activity;

- describe the nature of subjective experience and its place in physical activity;

- explain factors that influence our personal experiences of physical activity—what makes some activities enjoyable, some boring, and some frustrating;

- discuss how our individual preferences, tastes, dispositions, activity history, and personal circumstances contribute to our subjective experiences in physical activity; and

- describe the subjective experience of watching sports and how various factors affect us as spectators.

Elena Delle Donne may be one of the best female basketball players ever to step on the court. As the star player for University of Delaware's Blue Hens, she averaged 27 points and nearly 9 rebounds per game in 2009-2010. In 2011-2012 she averaged 28.1 points per game and was named Most Valuable Player of the Year by the Colonial Athletic Association. As a sophomore in high school she set a national record by making 80 consecutive free throws during competition and scoring 50 points in a single game. She won the Naismith Award in her senior year and was the star of the USA team when it won the World University Games. Many sportswriters have predicted that she will be a Women's National Basketball Association star within a few years.

Yet Delle Donne almost never played basketball in college. After signing to play at University of Connecticut, a perennial top contender for the NCAA crown, and living on the UConn campus for 2 days, she went back to her home in Delaware. She claimed she had lost her passion for basketball. After enrolling at University of Delaware, she joined the volleyball team and played well enough to be chosen for the all-conference rookie team. All the time she stayed away from basketball, going to only a few games and never touching a basketball. Then, after watching the women's NCAA tournament, she shook her head and asked herself, "What am I doing, not playing basketball?" After practicing in secret to regain her former skill level, she presented herself to the coach, and the rest is history (Werthheim 2012).

Elena's story is a very good example of how easily we can be misled by focusing only on the outward, physical experiences of physical activity. Often, unobservable feelings and emotions are at least as important. In most people's eyes, Elena Delle Donne's success in high school virtually assured that she would continue to play basketball in college, but her brilliant performances revealed little about her emotions or motivations. Examining physical activity in terms of performance characteristics or health benefits tells us little about what is happening inside the performer. If we are to become enthusiastic about a particular form of physical activity—whether basketball, aerobics, hiking, weight training, or taekwondo—something in the activity must contact our innermost selves in a way that causes us to return repeatedly to the activity. This relationship between physical activity and our inner lives—our thoughts and feelings—is the subject of this chapter.

You will notice that this chapter focuses primarily on physical activity experiences associated with sport and exercise (the competition and health spheres of experience). There are two reasons. First, our primary reason for playing sports or continuing to engage in exercise is that we enjoy these activities. Second, the focus is on activities in these spheres because traditionally they have formed the centerpiece of kinesiology. Kinesiologists are interested in physical activity of all kinds, but they are especially interested in sport and exercise.

Four Truths About Sport and Exercise

Elena wasn't the first talented athlete to quit a sport when she was at the top of her game. Tennis player Andrea Jaeger was ranked No. 2 in the world in 1983 when she quit the game at the age of 19, citing not only an injured shoulder but also the uneasiness she felt about making people unhappy when she defeated them on the court. Mary Hall McArver, a three-time all-conference basketball player at her high school and honorable mention All American was aggressively recruited by nationally competitive University of North Carolina but decided not to play after she arrived on campus. She told her coaches she wasn't enjoying the game. These incidences highlight the critical importance of subjective experiences, not simply in sport but in most forms of physical activity. Some people like to exercise; others, no matter how substantial the health benefits or material benefits, don't. Some like to dig in their garden, hike, or canoe; others avoid such activities at all costs. The subjective aspect of physical activity not only is fascinating but it underscores four fascinating truths:

- Physical activity is always accompanied by subjective experiences.
- Subjective experiences of physical activity are unique.
- We may not pay attention to the subjective experiences of physical activity.
- Physical activity will not be meaningful unless we enjoy it.

Physical Activity Is Always Accompanied by Subjective Experiences

The first truth is that physical activity of any sort is always accompanied by subjective experiences. Typically, when people talk about their "experience" with a particular activity, they usually refer to their physical, hands-on experience. Most of us consider an experienced runner as one who has spent much time running on the track, on roads, or on country trails. The reason we naturally equate experience with performing is that the sheer physicality of human movement rivets our attention. When Serena Williams glides effortlessly across the court and returns her opponent's backhand, her movement represents a concrete image, an *objective* event that we can observe.

But beneath these performances is an entirely different *subjective* world. Experiencing running is not just putting one foot ahead of the other in a mindless repetition of movements or simply a matter of your muscles contracting and your heart pounding in a mechanical fashion. The experience also involves feeling your racing pulse and the warmth of your body, the sense of energy being released in your muscles, the pain of fatigue. It includes the memories that the experience impresses upon you and the changes in the way you think and feel about the act of running and about yourself.

Anyone who was fortunate enough to have watched Michael Jordan play basketball at his peak knows that his performance could generate dramatic emotions—amazement, awe, and even disbelief. But Jordan's performances also affected his own subjective state. His play did not involve robotic movements executed with little feeling or emotion. After making an outstanding shot or pass, he would raise his fist in triumph, leap in the air, or laugh. Those gestures are external signals that physical activity also involves subjective experiences and that performing activities can touch us at a deep emotional level.

It is easy to underestimate the importance of the subjective side of physical activity. One reason is that historically, physical activity and kinesiology have been linked to the scientific

❚ The feelings associated with physical activity may be just as important as its physical benefits.

▶ The interior and sometimes mysterious aspect of our lives, in which we collect, recall, and reflect on the feelings and meanings that physical activity has for us, is the realm of subjective experience.

fields of biology and medicine. These areas tend to focus exclusively on objective qualities such as anatomical, physiological, and biomedical phenomena. The realm of the subjective is studied in the humanities or social and behavioral sciences, where intangible qualities such as values, feelings, and cognitions are examined. Even those working in the biological or health-related fields now recognize that a complete understanding of physical activity must take into account the response of the entire person—the intellectual, emotional, and spiritual capacities as well as those of the physiological systems. In fact, the physical features of activity merely set the stage for our subjective experiences. The late George Sheehan, a runner and a cardiologist who wrote extensively about the spiritual side of running, described it this way: "The first half hour of my run is for my body. The last half hour is for my soul. In the beginning the road is a miracle of solitude and escape. In the end it is a miracle of discovery and joy" (1978, p. 225).

Subjective experience refers to the entire range of emotions and cognitions, dispositions, knowledges, and meanings that we derive from physical activity. Sheehan found deep meaning in running on tracks and trails throughout his adult life into old age. McArver's experiences led her to stop playing basketball when she was still a teenager. This isn't to say that Sheehan's reaction was correct or that McArver's reaction was wrong. The stories merely illustrate that the feelings and meanings we associate with physical activity are very personal and often determine our level of engagement in it.

Subjective Experiences of Physical Activity Are Unique

The second truth about physical activity is that its associated subjective experiences, particularly in the case of sport and exercise, are unique. Whether physical activity is associated with work, education, or some other sphere of experience, how we feel when we move our bodies is different from how we feel when we are sitting or lying motionless. But not all physical activity stimulates the same subjective experiences. Executing a series of push-ups conjures up a distinctly different subjective experience than putting a golf ball. The feelings associated with playing racquetball are different from those associated with running on a treadmill, and all these are different from the feelings associated with typing on a keyboard, even though all require physical activity. What attract us are the unique subjective characteristics of sport and exercise. Would you work out in the gym, play sports, swim 5 miles (8 kilometers) each day, or run cross-country if those activities provided essentially the same subjective experiences as brushing your teeth, driving your car, or taking a final exam? Probably not.

> ▶ One of the primary reasons we seek out exercise and sport is that they supply us with unique forms of human experience unavailable to us in our everyday lives.

We May Not Pay Attention to the Subjective Experiences of Physical Activity

A third truth about physical activity is that we can easily overlook the subjective experiences that accompany it. Sometimes we engage in physical activity without pausing to ask ourselves what thoughts and feelings the activity generates or how the activity fits within the larger scheme of our lives. Even though our subjective experiences may be the underlying reason we engage in an activity, we sometimes become so involved in the competition or in achieving our goals that we lose sight of this important fact.

Listening to Your Body

We have said that sometimes we don't pay attention to the subjective experiences of physical activity. Here's a much different point of view from philosopher Heather Reid: "In contrast to such sedentary activities as taking tests or watching TV in which we try to ignore our bodies, sport forces us to *listen to our bodies*. . . . [Athletes] savor the exhaustion that follows a well-played game or a race well-run, even as they welcome the pain-numbing endorphins that miraculously quiet their bodies' screams midway through a hard effort. . . . What most athletes don't know but philosophical athletes do know is just how much can be learned from listening to their bodies. Listen closely enough to your body and you might discover a window to your soul" (Reid 2002, p. 52).

Physical Activity Should Be Meaningful

A fourth truth about physical activity is that unless we are attracted to the subjective aspects of physical activity—unless we discover something enjoyable in it—the activity is unlikely to become personally meaningful to us. The physical, tangible side of physical activity provides the raw material for the experience, but the subjective experience keeps us coming back for more. As we saw at the start of the chapter, accolades, recognition, prestige, and other factors were not enough to cause Elena Delle Donne to play basketball after graduating from high school. It was only after she went to a game and yearned for the meaningfulness she once experienced in earlier years that she came back to the sport. In sport, and in many other kinds of physical activity, unless you are able to invest the activity with the deepest resources of your mind and will, it can be hollow and meaningless.

If you have ever ice skated in solitude on a frozen pond or cycled along a country road on a brisk fall morning, you have probably enjoyed the subjective pleasures that physical activity can bring. The same subjective pleasures may come to you when you're struggling through a difficult wrestling practice in a hot, humid room or when your lungs are bursting in the final stages of a tough aerobics workout. Engaging in physical activity can touch us emotionally, mentally, and spiritually. When we allow this to happen, our physical activity experiences are more meaningful.

Why Subjective Experiences Are Important

Learning about subjective experiences in physical activity is important for undergraduate kinesiology students for several reasons. For one, subjective experiences can help clarify the bases of your career choices. You may be planning a career as an athletic trainer because you want others to have the same subjective experiences you did when you were in rehabilitation for a torn ligament. You may have decided to be an exercise instructor because you want to help others feel the self-confidence that comes from physical fitness. Or perhaps you have decided to become a physical education teacher or coach because of the good feelings that you experienced when participating in sport in high school.

Learning about subjective experiences can also help develop your skills as a physical activity professional. Physical educators who have experienced the feelings associated with the clumsy beginning trials of a skill can better serve their students; personal trainers and fitness leaders who have suffered through the painful early stages of a training regimen are better able to give their clients assurance and encouragement. Such knowledge can be useful in designing programs and interventions that help people understand their subjective experiences and meanings in physical activity.

But the most important reason to study and learn about subjective experiences is that how we feel and what we think before, during, and after we engage in a physical activity largely determine whether we will make that activity part of our lives. Being attracted to mountain climbing rather than canoeing, aerobics rather than running, or baseball rather than track and field may be traced in some part to the subjective experiences that these activities have evoked in us. This chapter explores how physical activity can affect us in ways that can't be measured by stop clocks, strength gauges, or competitive points: The focus is on internal dynamics rather than external performance.

The Nature of Subjective Experiences

Imagine that you have just run a half marathon in your best time ever. Imagine that after an extremely tough football practice you are sitting on a bench in the locker room, tired beyond belief but also surprised at how good it felt to endure the discomfort and pressure of that hot August practice. Imagine yourself reflecting on a mountain hike you took 10 years ago. You recall the dense forests you walked through, the musky smell of damp leaves and earth, the sound of birds chirping around you, the steep and winding hills you climbed, the treacherous descents, and the exhilaration of having exerted your body to its limits.

ACTIVITY 4.1

Rate Your Feelings
In this activity in your web study guide, you'll identify the emotional responses you have to various physical activities and the intensity of those responses.

These three short scenarios probably stimulate distinctly different feelings within you. In the first, your thoughts and feelings were likely to be intense and sharply focused on your subjective experiences while you were running the race. The sensations of a heightened pulse rate, of near-cramping legs, and burning lungs all may be indelibly impressed on your mind.

In the second scenario, you may again feel pride about having gutted it out and finished the practice, but you might also think that you learned a valuable lesson about yourself and your capacity to endure discomfort. Although these thoughts and feelings are less immediate than those evoked by the first scenario, they are still in part a direct response to the activity just performed.

In the third scenario, your thoughts and feelings are nostalgic, mellow, and far removed from the experience of performing any specific activity. Many other experiences have intervened in the years between the hike and your recollection. You could not have experienced this sense of meaning without having been deeply involved in the activity, but neither could you have discovered it without the mellowing and integration of memories in the time that has passed.

These scenarios illustrate how multifaceted subjective experiences can be—they can be intense and immediate, or they can be past experiences that you relive repeatedly. Two general kinds of experiences deserve special attention.

Immediate Subjective Experiences

When we engage in physical activity, our movements create immediate emotional and cognitive impressions. Our bodies are equipped with movement sensors called *proprioceptors*—sensory devices in tendons, ligaments, muscles, and the inner ear that are stimulated by physical actions. Proprioceptors provide us with information about the body's movements and position in space that is just as important as the unending array of visual and auditory information that often impinges on us when we move, maybe more so. Another sensory apparatus in the circulatory system picks up biochemical changes in the blood that affect our perceptions of fatigue and effort. Thus, each time we engage in physical activity, complex sensory signals flood our nervous system. These signals are vital to our performance of the activity, and they provide raw data that will be transformed into perceptions, feelings, and knowledge. Of course, we can choose to ignore many of these sensations, just as we can ignore the vast array of subtle colors in a beautiful painting or the nuances of sound in a symphony recording. Remaining open to the sensory information that accompanies physical activity is the first step toward appreciating its subjective dimensions.

> All physical activity is accompanied by sensations that we can convert to perceptions, emotions, and knowledge. To get the full experience of a physical activity, we must be open to the emotional and cognitive impressions that the activity provides.

Replayed Subjective Experiences

Although an attention-demanding activity may seem to yield little in the way of subjective experiences while it is being performed, one may discover much later that it has produced a lasting effect. Subjective experiences often endure in memory for months, years, or even a lifetime. In one of the scenarios that opened this section, you reflected on a hike you did 10 years ago and appreciated in a new and exciting way what it had meant to you. A person might experience the same feelings when she thinks about the day she earned a fitness award or about the early-morning high school exercise program that taught her how to control her weight and eat nutritious meals. The process by which we reexperience the subjectiveness of physical activity is called **self-reflection**. Self-reflection refers to reliving a past experience. It is as though we replay a videotape that includes not only visual but also kinesthetic, auditory, and other impressions from an earlier time. Because of the passage of time, self-reflection often enables us to place specific subjective experiences within a framework that makes them more meaningful.

> The subjective experiences that accompany physical activity may have a profound effect on us during and immediately after our performances. Reflecting on these experiences months or years later often helps us put them into a more comprehensive and meaningful frame of reference.

We can reflect on our physical activity experiences in various ways and for different reasons. Reflecting might be a means of summing up our total experience with the activity, accumulated over many years, to determine its value in our lives. Sometimes when we are learning a skill, it helps to focus on how the previous attempt felt so that we can adjust our movements. Sport psychologists often ask clients to replay experiences at times of anxiety as a way of calming them or giving them a boost in confidence. A tennis player, for instance, after losing the first set in a two-out-of-three match and then quickly falling behind in the second set, might replay previous successes that she has had under similar circumstances. If she can relive the feelings of the previous experience, she might be able to attain some of the same mental focus that led to her earlier victories.

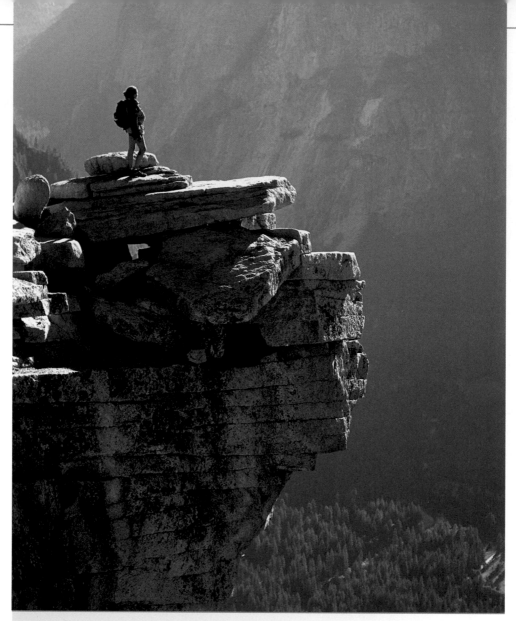

Some forms of physical activity naturally lend themselves to reflecting on past experiences.

Components of Subjective Experience

When exploring a territory as broad and expansive as subjective experience, it's helpful to break it down into its various components. In the following sections, we discuss sensations and perceptions, emotions and emotional responses, and various ways in which we gain knowledge through subjective experience.

Sensations and Perceptions

Sensations are raw, uninterpreted data collected through sensory organs. Our experiences help us interpret these raw sensations as meaningful constructs or **perceptions**. Physical activity automatically stimulates a barrage of sensations. These include the feelings of contracting our muscles and moving our bodies, as well as feelings of an increased heart rate, breathlessness, or muscle fatigue. If we attend to these sensations, they provide us with an "online" report of our inner states. Organized into meaningful information by the process of perception, these sensations allow us to make ongoing corrections in our physical activity.

For instance, perceptions of fatigue or discomfort usually form the basis of our decision about when to stop an exercise. Similarly, perceptions inform our judgment about how forcefully we are moving, how much resistance our movements are encountering, or, in the case of divers and gymnasts, how our bodies are oriented to the ground.

Sensations from outside our bodies also become part of the subjective experience. What we see or hear when we perform, even what we smell, all coalesce into a cohesive subjective experience. Football or soccer players who spend much time in close contact with the turf may automatically reflect on their playing experiences whenever they smell wet earth. For surfers, the sound of the ocean may trigger memories of riding the waves, and for boxers, the clang of a bell may prompt memories about their experiences in the ring.

Emotions and Emotional Responses

Perceptions during physical activity can elicit many different internal reactions. Sometimes physical activity increases our level of excitement and motivation and at other times may dampen it. We might become angry, annoyed, surprised, pleased, disappointed, or enthusiastic. Such reactions can stem from the feelings elicited by an activity, our impressions about the quality of our performance, or the outcome of a particular event. These subjective reactions are called emotions; they differ depending on the person and situation. For example, achieving a personal goal or performing an activity well may result in the emotion of elation or joy; failing to achieve a goal or performing poorly may prompt such emotions as disappointment or shame. Psychologists have identified more than 200 different emotions but have yet to agree about precisely how to define them. What seems clear is that emotions involve a complex combination of psychological processes including perception, memory, reasoning, and action (Willis & Campbell 1992).

Knowledge and Subjective Experience

When someone says she has gained knowledge, we usually interpret this to mean that she has learned facts, mathematical formulas, or story themes from great works of literature. This type of knowledge is called **rational knowledge**, and you will rely on it as you study the theoretical aspects of physical activity such as exercise physiology, biomechanics, or sport history or when you read this book.

A different type of knowledge derived from our subjective experiences in physical activity is equally important. **Intuitive knowledge** is a process by which we come to know something without conscious reasoning. Describing this type of knowledge or knowing precisely how we mastered it can be difficult. Intuitive knowledge gained from subjective experiences is usually personal in that it is knowledge about ourselves rather than about the exercise, the skill, or other people. Philosopher Drew Hyland (1990) identified three types of intuitive, personal self-knowledge associated with participating in physical activity: psychoanalytic self-knowledge, mystical knowledge, and Socratic self-knowledge (see figure 4.1). Because knowledge gained through

> Physical activity is always accompanied by the internal sensations of moving our limbs and bodies; changes in our physiological functioning; and sensations from outside our bodies including visual, auditory, touch, and smell sensations. These sensations can be organized into perceptions, which can give rise to emotions that we associate with particular physical activities.

Sources of knowledge from physical activity

Psychoanalytic self-knowledge

Mystical knowledge

Socratic self-knowledge

FIGURE 4.1 Three types of knowledge embedded in subjective experiences of physical activity.

subjective experience is closely associated with our motivations for participating in physical activity, we discuss each of these types of self-knowledge in detail.

Psychoanalytic Self-Knowledge

Psychoanalytic self-knowledge is knowledge about our deep-seated desires, motivations, and behavior. This form of self-knowledge relates primarily to the types of activities in which we choose to participate and the manner in which we pursue them. For example, when a young woman with an impoverished, disadvantaged background chooses to participate in a sport such as polo, golf, or squash, her choice may be a commendable effort to transcend social barriers. In contrast, it may represent an attempt to deny her social roots. When a young man chooses contact sports over noncontact sports, it may reveal a love of competition and physical contact—or it could represent "a precariously controlled desire to physically dominate or even hurt others" (Hyland 1990, p. 74). That many active sportspersons probably don't access this type of knowledge doesn't mean that it is not available to them. Some psychologists have suggested that participation in high-risk sports such as hang gliding, scuba, or motor car racing might be "propelled by personal needs to seek god-like experiences, which can activate an inflation of the ego" (Heyman 1994, p. 194).

The manner in which we choose to participate may also reveal psychoanalytic self-knowledge. Exercisers who are so concerned about the times of their runs or so intent on improving their performance that they can't enjoy their participation may have developed a narrow, single-focused pattern of behavior that may not serve them well in ordinary life, where they must meet many different responsibilities. A softball player who is a poor loser may have invested an inordinate amount of her self-worth in the game; the same may be true for a winner who gloats after the game is over. On the other side, someone who can't be serious about an exercise program or sport competition may be reflecting fear of failure. By not taking the game seriously, the player need not take failure seriously either. By carefully examining the activities in which we choose to engage, as well as our styles of involvement, we can learn a great deal about ourselves.

Mystical Knowledge

Mystical knowledge, termed *Zen self-knowledge* by Hyland (1990), refers to subjective experiences available to experienced performers only in rare and special circumstances. Mystical knowledge touches on a dimension not experienced in ordinary life. Some would call the experiences that give rise to this type of knowledge *transcendent* to suggest that the experiences took them out of the real world. These subjective experiences can be so powerful that memories of them remain with performers for years, sometimes for a lifetime. The most frequently cited type of mystical experience is called **peak experience**. Research suggests that peak experiences tend to come involuntarily and unexpectedly (Ravizza 1984). Time usually seems to slow down, and specific features of the environment sometimes stand out in sharp contrast to the background. Usually, athletes report being totally absorbed in the activity.

You might also have heard people speak of experiencing "flow" or "being in the zone." Both are quite similar to peak experiences. Psychologists Susan Jackson and Mihaly Csikszentmihalyi interviewed many elite athletes and asked them to describe their flow experiences.

Simon, an elite cyclist, talked about his last day in the 1,200-mile (1,931-kilometer) race known as the Tour de France: "My body felt great. Nothing, you feel like just nothing can go wrong and there's nothing that will be able to stop you or get in your way. And you're ready to tackle anything, and you don't fear any possibility happening, and it's just exhilarating. Afterward I couldn't come down, I was on a high. I felt like I wanted to go ride, up that hill again" (Jackson & Csikszentmihalyi 1999, p. 9).

Such experiences provide no knowledge of their own, but in some cases they can be windows through which performers claim to have a more extensive and inclusive world opened to them. Sometimes the experience can be feelings of awe and reverence to which the performer may attach a religious or philosophical interpretation. In other cases performers may claim to have gained new insights into themselves as beings connected to an all-encompassing universe.

Other Knowledge Gained From Physical Activity

Philosopher Scott Kretchmar (co-author of chapter 5) provides a different description of the knowing that comes from performing physical activity. He identifies nine types of knowledge (Kretchmar 1994, pp. 141–142):

- Health knowledge: knowledge of psychobiological factors that influence vitality and longevity
- Safety knowledge: knowledge about behaviors that influence physical well-being
- General performance knowledge: knowledge of factors that promote effective and efficient performance
- Specific performance knowledge: knowledge of factors that promote effective performance in specific sport, dance, work, or play activities
- Competitive performance knowledge: knowledge about one's own performance that is gained from competing with others
- Rules knowledge: understanding the rules (the "dos and don'ts") of a game
- Game spectator knowledge: knowledge about a game that enables one to be an intelligent spectator
- Sport history knowledge: knowledge about earlier events, persons, and social forces that influenced a particular activity
- Sport strategy knowledge: knowledge about techniques and tactics that can produce superior performance

Socratic Self-Knowledge

Socratic self-knowledge is an understanding of the difference between what we know and what we don't know. We may translate this into the realm of exercise and sport as knowledge of what we can and cannot do. Knowing our performance limits can help us operate within the confines of our skills and abilities. Coaches give testimony to the importance of Socratic self-knowledge when they say that athletes should "play within themselves."

Ignoring personal limits can lead to disastrous performances and even injury. Weightlifters who injudiciously attempt to lift beyond amounts not recommended for their weight class, or ski jumpers who challenge heights for which they have no preparation, risk not only performance failure but physical harm. At the same time, as we saw in chapter 3, we must test our personal limits or we'll never improve. The purpose of practice and training is to expand the range of our performance abilities systematically and gradually, thereby not only improving the quality of our performance but also reorienting our perceptions of our own limitations after taking into account our newly developed skill or performance capacity. One of the important roles of the personal trainer, coach, physical education teacher, physical therapist, or exercise leader is to help performers set realistic goals that will allow them to test the limits of their performances safely and productively.

Performing physical activities can be a source of knowledge, including knowledge about our motivations for engaging in activity, knowledge about different dimensions of reality, and knowledge of our personal performance capabilities.

Talking About Subjective Experiences

Subjective experiences, by nature, are private experiences. You cannot know precisely what the person riding the stationary bicycle next to you is experiencing; neither can that person know what you are experiencing. Because subjective experiences are interior by nature, the only way to compare our experiences with those of others is by talking about them. Some people have been able to describe their experiences quite eloquently. Yiannis Kouros, who could be classified as a legend in the world of ultrarunning, once explained what he felt when he ran:

> Some may ask why I am running such long distances. There are reasons. During the ultras I come to a point where my body is almost dead. My mind has to take leadership. When it is very hard there is a war going on between the body and the

mind. If my body wins, I will have to give up; if my mind wins, I will continue. At that time I feel that I stay outside of my body. It is as if I see my body in front of me; my mind commands and my body follows. This is a very special feeling, which I like very much. . . . It is a very beautiful feeling and the only time I experience my personality separate from my body, as two different things. (Kouros 1990, p. 19)

We have no reason to suspect that Kouros was exaggerating or fabricating his experiences, but it is impossible to validate his accounts. We can only know for certain what we experience; we cannot have the same certainty about the experiences of others.

If you've never before talked about your subjective experiences in physical activity, you may find it difficult at first. Most people are hesitant to share their deepest feelings—not only about physical activity but also about anything personal—with others, especially those whom they don't know well. You may worry that what you say about your subjective experiences will be misunderstood or will seem silly to others. Such feelings are normal, because most of us have had little practice sharing such information, and we have not been encouraged to do so. Unfortunately, coaches, personal trainers, aerobics instructors, and physical education teachers often are more concerned about the physical aspects of human performances than the subjective aspects. Thus, they rarely encourage students to talk about these aspects.

Even when we overcome our natural reluctance to talk about our subjective experiences, we often cannot find the right words to describe our feelings. Scott Kretchmar has suggested that words may not be able to adequately describe what we feel about physical activity. He pointed out that "the 'medium of exchange' in sport is 'feel'" and noted that "any meaningful distinctions in this realm typically outrun any verbal ability to refer to them" (1985, p. 101).

Nevertheless, words remain our primary medium for communicating our subjective experiences. Furthermore, struggling to put our feelings into words may help us come to grips with the deeper meanings that physical activity has for us. Sy Kleinman, a philosopher particularly sensitive to the meanings of physical movements, would agree: "Engagement in game, sport, or art, and a *description* of this kind of engagement enable us to know what game, sport, or art is on a level that adds another dimension to our knowing" (italics added; 1968, p. 31). Although you may feel inadequate to describe your physical activity experiences, such descriptions can be an important way to clarify to yourself why you engage in activity and what it means to you.

> ▶ In attempting to communicate our subjective experiences in sport and exercise to others, finding the correct words to express our feelings is usually difficult. This activity is important, however, because it helps us better understand the meanings that we find in physical activity.

Intrinsic and Extrinsic Approaches to Physical Activity

By now it should be apparent to you that how you feel when and after you engage in a physical activity can determine whether you return to it in the future. But our subjective experiences are not the only factor that motivates us to participate.

Practical Approach to Physical Activity

One of the most often-cited reasons for participating is the benefits that physical activity offers. Perhaps you recall listening to a lecture by your physical education or health teacher, who outlined these benefits and encouraged you to "exercise for the health of it." Maybe you recall listening to your coach's speech at the athletic banquet suggesting that sport is important because it builds character. The promise of a paycheck probably motivates the physical activity that you perform at work; the promise of returning to normal functioning motivates patients who adhere to a grueling program of therapeutic physical activity.

These are examples of **extrinsic approaches to physical activity**. One who has an extrinsic orientation to physical activity professes to engage in it primarily because of something other than the subjective experiences of the activity itself. The physical education teacher who argues for her program because it develops fundamental skills in young children and the recreational therapist who justifies shuffleboard as a means of psychological relaxation are approaching sport in an extrinsic manner, as is the businessman who works out at the fitness center with

> ▶ Physical activity usually is accompanied by external benefits, such as gaining health benefits from exercise, developing friendships through sport, or earning a salary through physical activity at work. Nevertheless, activities—especially those connected with sport and exercise—are more likely to become personally meaningful to us if we are attracted to the subjective experiences of the activity itself.

his boss as a way of impressing his supervisor and boosting his chances for a promotion. When we approach an activity from an extrinsic standpoint, we do not value it primarily for its subjective experiences; we value it because we believe it contributes to some more important end.

Physical activity produces many extrinsic benefits, some of them inevitable. The physiological benefits of physical activity, for example, are largely inescapable. When you pursue the subjective experiences of mountain climbing, you automatically reap the benefits of increased muscular endurance and strength. But we should never presume that the mere fact of these benefits makes them the primary reason for engaging in the activity.

Putting Thoughts and Feelings Into Words

Replay in your mind a notable physical activity experience—one that left an indelible impression on you, whether it occurred during exercise, sport, rehabilitation therapy, or another type of physical activity. Why did the experience make such an impression? Try to describe the experience in one word. Can you describe it in one sentence? In five sentences?

One way to think about the physiological, psychological, or sociological benefits of physical activity is to compare them to the nutritional benefits that we might gain after eating a meal at a gourmet restaurant. Although a nutritionist might be able to demonstrate that the meal served at the restaurant supplies us with a significant percentage of our daily dietary requirements, nutritional benefits are not the primary reason we go to such places. We are attracted not by the promise of improved health but by the unique subjective experience of eating at a first-rate restaurant. Whatever nutritional benefits accrue are merely icing on the cake. Likewise, we may think of many of the extrinsic benefits of physical activity—for example, health or character or new friendships—as merely icing on the cake. Physical activity, especially sport, exercise, and dance, are much more likely to become meaningful to us if we develop an attraction to their intrinsic qualities. Having said that, we should recognize that few people approach any activity in an entirely intrinsic or an entirely extrinsic manner; most of us use a combination of the two approaches.

Subjective Approach to Exercise for Health

People generally find it easier to appreciate the importance of subjective experiences when they play or watch sports than when they exercise. "We play sports for fun and enjoyment; we exercise for health," so the thinking goes. It is difficult to imagine being attracted to exercise by its subjective experiences when they often seem so painful. Wouldn't most of us prefer to have the improved strength, endurance, flexibility, longevity, and other benefits of exercise without the pain and discomfort required to attain them? Some of us might, but certainly not all. For many ardent exercisers and athletes, the discomfort is an indispensable ingredient in the satisfaction they derive from physical activity. Often, the subjective experience of performing rigorous exercise is an algebraic summation of both pleasure and pain that results in an overall sense of satisfaction and well-being. Psychologist Mihaly Csikszentmihalyi made this point in talking about the agony endured by a swimmer: "The swimmer's muscles might have ached during his most memorable race, his lungs might have felt like exploding, and he might have been dizzy with fatigue—yet these could have been the best moments of his life" (1990a, January, p. 4).

How do we know that subjective experiences of exercise are important to us? Two lines of evidence suggest this. The first is that although extrinsic factors may be the principal reason we initially undertake an exercise program, intrinsic, subjective experiences usually come to be the primary determinant over time. For instance, when people become committed to long-distance running, the prospective health benefits that may have prompted them to run

> Although exercise provides participants with numerous health benefits, many people may engage in it because of the unique subjective experiences that it offers.

in the first place often take a backseat to experiential outcomes such as feelings of achievement and self-development (Clough et al. 1989). Psychologist Michael Sachs' description of his feelings about running reflected this idea: "Running has become much more than a means to the end of getting in shape; it has become the end itself" (1981, p. 118).

We can observe the second line of evidence in our everyday lives in the way that we make qualitative distinctions between the subjective experiences of strenuous work and strenuous exercise. "Muscles," someone has said, "don't know the difference between basketball and burglary." If this is true, when we exercise specifically for health and fitness we shouldn't care whether the exercise comes in the form of using a stair stepper, running on a treadmill, riding a stationary bike, walking briskly behind a wheelbarrow, digging a ditch, or moving heavy stones. (In fact, getting our exercise through work would make more sense because at its conclusion we would have accomplished something!) But most of us *do* care. We make important distinctions between these forms of activity, and our decisions reflect these distinctions. Have you known someone who hires out the jobs of painting the house, mowing the lawn, or raking leaves and at the same time *pays dues* to a local fitness center for the privilege of doing even more grueling forms of exercise there? Measured from a purely physiological standpoint, the physical activity of work may have the same healthful benefit as the physical activity of exercise, yet we tend to avoid the former and eagerly seek the latter.

ACTIVITY 4.2

Subjective Exercise Experience Scale
Fill out the SEES in your web study guide to assess your psychological responses to exercise.

Clearly, then, the subjective experiences gained from exercising appear to be unique, different from work, and influential in our decisions to engage in exercise. Scientists have yet to identify precisely how the subjective aspects of exercise differ from those of work. How does exercising make you feel? One way to assess your subjective responses to exercise is to complete the Subjective Exercise Experience Scale (SEES) (McAuley & Courneya 1994) in activity 4.2 in your web study guide.

Internalization of Physical Activity

Have you ever decided to participate in an activity merely to please a friend or meet a class requirement? If so, the subjective experiences of the activity at first were probably incidental to your participation. But with time, your attitude toward the activity may have changed. Eventually, your decision to participate may have begun to center on something in the activity itself rather than extrinsic factors. Psychologists sometimes use the term **internalization** to refer to the gradual process by which something takes on intrinsic value, or passes from "a level of bare awareness to a position of some power to guide or control the behavior of a person" (Krawthwohl et al. 1964, p. 27).

The process of internalization can be summarized in five stages (see figure 4.2). The first time you watched Major League Baseball, hiked up a mountain trail, embarked on a weight training program, or went fly fishing, you were probably merely aware of the new stimuli and physical activity that surrounded you. Later, after doing the activity many more times, you may have responded to the activity with positive feelings. Eventually, you might have come to value the activity to the point of going out of your way to seek it out. With continued involvement you might have begun to conceptualize and organize the importance of the activity by talking about the characteristics of the activity you found most attractive. If you were further attracted to the activity, you reached the final stage, internalization. This stage occurs when your behaviors demonstrate a commitment to the activity and you are able to integrate your beliefs and attitudes about the activity into a comprehensive philosophy or worldview.

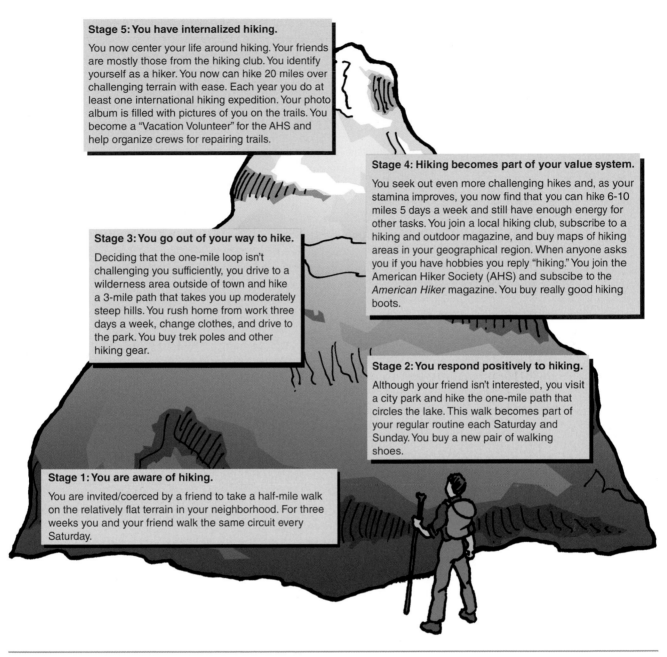

Stage 5: You have internalized hiking.

You now center your life around hiking. Your friends are mostly those from the hiking club. You identify yourself as a hiker. You now can hike 20 miles over challenging terrain with ease. Each year you do at least one international hiking expedition. Your photo album is filled with pictures of you on the trails. You become a "Vacation Volunteer" for the AHS and help organize crews for repairing trails.

Stage 4: Hiking becomes part of your value system.

You seek out even more challenging hikes and, as your stamina improves, you now find that you can hike 6-10 miles 5 days a week and still have enough energy for other tasks. You join a local hiking club, subscribe to a hiking and outdoor magazine, and buy maps of hiking areas in your geographical region. When anyone asks you if you have hobbies you reply "hiking." You join the American Hiker Society (AHS) and subscibe to the *American Hiker* magazine. You buy really good hiking boots.

Stage 3: You go out of your way to hike.

Deciding that the one-mile loop isn't challenging you sufficiently, you drive to a wilderness area outside of town and hike a 3-mile path that takes you up moderately steep hills. You rush home from work three days a week, change clothes, and drive to the park. You buy trek poles and other hiking gear.

Stage 2: You respond positively to hiking.

Although your friend isn't interested, you visit a city park and hike the one-mile path that circles the lake. This walk becomes part of your regular routine each Saturday and Sunday. You buy a new pair of walking shoes.

Stage 1: You are aware of hiking.

You are invited/coerced by a friend to take a half-mile walk on the relatively flat terrain in your neighborhood. For three weeks you and your friend walk the same circuit every Saturday.

FIGURE 4.2 Internalizing hiking.

As we progress from merely enjoying an activity to becoming engrossed in it, the activity can develop deep meaning for us, so deep that we incorporate it into our beliefs, attitudes, and personal identity. When this happens, we have internalized the activity.

Factors Affecting Our Enjoyment of Physical Activity

It is not altogether clear how internalizing an activity is different from learning to enjoy it. Clearly we aren't likely to internalize an activity that we don't enjoy. Perhaps the first thing that comes to your mind when asked why you work out regularly at the fitness club, go in-line skating, or play lacrosse is that the activity is enjoyable. This idea sounds simplistic, but the concept of enjoyment is complex. Let's focus briefly on this concept so that we can better understand it and, in turn, understand how enjoyment attracts us to physical activity. What is it about physical activity that causes people to enjoy it?

Although a long list of factors are probably involved, here we focus on three general categories: factors related to the activity, factors related to the performer, and factors related to the social context in which the activity is performed.

Factors Related to the Activity

If you reflect on the physical activities that you enjoy and ask yourself why you enjoy them, the first things that come to mind are probably specific characteristics of the activities. For instance, you might enjoy tennis because you prefer individual sports to team sports and enjoy a game that makes you concentrate. You might prefer aerobics to running because you enjoy moving your body to music. In both examples, the enjoyment comes from a specific quality of the activity itself. Csikszentmihalyi (1990b) identified several characteristics of activities that make them enjoyable. The three that we discuss here involve (1) balance between the challenges of the activity and the abilities of the performer, (2) whether the activity provides clear goals and feedback, and (3) whether the activity is competitive.

Evenly Matched Challenges

Imagine that two students have designed strength development programs using free weights. The first selects a weight that is very easy to lift (30 pounds [13.5 kilograms]), and each day he lifts this weight 10 times, never increasing the weight or the repetitions. Soon he becomes bored, realizes that he's not enjoying the activity, and quits. The second student decides to begin his weight training program by trying to lift 200 pounds (90.6 kilograms). Each day he comes to the gym and pulls with all of his might, but is unable to lift the weight off the ground. Soon he becomes frustrated and, like the first student, realizes that he's not enjoying the activity; he also quits.

Figure 4.3 shows that when our skills and abilities go far beyond the challenges of the activity, we usually experience boredom. In contrast, when success in the activity requires skills and abilities that we lack, we often experience frustration or anxiety. Csikszentmihalyi claims that "enjoyment appears at the boundary between boredom and anxiety" (1990b), which is another way of saying that enjoyment requires a delicate balance between the challenges of the activity and the skills and abilities of the performer.

Clear Goals and Feedback

Part of the fun we get from playing sports results from testing our skills and abilities against the challenges of the game. Shooting the ball into the basket and putting the ball into the cup are goals that we try to accomplish by executing a series of coordinated movements. Enjoyment stems not so much from accomplishing the goal (e.g., the golf ball's being in the cup) as from our having attempted and succeeded at putting the ball into the cup. Unsurprisingly, the enjoyment of attaining goals is greatest when the goals are difficult to attain. If we attained the goal at every attempt, our enjoyment would soon vanish.

If attempting to attain the goal of an activity is central to our enjoyment of it, it makes sense that lacking a clear idea of what that goal is can detract from our enjoyment. This can happen when we find ourselves in a game—say cricket—in which we don't understand the rules or purpose of the game. Exercising on a piece of equipment about which we know little may be unenjoyable because we don't have a clear sense of what we are supposed to be doing.

Some degree of feedback is also central to our enjoyment of physical activity. Imagine yourself shooting free throws. You may have a clear idea of the goal, but each time the ball leaves your hands, the lights go out and you never know whether the ball goes through the hoop. At first, the novelty of the situation might make it fun, but soon it would become dull.

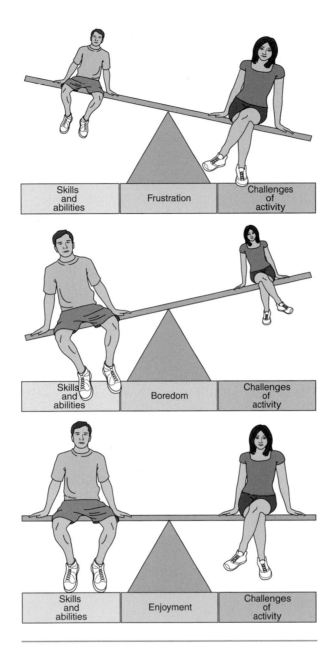

FIGURE 4.3 We are more likely to enjoy a physical activity when its challenges are balanced by our skills and abilities.

Knowing how we're doing relative to the activity goal is an important component of enjoyment. Activities such as table tennis or squash provide a lot of immediate feedback; others such as marathon running, in which the outcome occurs more than 2 hours after the gun has been fired, provide much less feedback.

Teachers, coaches, and trainers can play an important role in providing feedback to learners. After each trial or group of trials, the teacher usually comments on the major mistakes that performers made and provides recommendations for correcting these mistakes. This information, when coupled with encouragement, can affect performers' subjective experiences. For example, age-group swimmers are more likely to enjoy swimming when they perceive their coach as one who gives informative feedback along with encouragement following unsuccessful performances (Black & Weiss 1992).

We are more likely to enjoy physical activities when the challenges of the activity match our abilities, when the activity has clear goals and is followed by feedback, and when the activity is arranged in a competitive framework.

Competition

According to Csikszentmihalyi (1990a), competition often heightens enjoyment of physical activity, something we have already noted in chapter 2. **Competition** is not an activity per se but an organizing principle that frames physical activity within a larger purpose. Usually, it is a way of relating to other people by comparing your performances with theirs. It also may involve comparing your performances with a standard such as par on a golf course or a time on a racecourse. Just as we can add poetry to language, color to art, or melody to sound to make them more enjoyable, we can add competition to physical activity to make it more enjoyable.

One problem with adding competition to activities is that it can become an end in itself. Coaches and players who value winning above enjoyment of the activity or the basic individual and social values of friendship, caring, and cooperation are elevating competition to an end rather than a means. Csikszentmihalyi summarizes the issue:

> The challenges of competition can be stimulating and enjoyable. But when beating the opponent takes precedence in the mind over performing as well as possible, enjoyment tends to disappear. Competition is enjoyable only when it is a means to perfect one's skills; when it becomes an end in itself, it ceases to be fun. (1990a, p. 50)

Factors Related to the Performer

The second category of factors that influence our enjoyment of physical activities are those that lie within us. Whether we enjoy a particular type of physical activity depends a great deal on our dispositions and attitudes. **Dispositions** are short-term, highly variable psychological states that may be affected by a host of external factors. **Attitudes** are relatively stable mind-sets toward concrete objects that may be favorable or unfavorable.

Have you ever put the value of winning above the enjoyment of the activity itself?

Disposition

Csikszentmihalyi (1990b) identified three dispositions affecting enjoyment that are particularly relevant to physical activity: (1) how competent we feel in performing the activity, (2) the extent to which we are able to become absorbed in the activity, and (3) how much control we feel we have over the activity. Brief descriptions of each follow.

- *Perceived competency.* As a rule, we enjoy activities that we do well more than those that make us feel incompetent. Psychologists use the term *self-efficacy* to refer to how adequate we feel to perform a task. Research has shown that people who feel competent in exercise are more likely to adhere to a rehabilitative exercise program (Dishman & Sallis 1994), are more likely to engage in higher levels of intensity when participating in physical activities (Sallis et al. 1986), and are more faithful in attending exercise programs (McAuley & Jacobson 1991). One study has shown that aerobic dancers with higher ratings of self-efficacy reported enjoying the activity more, and they exerted more effort than those with low ratings of self-efficacy (McAuley et al. 1991).

- *Absorption.* We all remember times when we were deeply engrossed in an exercise or sport and other times when we simply went through the motions. When we surrender to an activity and become absorbed in it, we lose consciousness of ourselves as distinct entities apart from the activity. Csikszentmihalyi (1990b) found that when people lose a sense of themselves while performing an activity, feelings of enjoyment increase, partly because when we aren't self-conscious, we aren't as likely to judge our own performances. Most of us have higher expectations for our performances than we are able to achieve; thus, self-evaluation usually results in negative feedback, which is not enjoyable.

- *Perceived control.* Experiencing a sense of mastery or control over our environment is inherently enjoyable. Knowing that we can master a tough cross-country course, control the basketball (when dribbling, passing, or shooting), control our opponent in a wrestling match, or control our bodies in a difficult gymnastics skill adds to our sense of enjoyment of those activities.

> Everybody harbors certain temporary dispositions that affect his or her enjoyment of physical activity. Enjoyment tends to increase when we feel competent in the activity, when we become absorbed or lost in the activity, and when we feel we have control over our bodies and the environment.

At first glance, people who enjoy **sensation-seeking activities** such as hang gliding, parachute jumping, ski jumping, and rock climbing, in which so many factors are uncontrollable, would seem to contradict this principle. Yet, strange as it might seem, a sense of control is one of the factors that draw people to such high-risk sports (McIntyre 1992). In these cases, enjoyment appears to stem not from the presence of threatening forces but from feelings of being able to control these forces through elaborate preparation, training, and painstaking adherence to safety procedures. After all, if we had complete control of our bodies and the surrounding environment, sports of all kinds would be boring because we could always accomplish our goals. A degree of risk and uncertainty is fundamental to our enjoyment.

Attitudes

Along with temporary dispositional factors, more stable and enduring attitudes also can affect our enjoyment of physical activity. Just as individual preferences, tastes, and attitudes toward food, movies, clothes, friends, cars, and the like vary widely, so do attitudes toward different activities. Why we vary so much in our tastes for physical activity is largely inexplicable, but there is no question that we do. In the late 1960s, Gerald Kenyon (1968) and his associates designed a scale for assessing attitudes toward physical activity. They used the scale in their surveys of individual preferences for a wide range of activities—in relation to which people could be either participants or observers—and identified six categories of attitudes toward physical activity. Each represents a distinct preference for a particular form of physical activity based on "sources of satisfaction" that accrue to performers or observers:

- Physical activity as a social experience
- Physical activity for health and fitness
- Physical activity as the pursuit of vertigo

- Physical activity as an aesthetic experience
- Physical activity as a cathartic experience
- Physical activity as an ascetic experience

People who have a natural affinity for social experiences are likely to seek out physical activities that provide opportunities for social interaction.

Physical Activity as a Social Experience   Some forms of physical activity, such as team sports or group exercises, are intrinsically social events. That is, they normally take place in active social environments and involve interaction with other people. Team sports are the most social of sport activities, as are team aerobics or exercises done in health clubs where others are present. Other forms of physical activity such as rock climbing, surfing, or long-distance running incorporate less social interaction, although even those activities may provide opportunities for social engagement before, during, and after participation. Participants consistently identify social aspects of physical activities as positive experiences (Neulinger & Raps 1972), and youngsters engaged in youth sports (Wankel & Krissel 1985), as well as elite ice skaters (Scanlan et al. 1988), have indicated that developing friendships and experiencing social relationships are important sources of enjoyment. Thus, individuals who enjoy high levels of social interaction are more likely to seek out physical activities that maximize such experiences.

Physical Activity and Special Moments

The social context in which physical activity takes place often provides a setting for some of life's most special moments. In his book *Assignment: Sports*, noted sports reporter Robert Lipsyte describes a special moment related to him by the great Russian Olympic weightlifter, Yury Vlasov:

> At the peak of tremendous and victorious effort, while the blood is still pounding in your head, all suddenly becomes quiet within you. Everything seems clearer and whiter than ever before, as if great spotlights had been turned on. At that moment you have the conviction that you contain all the power in the world, that you are capable of everything, that you have wings. There is no more precious moment in life than this, and you will work very hard for years just to taste it again. (Lipsyte, 1970, p. 82)

What special moments within the social context of physical activity have you experienced?

Some people are attracted to activities requiring a great deal of strength and endurance, possibly because of the health benefits that such activities bring or because they enjoy pushing their bodies to their physiological limits.

Physical Activity for Health and Fitness   Some people value physical activity for the contribution that it makes to their health and fitness. Feeling in shape, knowing that you can meet any physical demand that you might face in the course of a day, can give you a sense of confidence and well-being that adds immensely to your quality of life. But, as we have seen, many people simply enjoy engaging in the activities that lead to the development of fitness. The unique sensations that accompany the physiological response to vigorous and sustained exercise cause many people to return again and again to running trails and gymnasiums.

Physical Activity as the Pursuit of Vertigo   Some people are attracted to a certain category of physical activity that presents an element of risk or thrill, usually through the medium of "speed, acceleration, sudden change of direction, or exposure to dangerous situations, with the participant usually remaining in control" (Kenyon 1968, p. 100). The thrill that comes from disorientation of the body in such activities is called **vertigo**. People pursue vertigo through sensation-seeking activities that reorient the body with respect to gravity, as in amusement park rides, free-fall parachuting, bungee jumping, downhill skiing, ski jumping, or any other activity that creates a sense of danger, thrill, and intense excitement.

We do not know why some individuals seek out such activities whereas others, just as deliberately, avoid them. Petrie (1967) suggested that those who pursue vertigo may tend habitually to reduce perceptual input so that they experience a specific event at a lower level of intensity than would someone who habitually augments (exaggerates) perceptual input. Involvement in vertiginous activities may also represent an attempt to compensate for tedious experiences in the workplace (Martin & Berry 1974), may be a form of stimulus addiction (Ogilvie 1973), or may be the result of complex sociocultural factors (Donnelly 1977). Whatever the cause, our interest in such activities, as both participants and spectators, clearly shows no indication of declining.

Physical Activity as an Aesthetic Experience We normally associate **aesthetic experiences** in physical activities with dance, but some individuals, as either participants or spectators, perceive sport—and to a lesser extent, exercise—as providing certain artistic or aesthetic experiences (Thomas 1983).

The aesthetic element is especially prevalent in gymnastics, diving, ice skating, and other sports in which grace and beauty of movement are primary considerations in awarding scores to competitors. The aesthetic element is less apparent in sports in which the outcome is usually viewed as more important than the manner in which the athlete moves (e.g., football, basketball, or baseball). Nevertheless, it is common to hear sport broadcasters refer to a well-choreographed football play, speak about the dancelike characteristics of a basketball player leaping through the air, or even describe the movements of a tennis player as "art in motion."

Exercise seems less likely to be a source of this type of subjective experience, although dance aerobics clearly contains a strong aesthetic component. Outdoor activities such as hiking, mountain climbing, or kayaking bring participants in contact with the beauty of the natural environment, and this element can add to the aesthetic experience of the activity. Women are more likely than men to engage in physical activities for aesthetic experiences, whereas men are more likely to value activities for their vertiginous, ascetic, and cathartic elements (Smoll & Schutz 1980; Zaichowsky 1975).

> People who find enjoyment in sensation-seeking activities also are attracted by the challenge of controlling the seemingly uncontrollable; people who find enjoyment in the aesthetic side of physical activity are likely to engage in gymnastics, diving, skating, and other movement forms that emphasize grace and beauty of motion.

Physical Activity as a Cathartic Experience Catharsis refers to a purging or venting of pent-up hostilities, either through attacking an enemy or some inanimate surrogate (object) in an aggressive fashion or through watching an aggressive event. Such a hypothesis owes to the work of behavioral physiologist Konrad Lorenz, who wrote in his best-selling book *On Aggression* that sport furnishes a healthy safety valve for "militant enthusiasm," which he viewed as the most dangerous form of aggression (Lorenz 1966). Scientists have since questioned this "hydraulic hypothesis" of aggression generally (Berkowitz 1969) and its application to sport specifically. It is no longer viewed as a credible hypothesis for explaining our attraction to sports. For example, if watching sports lowered levels of aggression in spectators, we would expect to find lower levels of aggression in spectators following their attendance at a game than before the game began, but research has shown just the opposite to be the case (Goldstein & Arms 1971).

But physical activity can be viewed as cathartic in another way. Scientists have discovered that a vigorous bout of physical activity can lower anxiety (Herring et al. 2011), thereby inducing a sense of relaxation and calm (but not necessarily reducing pent-up hostilities). Leisure theorists have long pointed to the value of leisure activities (including sport) in promoting feelings of rest and relaxation. Sport and exercise are novel activities that provide us with a change of pace. This change may, by itself, recharge our batteries by shifting our attention from problems and worries. Thus, the noontime racquetball player may derive a sense of enjoyment, relaxation, and calm simply because he has directed his energies toward a new task and has become absorbed in its features, not because he has purged hostilities.

> Participating in sport and exercise will not purge us of hostilities and aggression but may calm, relax, and refresh us.

Physical Activity as an Ascetic Experience Physical fitness and training programs often require us to undergo pain, sacrifice, and self-denial and to delay gratification. These

Hoffman

Sometimes participating in exercise and sport can be painful and can lead to uncomfortable subjective experiences. Participants do not always interpret these sensations negatively, and, indeed, the discomfort may be a source of attraction.

experiences are sometimes referred to as **ascetic experiences**. Elite athletes are accustomed to ascetic experiences—torturous training regimens intended to improve their capacity for performance. Ascetic experiences are also familiar to recreational athletes. Occasionally, you may see a runner, who by her attire and running style is obviously a novice, wheezing and groaning and from all appearances suffering. Fitness clubs specialize in ascetic experiences; patrons with pained faces struggle to complete their exercise routines while a personal trainer shouts encouragement as they work through the pain.

 ACTIVITY 4.3

Attitudes Toward Physical Activity
In this activity in your web study guide, test your understanding of the six main reasons people engage in physical activity.

Effect of Physical Activity Experiences on Family Conflicts

Sometimes the subjective experiences of physical activity can interact with other subjective experiences in ways that are beneficial to participants. For example, family conflict, including arguments and physical violence in the home, can be major factors in adolescent depression. In some cases, the negative emotions produced by family conflict lead not only to depression but to delinquency. Since participation in physical activity has been discovered to have a positive influence on emotional well-being, Sigfusdottir and colleagues (2011) wondered whether it also may help to buffer the sometimes harsh subjective experiences of adolescents brought on by family conflict. This study, involving 7,430 ninth- and tenth-grade students, was conducted by researchers in Iceland.

Hypotheses

Researchers proposed three hypotheses: (A) Family conflict increases the likelihood of depressed mood; (B) physical activity decreases the likelihood of depressed mood; and (C) the effects of family conflict on depressed mood are stronger when physical activity levels are low than when they are high.

Method

Among the questions asked of participants were whether in the past year they had (1) witnessed or heard severe arguments between their parents; (2) had severe arguments themselves with their parents; (3) witnessed physical violence at home; or (4) experienced physical violence at home. A standard scale was used to assess levels of symptoms of depression. A questionnaire was used to determine how physically active the students were.

Results

- Family conflict increased the likelihood of depressed mood among boys and girls, whereas participation in physical activity decreased the likelihood of depressed mood.
- The effects of family conflict on depressed mood were less on adolescents who were physically active than on those who were not.
- Physical activity was a stronger inhibitor of depressed mood among adolescents who live under conditions of family conflict at home than among adolescents generally, especially girls.

You may be surprised to learn that not everybody interprets the ascetic experiences of breathlessness, fatigue, and muscle soreness as unpleasant. For example, in one study, 12- and 13-year-old soccer players told investigators that they enjoyed not only winning, learning new skills, and playing with teammates but also "working hard" and "feeling tired after practice" (Shi & Ewing 1993). In another study (Wankel 1985), exercisers who did *not* drop out of the exercise program reported experiencing physical discomfort more frequently than those who *did* drop out. In this study, middle-aged men who reported experiencing fatigue because of vigorous exercise also reported *decreased* feelings of distress and *increased* feelings of psychological well-being because of the exercise. Thus, for some people, the discomfort of physical activity is attractive rather than unappealing.

Factors Related to the Social Context

The third category of factors that can affect our enjoyment of physical activity concerns the nature of the social context in which the activity occurs. Have you ever competed in an important sport contest with a large crowd looking on? How did that compare with playing the same sport on the playground or in intramurals? Do you feel different when you work out in a fitness center with many onlookers than you do when you work out alone? What about exercising with music playing in the background? Most people find these experiences profoundly different. One isn't any better than the other; they are simply different.

Spotlight on Research: Does Music Improve Exercise Performance?

Are you one of those people who plug into their iPod when they are on the exercise bike or treadmill? Most people who listen to music while exercising do so because it takes their mind off of the pain that usually accompanies hard workouts. But does listening to music improve your performance? Atkinson (2004) examined whether music played in the background would improve starting, finishing, or overall power during a 10-kilometer cycling time trial. The investigators also evaluated whether music affected heart rate and subjective responses (perceived effort) to the time trial. Two 10-kilometer time trials on a Cybex cycle ergometer were performed by 16 subjects, one with and one without "trance" dance music (a combination of forms of electronic music) being played in the background. Cyclists also completed a music rating inventory after each time trial in the music condition.

When music was played, cyclists completed the time trial in an average of 1,030 seconds; when music wasn't being played average time was 1,052 seconds, a statistically and practically significant difference. Interestingly, even though cyclists performed faster under the music condition, their perceptions of the effort required to cycle were higher under the music condition than under the no-music condition, suggesting that listening to music didn't make the exercise seem easier. Music brought about the largest increases in time trial performance and heart rate during the first 3 kilometers of the time trials. Participants rated the "tempo" and "rhythm" aspects of the music as more motivating than the "harmony" and "melody" aspects.

How participants feel when they are engaged in physical activity may depend on the social conditions surrounding the activity as much as or more than on the activity itself. The presence of others (parents, friends, strangers), the hype preceding a contest, players' feelings toward and relationship with an opponent, and the way media or coaches interpret a particular game to players can all affect how the players experience the sport. Running on a treadmill offers a different subjective experience than running through a park. Running when it is cold, dark, and raining offers a different subjective experience than running on a bright, warm summer day. Adding people to your exercise group changes the subjective experience of exercising, just as adding competitors can change the social climate and the subjective experience of playing a game. Whether these changes add to or subtract from your enjoyment depends on your individual preferences.

The social atmosphere created by the exercise leader can have a powerful influence on exercise class participants' enjoyment of the experience. Fox and colleagues (2000) reported on a study designed to test the effects of leadership style and group environment on enjoyment experienced by those attending a step aerobics class. The instructor, highly trained in aerobics, adopted either an "enriched style" that included high levels of interaction with class members (addressing them by name, providing positive reinforcement, orally rewarding effort) or a "bland style" (not addressing participants by name, avoiding conversation, not providing positive reinforcement, not rewarding effort). The investigators manipulated the group environment by planting undergraduate students in the experimental sessions. The planted students established an "enriched" environment in one exercise group and a "bland" environment in another. In the rich environment, the students introduced themselves to all participants, initiated casual conversation, were compliant with the instructor's wishes, and made positive remarks about the instructor. In the bland environment, the students did not introduce themselves to other members, did not initiate conversation, and were compliant but not enthusiastic or orally supportive of the instructor. Thus, the study established two leadership styles and two group environment conditions. The investigators examined the effects of each condition as well as combinations of the conditions on participants' enjoyment. They also examined the effect of these variables on participants' intentions to take another exercise class.

They found that social environment and leadership affected participants' enjoyment and their intentions to enroll in another exercise class. Ratings of enjoyment were higher for the enriched leadership style than for the bland leadership style. Similarly, ratings were higher for the enriched group environment than for the bland group environment. But the most powerful effect was found for the condition in which participants were exposed to an enriched leadership style and exercised in an enriched group environment. Ratings of enjoyment were, on average, 22% higher for this condition than the other conditions. Intentions to enroll in another class were highest in this condition as well (approximately 16% higher than in the other conditions). The conclusion is easy to draw: Both the style adopted by the exercise leader and the social environment in which the exercise occurs can have a major effect on participants' enjoyment.

Another way in which the social context can affect our enjoyment is by influencing our sense of perceived freedom. As you learned in chapter 2, leisure theorists have long known that we enjoy activities more when we are free to choose them than when we feel obligated to do them. In fact, Bart Giamatti, former president of Yale University and at the time of his death the commissioner of Major League Baseball, described leisure activity as "that form of nonwork activity felt to be chosen, not imposed" (1989, p. 22). This sense of freedom to participate is not simply a matter of being free from actual coercion; it is feeling free from any subtle coercive mental forces that instill a sense of obligation. Getting up early on Saturday morning to lift weights with a friend because she has been begging you for weeks to join her robs you of your sense of freedom, and you may have a far different subjective experience in this situation than you do when you lift weights because you want to.

> ► We never perform physical activities in a vacuum, and the social context that surrounds them can affect our sense of enjoyment. An example of social context occurs when we feel forced to engage in an activity rather than freely choosing to do it.

Watching Sports as a Subjective Experience

One of the most influential people in the development of modern physical education theory was Jesse Feiring Williams, a medical doctor turned professor who devoted his life to preparing generations of physical education teachers, coaches, and professors. Like many scholars of his generation, Williams (1964) was skeptical of sports watching because he believed that it demanded only "simple sensory responses" and did not require "expressive, cooperative skill activity" that could be related to the purposes of education. Although one might question Williams' assessment, it is obvious that the rise of sports watching that so worried him in 1927 has continued into the 21st century and, for the most part, shows no sign of stalling as we discover more sports to watch and more ways to watch them. Whether all of this sedentary sports watching has been beneficial to the public or not, it is clear that it is a source of vivid subjective experiences for wide segments of the human community. Because of the connection that sports watching has to the physical activity professions, we should examine, if only briefly, the factors that affect our enjoyment of and attraction to watching sports.

Watching a sport contest on the local playground is unlikely to bring about the subjective experiences that are evoked by a sport spectacle such as the Olympic Games, and neither may evoke the same subjective experiences that come from watching sports on television. Although we often watch our peers play sports in informal settings, **sport spectacles** attract most of our attention. Sport spectacles are staged competitions designed and promoted for audiences and intended to evoke an entire range of human emotions by virtue of their grandeur, scale, and drama. Unlike pickup games on the playground, sport spectacles are primarily entertainment.

Usually, we reserve the term *sport spectacle* for professional, collegiate, or international events that attract large numbers of spectators and have supporting casts of cheerleaders, bands, majorettes, officials, and the media. In spectacles, both players and spectators play defined roles; the players take on various roles associated with the competition, and the spectators take on various roles as consumers, supporters, or detractors. Thus, when spectators cheer, boo, do the wave, or sing the alma mater, they are participating in a spectacle, of which the sport contest is only one aspect.

Ways of Watching Sports

We can engage in spectating in many different ways, ranging from aligning ourselves with a team's performance and suffering extreme disappointment when the team loses or jubilation when it wins, to the other extreme of watching in a completely disinterested fashion.

Sport spectacles, such as the FIFA World Cup, play on our desire to experience drama, competition, hard work, and success.

> **Vicarious participation in sport** occurs when spectators imagine themselves performing the same activities as the athletes they are watching.

Vicarious Participation

Vicarious participation is a form of watching in which observers participate in the contest through the powers of imagination. If, while watching a sport event, you notice yourself tensing your muscles or adjusting your body position in accordance with an athlete's movements, you probably are participating vicariously in the activity. Usually, vicarious participators are fans who identify with a particular player or team and have a stake in the outcome of the contest. The word *fan* is rooted in *fanum,* the Latin word for "temple," which underscores a metaphoric link between religious zeal and the enthusiasm fans often have for their favorite teams and players. When their team wins, it is as though the fans have won; when their team loses, they experience profound disappointment.

Disinterested Sport Spectating

Sometimes we watch sport events without great emotional investment, a phenomenon known as **disinterested spectating**. We are likely to watch games in this manner when we care little about the outcome of the contest. Watching a contest between two teams we are not familiar with or watching a game that we have attended only because a friend invited us may result in disinterested spectatorship. Those who watch sports as part of their employment, such as reporters, referees, hot dog vendors, and sometimes coaches, often do so in a disinterested fashion.

Nevertheless, it would be a mistake to assume that disinterested spectators derive no pleasure from watching contests. For example, coaches who watch competitions between teams other than their own tend to observe athletes' performances in a detached, objective manner, yet they may enjoy and admire the performances of those athletes. In these cases, spectators watch in much the same way audiences watch a symphony or a ballet: Enjoyment comes not from the drama of competition but from observing the skillful actions of the performers.

Why Do You Watch Sports?

People watch sports for a variety of reasons. In an effort to determine those reasons, researchers have developed a Sport Fan Motivational Scale. Answers to each of the following questions about sport spectating can be sorted into different motivational subscales. To determine the underlying motivations for your own sport spectating (watching), indicate next to each item, using the 1 to 8 scale, how well the item describes you (1 = This is not descriptive of me; 8 = This is very descriptive of me). There are no right or wrong answers, but be honest in your responses. Remember, these items are about spectating, not sport participation.

____ 1. One of the main reasons that I watch, read, or discuss sports is that doing so gives me the opportunity to temporarily escape life's problems.

____ 2. One of the main reasons that I watch, read, or discuss sports is so that I can bet on sporting events.

____ 3. One of the main reasons that I watch, read, or discuss sports is that I get pumped up when I watch my favorite teams.

____ 4. One of the main reasons that I watch, read, or discuss sports is for their artistic value.

____ 5. One of the main reasons that I watch, read, or discuss sports is that I enjoy the beauty and grace of sports.

____ 6. One of the main reasons that I watch, read, or discuss sports is that I enjoy being physiologically aroused by the competition.

____ 7. Sports are enjoyable only if you can bet on the outcome.

____ 8. One of the main reasons that I watch, read, or discuss sports is that doing so makes me feel good when my team wins.

____ 9. One of the main reasons that I watch, read, or discuss sports is that doing so allows me to forget about my problems.

____ 10. Making wagers is the most enjoyable aspect of being a sports fan.

____ 11. One of the main reasons that I watch, read, or discuss sports is because most of my friends are sports fans.

____ 12. I enjoy watching sporting events because to me sports are a form of art.

____ 13. To me, watching, reading, or discussing sports is like daydreaming because it takes me away from life's hassles.

____ 14. One of the main reasons that I watch, read, or discuss sports is that I like to be with other people.

____ 15. I enjoy sports because of their entertainment value.

____ 16. I enjoy watching sports more when I am with a large group of people.

____ 17. I enjoy watching sports because it increases my self-esteem.

____ 18. I like the stimulation I get from watching sports.

____ 19. I enjoy watching, reading, or discussing sports simply because it is a good time.

____ 20. To me, sport spectating is simply a form of recreation.

____ 21. To me, my favorite team's successes are my successes and their losses are my losses.

____ 22. I like to watch, read, or discuss sports because doing so gives me an opportunity to be with my spouse.

____ 23. I like to watch, read, or discuss sports because doing so gives me an opportunity to be with my family.

Answers to these questions can be grouped into 8 motivation subscales. Sum your responses on the following items to determine the motivation subscale score. Total subscale scores reveal your dominant motivations for watching sports.

Motivation Subscale	Items
Group affiliation	11, 14, 16
Family	22, 23
Aesthetic	4, 5, 12
Self-esteem	8, 17, 21
Economic	2, 7, 10
Eustress	3, 6, 18
Escape	1, 9, 13
Entertainment	15, 19, 20

Adapted, by permission, from D.L. Wann, 1995, "Preliminary validation of the Sport Fan Motivational Scale," *Journal of Sport & Social Issues* 19: 377-396.

Factors Affecting Enjoyment of Watching Sports

If someone were to ask whether you like to watch sporting events, you might answer sometimes yes, sometimes no. Your answer depends on a number of factors. Few people enjoy watching their favorite team being trounced, and fewer yet are likely to enjoy watching a game that they know little about. Americans are unlikely to get excited about watching cricket or Australian rules football for the same reason that Australians and citizens of the United Kingdom are unlikely to enjoy American baseball. Our familiarity with sports and their importance in our culture often influence our attitudes toward them. Three factors play a major role in determining how much enjoyment we are likely to experience when watching a sport event: our knowledge of the game being played; our feelings toward the competing teams and players; and the extent to which the game entails a sense of drama, suspense, and uncertainty. Let's look a bit more closely at these factors.

Game Knowledge

Think about the sports you most like to watch. These sports are probably the ones that you know the most about. Knowledge about the game, what sport philosopher Scott Kretchmar (1994) calls **game spectator knowledge**, often determines our enjoyment as spectators. Game spectator knowledge is knowing about the game, including players, strategies, and competitive tactics. Only by having comprehensive knowledge of the activity can we fully appreciate the quality and significance of the athletes' performances. Such knowledge may come from reading about, watching, or participating in an activity, although it is not necessary to have played a sport to enjoy watching it. An individual who knows little about the game of lacrosse may enjoy watching a game without knowing the game's history, strategies, rules, or personalities. But it is unlikely that this person's enjoyment of the game will be as robust as that of the lacrosse fan who understands precisely what the players and teams are trying to do at every turn of the game.

Feelings Toward Competing Teams and Players

Think of the last time you watched a contest in which one of your favorite teams was competing against an opponent that, for one reason or another, you had come to dislike. What turn of events added to your enjoyment of the experience? Perhaps you enjoyed the experience more when "good things" happened to your favorite team (they played well) and when "bad things" happened to the team that you disliked (when they played poorly).

These connections between enjoyment and feelings about the competing teams are the focal point of the *dispositional theory of enjoyment,* which has been the subject of considerable research (Zillman et al. 1979) involving football, tennis, and basketball spectators. For example, researchers have discovered that spectators applauded the failed plays of the disliked team almost as much as they did the successful plays of the favorite team. Comparing reports by those who watched the Minnesota Vikings defeat the St. Louis Rams in a professional football game, the investigators found that those who both liked the Vikings and disliked the Rams reported maximal enjoyment, whereas those who both disliked the Vikings and liked the Rams reported maximal disappointment (Zillman & Cantor 1976).

Other research has shown that the enjoyment brought about after a fan's favored team wins can lead to elevated self-esteem and an overall boost in self-confidence. By contrast, diminished self-esteem and self-confidence have been associated with losses by one's favored team. Our enjoyment of sport spectating also tends to increase when our favorite team defeats an opponent of high rather than low quality. In addition, we tend to enjoy the game more when our favorite team plays much better than we expected it to (Madrigal 1995).

Human Drama of Sport Competition

The sense of drama, suspense, and uncertainty that often accompanies sport contests also enhances our enjoyment of them. Usually, this sense of drama requires that the competing teams be equally matched in talent and ability, although some of the best drama in sport occurs when the underdog overcomes great odds to defeat a heavily favored opponent. Games in which one team far outplays the other (as in the typical Super Bowl game) tend to be less enjoyable, even when teams are evenly matched.

Enjoyment also is related to the extent to which spectators perceive the contest to involve human conflict. Research by Zillman and his associates (1979) showed that when identical sport telecasts were viewed in the presence of broadcaster commentary that described the

Having a comprehensive knowledge of the players, rules, and competitive strategies adds to our enjoyment of watching sports, as do the feelings we harbor toward the participating teams.

game as either involving bitter rivals who felt intense hatred for each other or involving friendly opponents, the former condition elevated spectators' enjoyment. Rough and aggressive play (e.g., a higher number of minutes spent in penalty boxes by hockey players) has been found to increase spectator enjoyment, presumably because it demonstrates that players are sincere in their efforts to win.

These findings demonstrate that in spite of the enormity of the sport spectator business, we know relatively little about the subjective effects of sports watching on the collective psyche. Is watching sports a constructive use of our time? When the game is over, do we feel good about having watched it? Is it possible that sport spectating is a culturally hollow experience, little more than junk food for the human spirit? Is it sometimes a debasing experience? What changes might make sport spectating a more ennobling human experience? These important questions offer fertile ground for research by future kinesiologists.

Reflecting on Your Subjective Experiences as a Spectator

One way to examine your experiences in watching sports is to interrupt your viewing of an event periodically to record your feelings at that moment. You can easily do this at home when watching sports on television, but you can also do it in your seat in the arena or stadium. Set a timer or wristwatch alarm for 10-minute intervals. Each time the alarm sounds, quickly jot down what you are feeling, why you think you are feeling as you are, and what is happening in the sport event that contributes to your feelings. You might also briefly note any effects that the social environment (people you are watching the game with at home, or the stadium crowd) might have on your feelings. After the contest is over, take time to review your notes and reflect on the total experience. How would you describe the overall emotional effect of the experience? What kind of knowledge, if any, did you gain from the experience? Would you describe the time that you spent watching the event as meaningful? Have you gained insight into the effects that sports watching has on you? How would you evaluate these effects—largely desirable or largely undesirable?

Wrap-Up

Let's take stock of what we have learned about physical activity in this chapter. The theme of the chapter is that physical activity involves more than moving our bones and muscles or increasing the flow of blood through our circulatory systems. When we move our bodies, unique sensations arising from internal and external sources give rise to emotions, thoughts, and other inner states. We have termed these sensations, along with our interpretations of their meanings, the subjective experience of physical activity. Subjective experiences of physical activity are important, not only because they play a significant role in determining our activity preferences but also because they constitute the primary reason for engaging in physical activity in the first place. They also are keys to helping us learn about ourselves in physical activity environments.

The goal of most physical activity professionals is to help people develop an intrinsic orientation to sport and exercise, although in most cases participants' extrinsic orientation also plays a part. We can think about an intrinsic orientation as enjoyment. Anything that we value for its intrinsic qualities usually constitutes an enjoyable experience for us. Many factors can make sport and exercise more or less enjoyable, including those that lie within the structure of the activities themselves, those related to the dispositions and attitudes of the performers, and those associated with the social context in which physical activity is performed.

Finally, we examined briefly the subjective experiences associated with watching sports and learned that people have different ways of watching. Our subjective experience is affected by our knowledge of the game, our feelings about the competitors, and the drama of the competition, including its description by sport broadcasters. Because a significant part of our profession (sport management, athletic coaching, and administration) is geared toward athletic displays for spectators, we need to monitor the long-term effects on spectators carefully.

This brings to a conclusion our study of what it means to experience physical activity through performance and spectatorship. You are now ready to proceed with an in-depth study of the research knowledge about physical activity. As you become engrossed in the basic concepts of the scholarly subdisciplines of our field, do not forget that performing physical activity is an important source of our knowledge about it.

 ACTIVITY 4.4

Use the key points review in your web study guide as a study aid for this chapter.

Review Questions

1. Give an example of how an individual may internalize a daily run through the park.

2. List and describe three types of knowledge available to us from subjective experiences in physical activity.

3. What evidence exists to refute the notion that people don't like the sensations that accompany the hard physical effort required to exercise vigorously?

4. What types of physical activities might be chosen by a person who values physical activity as an aesthetic experience? As an ascetic experience? As a social experience?

5. What are the various ways in which one can watch sports? In your opinion, which ones add most to your knowledge of physical activity?

Scholarly Study of Physical Activity

Although learning through physical activity experiences is essential for a kinesiologist, a defining characteristic of a professional in this field is mastery of a complex body of knowledge about physical activity. Laypersons without any training in kinesiology who have been exposed to a variety of physical activity experiences may have learned much about physical activity and themselves. But it is unusual to encounter people who, in addition to having first-hand experience performing physical activity, possess in-depth knowledge of its theoretical aspects, including its sociocultural, behavioral, and biophysical dimensions, apart from those who have successfully negotiated a formal course of study in kinesiology. Some people who haven't formally studied kinesiology may think they know much about physical activity (and some, of course, may), but often their knowledge is incomplete, out-of-date, or simply untrue. As a kinesiology major, you—more than the typical person on the street, and more than your friends who are studying biology, chemistry, history, or some other discipline—need to possess a solid knowledge base about physical activity.

In this part of the book you will be introduced to seven areas of study, or subdisciplines, each of which deals with a different dimension of physical activity. Over the past 5 decades, these subdisciplines have become recognized as the traditional categories of kinesiology knowledge. Chapter 5 focuses on philosophy of physical activity, asking and answering questions about values, meaning, and ethics in sport and exercise. Chapter 6 is devoted to the history of sport and exercise. It also will provide you with a glimpse of how the field of kinesiology evolved. In chapter 7 you will examine relationships between sport and exercise and some of the pressing social issues of our day, such as racial and gender discrimination. Chapter 8 directs your attention to the performance of skilled movement and factors that affect it, including those related to human development. In chapter 9 you will study the psychological aspects of physical activity and learn how this important part of our lives affects and is affected by sport and exercise. Chapter 10 introduces you to the fascinating world of biomechanics, an approach to studying physical activity that will give you insights into the way forces affect and are affected by physical activity. Finally, chapter 11 reviews basic physiological aspects of physical activity, including the effects of exercise on muscular and cardiovascular systems and the physiological basis for training.

The knowledge reviewed here, and the hundreds of concepts, theories, and principles embedded in it, will teach you much about physical activity. You may already have recognized that the organization of the discipline of kinesiology mimics the typical organization of liberal studies. Studying philosophy of physical activity, history of physical activity, and sociology of physical activity builds on and extends a way of thinking that you will develop by taking courses in the philosophy, history, and sociology departments. Similarly, motor behavior and psychology of sport and exercise build on and extend basic knowledge of psychology; biomechanics builds on both biology and mathematics; and physiology of physical activity builds on general physiology.

Because each subdiscipline of kinesiology is, to some extent, an extension of another, older discipline (e.g., psychology, sociology, biology, history, and so on), your past experiences with these disciplines will help you understand the theories and terms used in the subdisciplines.

 AN INTEGRATIVE DISCIPLINE

Roundtable Discussion
Before you begin part II, go to your web study guide and listen to this insightful conversation among experts in the field.

As you embark on your study, you may wonder at times whether what you are learning about is really an organized area of study. At times, kinesiology seems to be merely a collection of bits and pieces of theory and concepts borrowed from the older, established parent disciplines. This notion can be troubling to those who prefer to think that kinesiology is an exclusive science that contains its own theories and scholarly methods. How the study of kinesiology should be organized, and the precise relationship between kinesiology and the parent disciplines, are subjects of ongoing debate among kinesiologists. The most popular view is that kinesiology consists of unique knowledge about physical activity drawn from other disciplines, knowledge that those disciplines have largely ignored. If biologists, for example, had concentrated on studying and teaching about the effects of exercise, perhaps the subdiscipline of exercise physiology would not have evolved. If philosophers, historians, and sociologists had focused their energy on sport and exercise, the subdisciplines of philosophy of physical activity, history of physical activity, and sociology of physical activity may not have evolved. Because knowledge about physical activity was judged important, and because other disciplines ignored it, the 1960s witnessed the rise of the subdisciplines of kinesiology. Thus, from its beginning, kinesiology has consisted of knowledge about physical activity gleaned from all the traditional disciplines and combined into a coherent discipline in its own right—the discipline of kinesiology. In addition, of course, the discipline of kinesiology includes unique knowledge gained from experiencing physical activity (see part I) and professional practice centered on physical activity (see part III).

Although the chapters in this section are jam-packed with information, they are a "once over lightly" survey of the scholarly subdisciplines. When you finish reading and studying them, you will not have mastered the contents. You will, however, have had a glimpse of what lies ahead of you in your program of studies, when you will have a chance to delve into each of these subdisciplines in greater depth.

 PART II PORTFOLIO

The Olympics: An Interdisciplinary Approach
When you've completed part II, go to the Part II Portfolio activity in your web study guide. There you'll consider the variety of ways in which professionals in various fields might investigate issues related to Olympic excellence.

chapter five

Philosophy of Physical Activity

Scott Kretchmar and Cesar R. Torres

chapter objectives

In this chapter we will

- examine the nature of philosophical thinking;

- describe how philosophy fits into the field of kinesiology;

- review the history of the philosophy of physical activity;

- describe the research methods in the philosophy of physical activity; and

- provide an overview of knowledge in this domain by focusing on three issues: the nature of the person, the nature and value of play, and the ethics of sport.

You play baseball for Smalltown High School. Your team just won the sectional tournament for the first time in the school's history and is now playing for the state championship. It is a three-game series with Largetown High School, and it takes place in a neutral ballpark. The series is tied 1-1. These are the two best teams in the state and the competition has been intense but fair. It has all come down to the last inning of the last game.

It is the bottom of the ninth and the game is tied. Your team is at bat with two outs and a runner on third base. A single, passed ball, wild pitch, or an error by a fielder would bring the runner (and the state championship) home. You notice, however, that your coach has called the next batter aside. He quietly gives the hitter the following instructions: "When the pitcher is in the middle of his windup, I want you to fall to the ground screaming loudly and holding your head—as if you had been shot." This, your coach says, should cause the pitcher to balk, which will allow the runner on third to score. There is nothing in the rules that makes this illegal.

The batter does as instructed and when the umpire calls the balk, he stops screaming and jumps up, all smiles. Your teammates rush out of the dugout to join the batter at home plate. However, you are unsure whether it is fitting to celebrate this victory. Something does not seem right. Is this the best way to win a baseball game, a championship? Should you question your coach? What if the coach argues that his action showed clever strategy and that the pitcher should have been prepared for it? Do you have a duty to voice your concern? If you do, should you say something now or later in the locker room? You wonder if anybody has answers to these kinds of questions. Maybe it all boils down to personal opinion—your coach's opinions against yours. What are you to do?

Maybe sports aren't your thing, so here's another example: You are a junior high school physical education teacher. (If you are headed into a health or medical field, you can substitute that setting.) You have several students (or clients) who are obese and clearly out of shape. You want to make a difference in their lives and get them involved in regular, healthful physical activity. Your goal is an ambitious one. It is nothing less than turning their lives around by promoting activity for a lifetime. You wonder how you will do it.

You decide to use a cognitive approach because they are old enough to understand both the importance of health and how activity is related to caloric expenditure and other health-related biological mechanisms. In short, you will motivate them to be healthy, and you will explain how activity combined with a good diet can get them there. You will have them set realistic goals and will introduce them to exercises that are designed to meet their individual needs.

You begin your semester with energy and enthusiasm. The students seem fairly responsive. Some even report a loss of weight, an increase in energy, and general satisfaction with your class. After several weeks of exercise, your own testing procedures show your students that their fitness measures are on the rise.

All too soon, however, the unit comes to an end. Your students are now on their own. You wonder whether they will continue to be active without the support provided by your program. You wonder about the way you chose to motivate them. You presented movement as prudent, as something of a duty. What if you had been more playful? What if you had tried to present movement more as a delight than a duty? What is more important—the health of your students or the various joys and delights of living? You wonder!

Situations like the first example presented in this chapter have been known to take place at different levels of baseball. On those occasions, pitchers often claim that base runners, batters, or coaches caused them to balk by taking unfair advantage of a natural reaction. When yelling or other startling events happen, pitchers are understandably unnerved. Most baseball rulebooks now identify such actions as "unnatural distractions," and they are expressly prohibited.

ACTIVITY 5.1

Critical Thinking Skills
Use your web study guide to develop arguments on both sides of this issue.

It is now illegal to distract opponents in the way described in the opening scenario. However, at the time of the incident, no rules were broken. In our version of this story, the coach underlined this fact as a partial justification for his plan. Some people, in fact, argue that *all* legal methods for winning games should be at any team's disposal. If this view is defensible, such tactics are simply "strategies," and their use is nothing more than an example of one coach outsmarting another.

On the other hand, some individuals regard these strategies as clear examples of poor sportsmanship. They are desperate measures designed to get another "W," tainted victories gained by means of cheap tricks. For those who see matters this way, such victories are not a source of pride but more occasions for embarrassment. Which side has it right? Can philosophy provide an objective argument to show whether these kinds of tactics are ethically appropriate?

The dilemma the teacher faced in the second example in the chapter opening is an issue for all of us in the physical activity professions. We love to move. We love to compete and dance and swim and bike. This is the play side of us. We see physical activity as a "jewel" that adorns our lives and improves them. From this perspective, movement is intrinsically valuable and should be presented so.

We also know the importance of health and fitness. The obesity epidemic concerns us. The declining health and vitality of many of our citizens who lead sedentary lives bothers us. This is the prudent, duty-bound side of us. We see physical activity as a tool that allows us to live longer and get more out of life. From this perspective movement is extrinsically valuable and should be presented so.

Once again we can ask, which side has it right? Should we be playmakers more than purveyors of duty? Or should we be no-nonsense health promoters more than frivolous fun-time activity leaders? Can philosophical methods help us reach reasonable and forceful conclusions on these questions? Even if it should be a mixture of the two, what is the right balance? Answers to these questions lay a foundation for central philosophical commitments that all movement professionals must make. They were identified in chapter 4 as among the most crucial issues we face—and indeed, they are. This is so because they color our curriculum, our interventions, our attitudes toward physical activity, the way we treat our students and clients—that is, virtually everything we do.

Why Use Philosophical Thinking?

The short answer to this question is that philosophical thinking is distinctive and valuable. Philosophy, in fact, has been called the queen of the sciences. This honorific description suggests that philosophy is both different and special. In this introductory section, we discuss the nature and value of philosophy and the role of philosophy as part of a cross-disciplinary team.

The Nature and Value of Philosophy

Philosophical thinking involves reflecting more than it requires testing, measuring, or examining things with any of our senses. Physiologists test physiological responses, for example, by putting people on treadmills to find out how they react to different conditions. Biomechanists measure the forces generated by different movement techniques or various kinds of equipment. Psychologists examine the subjective reports of individuals who have been interviewed, and historians analyze various documents for information about our past.

Philosophers also learn a great deal by living in the world and experiencing it. Thus, in some ways they also test, measure, and examine. But they use this information from daily living for later reflections. In other words, the tools they use are those related to different ways of reflecting, using logic, speculating, imagining, or, simply, thinking! Later in this chapter, you will have a chance to learn about and practice three of these ways of reflecting.

But here it is important to notice two things: first, that philosophical methods typically do not include the gathering of data from controlled experiments, and second, that various types of reflection are used instead.

A third issue is important as well. It has to do with the validity and reliability of philosophical reflection. Contrary to what some believe, and perhaps at odds with some of your own doubts, philosophy can produce results that are as valid and reliable as anything discovered by the physical sciences. Of course not all philosophy fits this description, but that does not mean that everything in philosophy is unreliable and thus little more than guesswork or opinion.

Let's take a look at four philosophical claims. They range from personal opinion (because they have low amounts of supporting evidence) to claims that seem patently true (because they have high amounts of supporting evidence).

- *Personal opinion.* Personal opinion is an indication of individual tastes or preferences. This is the lowest level of truth claim because it is difficult to rally persuasive arguments to support these judgments. Individual differences, diverse life experience, and contradictory but equally valid attitudes lie behind these judgments. Personal opinions, we might say, are true for that person but not necessarily for others. My claim that vanilla ice cream is the best flavor of ice cream is a reasonable one. But your own opinion that coffee-flavored ice cream is best is reasonable too. So we agree to disagree without any hope of solving the matter. You are entitled, as the common expression goes, to your own opinion . . . as long as it is a reasonable one.

- *Speculation.* Although speculations are not merely opinions, they are still very hard to support with evidence. When someone tells us why he or she believes a speculation is true, we may believe that the claim is plausible. But typically we still have reasonable doubts. Philosophical speculations like the claim that God exists or that love is the most important value in life could be true, but we may never know for sure. Also, it is unlikely that slam-dunk evidence will ever come along to support or refute such claims.

- *Probable assertion.* This is the kind of claim around which a considerable amount of evidence can be gathered. As we see later in the chapter, this evidence may speak to personal experience, logic, immediate insight or intuition, or objective observations of the world and human behavior. When we hear evidence for a probable assertion, we are likely to conclude that it is likely to be true. Or we can conclude that the claim is at least partly true. When we argue that all people are created equal, that all human beings should enjoy certain rights, a considerable amount of persuasive evidence can be brought to bear on that claim.

- *Truth assertion.* Some have argued that philosophy can never provide sufficient evidence for truth claims. But the same has been said in relationship to much research in science. For our purposes, however, we will regard truth assertions as those claims that are very difficult to question. When we claim that it is wrong to torture people, with rare, if any, exceptions, we think we are making an ethical claim that is self-evident. How could we be wrong? Or when we make the metaphysical assertion that compassion is not the same thing as hatred, it would be hard to dispute a powerful distinction that has affected all of us. Even if some characteristics are similar (e.g., compassion and hatred are both emotions), other aspects of these two things provide a foundation for conceptually telling them apart.

▶ Philosophers reflect more than they measure, and their reflections range from informed personal opinions to truth claims.

ACTIVITY 5.2

Making Value Claims
In this activity in your web study guide, you'll review four philosophical claims about values in physical activity, and you'll see if you can categorize each claim in terms of level of assertion (opinion, speculation, probable assertion, or truth assertion).

▶ Philosophy of physical activity involves reflection on both tangible and intangible objects in ways that provide insights into the nature and value of these phenomena. The validity and reliability of reflection range from opinions and highly speculative assertions to claims that are likely to be true. Philosophy is a member of a cross-disciplinary team, one that provides unique information that complements and is complemented by data gained from other sources.

Behind all this activity lies what many have called a love of wisdom. In fact, the word *philosophy* means precisely that—the love of wisdom. The pursuit of wisdom requires that we ask different questions and use different tools of research than those in other parts of our colleges and universities. The questions probe the very core of our profession. Why should I be a kinesiologist? What kind of good do kinesiologists bring into the world? Is a profession that focuses on the physical part of the person and movement as important as a profession that focuses on the mind, such as teaching writing or mathematics? Is it ethically acceptable to spend more time on a needy client and neglect one who seems to be in better health?

Answers to these questions cannot be found merely by using skin calipers, microscopes, petri dishes, or even surveys. These questions require that we reflect, examine with the mind's eye, use logic, and discern value. Philosophy, in short, asks questions and uses research tools that transcend those of other disciplines.

ACTIVITY 5.3

Interview with a Practicing Professional
In this interview in your web study guide with Wayne Sebastianelli, professor in Orthopaedics and director of Athletic Medicine at Penn State University, he shares his thoughts on philosophy and ethics in medicine.

Kinesiology: Cross-Disciplinary and Not Interdisciplinary

The methods of philosophy and its products are typically thought to complement those of the natural sciences and social sciences. Accordingly, we are used to thinking of our field as interdisciplinary, not cross-disciplinary. This means that kinesiology's component parts like physiology, biomechanics, and philosophy are relatively independent. Researchers in each area conduct their research on their own and then join hands after the fact in order to solve larger problems. In discussing these complementary, interdisciplinary arrangements, we can use the metaphor of putting jigsaw puzzle pieces together. Under this scenario, philosophy would produce its own puzzle piece and then take it to the joint meeting where all the diverse pieces would be put together.

But what if philosophy and other disciplines not only complement one another but are actually interdependent? What if philosophy cannot stand on its own but rather needs insights of cell biology, anthropology, and other areas of study in order to ask the right questions? And what if this is true of physiology, biomechanics, and sport psychology, too? What if they need the insights of literature, history, and philosophy in order to ask the right questions and design their empirical studies accordingly? If our research areas are actually interdependent, then we need a holistic representation of the spheres of physical activity, one that is a bit different than the traditional representation we use in other parts of the book (see chapter 1, for example). Although we can identify different areas of study listed in the big circle (see figure 5.1), none of them is entirely distinct. Thus, no lines divide them. They fade off and run into one another—philosophy to history to sociology, and so on. The satellites that are attached to the big circle depict the ways in which scholarship and application are related. The sphere nearest the big circle represents the sum benefit of our cross-disciplinary field of study. A reciprocal relationship then exists between the two other parts. One leads to participation in physical activity—in work, exercise, games, play, and sport. The other leads to the workplace and the practical problems faced there. The interaction among all spheres is two-way, suggesting that theory informs practice and, conversely, that practice informs theory.

No one discipline rules the roost in the big circle. We are mutually dependent on one another, and we are the mutual beneficiaries of one another's work. Likewise, for the satellites

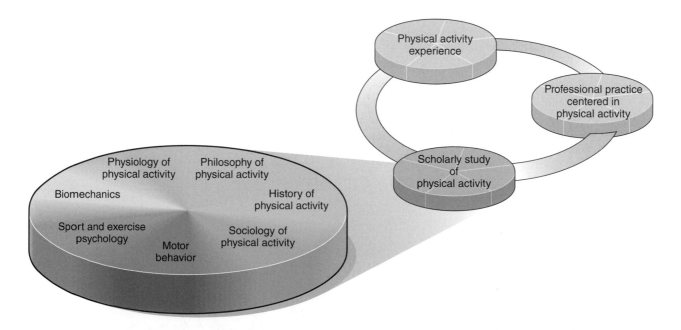

FIGURE 5.1 Holistic kinesiology encourages cross-disciplinary research, professional practice, and performance, as illustrated by the removal of hard lines and the opening of doors between the subdisciplines.

to the right, no circle dominates the others. Theory, performance, and practical application are mutually dependent on one another and are likewise the beneficiaries of each other's perspectives. This produces the holistic, democratic, and egalitarian notion of the discipline of kinesiology as described in chapter 1.

What Do Philosophers Do?

Philosophers pursue a goal that is no different than the goal of other members of the cross-disciplinary kinesiology team. In the broadest terms, it is simply to better understand the world and our lives in it. Specific to our field, the goal is to better understand human movement or, as it is identified in this text, physical activity. Philosophical thinking is needed to address a number of issues that are important for kinesiologists. Here is a list of some of them.

• What is the scope of our field? Is kinesiology responsible for all forms of movement? Only skilled movement? Should we focus on physical fitness? Or is our domain that of overall health? For those interested in cultural activities, which ones should be included under the umbrella of kinesiology? Sport? Games? Play? Exercise? Dance? Judgments about these things require that we do some groundwork in an area called **metaphysics**. This is the branch of philosophy that distinguishes one thing from another. The work here is largely descriptive. Before we can decide whether the scope of our field should include all movement or only skilled movement, we need a clear description of what these two things are. In metaphysics we ask, what is movement? What is skilled movement? What is health?

• How confident can we be as kinesiologists that our attitudes and perceptions about health, sport, obesity, and related matters are objective and valid? Are we confident enough in our understandings to export our conclusions to other cultures? Is obesity really an epidemic, or does it depend on one's perspective? Should obesity be conceptualized as inherently dangerous, or does it depend, for example, on how it is seen in different cultures? Judgments about these questions require that we do some groundwork in an area called **epistemology**. This is the branch of philosophy that examines how we know things and with what assurance we can claim that something is true. Thus, it has to do with bias, perspective, objectivity, socialization or enculturation, and other issues related to the strength of the conclusions we reach in kinesiology. In epistemology we ask, am I aware of all (or even most) of the attitudes and potentially skewed perspectives I bring to my reflections? Can I even become entirely objective, or objective enough to trust my conclusions?

• What really matters in kinesiology? Is it health? Pleasure? Knowledge? Skill? How does kinesiology contribute to the good life? Can a person really live a meaningful life without spending time outdoors or without possessing a variety of recreational skills? Should everyone be able to swim? Is kinesiology largely an auxiliary profession in which we teach skills, fix broken bodies, and promote health, or is it a liberal profession in which we celebrate play and other forms of freedom in movement? Should we promote life or quality of life? Judgments about these questions require that we do some groundwork in an area called **axiology**. This is the branch of philosophy that examines the value of things. It requires that we make judgments about the relative worth of things in the world (like health) and experiences (like pain-free living). In axiology we ask, what kind of life is worth pursuing? What components go into high-quality living? How do we rank values that are means to other good things against values that are ends in themselves?

• How should we behave as kinesiologists? What exactly do we owe our employers, our patients, clients, or students? When we sign a contract, what obligations do we have? What if we wake up some morning and don't feel like going to work? We are not sick but emotionally exhausted. Would it be acceptable to take just one day off? What if we see one of our colleagues doing something inappropriate or even illegal? Do we have a moral obligation to blow the whistle? What if blowing the whistle carries some risk to ourselves?

What do we do then? Judgments about these questions fall in an area called **ethics**. This is the branch of philosophy that examines how we ought to behave. It is different from historical, cultural, or so-called descriptive ethics that describe attitudes and how people do behave. Ethics focuses on right and wrong, good actions and bad actions, what morally responsible people do, and what irresponsible or selfish people do. In ethics, we ask how we should behave.

ACTIVITY 5.4

Practical Plug-In

In this activity in your web study guide, you'll meet a professional who is facing an ethical dilemma. You'll consider what he needs to know, and then you'll predict what his decision will be.

Goals of Philosophy of Physical Activity in Kinesiology

- To understand the nature and value of health and physical activity, particularly in the form of exercise, sport, games, play, and dance
- To understand how confident we can be about our claims in kinesiology
- To understand the most important values of physical activity and its contribution to good living
- To learn how we ought to behave in sport and in our professional lives as kinesiologists

---

History of Philosophy of Physical Activity

By the 1960s, American physical educators studying philosophy of physical activity were pursuing new directions by distancing themselves from educational philosophy and the single profession of physical education and using philosophical techniques to produce new insights about physical activity itself—especially sport. This approach attracted a few scholars in other disciplines as well. The two most important physical educators in this vanguard were Eleanor Metheny (*Connotations of Movement in Sport and Dance*, 1965; *Movement and Meaning*, 1968) and Howard Slusher (*Man, Sport and Existence: A Critical Analysis*, 1967), both of them at the University of Southern California and both active in doctoral education. Philosopher Paul Weiss (*Sport: A Philosophic Inquiry*, 1969), then at Yale, was also at the forefront of this interest in sport per se.

Metheny conceptualized movement as a source of insight and meaning for individuals. Slusher used existential perspectives to examine ways in which sport enhances our authenticity as human beings as well as our freedom and responsibility. Weiss was concerned with excellence in sport.

In the middle of the country, physical educator Earle Zeigler mentored a group of graduate students at the University of Illinois, while another important center of doctoral education developed at Ohio State University under the leadership of Seymour Kleinman. Zeigler wrote about philosophical schools of thought and prominent philosophers while extrapolating implications of their work for the field of physical education. Kleinman focused on philosophical analyses of movement, dance, Eastern movement forms, and the bodily dimensions of being human.

The Philosophic Society for the Study of Sport (PSSS), now the International Association for the Philosophy of Sport (IAPS), was formed in 1972 following discussions at the Olympic Scientific Congress in Munich, Germany, among Weiss, Warren Fraleigh from the State University of New York at Brockport, and German philosopher Hans Lenk. Fraleigh was the key organizer of the association, and he continued to be centrally involved as it developed. The PSSS almost immediately set about sponsoring a scholarly publication for the subdiscipline, and the first issue of the *Journal of the Philosophy of Sport* appeared in 1974 with Bob Osterhoudt as its editor.

Philosophy of sport grew slowly for at least three reasons. Philosophers and philosophy of sport classes were few and far between because many departments of physical education and kinesiology emphasized the physical sciences. Some schools that had multiple physiologists, biomechanists, and motor control specialists, for example, did not have even a single sport philosopher. Second, the cognate area of philosophy is uniquely conservative. Many parent discipline philosophers preferred not to be associated with a group that studied sport and thus stayed away from philosophy of sport meetings and chose not to publish in the *Journal of the Philosophy of Sport*. Third, early philosophers of sport worked hard to separate themselves from education and other applied fields. As a result, they produced very abstract literature that was often difficult to read and was perceived to have very little utility.

Much, however, has changed over the past 10 years, and this change is favoring the philosophy of sport. First, there is a greater appreciation for cross-disciplinary study and the role that philosophy, history, and the social sciences can play in this research. For example, new guidelines from the U.S. National Institutes of Health indicate that single-discipline research is often constrained and ineffectual and that collaboration across disciplines is to be encouraged. In addition, philosophers in parent disciplines are beginning to show up more regularly at philosophy of sport meetings, and the literature is improving significantly in both quantity and quality. Philosophy of sport, for example, now has two refereed journals. Finally, we have seen philosophy of sport spread to universities around the world. Whereas early activity in the 1960s and 1970s was confined largely to North America, groups now exist in Australia, Great Britain, the Scandinavian countries, the Middle European countries, South Korea, and Japan. Likewise, the philosophy of sport is growing, albeit slowly, in other parts of the world. An issue of the *Journal of the Philosophy of Sport* (Volume XXXVII, Issue 2, 2010) was devoted to the philosophy of sport in non-English-speaking countries.

Even with this positive activity afoot, the fate of sport philosophy in kinesiology departments in North America is uncertain. Because of the dominance of the biological sciences, the emphasis on large external grants, and the push for scientific specialization, among other factors, relatively few philosophy of sport doctoral programs remain in the United States. It is unclear where the philosophical leaders and researchers related to kinesiology will be produced in the years to come.

Research Methods for Philosophy of Physical Activity

Now that you know about the goals of philosophy of physical activity, as well as a bit of its history, let's look at how scholars produce knowledge in the subdiscipline. As you saw, philosophers use various techniques of reflection rather than measurement. Like all scholars, they ask questions. But unlike many other researchers, philosophers attempt to answer those questions by reflecting on experience, using logic, appealing to common sense, and deploying their conclusions in the everyday world to see how well they work. Nagel summarized the philosophical process by noting that philosophy "is done just by asking questions, arguing, trying out ideas and thinking of possible arguments against them, and wondering how our concepts really work" (Nagel 1987, p. 4).

You don't need to be able to conduct scientific experiments or work through mathematical proofs. You do, however, need to be able to think carefully and precisely, and you can improve this skill with practice, just as people improve writing skills or athletic skills by working on them. For some of us, reflection may be a relatively new experience. This is so because in our fast-paced, rapidly changing world, finding time for contemplation is sometimes difficult. And in our modern science-dominated world, measurement tends to be trusted more than reflection. (As one academic put it, "In God we trust. All others bring data!") Because of these sentiments, you may not have had sufficient opportunity to practice your skills of reflection.

In truth, the gap between those who measure and those who reflect is not as great as it seems. Most philosophers think that good empirical experimentation is crucial and that the gains made by the physical sciences have been both impressive and useful. Philosophers use scientific findings to put parameters on their own work. For instance, in the philosophy of mind, philosophers are very much aware of advances made in neurophysiology. Accordingly, philosophical claims about the nature of mind need to be consistent with the findings of brain science. It could be the case that intelligence is the product of an incredibly complex computer-like brain. Or perhaps it is better to model thinking on chaos theory and dynamic systems. Perhaps intelligence is a single generic capability. Or maybe we have multiple intelligences that have been developed in specific environments to solve specific problems. It is possible, in other words, that we have multiple IQs—one for spatial problems and another for motor problems, music problems, mathematical issues, language dilemmas, and so on—rather than a homogeneous aptitude that allows us to call ourselves "smart." Philosophers need to know these things and use them to constrain and direct their philosophical reflections on the so-called theory of mind.

> The gap between those who reflect and those who measure is not as great as some have thought. Both philosophers and scientists are affected by the realities of the physical world, on the one hand, and the force of ideas and meanings on the other. Both philosophers and scientists have to regularly modify their conclusions based on new studies, perspectives, data, and reflections.

However, scientists necessarily lean on philosophical methods and such intangibles as meaning. They reflect on various matters throughout the inductive scientific process. First, they employ values to determine what is worth looking at and what is not. They settle on issues that are "important," "significant," "of potential value for humankind," or "crucial for the advancement of science" in general or in their particular branch of scientific inquiry. Then, they reflect in formulating their problem, coming up with potential hypotheses, and developing a research design. They imagine different possibilities and use logic as well as empirical data in deciding to test one way or another. Ultimately, they reflect on their findings as the data begin to come in. They ask, "What do these data mean? Could they mean two or more different things? What are their implications for future study?" The meanings that come to mind in answering these questions are recorded as part of the findings and then are discussed even more thoroughly in terms of what the findings mean and their implications for future study.

If scientists are good at these kinds of value judgments and reflections on meanings, their work goes better. If they are not, their research never gets off the ground, and their findings are never published in reputable journals. In short, scientists who are unable to reflect skillfully and are not good at uncovering the right meanings associated with their data are not likely to be very good scientists.

Even though scientific and philosophical methods overlap in some ways, they are not the same. Reflective reasoning processes are central to philosophical research, and three of them are highlighted in this section of the chapter (see figure 5.2) (Kretchmar 1994, 1996):

- Inductive reasoning
- Deductive reasoning
- Descriptive reasoning

Inductive Reasoning

Inductive reasoning moves from the examination of a limited number of specific examples to broad, general conclusions. For example, you could use inductive reasoning to examine a metaphysical question on the nature of exercise. You would begin by developing a list of examples of exercise activities (e.g., participating in rhythmic aerobics classes, training for

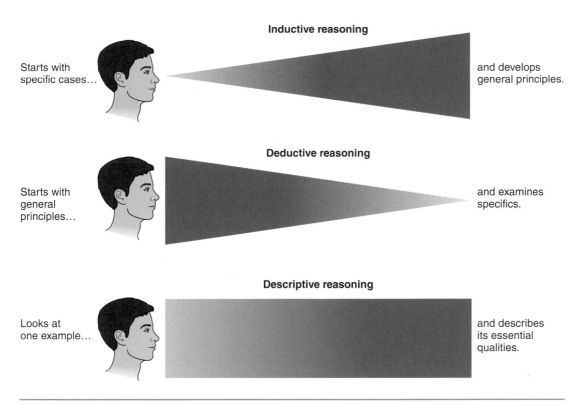

Inductive reasoning

Starts with specific cases... and develops general principles.

Deductive reasoning

Starts with general principles... and examines specifics.

Descriptive reasoning

Looks at one example... and describes its essential qualities.

FIGURE 5.2 Three types of philosophical reasoning.

strength to achieve better performance in football, lifting weights to develop muscles for a bodybuilding contest, mowing the lawn, walking to work because you don't own a car, taking a brisk walk at lunch each day to improve your health), deciding whether all of them are genuine examples of exercise, eliminating any that are questionable, and then identifying their general or common characteristics—in particular those features that distinguish exercise from activities like dance, sport, and play. You will need to be sure your examples are noncontroversial or "safe." Otherwise, your inductive conclusions are likely to be wrong. Also, you will need to be sure that your generalizations are neither too narrow (e.g., exercise requires the use of equipment) nor too broad (e.g., exercise requires a person who moves). When conclusions are too narrow, they are typically inaccurate. (Many types of exercise require equipment, but some do not.) When they are too broad, conclusions are unhelpful. (Sport, dance, and play require performers too, so this characteristic does not tell us much about what is unique to movement.)

Deductive Reasoning

Deductive reasoning starts with one or more broad premises to identify conclusions that follow from them. You might use deductive reasoning to determine whether a particular activity you happen to be observing is a sport. Let's say that you are attending your first chess tournament and are curious about whether chess should be identified as a sport. You develop a major and minor premise as follows: (Major premise) If it is true that sport requires the use of physical skills to solve its game problems, (Minor premise) and if it is true that chess does not require physical skills in order to achieve checkmate, (Conclusion) then it necessarily follows that chess is not a sport. Of course, the validity of this reasoning depends on the accuracy of the two premises. Someone could argue with the first premise and suggest that because chess is competitive and is a game, it deserves to be identified as "sport" even though physical skills are not required.

Descriptive Reasoning

Descriptive reasoning involves looking at one example of an event and describing its essential qualities. For example, you might imagine slightly altered forms of basketball and then, through a process of descriptive reasoning, determine whether these are still the same game. What do you think about the following variations on the sport of basketball? Would each one still be basketball?

- Changing the goal of the game from winning to achieving a tied score at the end of the contest

- Eliminating all restrictions on the means of scoring a basket, thus making it legal to use a stepladder under the basket or stand on a teammate's shoulders

- Playing on an outdoor court rather than in an indoor facility

- Playing a computer version of the game, thus eliminating the need to move around on the court

Negative answers to any of these variations should help you zero in on what is central to this activity. If the variation does not eliminate basketball, then the variation in question is probably not crucial to the identity or integrity of the game. An important research technique in which descriptive reasoning is often used is **phenomenology**. Phenomenologists, as Edmund Husserl noted, are guided by the requirement that they "return to the things themselves." This means that they should get back in touch with their experiences and examine those experiences to determine what exactly they mean. Variations help phenomenologists distinguish meanings that are central from those that are peripheral, just as we saw with the basketball example. Playing outdoors versus indoors does little to change the phenomenon of basketball. But if people use ladders to score baskets, something more central or essential to the game is lost.

> Inductive reasoning begins with specific cases to develop broad, general conclusions. Deductive reasoning begins with broad factual or hypothetical premises in order to determine more specific conclusions that follow from them. Descriptive reasoning begins with one example of some phenomenon and then varies it to see how dramatically it changes. Change (or its absence) allows a more accurate description of the central characteristics of the item being examined.

Applying Inductive, Deductive, and Descriptive Reasoning to Philosophical Questions

Many philosophers use inductive, deductive, and descriptive approaches jointly in the process of researching a single problem. To demonstrate this, we use the second issue introduced at the start of the chapter, the one that raises questions about the sufficiency of extrinsic, rational, duty-oriented methods for promoting persistence. It also raises the possibility that play should be used as a complementary or alternate method for keeping kids moving. See whether you can use each method to make some headway. We will try to get you started on each one:

- For inductive reasoning, develop a list of examples of individuals at play—that is, people who are fully engaged in some activity in the spirit of play. Once you have your set of examples in place, ask yourself the following question: What is common to all of these—that is, what general features make all these examples of play and not of work, labor, or duty?

- For deductive reasoning, start with premises: If it is true that play promotes future participation because it has the qualities of (fill in this blank with the qualities you identified using inductive reasoning), and if it is true that calisthenic exercises typically lack these qualities, then it follows that . . . (it is up to you to draw the most accurate conclusion based on the content of premises 1 and 2).

- For descriptive reasoning, let us suppose that you picture 3rd graders in the following ways:

  ○ They are involved in make-believe play in a sand pile behind the school. They are making "mountains," "rivers," and "towns." It is difficult but interesting. They are focused on the process.

> ○ Now imagine the same children involved with the sand doing anything they want. There seems to be no particular goal, no objective. They are just "playing" or, as we say, "fooling around."
>
> Now ask yourself the question: Was anything important lost when the project was varied from making something to fooling around aimlessly? What was lost? Was this loss important to the spirit of play?

Overview of Knowledge in Philosophy of Physical Activity

It is now time to look at several important topics in the philosophy of physical activity. Although scholars have produced a complex, broad literature related to human movement, here we examine only a few examples of the insights produced by this fascinating subdiscipline. Specifically, we examine

- the nature of the person—particularly the mind–body relationship,
- the nature of sport and competition and its relationships to work and play,
- the values promoted by physical activity, and
- ethical values and sport.

The "Person Problem" (How to Understand Mind and Body)

Some sport philosophers believe that you should begin work on understanding human movement with the adjective (*human*) before you get to the noun (*movement*), a pattern we follow here. We look at who moves before we get to the nature and value of movement in forms like play, games, sport, and competition.

Good reasons exist for dealing with philosophical knowledge in this sequence. It can be argued that we must understand the person before we can understand how and why he or she is attracted to games and play. For practitioners, the same logic holds. If you are going to use play or exercise as an intervention to promote human welfare, you must first know what it is to be human. The subsequent intervention, in short, must be appropriate for the individual. So it is with the individual that we should begin our analysis.

Philosophers have taken many positions on what human beings are. Human beings have been called social animals, symbolic creatures, sexually driven individuals, the product of economics, composites of two elements (mind and body), composites of three things (mind, body, and spirit), or simply complex machines and thus little more than a complex collection of atoms. Although each of these positions may merit some attention, we focus on only three of them in this chapter: materialism, dualism, and holism. If you understand these three, you will have a reasonably good grasp of fundamental positions that can be taken on the nature of the human being. And you should have a good grasp of the implications of these positions for kinesiology and the movement professions.

Materialism

Perhaps the simplest and most straightforward position on human nature is captured by **materialism**. Philosophers call it a monistic position because the person is said to be made of only one thing—namely, atoms. In fact, all that truly exists in the world is atoms and void. The human being, including the body, the blood, the wiring, the brain—all of the human being—is nothing more than a complex machine. The brain can be likened to an incredibly sophisticated computer. The body is a combination of very complex lever systems, motors, wiring, and plumbing.

Consciousness for materialists is often thought of as a kind of a sideshow. Although these philosophers may acknowledge that subjective experiences are real, materialists do not believe that they have any power or efficacy. Ideas and values are a product of brain states, and brain states, now and in the future, make things happen.

This is a no-nonsense kind of philosophical position. What you see is what you get! The person appears to be an impressive material object . . . and is! Nothing more. Nothing less. You and I do not need to rely on spooky substances like mind or spirit to account for human behavior. Neither should we discount the considerable capabilities of this powerful machine. Once we gain the ability to measure and understand all parts of the organism, all human capabilities will be accounted for in terms of atoms and void.

Materialistic accounts of persons were popular as early as the time of Plato and before, but they became ever more prominent during the period of the scientific revolution in the 15th century. The careful skepticism inherent in scientific inquiry, the emphasis on measurement, the onset of the Copernican revolution and the realization that the Earth is not the center of it all, the rise of Newtonian physics and the increased appreciation of causal relationships, the invention of the microscope—all of these things and others encouraged us to think of persons as simply one kind of physical object among others.

Many physical educators and exercise scientists adopted the assumptions and perspectives associated with materialism and began to study the person as a mobile machine. Physiologists, biomechanists, motor control and motor learning experts, sport psychologists, and even some sociologists and philosophers of sport began looking for the holy grail of materialism, often referred to as "underlying mechanisms"—those multiple physical causes that, when pasted together, would explain the whole.

Dualism

How we deal with people—how we cure them, educate them, or teach them sport skills, for example—is affected by what we think a person is. If we labor under false impressions of human nature, the risk of inappropriate, ineffectual, or even dangerous interventions increases. Or, to put it another way, we are unlikely to understand health, movement, the body, games, and play unless we understand what a person is. That question lies at the very heart of our field.

Some philosophers, although appreciating the simplicity and honesty of materialism, doubt that consciousness can be cast aside so quickly and easily. In other words, they doubt that human behavior can be accounted for by looking only for underlying mechanisms or only for material causes. Although acknowledging that human beings have a body that works very much like a machine, these philosophers argue that people are also endowed with a mind, with consciousness, with a steady stream of very concrete emotions and ideas. Moreover, these subjective states have power. Ideas of love and hope and justice, for example, can (and do) change our behavior.

When these philosophers are asked how many parts go into the production of a human being, they answer "two"—mind and body. Thus, instead of being monists like the materialists, they are identified as dualists. René Descartes, perhaps the most articulate and forceful of the modern dualists, said that a person was composed of two substances, one that was extended in space (body) and one whose nature it was to think (mind).

Dualism is an attractive philosophical position. It is attractive because our intuitions and daily experiences suggest that our thoughts must count for something. We have an idea, we draw up a plan reflectively, and then we act on it. We have an exciting idea, and everything from our heart rate to our outlook on life is affected. Dualism, in short, can be attractive because it takes ideas seriously.

Many physical educators and exercise scientists are dualists because they believe that consciousness is real and that nonphysical ideas somehow invade the world of motor behavior—of swimming across pools and catching baseballs and performing pirouettes. Many have expressed their respect for both of our human elements and their close interrelationships, as did Juvenal when he argued for "a sound mind in a sound body" or Jesse Feiring Williams when he proposed an idealistically flavored "education through the physical."

Holism

A third position—holism—has received a considerable amount of attention more recently. In some ways it can be considered a middle position between monistic materialism and

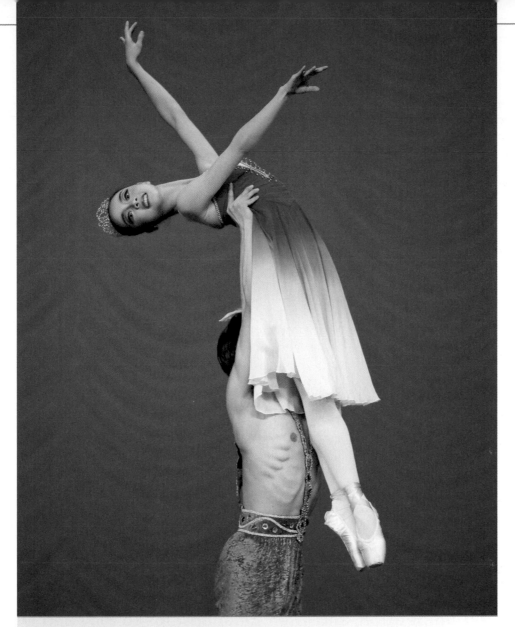

▍From a dualistic point of view, our ability to complete physical tasks is the result of our ability to put thoughts into action.

mind–body dualism. Holism takes both the mechanistic and thoughtful sides of our human nature seriously but refuses to separate them. In other words, it argues that our physical nature, all the way from our anatomy to our genes, is always shaped and influenced by emotions and ideas. Likewise, our subjective experiences are always shaped and influenced by our physicality. In short, we never get one without the other. Thus, neither the physical nor the thoughtful side of our nature has any independence. We are thoroughly ambiguous—whether one is looking at an injured knee or at a child playing soccer. Both phenomena are a composite of "bodymind"!

ACTIVITY 5.5

Practical Plug-In

In this activity in your web study guide, you'll consider what a high school physical education teacher needs to know to solve the problem of lack of motivation among her students.

Philosophers like William James, John Dewey, and Maurice Merleau-Ponty are frequently identified as champions of holism. They all claim that it is unhelpful to think of a person in terms of parts or dichotomies like mind and body. They all believe that intelligence works at many levels, from mere nonverbal reflex behavior to creative action to reflection and verbalizations. Thus, behavior is more or less mechanistic or more or less creative. But no behavior is simply mechanical, and no actions are the result of an independent mind giving instructions to a dependent body. Once again, all behavior is ambiguous.

 ACTIVITY 5.6

Interview with a Practicing Professional
In this interview in your web study guide, Jack Hogan, a former teacher, coach, and athletic director, discusses why it's critical for educators to develop a guiding philosophy in their professions.

Kinesiologists and physical educators have seen the relevance of such a position for their own work. Among other things it has expanded the significance of physical activity as something that counts as creative, intelligent behavior in its own right, without the addition of theory or other propositions. Holism has increased the significance of the body in understanding human nature and behavior. Holists, unlike materialists and dualists, do not say that we *have* a body. Rather, they argue that we *are* our bodies. Or more accurately, as some Zen Buddhists like to say, we are simply bodymind! Finally, holism has expanded the kinds of interventions that are designed to improve health or repair injuries. Rather than working on the plumbing (our cardiovascular system), or fixing the hinge (e.g., one of our knees), or manipulating the computer (our brain), we can intervene at multiple levels—from our embodied ideas and aspirations to our idea-shaped cells. Holistic medicine, for example, is now taken seriously in many quarters. A bad back, for example, can be addressed by any number of interrelated therapies, such as meditation, yoga, chiropractic manipulations, and surgery.

Sport, Competition, and Their Relationships to Play and Duty

Sport is a game in which motor skills are tested. Its rules specify a goal to be achieved and limit the means that participants can use to reach the goal. These rules exist to create and make possible the game.

Sport is virtually everywhere today. It is not surprising, then, that it occupies a prominent place in physical education programs across the United States as well as in the lives of millions of citizens. As a kind of game, sport has a unique structure, one that can be experienced in multiple ways. Here you will look at its structure, its connection to competition, and both its dutylike and its playlike potential. You will see that at its best, sport allows us to test our capabilities while thoroughly enjoying the doing.

Games and Sport

Games are like no other activities in the world. They are created by a set of rules that specify a goal to be achieved and limit the means that participants can use to reach the goal. These rules exist to create and make possible the game; they would be absurd in ordinary life (Suits 1978). In other words, games are artificial problems (Kretchmar 2004; Morgan 1994). The built-in inefficiencies structured by the rules of games are different from the inefficiencies encountered in many other realms of life—for example, education, politics, and business—in which limits on efficiency come mainly from ethical constraints. For instance, we cannot copy someone else's writing and call it our own. That is an act of plagiarism, and it is a serious breach of ethics. Game rules, however, handicap us in gratuitous or unnecessary ways that have nothing to do with ethics. Think about the ordinary activity of shopping for groceries. This activity could become a game, for example, if you restrict yourself to buying products whose brand names start with a vowel. You could make your game even more complex by giving yourself only 30 minutes to complete your shopping. This method is not the most efficient way to shop for groceries, but it does turn shopping for groceries into a game. Nearly any activity can be converted into a game . . . even reading this textbook!

In light of this definition of games, we can describe sport as a game in which motor skills are required to reach its goal. In National Football League football, for example, the objective is to get the ball over the goal line. In competitive situations, the objective is not just to do that but to do it more times than the other team does. Furthermore, the rules prohibit certain actions that might otherwise increase participants' ability to accomplish the goal, such as the following:

- Offensive linemen are limited in the ways in which they can block opponents.
- Only certain players are eligible to receive forward passes.
- A player who is advancing the ball must stay within the boundaries of the field.
- Each team may have only 11 players on the field at one time.
- Teams may not use mounted horsemen to outrun their opponents.
- Players may not use brass knuckles or clubs to beat their opponents into submission.

The built-in inefficiencies structured by the rules of football not only make it more difficult to attain its goal but also provide the sport with its distinctiveness. A sport is a game primarily meant to test our motor skills.

Significance of Rules   Rules can be thought of as formal types of game cues. They tell us the structure of the test, that is, what should be accomplished and how we should accomplish it. In this sense, rules create a problem that is artificial yet intelligible. Only within the rules of a game such as basketball or baseball do the activities of jump shooting and fielding ground balls make sense and take on value. It is precisely the artificiality created by the rules, the distinctive problem to be solved, that gives sport its special meaning. That is why getting a basketball through a hoop while not using a ladder or pitching a baseball across home plate while standing a certain distance away becomes an important human project. It appears that respecting the rules not only preserves sport but also makes room for the creation of excellence and the emergence of meaning. Engaging in acts that would be considered inconsequential in ordinary life also liberates us a bit, making it possible to explore our capabilities in a protected environment. To keep that liberation and exploration appealing, we often make rule changes to recalibrate a test that has become either too easy or too difficult. We like our sports to more or less match our ever-changing potential.

Significance of Skills   Although sport can be understood in abstract terms, the actual attainment of the goal requires the deployment of a variety of motor skills. For instance, successful participation in soccer requires proficiency in kicking, passing, running, marking, and tackling, among other motor skills. Indeed, soccer expertise involves a blend of these motor skills. Rather than being generic in nature, soccer skills require specialization. The ways soccer players kick the ball are markedly different from the ways rugby or football players do, and all of these are different from how we kick pebbles down the street or sidewalk. This is so because sport skills develop out of the relationship between the goal, on one hand, and the means allowed and prohibited to pursue it, on the other.

These relationships are established in the rulebooks of different sports (Torres 2000). The kicking skills developed in soccer are the result of players' inventiveness in negotiating the test established by the rules, which prohibits the use of their hands, allows for passes in any direction, and does not impose a time limit on ball possession, among many other stipulations. It should be clear that the rulebook of each sport indicates the set of motor skills that this unique game is designed to test. Thus, a sport's set of motor skills provides that sport with idiosyncratic characteristics that differentiate it from other sports. Moreover, motor skills are very important because they represent the standards of excellence by which players evaluate their performance.

I What rules do you see being followed in this tennis match?

Significance of Competition Any sport, or any game for that matter, can be organized both competitively and noncompetitively. This is easier to see in the so-called parallel tests such as bowling, archery, or track events. You can bowl by yourself, or you can test how fast you can run the 100-meter dash in the absence of competitors. This concept is more difficult to see in sports in which the artificial test is provided, in part, by others. In soccer, wheelchair basketball, and tennis, for example, you facilitate the test of another test taker who at the same time facilitates the test for you. This, however, does not mean that both parties are automatically engaged in a contest. Think about scrimmages or games between parents and their children. Although fully engaged in their respective tests, these people are not competing. What is missing in these examples of parallel and simultaneous testing is the central element of competition—that is, the attempt to determine athletic superiority.

Thus, competition requires that two parties share the same test and that they commit themselves to surpassing each other's performances (Kretchmar 1975). When individuals enter the world of competitive sport, they enter a community that defines itself in terms of the unique test they share. Thus, in competition, players learn not only whether they are able to do well on the test provided by the sport but also how they fare in comparison to their opponents. Testing evaluations and competitive comparisons here go hand in hand.

At one level, competition determines winners and losers. A more encompassing view of competition highlights the value of performance and cherishes process as much as the outcomes that result from the play (Torres & McLaughlin 2003). Excellent play, in other words, transcends mere wins and losses and can even become the central value of competition (Simon 2010).

Play and Duty in Sport

We have seen that sport is a goal-oriented activity in which we accept rules—imposed limitations on how we can go about attaining that goal. This is done primarily to test our proficiency in a given set of motor skills. We have also seen that this activity can be organized just to evaluate our proficiency in solving the test or, in addition, to compare our skills to those of an opponent. That is, sport is not necessarily competitive.

A second distinction is important as well. Sport may be encountered as a chore or as play. In what follows we explore the differences between dutylike and playlike modes of experiencing sport and physical activity.

Dutylike Sport   One of the most common arguments used for including sport and physical activity in educational settings and used to justify personal involvement in these activities focuses on their power to bring about beneficial effects. That is, sport and physical education are presented as valuable because they help us advance in concrete and desirable ways, from improving our health and teaching civil values to fostering national pride and combating sedentary living and obesity. This account is clearly based on utility, a rationale for activity that many professionals in the field have emphasized for years. When sport and physical activity are articulated in this way, it is likely that we will approach them as tools, as activities that we must do. In short, they become a form of duty.

Regarding sport and physical activities as chores that produce external rewards tends to diminish their intrinsic worth. The more we remind ourselves that our participation in sport and physical activity produces good results, the less we are prone to cavort in their intrinsic meaning and value. Have you ever wondered why time and again we use the word *workouts* when referring to our sport and physical activity experiences? This says something about what we expect them to be like—namely, work! Wouldn't it better to describe instances of physical activity and sport as "playouts"?

By stressing that physical activity is "good for you," dutylike sport and physical activity might not offer the most effective strategies for motivating people to get up and move. This is true because we tend to avoid whatever we *live* as a monotonous duty. Even the most committed and cautious among us would welcome the addition of some delight and sensuosity to his or her workout. Dutylike sport and physical activity serve the professional need to defend against the charge that they are frivolous and unimportant. In the process, however, they become not only a "workload" that we often cannot wait to escape but also activities that are only indirectly valuable. Their value lies in what they produce, not in what they allow us to be and to experience.

Playlike Sport   If you asked a sample of average people how play ranks in comparison to education, religion, family, business, or government, most would probably place it close to the bottom. This is so because play is normally presented as trivial, infertile, and senseless—an activity reserved mostly for children and retired people. In the face of the delightful and mesmeric power of play, it is paradoxical that many still want to confine it to childhood

> Dutylike sport and physical activity emphasize instrumental effects and their capacity to bring about beneficial effects such as health, moral values, and nationalistic sentiments. When the focus is on external rewards, sport and physical activity are more likely to be encountered as a form of duty. On the other hand, playlike sport and physical activity are self-contained. As a way to live these activities, play accentuates their intrinsic values and ends. At play, sport and physical activity are experienced as delightful, absorbing, and meaningful, and they allow us to show our human potential.

❚ When we're forced to complete a task, we often can't find the joy in that movement.

153

and retirement and consider it at best an ornamental aspect of life. Are we adults frightened of play? Do we really understand it?

Although found during engagement in an activity, play is not the activity itself. Rather, play is the manner in which we approach, embrace, and experience the activity (Meier 1980). Play requires an **autotelic attitude** in contrast to all forms of instrumental or utilitarian orientations toward the world. This means that while at play, we are interested not in the payoffs that might potentially come from our participation in the activity but rather in its intrinsic values and ends. To put this in a different way, at play the doing is what really counts. For example, a playlike encounter with soccer is experienced as utterly absorbing. Even if extrinsic rewards and goals were of concern when the activity started, they recede into the background as we take great pleasure in just embodying our soccer skills. In those moments, soccer is lived as "good for nothing" and as an activity that we like for what it is and allows us to feel. At play we simply let go.

Playlike sport and physical activity, and for that matter any activity lived in a playful spirit, become something we *want* and *like* to do . . . sometimes desperately! We often forget about time, the necessities of life, and duties as we surrender to the magnetism of the playground. Play, then, is a way of embracing the world that includes a reciprocal relationship between a willing player and an inviting playground, where the value of the activity is experienced as intrinsic to itself (Kretchmar 2005; Torres 2002). This means that we can predispose ourselves to find play but cannot guarantee it. And, as professionals, we can help others predispose themselves to find play in any variety of potential playgrounds—but we cannot force the experience on them. We can set an example by respecting play, by being patient in waiting for the call of play in likely playgrounds.

As you undoubtedly know from your own experiences, playlike sport is characterized by freedom, spontaneity, creativity, sheer joy, and personal meaning. You are also aware that it is fragile and temporary. Playlike sport cannot be dictated but, as already suggested, can be coaxed out of hiding more often through ingenuity and care. Don't you think that fostering playlike sport and physical activity can enhance our chances that people will fall in love with these activities and stick with them for the rest of their lives?

Playlike elements are potentially found in both competitive (from the elite level all the way to recreational leagues and low-key youth sports) and noncompetitive sports. Many casual and professional athletes are clearly drawn to the process of doing the activity, an internal focus, rather than just to its instrumental value. Similarly, play does not recognize differences of age, gender, race, religion, socioeconomic status, or motor skill development. It is not the exclusive property of any particular group. While alive we are always at risk of finding or being found by play, or both (Kretchmar 1994). The magnetism of play and the way we live an activity while at play make it an essential component in our lives. No wonder some have recommended that we live our lives fostering as much play as possible (Lasch 1979; Morgan 1982; Pieper 1952).

Two Potent Combinations

First, the combination of sport or physical activity and play offers powerful incentives to get us out of our chairs and into the realm of movement. When the doing becomes intrinsically meaningful and sensuously enchanting, we are more likely to continue with the activity. The artificiality of sport seems to be especially attractive. We love to solve problems created for the sole purpose of discovering whether they can be solved.

Second, if the playful sporting test is appealing, many of us encounter yet another layer of meaning when competition is added to the mix. This combination offers not only the uncertainty and tension of discovering our chances in the sport but also the ambiguity and drama of learning how we will fare in comparison to our opponents. When competitive sport is experienced playfully, we can focus on the doing, develop friendships with fellow sportspeople, and aim for excellence, not just victories.

Creating Movement Playgrounds: How is it Done?

While teaching an introductory unit on skateboarding, what might you do to enable your students to find play? What allows these strategies to generate play? (Please be sure to address, at minimum, the issues of skill development, repetition, feedback, and the amount of time devoted to this unit.) Can your students play in ways that are compatible with, or that even advance, your goals as a teacher? If this is possible, think of some examples.

Values Connected With Physical Activity

Given past and current scandals, what do you think about rewarding players or teams for being good sports by giving extra points at the end of games based on referees' judgments of players' behavior? Did you ever wonder what your responsibility would be if you were to discover that a teammate had been cheating? Should you tell your coach? League officials? Would it be acceptable just to keep quiet about it? Are there strong reasons to eliminate dodgeball from school physical education programs? Will physical activity become antiquated if science and technology solve most of our health-related problems? The search for answers to these questions require that you examine your values and develop persuasive arguments to defend them. The techniques of reflection discussed earlier in the chapter will help you in doing so.

All of us use values—our conceptions about the importance of things—to make decisions, both in personal and in professional matters. Values can be moral or nonmoral. Moral values refer to our character and how we ought to behave; nonmoral values refer to objects of desire such as happiness, ice cream cones, and good health. In this section we look at several central nonmoral values promoted by the field of physical activity. Then we examine the connection between ethical values and sport—a fascinating topic that has attracted a considerable amount of philosophical study. Philosophers of physical activity have also studied how aesthetic and sociopolitical values relate to sport. Although there is not space to discuss these relationships here, you'll look up additional information about them in the activity in your online study guide.

Values Promoted by the Field of Physical Activity

Nonmoral values are things that people consider desirable. Your values help you define the good life and steer a course toward achieving it (Kretchmar 2005). Are there values that are especially central in physical activity, and if so, what are they? In his analysis of major values promoted by physical activity programs, Kretchmar (2005) identifies four:

- Health-related physical fitness
- Knowledge about the human body, physical activity, and health practices
- Motor skill
- Activity-related pleasure or fun

Each of these values supports a different approach to developing physical activity programs. Physical fitness provides direction for those who are most interested in health and appearance. From this perspective, physical activity professionals should be primarily concerned about such things as assessing fitness, measuring fitness changes resulting from physical activity, and promoting physically active lifestyles.

Knowledge is the primary value for those who believe that information is a precursor to improved behavior—for example, for staying healthy, looking good, and performing physical skills better. This view would encourage more research in scholarly subdisciplines of

> Each of the four values promoted by physical activity programs—fitness, knowledge, skill, and pleasure—would lead to a different sort of program if used as the central guide or priority for planning and intervention.

155

kinesiology. Physical activity professionals should become well grounded in these subdisciplines on the assumption that having more of this knowledge is good for its own sake and also will make them better practitioners.

Motor skill is valued by those who like to move well and who know the joy of proficiency. Some might rank this goal highest because the pleasures of skilled movement might encourage people to stay physically active and thus improve fitness. Intrinsic motivations, in other words, lead to extrinsic gains. For those who highly value motor skill, activity professionals should concentrate on teaching appropriate movement techniques, for example, in school and college physical activity classes, focusing on particular sports such as golf or softball.

Valuing pleasure resonates well with people's conceptions of the good life. Fun and enjoyment are important for just about everyone. Enjoyment grows when a situation is complex enough to present stimulating challenges but simple enough so as not to overwhelm people's capabilities (Csikszentmihalyi 1990). We become bored if things are too easy and anxious if the situation is too difficult. Physical activity professionals who see the world from this perspective should make sure that games are simpler for beginners and more complex for advanced players and that the games are suited to the motor skill level of the players. We have already seen that competition notches up complexity even further, and it probably increases fun as long as the challenges do not overwhelm the players. Another way of varying the complexity is to make sure that opponents are evenly matched.

ACTIVITY 5.7

Using Values in Program Development
Explore the approaches of four professionals, as well as your own, in this activity in your web study guide.

The four values discussed here—fitness, knowledge, skill, and pleasure—do not exist in isolation from one another. All are important in physical activity, and each one influences the others (Kretchmar 2005). Consequently, the main task we face is not to pick one or two values to the exclusion of the others but to weigh their relative merits and rank them. We can use this ranking as a guide for setting priorities in physical activity—priorities for experiencing it as performers and spectators, for studying it, and for engaging in professional practice centered on it.

To illustrate the latter, professionals might use their value rankings to prioritize the topics they cover in discussions with people who control purse strings (e.g., city government officials, school boards, representatives of private foundations that donate to worthy causes) in an effort to encourage more spending for physical activity programs. If fitness is emphasized, then the profession becomes largely a means to an end or, as Kretchmar (2005) argued, an "auxiliary profession," albeit a very important one. If values like skill and knowledge are emphasized, the field takes on a liberal arts character. Movement, in other words, contributes directly to the good life by freeing us to move well and to understand. Kretchmar clusters a kind of knowledge (wisdom- or encounter-understanding) with skill as his own choice for the two highest-priority values. Fun, fitness, and theoretical knowledge are ranked lower. This profile presents physical education as a liberal arts subject matter, one that is aimed more at improving the quality of one's life than at meeting various day-to-day health and recreation needs. His thoughts about the intrinsic and extrinsic values associated with each value are important in his rationale for his rankings. He cites three criteria in his rationale:

- Intrinsic values are better than extrinsic values because intrinsic values have immediate, direct worth, whereas extrinsic values lead only indirectly to an intrinsic value.

- Among intrinsic values, those involving broader satisfaction and contentment are better than those involving more limited pleasure.

- Among intrinsic values involving broader satisfaction, those involving longer-term meaning and coherence in our lives are better than those involving more short-term or immediate pleasure.

Ethical Values and Sport

Given that in elite levels of sport, so much rides on winning and securing large amounts of money, athletes, coaches, and administrators may be tempted to give less attention to questions of morality. Cheating, shaving points, fixing games, using performance-enhancing drugs, and engaging in different kinds of craftiness often make the headlines. Sometimes these headlines are about misbehavior in youth sport and recreational leagues. It is not unusual to hear about dishonest behavior, brawls, and verbal abuse in these settings. It seems that in virtually all sporting environments, our capacities for moral reasoning are severely tested.

As intimated earlier in the chapter, ethics is about determining what is right and wrong and what ought and ought not to be done. To put it another way, ethics is concerned with questions about how we should live our lives. In sport, this means formulating defensible standards of behavior. While thinking ethically you must make an effort to be impartial, consistent, and critical. As you know, our emotions, assumptions, and tendencies to give ourselves a break can bias our thinking. The moral point of view is universal, one in which you do not count more than anyone else. Thus, bending the rules simply because "everyone does it" or intentionally injuring the best opposing player in a soccer match to send her to the sidelines for the rest of the game is difficult to justify. On the basis of universality, we would not want to be the recipient of such actions, and, because no one counts as more important than anyone else, it would not be right for us to prohibit rule bending or intentional harm but exempt ourselves from these restrictions. Consequently, societal understandings about right and wrong and the force of self-interest do not provide solid foundations for ethics.

In sport, decisions about right and wrong actions require that we keep in mind our conceptual analysis of competition. The more we understand the nature of competitive sport, the clearer and stronger our deliberations will be about what actions by athletes, coaches, and administrators would promote its health and well-being. These deliberations will provide principles about how these folks ought to behave. Sometimes ethical considerations in sport also include spectators and other consumers of sport. The goal is to produce the best possible world of sport for everyone.

Ethics and the Use of Performance Enhancers

Although athletes have used performance-enhancing drugs for a long time, these drugs became more prevalent in the late 1950s, when anabolic steroids came on the scene. Their use has been on the rise ever since (Todd 1987). Although precise figures are difficult to find, data from the U.S. National Football League show a large number of suspensions since 1982, when it implemented a policy of suspending players who tested positive for drugs (Boeck & Staimer 1996; Mosher & Atkins 2007). In this section we look at the use of performance enhancers—drugs as well as other things—from ethical perspectives (Osterhoudt 1991; Simon 2010).

The main ethical problem is that performance enhancers give athletes an unfair advantage over opponents. You might ask, then, which substances improve performance unfairly, and which substances improve it fairly? Your first answer might be that natural substances (e.g., food, testosterone) are fair, whereas unnatural substances (e.g., steroids, amphetamines) are not.

(continued)

> ### Ethics and the Use of Performance Enhancers   *(continued)*
>
> The distinction between *natural* and *unnatural*, however, is problematic. Technological breakthroughs in facilities and equipment—long-body tennis rackets, artificial track surfaces, sophisticated basketball shoes, and better-engineered racing bicycles—are unnatural. Yet performance enhancers such as these would generally be considered fair, even though limits are sometimes placed on technology (e.g., golf balls and baseballs could be engineered to travel farther, but limits are placed on this attribute). However, although testosterone is a natural substance, its use is typically considered unfair when it is taken in doses beyond what a person might produce in his or her own body (Brown 1980).
>
> One solution is to create a level playing field by giving all competitors access to whatever performance-enhancing aids they want. Another possibility is to distinguish between performance enhancers that increase a body's capabilities and those that just make it possible to use more effectively whatever capabilities the body already has (Perry 1983). But neither of these options, for a variety of reasons, has proved satisfactory.
>
> Another view suggests that performance enhancers should be prohibited if they put the health of an athlete or others at risk or if they are shown to make athletes dangerous role models for young children. But this approach raises serious questions about the rights and freedoms of athletes to make informed choices about their own health. It raises equally complicated questions about where, when, and under what conditions adult freedoms should be curtailed for the well-being of less mature or less well-informed individuals.
>
> Finally, there is the matter of privacy and individual rights. When is drug testing acceptable, and when is it an ethically unwarranted intrusion on an athlete's freedom as a human being? If we test athletes for recreational drugs or illegal drugs that don't enhance performance, some would argue that we are invading their privacy in unacceptable ways (Thompson 1982).
>
> Clearly, difficult ethical issues are connected with athletes' use of performance enhancers. Uncertainties remain concerning which aids are morally unacceptable, although performance enhancers that present health risks to athletes seem to be viewed as less acceptable than other performance enhancers. Additional investigations are needed to examine the ethics connected with using different kinds of aids to enhance sport performance.

When considering competitive sport as a reciprocal challenge in the pursuit of excellence and when using the moral guideline of creating the greatest good for the most people, we can generate the following behavioral guidelines:

- Follow the rules of the sport not simply because they are the rules but rather because they are the foundations of the artificial problem you find special. Avoid cheating, which alters, and even destroys, the sport and vitiates the legitimacy of results.

- Respect your opponent as someone who not only facilitates the contest but also makes possible the creation of athletic excellence. See your opponent as a partner who belongs to the same community of contestants and shares your interests and passion.

- Strive to bring out the best performance in one another. This honors each person's motor skills and the sport's standards of excellence.

- Recognize and celebrate athletic excellence, your own as well as your opponents'.

- Seek opponents who are close to you in ability and who compete with personal resolve to win without slacking off during the contest.

- Care about your opponent's well-being as much as your own. Your opponent is integral to the contest, and a victory is fully meaningful when opponents are at their best.

Winning Strategies

What kinds of strategies do you believe should be acceptable in sport? What about flopping in soccer to get a foul, intentionally fouling opponents in basketball in order to stop the clock, or intimidating a batter in baseball by intentionally throwing a pitch toward his or her head? What principles can you develop to distinguish between acceptable and unacceptable sporting actions? Can you come up with ideas to discourage morally unacceptable strategies? Should we levy harsher penalties? Use incentives? Provide more education?

- Remember that how you play says as much about you as an athlete as the scoreboard does.

In the high school baseball scenario at the beginning of the chapter, you saw that a dubious strategy cast doubt on the value of the victory and on any excellence shown by the winning team. Was this a tainted victory? It did not appear to be so for the coach, who believed, wrongly, that distracting pitchers with startling sounds and gestures to make them balk did not violate any rules and was not an instance of cheating. For him, it was simply clever and effective strategy.

However, even if distracting pitchers to make them balk was not cheating, it still raises moral questions. This action subverted the central purpose of competitive sport in several ways. First, the coach's strategy clearly undermined the set of motor skills that shape baseball. (Is yelling at pitchers to distract them and cause a balk a baseball skill? Is this a skill at which we want our players to excel?) Second, this action disregarded the standards of excellence that define baseball. (Was this an example of baseball at its best?) Third, by using this action, the coach showed a greater concern for winning than for the way in which winning was achieved. Perhaps even more important, his players were treated merely as means to an end and were not given a chance to demonstrate legitimate baseball superiority. Similarly, opponents' interests were not taken into consideration. In short, startling pitchers to make them balk diminished the victory by trivializing athletic competition, subverting baseball excellence, and not considering the good of everyone.

> Ethical reflection in sport typically requires that we critically examine societal understandings about right and wrong and tendencies that favor self-interest. When evaluating the best course of action in a given situation, we must also consider what is good for sport and the interests of everybody.

Wrap-Up

Philosophy offers not just a body of knowledge but, more important, an intriguing way of thinking about reality. Philosophy of physical activity helps us reach new insights about various forms of physical activity and what it means to move and move skillfully. It also includes explorations of the structure of games, sport, and play and their relationships as well as knowledge that can be derived from experiences with physical activity. Values connected with physical activities are of much concern in the subdiscipline. Philosophy of physical activity encourages using reasoning skills and insights to positively influence people's lives as much as your own. For example, you might bring your knowledge of play to bear on important decisions that you have to make to motivate people to exercise. Similarly, your ethical analyses might help you decide the best policy concerning the use of performance-enhancing drugs in a league for which you serve as the executive director.

If you find the research methods and knowledge in this subdiscipline interesting, take time to leaf through some of the texts on the philosophy of physical activity as well as the *Journal of the Philosophy of Sport* and *Sport, Ethics, and Philosophy*. You might enjoy an introductory course on the topic. You'll learn to think more carefully and precisely about the nature of things, and this will help you be a better performer, scholar, and professional.

More Information on . . . Philosophy of Physical Activity

Organizations

British Philosophy of Sport Association

European Association for the Philosophy of Sport

International Association for the Philosophy of Sport

Journals

International Journal of Religion and Sport

Journal of the Philosophy of Sport

Quest

Sport, Ethics and Philosophy

 ACTIVITY 5.8

Use the key points review in your web study guide as a study aid for this chapter.

Review Questions

1. Describe the main goal of philosophical study of physical activity and the four major kinds of issues it most commonly tackles.

2. Describe the major change in the philosophy of physical activity that occurred in the 1960s.

3. Describe the three reasoning processes that are central research tools in philosophical studies of physical activity.

4. Discuss the concept of blended unity of mind and body, its implications for school physical activity programs, its implications for research on physical activity, and its implications for the well-being of competitive athletes. Why is this an improvement over dualistic views of mind apart from body?

5. Explain the relationship between rules and skills in sport. Why is this relationship relevant to competition?

6. Discuss dutylike play, sport, and physical activity and their implications for how people relate to and experience these activities.

7. Elaborate on the four values promoted by the field of physical activity and their implications for designing physical activity programs.

8. Provide five examples of morally defensible sport contests. What makes them so?

chapter six

History of Physical Activity

Richard A. Swanson

chapter objectives

In this chapter we will

- explain what a physical activity historian does,

- describe the goals of history of physical activity,

- describe how the subdiscipline of history of physical activity developed,

- explain how research is conducted in history of physical activity, and

- explain what research tells us about physical activity in American society from the Industrial Revolution to the present.

A Survey of American History course will tell you about some rather strict rules in New England Puritan society in the 17th century regarding use of time away from work, Sabbath behavior and practices, and some forms of play and amusements that were banned altogether. Many of these rules were made into laws in most of the American states following the Revolutionary War and were still in effect by the late 19th and early 20th centuries. But churches, YMCAs, and YWCAs were sponsoring sport competitions and even building gymnasiums and playfields by the 1870s and 1880s. You can see photographs of some of these facilities and activities in books and on the Internet, showing children and young men and women participating in exercise classes and playing baseball on a lot next to a church. Rules for the first game of basketball were developed by a young instructor at the International YMCA Training Institute in Springfield, Massachusetts, in 1891. Even in the 1920s, some religious publishing companies were producing folk and square dance records for use in church youth recreational programs.

What happened to change these beliefs and practices so rapidly over a 50- to 60-year period? What societal influences played a role in changing religious behaviors so profoundly? What biblical understandings and interpretations had to be modified to allow such an alteration in practice and teaching? Were there theological differences over this issue within and between religious denominations and congregations? Did the leaders in various denominations lead this evolution, or did the secular culture influence the behaviors of otherwise devout church members? Was such a change positive or negative for the church? Was it positive or negative for society?

The scenario presented here describes one aspect of the historical relationship between two societal institutions found in most, if not all, societies in the world: (1) organized play, sport, and exercise, which in this text we place under the umbrella-like term *physical activity,* and (2) religion. Historians also examine relationships between physical activity and other social institutions such as economics, politics, business, and education, to name a few. The same is true of music and art historians, labor historians, medical historians, and military historians. The history of humankind is too complex to be understood through only one lens such as the political development of a nation. In this chapter you will begin to see many ways in which physical activity has been acted upon by the larger society as well as how it has affected and influenced other areas of our communal life as a nation and world. Likewise, we trust that you will begin to appreciate the exciting array of opportunities that exist for those who choose to become physical activity historians as well as the many ways in which a knowledge of our professional history can be used by kinesiologists in other subdisciplines.

Why Use History of Physical Activity?

Consider what life would be like if you didn't have a memory. You wouldn't be able to recall the 10 years that you spent training with a swim team, the pickup basketball games you played at your local neighborhood courts, the time you broke your leg falling off your bike and went to an exercise rehabilitation program for several months, or the details about the fortunes of your favorite professional football team. Memories give you insights about how things came to be the way they are. No one has perfect memory, and no one can predict the future with total accuracy, but our recollections help us act intelligently and develop reasonable plans for the future.

Goals of History of Physical Activity

- To identify and describe patterns of change and stability in physical activity in particular societies or cultures during specific periods
- To analyze patterns of change and stability in physical activity in particular societies or cultures during specific periods

History offers broad and detailed insights that go far beyond our own memory. It gives us the opportunity to develop a more extensive "memory" than we could ever acquire independently. History consists of a vast collection of information—mostly events that occurred before we were born, often in geographic regions and societies different from our own. The ancient Greek Olympic Games, 18th-century peasant ball games in Europe, and 20th-century American basketball all influenced the physical activities that we take part in today. Studying history gives us windows to the past and magnifying lenses to look closely at things that we find especially interesting. It helps us understand how and why our current physical activities are structured the way they are, allows us to compare them with physical activities from earlier periods, and gives us the tools to look toward the future from new vantage points.

A kinesiologist who is knowledgeable about history of physical activity would probably interpret such things as a fitness trend, the popularity of spectator sports in colleges and universities, or the availability of opportunity in sport across racial, ethnic, and gender lines quite differently from someone who lacks this information. A person who has studied history of physical activity would look at the opening scenario and have answers for the questions that we asked. For example, the historian would know that Protestant Christianity in the United States underwent a gradual but radical change in attitude toward recreational physical activity in the latter half of the 19th century. As organized sport and exercise were incorporated into the missions of many congregations and approved by many clergy members, this led to the acceptance of physical education in the public schools and the development of parks and recreation departments in cities across the United States in the early 20th century.

@ ACTIVITY 6.1

History of Women's Pro Basketball
Use this activity in your web study guide to document the development of professional basketball for women.

Although historians of physical activity are academic specialists in their own right, with specific goals that help to guide their research, the fruit of their labors often finds application in a variety of other professional roles. For instance, many of the modern exercise machines found in college recreation centers, athletic training facilities, and commercial fitness centers are based on models developed by an early American physical education leader, Dudley Allen Sargent, director of the Hemenway Gymnasium at Harvard University in the late 19th and early 20th centuries. Later, exercise specialists and engineers took the ideas and principles upon which Sargent had based his inventions and adapted them to contemporary knowledge and construction materials unavailable in an earlier time. Some of Sargent's ideas were, in turn, based on principles developed by earlier inventors that he undoubtedly learned of through the work of historians. Athletic coaches often create new playing strategies and training techniques by adapting those of an earlier generation of coaches whose work they have studied.

Knowledge of the history of sport, games, and exercise likewise gives the creative physical education teacher another tool with which to build interest in the lessons being presented by helping students discover the development and evolution of an activity. Scholars and scientists doing research in any of the other kinesiology subdisciplines described in this book also rely on the work of earlier researchers in their field and often use the work of physical activity historians who have analyzed this past work and summarized it within a broader historical context. One example of the latter would be examining how the first generation of scholars in a given field resolved any ethical considerations related to the use of human or animal subjects and whether those issues have changed as the field has matured. An understanding of past practices can help one avoid earlier mistakes, make adaptations that work within a new scheme, and gain insights that sometimes lead to entirely new ways of thinking about a specific problem.

Knowledge of history also helps us make better decisions today. A good example of this would be decisions by sport organizations to ban players, coaches, and managers from contact with gamblers. Major League Baseball was the first to adopt such a policy following the 1919 World Series, when players on the heavily favored Chicago White Sox were accused of accepting money from gamblers in exchange for purposely losing the series to the Cincinnati Reds. The public outcry about the loss of the integrity of the game threatened to reduce public confidence in the honesty of the game, which, in turn, could result in a decrease in ticket sales. With that lesson in mind, other sport organizations such as the National Football League, the National Basketball Association, the National Hockey League, the National Collegiate Athletic Association, and the U.S. Olympic Committee adopted similar policies. The memory of the 1919 World Series scandal has informed decisions by these organizations to fine, suspend, and dismiss player and management personnel for associating with gamblers in any manner.

American society's current interest in exercise, health, and fitness is a phenomenon that's hard to miss. Exercise studios and health clubs dot the landscape, and videos, books, and magazines on these topics are available in stores in almost every neighborhood shopping center. But if you asked teenagers or people in their 20s about the origins of this trend, most would probably draw a blank. Ironically, at the same time that America seems absorbed in the idea and the practice of exercise, health, and fitness, the nation is faced with rising obesity rates in the youngest of children as well as the adult population. Kinesiologists, including some historians, as well as medical professionals are seeking to understand the origins and current conditions creating this perplexing issue.

Kinesiologists know that interest in exercise picked up in the 1950s because of a variety of influences, such as Cold War fears, President Eisenhower's heart attack, and evidence suggesting that American children were less physically fit than their European counterparts were. Kinesiologists also know about an earlier period in American history when exercise was also prominent—the late 19th century (Park 1987a).

Kinesiologists can point out that American fascination with exercise has fluctuated over the years. This information is important: Is the current, relatively high level of interest in exercise likely to decrease in the future, in line with the fluctuating pattern that we've seen in the past? If so, this decrease will have serious negative implications for the health of many Americans. Can we do anything to prevent this possible downward cycle? Or do we now have such strong scientific evidence confirming the health benefits of exercise that we will break the roller-coaster pattern of the past? We could not even raise these questions without having an understanding of history of physical activity.

Your previous physical activity experiences, academic studies, and even observations of professional practice have already given you some insight into history of physical activity. From experience in competitive team sports such as basketball, volleyball, and soccer, you may have noticed that these sports are incredibly popular in Western societies and have in recent years experienced tremendous growth in popularity in Asian and African countries. If you have ever wondered how this came to be, you have been on the verge of historical inquiry. Or perhaps you have noticed gender inequities in physical activity and wondered what historical events led to this situation. If you are a man, maybe at some time you have feared being teased for your lack of athletic ability. If you are a male athlete, you may have noticed uneasily how much more financial support you received to develop your athletic talents in relation to your female counterparts. If you are a woman, perhaps at times you have felt strangely out of place in a competitive environment, or maybe limited financial support has been available to help you improve your athletic skills. In either case, if you have ever wondered why such situations exist, you were on the verge of delving into history.

From previous scholarly study, you are probably familiar with a lot of historical information that will help you understand the history of physical activity. Most of the historical research on physical activity deals with ancient Greece and Rome, Europe and Great Britain, and North America. Knowing something about the overall history of these societies will make history of physical activity easier to learn and remember. For example, if you know that nationalism and national patriotism swept through Europe in the 19th century, then it's not surprising to learn that in countries such as Germany and Sweden, people thought that special gymnastics systems could help build a healthy, strong citizenry that could keep their nations strong. Making this connection helps you remember when, and at least one reason why, these gymnastics systems came into being.

Finally, from watching physical activity professionals in action, you may have observed some history in the making. Perhaps you know people who work as athletic trainers, sport marketers, personal trainers, and sport physical therapists. These professionals are part of an important historical trend. Not long ago, career opportunities for people in the physical activity field were limited to professors, physical education teachers, and coaches. But in recent years, professional opportunities have greatly expanded, and this is a historical phenomenon.

As your knowledge of history of physical activity broadens, you will see that it is related to both philosophy of physical activity and sociology of physical activity. As you can see in figure 6.1, all three are part of the sociocultural sphere of scholarly study—that is, the sphere of research that focuses on the social and cultural aspects of physical activity.

> History of physical activity teaches us about changes as well as stability in the past, and this helps us understand the present and make reasonable decisions for the future.

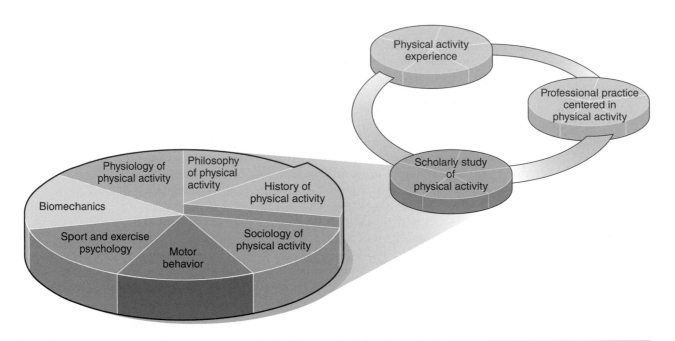

FIGURE 6.1 History of physical activity and the other scholarly subdisciplines.

What Do Historians of Physical Activity Do?

Physical activity historians have two main goals, as outlined at the beginning of the chapter. The first is to identify and describe patterns of change and stability in physical activity in particular societies or cultures during specific periods. Historians provide an enormous number of factual details to give readers a thick, robust description of such things as what it was like to be involved in the particular physical activities being studied, the values and attitudes that people had about those activities, people's broader lives, general societal values and attitudes prevalent at the time, how people's physical activities fit into their overall lives, and how they changed their physical activities. Because written historical descriptions require many pages, we can't include examples here.

Sometimes changes occur slowly, giving us the impression that things have stayed the same. For example, the rules, strategies, styles of play, and equipment in many of our sports

Comparing Sport: Then and Now

A lot of movies show people playing sport a long time ago. Here are just a few examples:

Cinderella Man (boxing)

A League of Their Own (female professional baseball)

Field of Dreams (baseball)

Hoosiers (basketball)

Seabiscuit (horse racing)

Kansas City Bomber (roller derby)

Breaking Away (bicycling)

Leatherheads (football)

There are many others. Think about film coverage that you have seen showing people playing a sport a long time ago. Compare this with the way this sport is played today.

changed a lot during the 20th century. Because these shifts occurred relatively slowly, however, the games remained recognizable. At other times, changes take place much more rapidly and dramatically. For example, improvements in bicycle design in the United States more than a hundred years ago contributed to a rapid rise in bicycling—the bicycle craze in the 1890s (Hardy 1982).

The other main goal of physical activity historians is to analyze patterns of change and stability in physical activity in particular societies or cultures during specific periods. Most historians go beyond description to demonstrate relationships and influences in an effort to explain why things occurred. They look for individuals, groups, events, and ideas that helped bring about changes or contributed to maintaining stability. For example, Hardy's *How Boston Played: Sport, Recreation, and Community 1865–1915* (1982) focuses on the city-building process that took place in Boston over several decades after the Civil War. During that time, the population of the city mushroomed as European immigrants and people from the American countryside arrived to take jobs created by the Industrial Revolution. The growing population led to major problems such as overcrowded tenements, poor health, neighborhood destruction, crime, ethnic conflict, and political and moral disorder—all serious threats to traditional ways of life. Hardy believed that Bostonians were looking for ways to revitalize their sense of connectedness to one another, and sport and recreation played an important part.

Drawing on what happened in Boston during this period, Hardy created three categories of ways in which Bostonians responded to the problems in their city: escape from the city and its problems, reform of the city and its problems, and accommodation to new forms of city life through developing a renewed sense of group identity (pp. 197-201). Hardy used these categories to examine the complex ways in which people sought community through building new playgrounds and parks, expanding sport and exercise programs in schools and universities, joining community-based sport clubs, riding bicycles, and conferring hero status on top athletes.

> Most physical activity historians are college and university faculty members; through their research, teaching, and service, they pursue the two main goals of the subdiscipline.

Understanding how this earlier generation of Bostonians used physical activity to benefit the entire city could well prompt modern-day community leaders to work with physical educators, recreation specialists, exercise consultants, landscape architects, land developers, and urban planners to create new and exciting opportunities for current and future generations of citizens.

Historians who study physical activity are usually college or university faculty members, but a handful are librarians, journalists, consultants for book publishing companies, archivists in charge of special collections of documents, or museum curators. We focus here on faculty members. They engage in typical activities consisting of teaching, research, and professional service. They often teach broad survey courses on history of physical activity, and some of their classes focus more narrowly on topics such as the ancient Olympic Games, physical activity in colonial America, or sport and ethnic relations in the United States. Examples of professional service include giving a community presentation about the history of a region's Minor League Baseball team, participating in an educational television panel to discuss the history of gender relations in American sport, and helping a city's nonprofit historical society put together a museum display about the history of rehabilitative exercise.

Libraries are perhaps the most important research tools for historians. If you want to study the physical activities of African American slaves before the Civil War in the United States, for example, you could locate library sources such as records of interviews with former slaves who were asked to recall their experiences under slavery, published accounts from white slaveholders written during the time of slavery, and accounts from northern whites who visited plantations where slaves worked (Wiggins 1980). Most historians take great care to develop solid working relationships with librarians. Sometimes library staff act almost as gatekeepers—deciding how much access you should have to rare and valuable materials.

Historians of physical activity sometimes apply for research money from their own colleges or universities as well as from outside organizations that specialize in supporting historical research. These historians often use the money to travel to libraries in distant cities or pay research assistants to help collect information. When asking for research money, they must write clear, persuasive proposals. Funding organizations usually require applicants to describe such things as the topic they want to investigate, why it is important, the sort of information they will collect, the sources where they'll find this information, what makes them competent to conduct the study, how much money they'll need, and what they'll buy with it.

History of Kinesiology and Physical Activity

A handful of American scholars studied history of physical activity in the early 20th century, and even in the late 19th century at least one physical educator viewed this as important. Even so, it wasn't until the 1960s that the subdiscipline began to develop a recognizable identity in North America.

One of the earliest American reports to emphasize history of physical activity was Edward Hartwell's *On Physical Training* (1899), which included information about ancient Greece, Europe and Great Britain, and the United States (Gerber 1971). Hartwell believed that professionals who would be charting a course for the future should know about the past, and as early as 1893 he pointed to the need for a textbook on history of physical education. Several decades would pass, however, before such texts appeared.

Historian Frederic Paxson wrote a classic article in 1917 titled "The Rise of Sport." Drawing from the well-known frontier thesis developed by his graduate school mentor Frederick Jackson Turner, Paxson argued that the disappearance of the American western frontier spurred the increased popularity of sport because it served some of the same escape or release functions as the frontier did (Pope 1997, pp. 1-2; Struna 1997, pp. 148-149). This safety valve function and Paxson's belief in the gradual, inevitable social evolution of sport were central ideas in the research of many sport history scholars for decades to come.

In the 1920s and 1930s, physical educators published several textbooks on the history of physical education as well as articles in professional journals on the development of physical education programs in schools and colleges. These books and articles probably helped lead college and university physical education departments to recognize the importance of history of physical activity.

▌ Sport historians look at where we've come from to help us see where we're going.

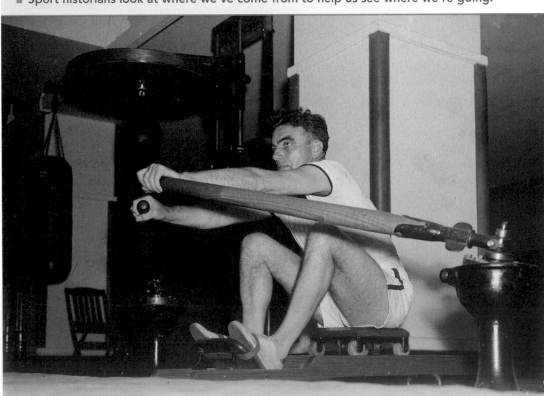

Seward Staley (1937), a professor of physical education at the University of Illinois, opened the door wider when he suggested that courses on the history of physical education should include a focus on the history of sport. Following World War II, graduate students began studying sport history at the University of Illinois under his direction (Struna 1997).

In 1940, historian Foster Rhea Dulles wrote his widely read *America Learns to Play: A History of Popular Recreation*. Dulles went beyond Paxson's safety valve and evolutionary perspectives to point out more positively that popular 20th-century American recreational pursuits—including sports—were well liked because they were satisfying to the American public (Struna 1997). Furthermore, a broad range of people had access to them, indicating to Dulles that American democracy had come of age (Pope 1997).

After World War II, historian John Betts completed an important doctoral dissertation titled "Organized Sport in Industrial America" (1952). He stressed that sport played an active role in society by helping bind people together, and he highlighted the ways in which late 19th-century entrepreneurs harnessed a variety of new manufacturing processes, new forms of transportation, and new forms of communication to shape sports into profitable commercial enterprises (Pope 1997; Struna 1997). In sharp contrast to Betts' active model, a passive model of sport was offered a decade later by journalist Robert Boyle in *Sport—Mirror of American Life* (1963). Boyle viewed sport as a mirror that simply reflected society. Both Betts' and Boyle's works had considerable influence on the subsequent work of sport historians.

The ranks of sport historians began to grow in the 1960s, but no scholarly association was in place to help the subdiscipline advance. The first attempt to remedy this occurred in 1962 when Marvin Eyler, Seward Staley, and Earle Zeigler worked to develop a section for sport history in the College Physical Education Association (soon renamed the National College Physical Education Association for Men [NCPEAM]), a professional organization for college and university physical education faculty members and administrators. Because this was the only formal organization for sport historians at the time, the NCPEAM *Proceedings* became an important published record of their work (Berryman 1973; Struna 1997).

Scholars in the discipline of history became more interested in studying sport when Eugen Weber, a prominent historian, presented "Gymnastics and Sports in Fin-de-Siècle France: Opium of the Classes?" at the 1970 meeting of the American Historical Association (AHA) and published it the following year in *American Historical Review* (Weber 1970; Berryman 1973). The program of the 1971 AHA meeting included an entire section on sport history. The North American Society for Sport History (NASSH) was established in 1972, and it was soon recognized as the main scholarly association for sport historians in the United States and Canada. The North American Society for Sport History was designed to attract a broader scholarly membership including women, Canadians, international scholars beyond North America, and sport historians from disciplines beyond physical education (Struna 1997). It began publishing *Journal of Sport History* in 1974, and a report issued in 1985 showed this publication to be the seventh most frequently cited history journal among hundreds in existence (Struna 1997).

Research Methods for History of Physical Activity

Research in sport history expanded in several directions during the 1970s and 1980s (Pope 1997; Struna 1997). Scholars began using two new analytical frameworks, or sets of general concepts or ideas, to make sense of the historical information they collected. The first framework, **modernization theory**, emphasized that the rise of modern sport occurred during the Industrial Revolution as American society shifted away from agricultural and local economies toward city-based industries rooted in science and technology. Sports changed from relatively unspecialized games to highly organized contests involving many rules and specialized playing positions. Allen Guttmann pioneered the use of modernization theory in his book *From Ritual to Record: The Nature of Modern Sports* (1978).

ACTIVITY 6.2

Interview with a Practicing Professional
In this interview in your web study guide, Jim Sumner, a sport historian, author, columnist, and contributor to a number of sport-related publications and websites, talks about how he turned his interest in sport history into a varied and meaningful career and what he finds most exciting and challenging about his work.

The second conceptual framework emphasized **human agency**, suggesting that people were actively involved in developing or "constructing" their own sports. Research focused on the details of how this occurred (Pope 1997; Struna 1997). One of the earliest works to use this approach was Stephen Hardy's *How Boston Played: Sport, Recreation, and Community 1865–1915* (1982). Hardy pointed to many local struggles among middle-class and working-class groups—often tinged with ethnic distinctions—that occurred as Bostonians went about structuring their sports and recreational activities.

Other scholars expanded the focus on human agency to look at gender differences. For example, in *Cheap Amusements: Working Women and Leisure in Turn-of-the-Century New York*, Kathy Peiss (1986) examined working women's culture between 1880 and 1920, emphasizing their newfound recreational activities outside the home in dance halls, nickelodeons, social clubs, and amusement parks.

An important research direction inaugurated in the 1980s dealt with exercise and health (Struna 1997). This trend was sparked by the American public's growing interest in exercise and physical fitness in the late 20th century and by the expansion of advanced course work on exercise in college and university kinesiology and physical education programs. One of the earliest studies was James Whorton's *Crusaders for Fitness: The History of American Health Reformers* (1982). Whorton showed the role of exercise among the many practices advocated in 19th- and early 20th-century reform movements aimed at improving the health of the American public.

In the 1990s, physical activity scholars continued to use the approaches to historical research that had been pioneered in the 1970s and 1980s. In addition, scholars increasingly linked their work to research in other social science disciplines, especially anthropology, economics, and sociology (Struna 1997). This approach has continued into the second decade of the 21st century.

Building on a small base of scholarly knowledge gathered earlier in the century, the history of physical activity subdiscipline grew in size, sophistication, and scope from the 1960s onward. Scholars initially focused on sport, but since the 1980s, exercise and physical fitness have been receiving greater attention.

Remembering what you've learned about the goals of history of physical activity, as well as the chronology of the development of the subdiscipline, let's look at how scholars produce this knowledge. How do they find out what happened, who was involved, when and where things took place, and why? The process has similarities to what detectives and attorneys do when they hunt for evidence that helps reconstruct important details of a crime. It involves locating evidence; critiquing and examining evidence; and piecing it together in a coherent, insightful framework that explains how and why things occurred. We can divide this process into three stages: finding sources of evidence; critiquing the sources; and examining, analyzing, and synthesizing the evidence.

Thinking From a Historical Perspective

Scholars in the discipline of history had little interest in studying sport history until the 1970s. Why did these historians ignore the topic for so long? What might have been their concerns? Can you think of any challenges or roadblocks they might have faced in conducting historical research in the field? What about sport and activity itself might have made historians slow to identify sport history as a viable research topic?

Finding Sources of Evidence

Think about the many different physical activities in current society. If historians 200 years in the future wanted to study these, what sorts of items would be useful? All kinds of things come to mind: aerobic workout videos; popular books, articles, and magazines about exercise and physical fitness; business records of professional sport franchises; videos of televised sporting events; school physical education curriculum guides; videos of dance concerts; policy statements from the National Collegiate Athletic Association; and the 1996 U.S. government report *Physical Activity and Health: A Report of the Surgeon General*. The list could go on and on. For instance, the digital age has brought us countless Internet sites devoted to health, fitness, exercise, and organized sport from youth activities through professional and masters-level competition. As never before, sports fans as well as people seeking information on health and exercise are able to blog about their interests and instantly express their own reactions to what they read via "Comment" to online newspaper and magazine articles and by means of e-mail and Twitter accounts. These digital resources will be a rich resource for historical researchers in the future. However, just as with all traditional primary and secondary resources, the hardest task for the historian will be testing the validity and reliability of the data. In addition, websites come and go, as do e-mail records. Uncovering this "lost" data may be extraordinarily difficult if not impossible.

These items are all primary sources of information, meaning that they were produced in the society and period being studied. But where would scholars find them? We can only guess where these things might be stored in the future, but we know where historians look today—libraries, archives (storage facilities for documents from important people and organizations), and private collections (Struna 1996a). Sometimes researchers interview people to tap their memories of life in earlier times.

Materials are scattered throughout the country. For example, the Avery Brundage Collection (Brundage was president of the International Olympic Committee from 1952 to 1972) is at the University of Illinois. The collection contains items such as letters to and from Brundage, books, pamphlets, scrapbooks, Brundage's notes on various topics, speeches given by Brundage, and policies from the International Olympic Committee and other sport governing bodies (Guttmann 1984). Copies of *Boston Medical and Surgical Journal* from the 19th century are located at the National Library of Medicine in Bethesda, Maryland, as well as at a library at the University of Chicago (Struna 1996a). Local historical societies are often rich sources of information, as are archives located in the library at your own college or university.

ACTIVITY 6.3

Primary Source Material
Distinguish between primary and secondary source material in this activity in your web study guide.

Historical reports written by people in later periods—secondary sources—are useful for finding primary sources. Examples of secondary sources include a journal article written in the 1980s about African Americans who participated in intercollegiate athletics in the 1930s and 1940s, or a book-length report produced in the 1990s on physical activity in the American colonies. Authors include detailed notes pointing to the nature and location of specific information (primary sources and other secondary sources) used in their investigation. Most libraries have electronic databases to help you locate sources, and some of these are accessible through the Internet.

Critiquing the Sources

After you locate primary sources, you must scrutinize them carefully for authenticity and credibility. Sources that are authentic and credible make your historical research believable. Let's concentrate on written documents such as newspaper articles, books, and letters. Determining authenticity involves such matters as who wrote the piece and when it was written. You might learn, for example, that a document was written many years after the events it describes took place or that the author listed on the document was not the person who wrote it. For example, if a letter dated in the 17th century uses the phrase "didn't get to first base," you can be sure that the piece was written much later because this term comes from American baseball, a game that didn't exist in the 17th century (Shafer 1980, p. 131). If ink used to publish a report wasn't manufactured until several years after the date listed, then you know that the piece was written later. (Such a determination would require a technical specialist.)

Determining the credibility of written documents helps you interpret them. The *rule of context* encourages you to make sense of a document's language in relation to what the words meant to people in the society in which it was produced. The *rule of perspective* requires you to examine an author's relationship to the events that he or she describes and how the author obtained the information. The *rule of omission* or *free editing* reminds you that you always have partial records of events—not the complete events themselves; locating multiple sources to add more details to your evidence is important (Struna 1996a).

Examining, Analyzing, and Synthesizing the Evidence

> Historical research involves finding sources that contain evidence about past events, critiquing the sources for authenticity and credibility, and analyzing the data contained in the sources to learn how and why things happened.

After you've located authentic and credible sources, you should examine the evidence to find information that addresses the tentative hypotheses or research questions that you established at the beginning of your investigation. The goal is to describe events in detail and then analyze them to learn how and why things took place. Historians accomplish this by placing events in an analytical framework that uses (1) trends and relationships in the events themselves or (2) theoretical models from the social sciences. You have already seen examples of analytical frameworks—the reestablishment-of-community framework, the modernization framework, and the human-agency framework. There are probably as many approaches to analysis and synthesis as there are researchers. The main goal is to piece together the evidence to gain a detailed understanding of what happened, as well as how and why (Shafer 1980; Struna 1996a).

✔ ACTIVITY 6.4

Credibility of Sources
Differentiate among context, perspective, and omission in this activity in your web study guide.

ACTIVITY 6.5

Practical Plug-In
A sport historian is hired to help a university decide whether to take its athletic program to the next level. In this activity in your web study guide, you'll consider what the historian needs to know and do to facilitate the decision-making process.

Overview of Knowledge in History of Physical Activity

Now that you know what historians of physical activity do, let's look at some of the fascinating information they've discovered. The body of knowledge is large and complex; it deals with many societies worldwide and covers many different periods. Here we touch on a few of the highlights.

If you were a boy living in ancient Greece, for example, during the 6th century BCE in the city-state of Sparta, you would have received more extensive and harsher physical activity training than you would have in the city-state of Athens, where you would have received a more balanced physical and intellectual education. In both places, however, your physical activity instruction would have had the goal of preparing you for warfare. And although the ancient Greeks considered the Olympic Games similar to warfare in competitiveness, this prestigious event for full-fledged citizens was distinct from the brutal gladiators' contests—mostly among slaves and convicts—staged several hundred years later in the Roman Empire. And the Roman contests were very different from the informal, rough team ball games played by European peasants in the Middle Ages and also distinct from high-profile women's gymnastics in the Olympics of the late 20th century. We haven't even mentioned other forms of physical activity such as dance and exercise. Understanding the ways in which physical activities have fit into societies around the world since the beginning of history is indeed a monumental task.

In North America we know a lot about the history of physical activity in Western civilizations—especially ancient Greece and Rome, Europe, Canada, and the United States. We must recognize, however, that people in Africa, Asia, Australia and New Zealand, the Indian subcontinent, Central and South America, the Caribbean region, and the Pacific Islands developed rich physical activity traditions. These include a variety of martial arts in Asian societies, sumo wrestling in Japan, ball games played in high-walled courts in native Central American civilizations, an African martial art brought to the Atlantic and Caribbean coasts of South America, and a sport in Afghanistan involving hundreds of mounted horsemen who simultaneously try to gain possession of a goat or calf carcass. Historical information exists about all of these. In this text, however, we focus on the history of physical activity in North America.

British and European colonists began arriving in North America in greater numbers beginning in the early 1600s, swelling an earlier trickle. Initially, people worked hard to establish themselves in unfamiliar and often harsh environments. These hardships, coupled with strict religious prohibitions against idleness and amusements in some areas of the colonies, meant that participation in sport and games was limited. Nevertheless, people sometimes engaged in their own versions of traditional British and European sport and recreational activities. Participation increased as the colonists became more established and religious sanctions were lifted. Activities included horse racing, fishing, hunting, sailing, boat racing, golf, team ball games, sleigh races, skating, foot racing, boxing, wrestling, animal baiting (in which a wild animal such as a bear was tied by a chain and attacked, often by dogs or rats), cockfighting, and billiards. Some of these were commercialized and yielded profits for promoters; gambling was also common (Baker 1988; Lucas & Smith 1978; Struna 1996b).

We'll look more closely at physical activity in North America during three periods:

- 1840-1900: Industrialization, growth of science and technology, immigration, urbanization, democratization, and westward expansion characterized North American society; the American Civil War occurred.

- 1900-1950: The growth of consumerism, immigration, and democratization characterized North American society; a major economic depression and two world wars occurred.

- 1950-2012: The growth of electronic communication, the growth of global trade, immigration, and democratization characterized North American society; the Cold War, Korean War, the Vietnam War, the first Iraq War, and the Afghanistan and Iraq wars of 2001-2012 occurred.

For each of these periods, we'll examine

- participation in physical activity,
- physical activity professions, and
- scholarly knowledge about physical activity.

1840-1900: Industrialization and Westward Expansion

From the mid-1800s to 1900, the economy, population, and geography of the United States and Canada expanded. New developments in science and technology fueled the growth of business and industry. As city populations mushroomed, major sanitation problems, crowded living conditions, lack of space for physical exercise, and unsafe factories threatened everyone's health. The advent of more leisure time sparked an interest in building city parks and playgrounds as well as in developing more commercial entertainment and amusements such as spectator sports. Many new schools and colleges opened, with both physical training and student-run intercollegiate sport programs. The citizenry started buying more products in response to newspapers, magazines, and eventually movies that offered glimpses of the good life. Westward expansion continued, highlighted by the completion of the transcontinental railroad in the United States. Massive numbers of European and British immigrants arrived, along with Asians on the West Coast. The Civil War (1861-1865) exacted a terrible toll of death, destruction, and disruption, but it led to legal affirmation of the ideal of racial equality, an important goal in American society ever since.

Participation in Physical Activity

Liberal religious and philosophical currents had begun to flow together in the early decades of the 19th century, emphasizing that a human being's body, mind, and soul were an inte-

▌ Public spaces, such as skating ponds in New York City's Central Park, provided residents of crowded cities a place to participate in healthy activities.

grated whole. This perspective promoted the importance of keeping one's body healthy to achieve peak mental functioning as well as the highest possible level of moral reasoning. Some believed that good health practices indicated a morally righteous person in the eyes of God. This thinking, combined with squalid conditions in overcrowded cities and a nationalistic desire to make the United States self-sufficient, led to widespread efforts throughout most of the 19th century to improve people's health. Proper diet and exercise were among the recommended practices (Berryman 1989; Whorton 1982).

Physicians, writers, and teachers, addressing primarily the middle class and social elites, called for more vigorous physical activity—especially for boys and men. These authorities coupled their notions of the importance of health reform with ideas from England about the value of sport for developing moral character and manliness. As the century progressed, physical activity grew increasingly attractive and popular.

Many professionals also recommended moderate exercises for improving the health of girls and women. Straining too hard in exercise or competitive sports, however, was usually considered unladylike and dangerous for women's physical well-being and future childbearing capabilities (Vertinsky 1990). By the end of the 19th century, some girls and women were involved in vigorous physical pursuits.

European systems of gymnastics, especially those from Germany and Sweden, gained an important American foothold during this time. German cultural societies called *Turnvereins* eventually offered well-equipped gymnastics facilities in many North American cities. In the United States alone, there were 148 of these by 1867 serving 10,200 members (Swanson & Spears 1995, p. 128). The centerpiece was exercise on heavy equipment such as parallel bars, vaulting horses, and hanging rings. By contrast, the Swedish system involved calisthenics and exercises with lighter equipment such as dumbbells, long wooden rods, and weights. Elements from both systems, as well as other exercise programs developed by Americans, grew increasingly popular in YMCAs, YWCAs, and city parks and playgrounds.

Sports were popular with many of the new immigrants who arrived in the United States and Canada from 1840 onward. They played American baseball as well as American and Canadian football, which evolved from rugby and soccer later in the century. Youngsters often learned American sports and games at local playgrounds, and many civic leaders hoped that immigrants' participation would speed their assimilation into American society. But immigrants also sought to keep the cultures of their homelands alive through participation in physical activities. Many belonged to special clubs in which sport, exercise, and other forms of recreation were often prominent. Examples in the city of Boston included the Boston Turnverein, featuring gymnastics and catering to German immigrants; the Caledonian Club, devoted to Scottish culture and including an array of sports and games such as caber toss, races, and pole vault; the Irish Athletic Club, Irish-American Athletic Club, Ancient Order of Hibernians, Boston Hurling Club, and Shamrock Hurling Club, all focusing on Irish sports and games, including oarsmen's regattas, stone throwing, and hurling; and cricket clubs formed by West Indian immigrants (Hardy 1982, pp. 136-138).

Other immigrants took part in sports and amusements that, at the time, were considered less than respectable by members of mainstream society. These included prize fighting, billiards, bowling, cycling, and wrestling. The Irish dominated prize fighting for several decades, and later in the century boxing gyms could be found in African American neighborhoods (Gorn 1986). After the Civil War, African Americans found their way into mainstream sports—professional baseball and horse racing, for example—but by 1890, racial prejudice forced most African Americans into segregated leagues and organizations.

In school and college physical activity classes, gymnastics exercise was preeminent. The Swedish and German systems were common, as were exercise programs developed by Americans that centered on calisthenics and lightweight equipment. The primary focus of such programs was health through fitness as promulgated by such medically trained early physical education leaders as Dr. Edward Hitchcock (Amherst College), Dr. Dudley Allen Sargent (Harvard University), and Dr. Delphine Hannah (Oberlin College). Sports were not usually included in classes but instead existed as extracurricular activities. Originally low-key intramural activities, sports were a rich supplement to often-stodgy exercise classes (Rudolph 1962; Smith 1988). Student-run intercollegiate competition appeared about midcentury and gathered steam in the 1870s. The most common intercollegiate sports were American football, crew, baseball, and track and field.

College and university administrators believed that intercollegiate sports could be helpful for recruiting new students, attracting alumni contributions, and promoting a sense of community among an increasingly diverse student body. According to thinking that arrived on North American shores from England, sport could also develop leadership, vigorous manliness, and upstanding moral character, all qualities considered desirable for college men. For all these reasons, athletics were believed too important to be left in the hands of students. Campus administrators, faculty, and coaches gradually took control.

During this era fewer sport opportunities were available for college women. They engaged in occasional intercollegiate competition and somewhat more frequent intramural competition. Many female physical educators, however, began including sport activities in their classes in the 1870s and 1880s (Swanson & Spears 1995). Women who wanted to compete faced societal concerns about overly strenuous and highly competitive physical activities for women and girls. They avoided sports in which injuries were more likely (football, for example). Basketball, tennis, golf, and baseball were popular. Women's sports usually took place in secluded locations with only a few spectators present. A delicate balance had to be struck between making competition available to college women and keeping the activities acceptable to faculty and campus administrators (Park 1987b).

Beyond schools and colleges, by the 1860s many amateur baseball teams were on their way to becoming professionalized; common practices by that time included charging admission and paying players. Baseball had an avid following, and newspapers gave it the most coverage of any sport (Adelman 1986, p. 174). In the 1880s, the most popular sports were horse racing, baseball, and prize fighting.

American athletes also competed in a new amateur event developed by Europeans and first held in Athens, Greece, in 1896—the modern Olympic Games. The United States fielded a team of 10 male track and field competitors—all collegians or ex-collegians—who participated in the fledgling Games; these competitors chalked up nine victories in 12 events, along with five second-place finishes. Three other American athletes, two pistol shooters and one swimmer, also competed in these first Olympic Games, resulting in two additional victories in the sharpshooting events.

> ► American interest in physical activity—for example, gymnastics exercises, baseball, football, crew, track and field, horse racing, boxing, bicycling, and less well-known sports tied closely to people's ethnic origins—increased throughout the 19th century.

Curriculum Changes in Physical Education

In the late 19th century, physical education programs emphasized health and the activities focused on calisthenics; sports were not usually included in men's physical education classes. Why do you believe there was such an emphasis? What types of activities have your physical education classes in high school and college included? What do you think caused the change between what was offered in the 1890s and what is offered today?

Physical Activity Professions

An identifiable physical activity profession didn't appear in the United States until late in the 19th century. Nevertheless, a variety of practitioners in earlier decades focused at least some of their work on physical activity. These practitioners included physicians, successful athletes, journalists, educators, ministers, health reform advocates, business entrepreneurs, and a handful of European gymnastics specialists who immigrated to the United States. Some of them wrote about the physical, intellectual, and moral benefits of exercise and sport. Some developed exercise programs in schools and colleges. Others worked to establish various European gymnastics systems on American soil. Some wrote popular self-help manuals on achieving healthy lifestyles that included how-to information on exercise routines. A few

■ Aaron Molyneaux Hewlett was an early professional in the physical activity world, serving as the first director of the Harvard College Gymnasium.

became professional athletes in sports such as horse racing, pedestrianism (long-distance walking), boxing, and baseball. Some tried to sort out the best training techniques for athletic success. Some became coaches, and some bankrolled professional sports.

By the 1880s, a climate of intense interest in exercise and sport had developed in the United States. Numerous professions were beginning to organize in response to the growing need for people with specialized knowledge and skills in technologically based businesses and industries. Efforts were under way to develop more stringent educational standards for new practitioners.

In this atmosphere, the physical education teaching profession took root (Park 1987a), the first recognizable physical activity profession in the United States. In 1885, the American Association for the Advancement of Physical Education was formed by about 60 people who wanted to have an organization that would promote the new profession. After several name changes, it is today known as the American Alliance for Health, Physical Education, Recreation and Dance, an important professional association for practitioners in several fields.

Several physical education teacher training programs were inaugurated, primarily in the northeastern United States, and they varied in length from a few weeks to about 2 years. Some of the most famous were the Sanatory Gymnasium (opened near Harvard University in 1881 by Dudley Sargent and renamed the Normal School of Physical Training in 1894), the Harvard Summer School of Physical Education (opened in 1887 by Dudley Sargent), and the Boston Normal School of Gymnastics (opened in 1889 by Mary Hemenway and Amy Morris

▶ The earliest identifiable American physical activity profession—teaching physical education—was established in the late 19th century during a period of high interest in physical activity among the general public.

Homans). Around the turn of the century, several 4-year bachelor's degree programs were initiated for the preparation of physical education teachers.

Scholarly Knowledge About Physical Activity

Scientific discoveries in the 19th century produced much new information about human anatomy and physiology. By the 1840s, improved microscopes permitted more detailed studies of phenomena such as oxygen transport in blood, energy transformation in muscles, and the anatomy and functioning of the nervous system (Park 1987a, p. 144). Later in the century, scholars who wanted better understanding of the biological effects of physical activity on the human body occasionally used such information. Educators also sometimes incorporated this knowledge into physical education teacher training programs.

▶ Scholarly knowledge about physical activity—mostly information from the biological and physical sciences—became important in a few professional teaching training curriculums in the late 19th century, but in most programs this subject took a backseat to learning physical activities and practical knowledge about how to teach them.

For example, Edward Hartwell, an associate in physical training and director of the gymnasium at Johns Hopkins University, published an article in 1887 titled "On the Physiology of Exercise." George Fitz, an instructor and researcher in physiology and hygiene at Harvard University, established a 4-year undergraduate program called Anatomy, Physiology, and Physical Training. Not far from Harvard University, the Boston Normal School of Gymnastics—a 2-year physical education teacher training institution that attracted mostly female students—offered a curriculum focused on Swedish gymnastics that also included a variety of basic science courses taught by well-known scholars from major universities around Boston. This excursion into the biological sciences by early leaders in physical education demonstrates the value that they placed on developing a scientific base for the emerging physical education teaching profession (Gerber 1971; Park 1987b).

Despite this new interest in the sciences, professional programs focused on learning physical activities and how to teach them outnumbered scientific curriculums. By the end of the 19th century, the emphasis was clearly on the professional programs as well as on the positive social values that students could learn through participating in play and sport. Investigations into the biological mechanisms underlying physical activity were not completely curtailed, but it was not until the 1960s that scientific information again moved to center stage in college and university physical education curriculums (Park 1987a).

1900-1950: Consumerism, Immigration, Democratization

In the first half of the 20th century, American industry perfected mass production and churned out a wide variety of consumer goods from automobiles to canned food to refrigerators. With readily available products, entrepreneurs stepped up their advertising to encourage the public to become enthusiastic consumers. Big-time intercollegiate sports and professional baseball and football leagues became major parts of the growing entertainment industry and benefited from the rise of this consumer culture. Many new immigrants—especially from southern Europe—arrived on American shores, adding more ethnic diversity to an already multifaceted population. American society made progress toward the democratic ideal of equal opportunity and social justice: Women won the right to vote, and African Americans were hired in greater numbers for federally funded jobs. Full achievement of social equity, however, remained distant and elusive. The severe economic depression that began in 1929 left personal, lifelong scars on massive numbers of Americans who suffered through it. Many other horrible and unforgettable disruptions occurred during World War I (1914-1918) and World War II (1939-1945).

Participation in Physical Activity

Competitive sports were in the limelight during the first half of the 20th century. Professional baseball, professional boxing, horse racing, and American collegiate football—all men's sports—were especially popular with fans. Professional baseball's American League and National League joined forces in 1903 to keep player salaries in check and reduce competition for fan loyalty. American professional football and basketball gained a degree of prominence by the 1930s, although they were still fledgling enterprises. World War II temporarily put a damper on men's professional athletics because most healthy young men entered military service. This spurred Chicago Cubs owner and chewing gum magnate Philip K. Wrigley to organize the All-American Girls Professional Baseball League, which played in the Midwest from 1943 to 1954.

American football dominated intercollegiate athletics during this era. Major college games received widespread news coverage, and they were occasions for lively partying and revelry. By 1903 Harvard University had a stadium that could seat 40,000 fans, and by 1930 seven of the "concrete giants" on college campuses could hold more than 70,000 people (Rader 1990, p. 182; Smith 1988, p. 169).

Competitive sports for boys were common in high schools and elementary schools, on playgrounds, and in youth service agencies such as YMCAs. By the 1930s, elementary school competition began to decline when physical education teachers complained about the high stress of competition. Many people still wanted competition for youngsters, however, and new programs designed by city recreation departments and organizations such as Little League Baseball, Inc. (formed in 1939) and Biddy Basketball (formed in 1950) filled the void.

The Golden Age of Sport: The 1920s

The 1920s, sometimes referred to as the "Roaring Twenties," was a decade of change and excess. Whether one measures the change in terms of clothing fashion, political opportunity, technology, or popular culture, there is no doubt that a seismic shift in American society took place. Change was apparent as women exercised their newly acquired right to vote as well as to raise the hemlines of their skirts. Excess might be measured by stock market and real estate speculation, as well as by personal spending as America became increasingly a consumer nation. Sport became one of those products consumed by a people hungry for entertainment, be it on the stage, on the silver screen, on the radio, or in the stadium. Where once the titans of business and industry were the childhood heroes and role models, now the Thomas Edisons, Henry Fords, and John D. Rockefellers were forced to make room for sport figures such as Babe Ruth and Gertrude Ederle and movie stars like Charlie Chaplin and Mary Pickford. The middle class was growing; times were good for larger numbers of Americans, and increasing numbers had more leisure time to watch and follow their favorite teams and individual sport stars.

Although earlier decades had their individual athletic heroes, such as the boxer John L. Sullivan in the 1880s and baseball pitcher Cy Young in the 1890s and early 1900s, the 1920s brought them in greater abundance and across a wide array of sports. Babe Ruth and Lou Gehrig (baseball), Bill Tilden and Helen Wills (tennis), Bobby Jones (golf), Gertrude Ederle and Johnny Weissmuller (swimming), Red Grange (football), and Jack Dempsey and Gene Tunney (boxing) were representative of this sudden explosion of athletic excellence and public interest in sport. To accommodate demand, larger and grander stadiums were constructed by both professional baseball clubs and college football programs. New York's Yankee Stadium was opened in 1923 with more than 50,000 seats to meet the demands of fans flocking to see the great Ruth hit his astounding home runs. New college football stadiums such as those at Ohio State University (1922) and the University of Illinois (1923) exceeded 60,000 seats to handle the masses of people who arrived by train and private automobile from near and far.

Fueling this phenomenon was a media industry just beginning to realize the potential revenue available through the exploitation of commercial sport. Newspaper owners had begun to recognize this around the turn of the 20th century, when the sports page made its debut in many large cities to meet the growing demand for Major League Baseball coverage. By the 1920s, many of these big-city papers began hiring talented writers such as Grantland Rice (*New York Herald Tribune*), John Drebinger (*New York Times*), Shirley Povich (*Washington Post*), and Paul Gallico (*New York Daily News*), who further

(continued)

The Golden Age of Sport: The 1920s   (continued)

popularized the stars of the sports world through their coverage of the contests and feature stories on the athletes. The introduction of commercial radio in 1924 brought a whole new dimension to sport coverage, and by the end of the decade, announcers such as Graham McNamee were recreating football and baseball games for listeners in millions of homes across the nation. Contrary to the initial fears of college administrators and baseball club owners, the radio brought in countless numbers of new fans whose interest led them to purchase tickets to see in person what they had heard on the airwaves.

In the mid-1930s, writer Paul Gallico, looking back on the tremendous growth of interest in sport during the 1920s, dubbed it "The Golden Age of Sport." Today we might refer to it more accurately as "The *First* Golden Age of Sport," since we have certainly been in a similar "sport-crazed" era for the past 4 decades or more. With the arrival of cable television and its numerous dedicated sport networks beginning in the 1980s, the publication of sport-specific magazines and almanacs, the construction of newer and ever more grandiose stadiums and arenas, and the creation of new and increasingly extreme competitive activities, it is fair to say that we are today in an unprecedented age. Nevertheless, we can trace the roots of today's sport mania to that amazing decade some 90 years ago.

Girls' and women's sports were subdued in comparison to major boys' and men's events with their accompanying publicity and hoopla. A few women achieved national and international success in sports such as tennis, basketball, golf, long-distance swimming, and track and field. For the most part women trained and competed outside the educational system in settings such as private country clubs and industry-sponsored leagues. For example, in 1930 Mildred "Babe" Didrikson was offered a job as a stenographer at Employers Casualty Insurance Company in Dallas so that she could play for the company's Golden Cyclones basketball team. The company also sponsored her to compete at the Amateur Athletic Union's national track and field championships in 1932. Her phenomenal success led her to the 1932 Olympics, where she won two gold medals and a silver (Guttmann 1991).

In schools and colleges during the 1920s, female physical educators turned toward intramural sports and low-key "extramural" competition for girls and women. They wanted to avoid the stress of high-level competition and to encourage all girls and women to participate. Although a few institutions continued to sponsor elite sports for females, "fun and games" became the goal at most schools and colleges, greatly curtailing the development of elite female athletes in educational settings. Things didn't begin to change until the late 1950s.

Public interest in competitive sport was so pronounced in the early decades of the 20th century that sports replaced gymnastics exercises as the centerpiece of most school and college physical education curriculums. A large number of physical education teachers believed that student participation in sport would help develop high moral character and other qualities needed by a good citizen. And students seemed more interested in physical activities that were sparked with competitive excitement, compared with the tedium of repetitious, traditional gymnastics exercises.

But exercise did not completely disappear from physical education classes. Teachers sometimes replaced the old gymnastics systems with expressive movements designed to communicate ideas and feelings or with sport-related exercises such as shooting baskets or batting baseballs. By the 1920s, "corrective" physical education classes offered special exercises to students with posture, fitness, and health challenges (Van Dalen & Bennett 1971, pp. 460, 461, 468).

The U.S. military recognized the need for physical training during World War I, when approximately one-third of U.S. draftees were initially declared unfit to serve. Moreover, recruits with musculoskeletal incapacities such as flat feet and backaches deluged overseas hospitals. Recreational sports—organized by about 345 military athletic directors—were widespread and popular with the troops (Murphy 1995; Rice et al. 1969).

Physical education teachers on the home front added military training such as marching drills and calisthenics to school and college physical education programs. Recognition of the poor physical condition of the troops served as a wake-up call after the war, spurring the expansion of school and college physical activity programs. By the end of the 1920s, people could also exercise by following an instructor on the radio.

Racial discrimination eliminated most African Americans from the highest levels of competitive sport by 1900. In spite of this, a few were active in boxing (Jack Johnson and Joe Louis were heavyweight champions); a small number participated in intercollegiate athletics at predominantly white schools as well as at predominantly African-American colleges; a few competed in the National Football League until 1933, when racial barriers were raised; and a handful took part in the Olympic Games (track star Jesse Owens won four gold medals in the 1936 Berlin Games). African American baseball players had no choice but to play for African American teams until Jackie Robinson signed with the Brooklyn Dodgers organization in 1945 and played on one of their minor league teams in 1946; he moved up to the Major League with the Dodgers in 1947.

Cultural heritage influenced the sport involvement of people in many racial and ethnic minority groups. For example, baseball, basketball, and boxing served as a middle ground for second-generation Jews, providing at the same time a place to celebrate Jewish heritage and a place to assimilate into mainstream American society (Levine 1992, pp. 3-25, 270-274). First-generation Japanese immigrants who arrived early in the 20th century often used traditional perspectives from their homeland to make sense of American sports, sometimes recalling samurai (historic Japanese knight-warriors) principles of courage and honor (Regalado 1992).

World War II renewed people's interest in exercise. After the war, educators expanded many school and college physical activity curriculums in response to weaknesses observed in the fitness of wartime military recruits, and sports were again at center stage (Swanson & Spears 1995).

> Sports—in elite-level competition, community recreation programs, and school and college physical education classes—were Americans' favorite physical activities in the first half of the 20th century. Americans paid less attention to exercise, although interest picked up during the two world wars in order to improve physical fitness.

Sport and Ethnicity

Think of as many examples as you can of groups of people with a particular ethnic heritage getting together to

- play or watch sports associated with their own cultural heritage and
- play or watch sports that are widely popular in the United States.

Is there any overlap between your two lists? If so, why do you think there are such overlaps?

Physical Activity Professions

Bachelor's degree programs in physical education increased in the early 1900s, expanding to about 135 by 1927 (Park 1980). Master's degrees were available just after the turn of the century, and doctoral programs began in the 1920s. Undergraduate programs focused on training physical education teachers, continuing the trend that had started in the late 19th century. Graduate programs offered advanced training for teachers as well as the academic preparation to become a college or university faculty member. The quality of these programs improved as research on physical activity picked up in the late 1920s.

Despite the popularity of sport during the first half of the 20th century, college and university physical education curriculums didn't include much course work to prepare students for the occupations of coach or athletic trainer. Coaches most often came from the ranks of successful athletes. If they worked in a school or college, they were often required to teach physical education classes as well, so they earned the necessary college degrees and professional teacher certifications.

The few athletic trainers on the scene during this period had little formal education in health care. In colleges and universities, they often began as gymnasium jacks-of-all-trades with custodial responsibilities as well as other duties such as repairing equipment and facilities, laundering clothes, driving the team bus, and maintaining outdoor playing fields. They sometimes met their counterparts from other institutions at intercollegiate athletic events and shared training techniques (Smith 1979). Because elite sports were primarily for boys and men, most coaches and athletic trainers were men.

The Cramer Company, founded in 1922, identified a commercial niche and began selling liniment for sprains as well as other products for athletic training. In 1933, Cramer inaugurated the *First Aider,* which soon became a popular newsletter aimed at helping high school coaches understand training techniques. For many years Cramer also sponsored educational seminars for athletic trainers. The profession grew, but it wasn't until 1950 that a professional organization was founded—the National Athletic Trainers' Association.

The outbreak of World War I sparked interest in physical therapy. By the end of the war, U.S. military reconstruction aides—an entirely female corps of therapists trained to help the war wounded—were receiving training in massage, anatomy, remedial exercise, hydrotherapy, electrotherapy, bandaging, kinesiology, ethics, and the psychological effects of injuries. In 1921, the reconstruction aides formed the American Women's Physical Therapeutic Association, the organization that eventually became the present-day American Physical Therapy Association, the main professional association for physical therapists (Murphy 1995, pp. 54, 71).

Teaching physical education continued to be the main profession for which students were prepared in college physical education programs during the first half of the 20th century; bachelor's degree programs increased in number, and master's and doctoral programs came on the scene.

Scholarly Knowledge About Physical Activity

At the beginning of the 20th century, a number of scholars were investigating topics such as neuromuscular fatigue, the vascular effects of exercise, kinesiology (today called functional anatomy or biomechanics), body measurements and proportions, the psychological aspects of play, and the history of physical education and sport (Park 1981). Not until the late 1920s, however, did research on physical activity gain much visibility. Faculty members in physical education as well as other disciplines conducted investigations. The prestigious Harvard Fatigue Laboratory, focused on research in exercise physiology, opened in 1927. *Research Quarterly,* a scholarly journal devoted to physical activity, began in 1930. One issue in 1934 contained the following topics: "physiology of respiration, reflex/reaction time, measurement of motor ability, effects of temperature on muscular activity, and test construction" (Park 1980, p. 5). Because of the importance of teacher training in college and university physical education departments, physical educators often did research that could be applied to teaching. For example, studies of techniques for assessing motor skills could be used to help teachers develop ways to evaluate student progress.

▶ Research on physical activity started to expand in the late 1920s. Most physical educators studied topics relevant to teaching physical education; a few examined other aspects of physical activity.

A handful of physical educators began conducting research on aspects of physical activity not centrally applicable to teaching, including motor ability, motor capacity, physical fitness, and exercise physiology. Research on these topics continued to increase in the 1940s and 1950s, but it was not until the 1960s that such research mushroomed.

1950-2012: Electronic Communication and Globalization

In the second half of the 20th century, electronic communication expanded at a breathtaking pace. Television—in its infancy in 1950—became a common fixture in most homes in fewer than 2 decades. Digital computers developed from massive, room-sized arrays of vacuum tubes with limited capabilities to tiny but enormously powerful collections of memory chips and electronic displays. Communications satellites were placed in orbit, greatly simplifying intercontinental communication. Air travel became faster and easier. As in the past, each of these technological advances affected sport and exercise in numerous dimensions. Global trade expanded, accompanied by greater worldwide political and economic interdependence. New immigrants continued to arrive in the United States—especially from Asia and Latin America. Progress was made toward the American democratic ideal of equal opportunity and social justice: Federal civil rights legislation mandated greater equality in education, business, and housing, and professional and intercollegiate sport became more racially integrated during this period. Nevertheless, complete achievement of this goal remained elusive. The sometimes-frightening Cold War between communist and capitalist countries was a worldwide reality for more than 40 years. The Korean War (1950-1953) and the Vietnam War (1960s-1975) were hot, disruptive episodes in the Cold War. Communism unraveled after the fall of the Berlin Wall in 1989, leading to vast global changes in political and economic relations. In the early 1990s Iraq invaded Kuwait, and the United States and many allies responded by driving the Iraqi armed forces back and then imposing a decade-long air blockade of Iraq. Ethnic conflict between Muslims and Christians led to civil war in the Balkan countries of the former Yugoslavia, and the city of Sarajevo, site of the 1984 Winter Olympic Games, was extensively damaged. The September 11, 2001, attack on the World Trade Center in New York City led to a decade-long war in Afghanistan and Iraq. In 2011 and 2012, several middle eastern nations experienced revolts by ordinary citizens that overthrew long-standing dictators. Modern electronic communication via e-mail and cell phone texting played a major role in the organization of these citizen protest movements.

Participation in Physical Activity: 1950-2012

A dramatic increase in health-related exercise began in the United States in the early 1950s and continued for several decades. A number of events contributed to this trend. The Korean War again underscored the importance of physical fitness for military preparedness. The USSR's 1957 launch of Sputnik—the first earth satellite—highlighted the continuing threat of communism. A highly publicized 1953 report showed that more than 50% of American children failed a strength and flexibility test, compared with a failure rate of less than 10% by children from European countries. After President Eisenhower's heart attacks in the mid-1950s, people were well aware that his doctor prescribed exercise for recovery, a radical idea at the time. And in the early 1960s, President Kennedy continuously demonstrated an active lifestyle—touch football, tennis, swimming, sailing, horseback riding, badminton, and general exercise (Rader 1990, p. 242).

By one estimate, commercial health clubs increased from 350 in 1968 to more than 7,000 in 1986 (Rader 1990, p. 243). By another, the number of health clubs and physical fitness centers was about 24,000 in the late 1980s (Eitzen & Sage 1993, p. 390). Marathon races throughout the United States increased from fewer than 25 in 1970 to more than 250 in 1996. Between 1980 and 1996, road races covering shorter distances increased from 4,100 to more than 20,000 (Eitzen & Sage 1997, p. 316). Adult membership in commercial and nonprofit health, racket, and sport clubs reportedly grew by a whopping 51% between 1987 and 1996, increasing from about 13.8 million to about 20.8 million (Carey & Mullins 1997). By 2008, one industry survey reported a total of 29,636 fitness centers in the United States with 42.7 million club memberships (IHRSA 2008).

The proportion of children and adults who said that they did something daily to keep physically fit increased from 24% in 1961 to 59% in 1984. Conversely, adults who said that they had a sedentary lifestyle decreased from 41% in 1971 to 27% in 1985 (Blair et al. 1987; Ramlow et al. 1987; Stephens 1987). But recent surveys conducted by the U.S. Centers for Disease Control and Prevention using more precise criteria for vigorous physical exercise indicate that in 2004, more than 50% of American adults 18 years of age and older were not engaging in the recommended amount of activity. Moreover, the level of engagement in vigorous physical activity in leisure time was significantly less for females, ethnic minorities, and rural populations. In 2005, less than one-third of adults 18 years of age and older engaged in regular leisure-time physical activity. Adults in families with incomes above twice the poverty level were more likely to engage in regular leisure-time physical activity than adults in lower-income families (34% compared with 20%-22%; age-adjusted) (National Center for

■ How would a sport historian determine what led to the increase in participation in Little League Baseball?

Health Statistics 2007). Four years later, these figures were basically unchanged (National Center for Health Statistics 2011).

Americans' well-established fascination with sport grew more widespread—both for participants and for spectators—in the latter half of the 20th century. For example, between 1950 and 2003 the number of youngsters worldwide competing in Little League Baseball increased by more than 140 times, from 18,300 to approximately 2,994,000 (Little League Baseball, Inc. personal communication, April, 2004). However, from 2000 to 2009, the number of children and youth ages 7 to 17 years participating in baseball fell 25% (Futterman 2011). This decrease may be due, in part, to the increase in participation in other youth sport programs in football, soccer, and lacrosse. Youth tackle football participation increased 21% between 2000 and 2009; some "studies suggest more people now play soccer in the U.S. than baseball, and lacrosse participation has more than doubled" between 2000 and 2011 (Futterman 2011). According to US Lacrosse, participation of youths under age 15 increased from 125,000 in 2001 to 297,271 in 2009, a gain of 137.8% (US Lacrosse 2011).

Elite athletic competition for girls and women reappeared in high schools and colleges beginning in the late 1950s. In 1972, the U.S. Congress approved Title IX of the 1972 Education Amendments to the 1964 Civil Rights Act, requiring equal opportunity for males and females in educational programs—including athletic programs. Between 1971 and 2007, participation in athletic competition among high school girls increased almost 11-fold. Similar dramatic increases occurred in the numbers of female players in colleges and universities (National Federation of State High School Associations 2012).

In the 1950s African Americans trickled back into elite professional athletics, and by the 1960s this trickle grew to a fast-moving stream. African Americans constituted about 10% of the players in professional baseball, basketball, and football in the late 1950s. These proportions increased in the 1960s, and by 1994, 77% of professional basketball players were African American, along with 65% of professional football players and 18% of professional baseball players (Eitzen & Sage 1997, p. 264; Leonard 1998, p. 212). By 2006 the number of African American players in Major League Baseball had fallen to an all-time low of 8.4%, although the number of front office personnel, coaches, and managers had significantly increased (Lapchick 2007). That number increased to 10% in 2010 but fell back to 8.8% by

2012 (Lapchick 2012). There is a consensus among many that the attraction of football and basketball has been largely responsible for the decreasing numbers of African American youth playing baseball in recent years. To counteract this trend, Major League Baseball has begun programs designed to sponsor youth baseball programs in many of the inner cities of America. In 2012, Latinos constituted 27.3% of Major League Baseball players, and the number of Asian players on major league teams steadily increased during the first decade of the 21st century to 2.1% (Lapchick 2012).

The result of this growth in high-level sport competition was an enormous increase in levels of performance. Between 1896 and 1956, Olympic track and field performances (times, distances, heights) increased 130%. Between 1908 and 1948, swimming times improved an average of 2.2% every Olympiad. However, since the middle of the 20th century, the rate of improvement in measureable world records in most men's track and field, swimming, and weightlifting events has slowed considerably, and although records continue to be consistently broken, the rate of improvement is slight each time. In contrast, the rate of improvement in women's events in these same sports over the past 40 years has been very high, undoubtedly speaking to the increased access to training, facilities and high-level competition. A recent study by a group of Italian scientists suggests that because "the performances of athletes are the product of genetic endowment, hard work and increasingly, the contribution of science," the mere willingness to work harder will no longer be sufficient to break world records. They conclude: "Future limits to athletic performance will be determined less and less by the innate physiology of the athlete, and more and more by scientific and technological advances and by the still evolving judgment on where to draw the line between what is 'natural' and what is artificially enhanced" (Lippi et al. 2008, p. 14).

Enthusiasm for outdoor physical recreation skyrocketed during this period. Between the mid-1960s and the mid-1970s, the increase in interest was so rapid that it sometimes outstripped manufacturers' ability to supply people with proper clothes and equipment. Cross-country skiers numbered about 2,000 in 1964 compared with 500,000 in 1974. Backpacking at Kings Canyon and Sequoia National Parks in California increased 100% between 1968 and 1971. Visitor days at New Hampshire's White Mountain National Forest increased from 234,000 in 1968 to 482,000 in 1972. Bike path mileage in the United States went from almost none in 1965 to more than 15,000 miles (24,140 kilometers) in 1974 (Wilson 1977). By 2011, The Adventure Cycling Association's Routes & Mapping Department had linked together some "40,974 miles of rural roads to create low-traffic bike routes through some of the nation's most scenic and historically significant terrain." This TransAmerica Bicycle Trail project was begun in 1976 (Adventure Cycling Association 2012). By the first decade of the 21st century, participation continued at a high level, although there were signs of changing patterns of activity choices. In its 2010 Participation Report, the Outdoor Foundation of the Outdoor Industry Association found that although the economic recession forced many American families to decrease the amount of money spent on outdoor activities, more than half of those surveyed continued their outdoor recreation spending at or above previous levels. The five most popular areas of outdoor recreation activity for ages 6 through 24 were running, jogging, and trail running (20.9 million participants, 26.2%); biking (road biking, mountain biking, BMX) (17.5 million participants, 21.9%); camping (car, RV, backyard) (16.4 million participants, 20.6%); fishing (freshwater, saltwater, fly) (15 million participants, 18.8%); and hiking (10.3 million participants, 12.9%). The five most popular outdoor recreation activities for adults age 25 and older were in the same categories but in a slightly different order: fishing (freshwater, saltwater, fly) (31.2 million participants, 15.1%); running, jogging, trail running (30.6 million, 14.8%); camping (car, backyard, RV) (26.1 million, 12.7%); biking (road biking, mountain biking, BMX) (25.5 million, 12.4%); and hiking (24.2 million, 11.8%) (Outdoor Foundation 2012).

Americans also increased their spectator involvement. Attendance at Major League Baseball games was more than four times greater in 2006 than it was in 1950 as the number of teams increased from 16 to 30. Professional football attendance mushroomed to just under 12 times its earlier size during the same period although it dropped by about 5 million by 2010. Attendance at National Basketball Association regular season games rose from 1,296,973 in

1959-1960 to 21,302,573 in 2010-2011. These increases far outstripped a more modest U.S. population growth over the same period (U.S. Bureau of the Census 1966, 2012).

Although soccer is a preeminent spectator sport throughout most of the world, it didn't gain much of a foothold in the United States as an entertainment event until the 1990s. The final rounds of the 1994 men's World Cup were held on American soil, and in 1996 entrepreneurs developed a relatively stable men's professional league—Major League Soccer—with teams that held special allure for Latino spectators. In 1991, the United States won the first Women's World Cup with hardly anyone in the nation noticing. After losing to China in 1995, the Americans won the gold medal at the 1996 Olympic Games before the largest crowd to ever witness a soccer game on U.S. soil. The following year, this same team again won the World Cup before 100,000 spectators in the Rose Bowl and a worldwide television audience numbering in the millions. These successes led to the creation of the professional Women's United Soccer Association (WUSA), which was moderately successful but fell into financial exigencies following the 2003 season. Nevertheless, the league did much to promote women's sport, especially soccer, in North America. In 2009, a new league, Women's Professional Soccer (WPS), began play, fielding seven teams during the first two seasons and six teams for the 2011 season.

✔ ACTIVITY 6.6

Timelines
Match historical events with the time periods during which they occurred in this activity in your web study guide.

> The enthusiasm of Americans—both as direct participants and as spectators—for a widening array of sports and exercise mushroomed in the second half of the 20th century.

By the early 1960s, Roone Arledge, head of sport programming at ABC, was incorporating the highly innovative and popular instant replay technique into sport broadcasts. He unveiled additional novel production technologies—for example, close-ups of players and fans, multiple cameras, and directional microphones—in 1970 with the inauguration of ABC's *Monday Night Football* (Gorn & Goldstein 1993, p. 239). These innovations made televised sports much more exciting to watch and drew new spectators. The arrival of all-sports cable channels, beginning with ESPN in the late 1970s, wedded sport and television in a 24-hour, 7-day-a-week relationship. The commercial success of sports on television—primarily men's team sports—provided vast amounts of money for professional and big-time intercollegiate athletics.

The Internet took off with phenomenal growth in the 1990s. By 2007, anyone with a computer attached to high-speed Internet service had instantaneous access to worldwide video broadcasts of sport events as well as news and discussions about sport through electronic chat rooms and websites.

Physical Activity Professions

At midcentury, most American college students who majored in physical education—the name used almost universally at the time—became high school teachers, and some also coached. Things began to change in the mid-1960s with the development of kinesiology—the body of knowledge about physical activity that included numerous scholarly subdisciplines. It soon became apparent that many careers other than teaching and coaching were centered in physical activity. Let's look at a few examples.

With the growing American interest in exercise and physical fitness, it is little wonder that new health clubs, cardiopulmonary rehabilitation programs, worksite health promotion programs, and health maintenance organizations began offering physical activity programs. People with specialized training in exercise began working in a variety of jobs such as cardiopulmonary rehabilitation, personal training, aerobics instruction, and strength training. Some of them rapidly advanced to management and business ownership roles.

Attitudes of Adults Toward Exercise Have Changed

Fifty or 60 years ago, most Americans perceived regular participation in exercise and sport to be the exclusive province of young people engaged in high school, college, or professional sport. Today, large numbers of people pursue such activities throughout much of their lives. What historical events and social changes in our society have influenced this shift in attitude and practice over the past several decades? How have you seen this shift in attitude and practice within your own family or circle of acquaintances?

The National Athletic Trainers' Association (NATA) was founded in 1950 to support the increasing number of specialists who worked to prevent and treat athletic injuries. The membership was almost entirely men until 1972, when Title IX (requiring equal opportunity for males and females in educational settings) became law, encouraging women to enter the field in greater numbers. As of 2011 there were about 325 NATA-accredited degree programs—a mixture of master's and bachelor's programs—and more than 34,000 NATA-certified athletic trainers plus another 6,000 noncertified athletic trainers who began practicing prior to the establishment of a certification process and are allowed to continue practicing under a "grandfather" status (National Athletic Trainers' Association 2011).

At midcentury there was little by way of standard educational preparation for most careers in sport management and leadership. As you already know, the sport industry experienced enormous growth during the second half of the 20th century. As a result, there were more

Professionals such as this massage therapist help people lead healthier lives.

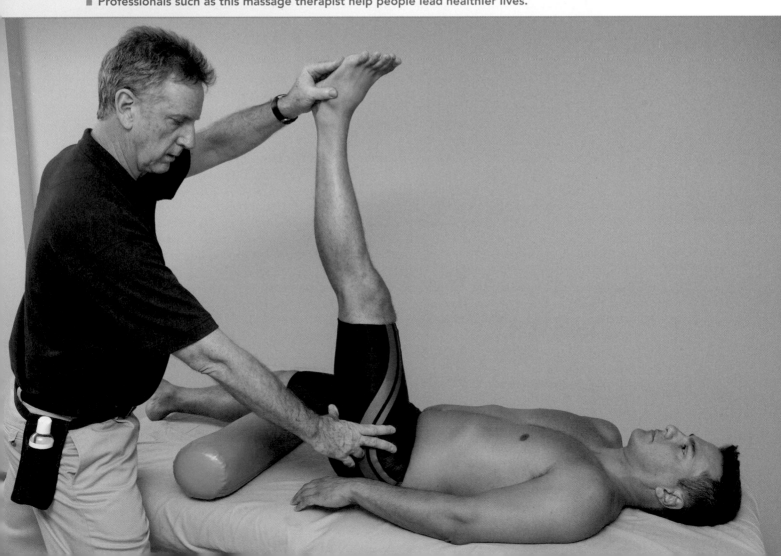

jobs in sport settings, and the variety of jobs was wider. Job opportunities were available, for example, with intercollegiate athletic departments, professional athletic teams, city stadiums, and participant-oriented nonprofit organizations such as Boys & Girls Clubs and YMCAs. Some jobs didn't even require a college degree (e.g., ticket seller in Minor League Baseball, YMCA program leader), whereas some required extensive high-level leadership experience or advanced degrees (e.g., big-time intercollegiate athletic director, city arena or stadium manager). This unevenness made it difficult to accredit degree programs or certify practitioners.

Nevertheless, in 1987 the National Association for Sport and Physical Education (affiliated with AAHPERD, the American Alliance for Health, Physical Education, Recreation and Dance) and the North American Society for Sport Management jointly published curriculum guidelines for bachelor's and master's degree programs in sport management. These were revised in 1993 and published as voluntary accreditation standards. By 2011, more than 300 colleges and universities offered undergraduate or graduate degrees in sport management (North American Society for Sport Management 2012).

▶ Most college students majoring in physical education in the middle of the 20th century entered the same profession—physical education teaching; by the end of the century, students had a wide array of physical activity careers to choose from.

Beginning in the 1960s, physical therapy continued to grow as a profession. The number of physical therapy jobs in the United States in 1960 was estimated to be 8,000; this grew almost 25-fold by 2011 to 198,600 (U.S. Bureau of Labor Statistics 1961, 2011). In the 1950s, students were trained in undergraduate physical therapy degree programs or in older, hospital-based certificate programs designed for people who had completed bachelor's degrees in other fields. By the late 20th century, hospital-based programs were phased out, undergraduate programs were in decline, and master's programs were on the rise. In 1981, accredited physical therapy curriculums included 100 undergraduate degree programs, seven certificate programs, and eight master's programs (U.S. Bureau of Labor Statistics 1982, p. 166). In 2009, there were 212 accredited physical therapy programs in the United States, including 12 offering the master's degree and 200 offering the doctorate (U.S. Bureau of Labor Statistics 2011). Since 2002, all physical therapy programs seeking accreditation are required to offer degrees at the master's level or above (American Physical Therapy Association 2012.). Increasingly, the doctor of physical therapy (DPT) is becoming the accepted entry-level degree.

Scholarly Knowledge About Physical Activity

Scholars developed new knowledge about physical activity at a feverish pace during the final 35 years of the 20th century, a trend that is continuing today. You will learn a lot more about this in the chapters of this text on the other subdisciplines—each has a section on history. Here we focus on the big picture.

▶ Beginning in the 1960s, the discipline of kinesiology grew rapidly and branched into numerous scholarly subdisciplines.

By the 1950s, researchers had already produced a lot of information about physical activity. Problematically, knowledge was scattered throughout many different research journals, and no one had taken the time to make it easily accessible. Warren Johnson, professor of health education and physical education at the University of Maryland, attempted in 1960 to remedy this by publishing a book titled *Science and Medicine of Exercise and Sports*. The book contained 36 essays summarizing current knowledge. Another early attempt at organizing research knowledge appeared in the May 1960 supplement to *Research Quarterly*, edited by several scholars headed by Raymond Weiss of New York University; each paper summarized research concerning a different aspect of physical activity (Karpovich et al. 1960).

Calls from politicians and educators in the early 1960s for better quality control in college and university teacher education programs resulted in scrutiny and criticism of physical education curriculums because they were mostly teacher education methodology programs at the time. In response to this, physical education faculty began charting new directions for their discipline. Franklin Henry, a physical education professor at the University of California and top-notch scholar in exercise physiology and motor learning, published a paper outlining his views on the nature of the discipline titled "Physical Education: An Academic Discipline" (1964).

Henry envisioned a body of knowledge about physical activity that would draw from "such diverse fields as anatomy, physics and physiology, cultural anthropology, history and sociology, as well as psychology" (p. 32). He also believed that the discipline should be cross-disciplinary. By that he meant that it should bind together knowledge from various disciplines to form a distinct, new body of knowledge concerning human movement.

Henry's vision of the discipline spurred rapid changes during the rest of the 1960s and 1970s. A nucleus of physical educators who were already heavily engaged in research began pushing to make scholarship more central in university physical education departments. They continued their own research and trained a new cadre of doctoral students who expected to be centrally involved in inquiry throughout their own careers. They communicated frequently with scholars in other disciplines—for example, physiology, medicine, psychology, history—who were also studying physical activity. This exchange led to rapid growth in the size and scope of the body of knowledge.

The knowledge also became more specialized, and this soon led to the formation of subdisciplines. By 1970, the earliest ones were clearly identified, and most of the others were on the drawing boards. You already know the names of the subdisciplines and how they fit together to form the scholarly source of knowledge in the discipline. Throughout the rest of the 20th century, scholars formed many new associations and worked with publishers to inaugurate a host of new research journals to support the growing subdisciplines. In the 1980s and 1990s, the amount and quality of research continued to increase, and the subdisciplines became stronger and more identifiable (Massengale & Swanson 1997; Park 1981, 1989).

Wrap-Up

History extends your "memory" with fascinating, evidence-based stories about how and why physical activities such as exercise and sport came to be shaped the way they are. Knowledge of the past gives you important, broad understanding about the present that you can use to make better-informed personal and professional decisions for the future. As demonstrated in this chapter, historical knowledge is used by kinesiology practitioners in a variety of ways across many specialties and professions. Since the 1970s, scholars have gone beyond earlier descriptive approaches to put more emphasis on analytical frameworks that help answer questions about why things happened as they did. So far, this subdiscipline has focused mostly on sport, but since the 1980s scholars have given greater attention to exercise.

If you are intrigued by what you have learned here, you might have fun venturing into sport history textbooks, taking an introductory course in sport history, looking through the book-length research studies mentioned in this chapter, and learning through the major journals such as *Journal of Sport History* and *International Journal of the History of Sport*. This interesting and provocative subdiscipline will give you a new outlook that will enrich your understanding of physical activities as you step into the future.

More Information on . . . History of Physical Activity

Organizations

American Historical Association (AHA)

Australian Society for Sports History (ASSH)

British Society of Sports History (BSSH)

International Society for the History of Physical Education and Sport (ISHPES)

National Association for Kinesiology in Higher Education (NAKHE)

North American Society for Sport History (NASSH)

Organization of American Historians (OAH)

Journals

American Historical Review (AHA)

Journal of American History (OAH)

Journal of Sport History (NASSH)

Olympika: The International Journal of Olympic Studies

Quest (National Association for Kinesiology in Higher Education)

Sport History Review

Sport in History (BSSH)

Sporting Traditions (ASSH)

 ACTIVITY 6.7

Use the key points review in your web study guide as a study aid for this chapter.

Review Questions

1. List and discuss the goals of history of physical activity.

2. List and discuss three ways in which a kinesiology practitioner might use a knowledge of physical activity history in her or his area of specialization.

3. Describe participation in physical activity in the United States during the following three periods:

   1840-1900

   1900-1950

   1950-2012

4. Describe professional practice centered in physical activity in the United States during the following three periods:

   1840-1900

   1900-1950

   1950-2012

5. Describe scholarly knowledge about physical activity during the following three periods:

   1840-1900

   1900-1950

   1950-2012

chapter seven

Sociology of Physical Activity

Katherine M. Jamieson and Margaret Carlisle Duncan

chapter objectives

In this chapter we will

- explain what a sociologist of physical activity does;

- identify the goals of sociology of physical activity;

- discuss how sociology of physical activity came into being;

- explain how research is conducted in sociology of physical activity; and

- examine what research tells us about inequitable power relations relevant to physical activity, especially gender relations, ethnic and racial relations, and socioeconomic relations.

You're having coffee with your roommate at the local bookstore, and a couple of people at the next table are having an intense conversation. You can't help overhearing, and you're interested, because they're talking about the beginner soccer program their kids are in. As you listen, you gather that many of the coaches in the program are parents who are volunteering their time and expertise. Your neighboring coffee drinkers are complaining to each other about how the parent-coaches seem to be very competitive and regimented in the ways they organize team sessions and interact with participants. You gather that the head coach in particular seems extremely competitive, having been overheard speaking ill of the less-skilled participants and calling the participants "losers" unless they work hard enough to beat all the other teams. You hear the complaints that their kids are not having fun and are constantly afraid of getting in trouble. Hearing these people, who are clearly frustrated, talk about what's going on in their soccer program makes you start to wonder about what ought to be the focus of such programs. Should the emphasis be on learning new skills, having fun, and making friends? Or should it be on winning? You start to think about the origin of the coaches' attitudes and beliefs. Is it appropriate for children to be taught to strive for excellence at such a young age? Is this how the win-at-all costs sporting mentality begins? You wonder what you would do if you were in the place of these people. Would you speak to the coaches or to the sport director at the community center? Would you withdraw your child from the program? Would another local soccer program be better, or would you encounter the same behaviors?

It would be difficult to think of physical activities that do not somehow involve interaction with other people. Try to imagine it—even if you exercise at home, how might you still be interacting with others? There are many kinds of social interaction, especially in this highly technologized world, where it is nearly impossible to escape the ideas, opinions, and behaviors of other people, many of whom you will never even meet in person. Without our varied opportunities for social interaction, our physical activities would be very different. Most of us would probably engage in basic movements such as walking, running, and jumping, but if in the process we never interacted with others—never watched them, never communicated with them, never played with them—each of us would probably develop a unique style of moving that would be unfamiliar to anyone else.

In such an absurd, asocial world, sport would be impossible; without communication we would have no way to decide on game rules. People also wouldn't have a mutually shared concept of exercise (i.e., physical activity done to improve human functioning or appearance), and they probably wouldn't be consciously aware that people's health can improve with moderate physical activity on a regular basis. Some individuals would probably do strenuous physical activities such as running fast or lifting heavy loads, and if they did these things often enough their health would probably improve. But other people would never learn about this idea. Exercise, sport, dance, and many other specific forms of physical activity have been brought into being, and are continuously being modified, through social interaction; without it, these activities wouldn't exist.

Why Use Sociology of Physical Activity?

Our social life is made up of mutually influential social practices (everyday behaviors) and shared beliefs; these are so commonplace that we rarely think about them. Whether we're playing soccer, coaching a basketball team, rehabilitating a broken leg, cheering for our favorite football team, or working out at a local gym, our social practices as athletes, coaches, exercisers, and spectators are influenced by commonly held beliefs in our society. And in turn, it is through our social practices that we come to have beliefs or understandings that we share with others. Yet few of us spend time analyzing the social arrangements—social practices and shared beliefs—that underlie physical activity. When we do look more closely, though, we gain insights about ourselves and our culture. It is the job of sociologists of physical activity to give this a closer look.

Goals of Sociology of Physical Activity in Kinesiology

- To look at physical activity with a penetrating gaze that goes beyond our common understanding of social life
- To identify and analyze patterns of change and stability in physical activity
- To critique physical activity programs in order to identify problems and recommend changes to enhance equality and human well-being

---

Although the subdiscipline of sociology of physical activity focuses primarily on sport, there has been an increasing interest in exercise and in the ways in which people in our society conceive of the human body. By exploring how we, as a culture, view our bodies, we may come to a renewed understanding of what it means to be human. Sociology of physical activity, then, is a set of intellectual tools that one may develop in order to look at human physical activity as more than physical—it is cultural as well. That is, a process of intellectually unpacking patterns, beliefs, values, and power relations will reveal human physical activity as

Sociology of physical activity focuses on the shared beliefs and social practices that constitute specific forms of physical activity (e.g., sport, exercise). Accordingly, sociological information adds to the breadth of knowledge of a well-educated kinesiologist.

a crucial organizing component of all social systems. These tools include an understanding of theoretical perspectives, knowledge about research methods, and the ability to identify and think through social problems as they relate to social spaces of human physical activity.

Additional course work in the humanities and social sciences will enhance your understanding of principles of sociology and, specifically, the sociology of physical activity. You don't have to know all these disciplines in depth, but the more information you have, the more insightful you will be about the social arrangements underlying physical activity. For example, if you know a little about the general nature of socialization—the process of learning about your own society and how to get along in it—you will find it easier to understand how societal members become *socialized* into dominant sports or into subcultures centered in activities such as surfing, triathlons, mountain climbing, skateboarding, rugby, and exercise related to social or political causes (e.g., Race for the Cure).

As you get to know more about socialization, however, you will likely notice details and nuances of physical activity involvement that might lead you to ask more complex questions. For example, rather than being content to think that socialization about participating in physical activity flows in only one direction—from parents to children—you might become interested in the extent to which youngsters contribute to socializing their parents into participating in physical activity.

Even your observations of professional practice have lent you some insight into sociology of physical activity. Perhaps you have noticed that there are few female athletic trainers for American football teams, that many people cannot afford to belong to a country club, that fitness professionals often promote achievement of a dominant body ideal, and that jobs in college and professional coaching remain largely held by white males. All these observations are ripe for sociological study, and sociologists of physical activity have already documented and analyzed many of them.

Sociology of physical activity joins philosophy and history of physical activity as a subject that deals with the sociocultural side of sport and exercise. Like philosophers of physical activity and historians of physical activity, sociologists of physical activity use their expertise to study the social life and patterns associated with physical activity. By examining physical activity from yet another standpoint, sociologists provide insights into our understanding of the human experience of being physically active. Figure 7.1 shows how sociology of activity fits into the broader context of physical activity studies.

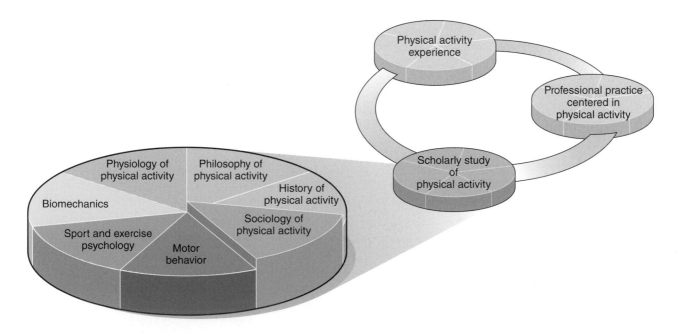

FIGURE 7.1 Sociology of physical activity and the other scholarly subdisciplines.

Let's look again at our opening scenario to see what advice a well-educated kinesiologist with knowledge of sociology of physical activity might give to the parent who questions the current competitive model as the best model for children. This advice would probably be different from that of a layperson without such training. Kinesiologists with a background in sociology are aware of the dominant competitive model of sport as well as the kinds of daily human interactions that make particular ways of organizing and engaging in physical activities more socially valued than others. Well-informed kinesiologists also recognize that there are varied social and cultural contexts for physical activity and that because of this, different forms of sport, exercise, and recreational activities may hold diverse values across time and cultural groups. Recognizing this diversity of social and cultural values applied to physical activities, organizers of many physical activity programs have proposed alternative models, especially for young children. In the opening scenario, a sociologically informed kinesiologist may apply this knowledge to advocate a less outcome-centered, more process-centered sporting experience for kids, one that promotes skill building, self-awareness, and responsibility for one's actions.

More specifically, the sociology of physical activity in kinesiology typically advances three main goals. The first is *to look at physical activity with a penetrating gaze that goes beyond our common understanding of social life.* This is a habit that must be developed, and acquiring it involves having a dash of skepticism about common assumptions. If you remember that the complex workings of social life are usually not readily observable, it will be easier to form the habit of looking further into your own physical activity experiences, knowledge, and professional practice. For example, on the surface it appears that sport and exercise are personal choices available to all societal members, but if you look more closely you'll find that a host of external social conditions influence the physical activity choices that people make.

To illustrate, in the United States, becoming an Olympic competitor in sports such as figure skating, gymnastics, or speed skating often requires a number of resources, including societal value for that particular sport; development of experts and training facilities in that sport; financial commitment of the athlete and family throughout training; a formal set of rules, guidelines, and benchmarks for elite performance in that sport; and enough international interest that the sport remains an Olympic event. Kinesiologists who lack a sociological understanding of physical activity might misinterpret trends in Olympic participation as indications of natural interest or willingness to work hard among particular groups (e.g., men more than women; whites more than other racial-ethnic groups). But a kinesiologist who is able to perceive what some sociologists refer to as "chains of interdependence" will be much better able to respond to the physical activity needs of an ever-changing, diverse societal membership.

Building on this sociological lens is the second goal of the subdiscipline, *to identify and analyze patterns of change and stability in physical activity.* Once you start looking beneath the surface of social life, your goal is to identify changes as well as ongoing regularities. In keeping with our example of Olympic involvement, sociological analyses indicate that the Olympic Games continue to reflect the sporting traditions of the most developed nations. Despite Pierre de Coubertin's desire to revive a modern Olympic movement that would truly be about cross-cultural exchange, the Games continue to privilege large, economically advanced countries in sporting displays of nationalist propaganda. Yet change has occurred: More women and more countries currently participate in the Olympic Games than in earlier times (see table 7.1); new sports have been added to the Games; professional athletes are now allowed to compete in particular sports; and environmental issues are often part of planning discussions for Olympic venues. Whether you agree with the current state of affairs in the Olympic Games is not the issue for sociologists; rather, it is crucial to be able to identify shifting trends and stabilizing forces in the physical activity experiences of societal members.

✔ ACTIVITY 7.1

Goals

In this activity in your web study guide, you'll review three scenarios and identify which goal each represents.

TABLE 7.1 Number of Women's Events and Female Athletes in Olympic Games

Year	OLYMPIC SUMMER GAMES				OLYMPIC WINTER GAMES			
	No. of women's events/total	% women's events	No. of female athletes	% female athletes	No. of women's events/total	% women's events	No. of female athletes	% female athletes
1980	50/203	24.63	1115	21.5	14/38	36.84	232	21.7
1984	62/221	28.05	1566	23	15/39	38.46	274	21.5
1988	86*/237	36.28	2194	26.1	18/46	39.13	301	21.2
1992	98*/257	38.13	2704	28.8	25/57	43.86	488	27.1
1994					27/61	44.26	522	30
1996	108*/271	39.85	3512	34.0				
1998					31/68	45.58	787	36.2
2000	132*/300	44	4069	38.2				
2002					37/78	47.4	886	36.9
2004	135*/301	44.9	4329	40.7				
2006					40/84	47.6	960	38.2
2008	137*/302	45.4	4637	42.4				
2010					43/89	48.3	1044	40.7
2012	141*/302	46.7	4862	44.4				

*Including mixed events.

Data from International Olympic Committee, 2011, *Women in the Olympic movement.*

Given that you have learned to see physical activity in new ways and are able to identify patterns of change and stability, you are now ready to engage the third goal of sociology of physical activity: *to critique physical activity programs in order to identify problems and recommend changes to enhance equality and human well-being.* As you have probably already guessed, sociologists point to the constant stream of social interaction that goes on in any society as the most important source of societal problems, although they don't deny that psychological and biological factors have important influences in our lives. Sociologists study social interaction and, more important, they evaluate it—pass judgment on its contributions to overall human well-being and equitable social relationships (e.g., relationships among racial and ethnic groups, men and women, or people at different socioeconomic levels). Returning to our Olympic Games example, sociologists have critiqued the Games as an exclusionary physical activity space that privileges societal members who happen to be from the best-resourced countries, teams, and families.

Additionally, as already mentioned, the Olympic Games typically require host cities to use public resources, land space, and human resources to build the facilities necessary to host a sporting event that will last only 2 weeks. Some suggest that hosting such an event will create a local interest in sport and exercise, thereby increasing the physical activity of societal members. Sadly, there has been no evidence of this, yet there has been evidence that watching the hypermediated (i.e., multiple forms of ongoing media attention), flashy, entertainment-oriented Olympic Games may lead to increased television viewing even after

Sociology of physical activity in kinesiology looks beneath the surface of social life to see things from different angles; kinesiologists with this knowledge are able to point out social problems in an effort to encourage people to make changes that may enhance human well-being and promote equitable social relationships.

■ Looking at this photo as a sport sociologist, what assertions might you make about the value, role, and access to physical activity in this social space?

the Games have ended. Clearly, for kinesiologists with a sociological understanding, these consequences of the Olympic Games would be considered problematic and would move them to modify the Games toward more beneficial outcomes for all societal members, not only those who compete in or profit from the Games.

In keeping with the third goal, sociologists of physical activity are not oriented toward supporting or reinforcing status quo physical activities such as present exercise programs and competitive sports. For example, a former Olympic athlete-turned-coach may start an Olympic training center where there previously was none; a biomechanist might design an exercise machine that meshes with the human body to improve elite performance. Both of these are examples of working within the status quo—working to improve performance in activities as they are currently structured. A sociologist might advocate changing the structure itself by questioning the desirability of fierce competition between nations or treatment of the human body like a piece of machinery that needs to be meshed with other pieces of machinery.

What Do Sociologists of Physical Activity Do?

Sociologists of physical activity are usually faculty members in colleges or universities whose teaching, research, and service advance a scholarly and practical understanding of the complex social and cultural contexts in which societal members experience physical activity. They may teach an introductory course on sociology of physical activity and offer courses on specialized topics such as the Olympic Games; interscholastic and intercollegiate athletics; or the ties between physical activity and phenomena such as gender relations, violence and aggression, or the mass media (e.g., television, magazines, the Internet).

Sometimes sociologists of physical activity combine teaching with professional service by encouraging their students to assist in physical activity programs located in nearby neighborhoods. This helps local communities and fosters a sense of civic responsibility in future kinesiology professionals. Faculty service activities might also include consulting with a community group about the role of physical activity in agendas for community-wide change (e.g., immigrant settlement issues and cultural forms of physical activity; city planning and physical activity space, access, and use priorities). Given these examples of scholarly engagement, it should be clear that sociologists of physical activity cannot carry out their research in a typical laboratory. Even if it were possible to bring a "slice of life" (e.g., fans watching a National Football League [NFL] football game, an aerobic exercise class) into a laboratory for study, in most cases such analyses would not be useful. An unrealistic result would be obtained because when you manipulate such an event, you alter the very thing that you're trying to study: You would be studying an aerobic exercise class in a laboratory rather than an aerobic exercise class in ordinary life. Sociologists usually prefer to study their subjects in the field—for example, professional golfers on the Ladies Professional Golf Association tour, a television production crew covering a downhill ski race, the political processes that underlie the preparation of elite Canadian athletes for international competition, or preferred physical activity practices among immigrant groups.

As part of this desire to understand everyday life and physical activity engagement, sociologists of physical activity spend a lot of time making contacts and gaining the confidence of the people and communities with whom they wish to collaborate for the purpose of developing new knowledge in sociology of physical activity. Regardless of the field setting, the quality of such relationships has an important bearing on the quality of the research. As such, sociologists of physical activity seldom rely on a physical laboratory space but rather see all sport settings in the social world as their laboratory.

History of Sociology of Physical Activity

▶ Since the 1970s, sociology of physical activity has expanded as scholars have explored a wider variety of topics, especially those concerning sport and socioeconomic, gender, racial, and ethnic inequalities; globalization and regional and national differences; exercise and societal conceptions of the human body; disability and ability; and obesity.

The subdiscipline is relatively young. It began to take shape in North America in the mid-1960s with the notable influence of English physical educator Peter McIntosh's 1963 book *Sport in Society*. His analysis of the social significance of sport encouraged several North American scholars to channel their careers toward this emerging area (Sage 1997). Subsequently, in 1964, the International Committee for Sport Sociology was founded, and in 1966 it inaugurated a scholarly journal, *International Review of Sport Sociology,* which was joined in 1977 by *Journal of Sport & Social Issues.* In 1978, the North American Society for the Sociology of Sport was formed under the leadership of physical educators Susan Greendorfer and Andrew Yiannakis. The society held its first annual conference in 1980 and began publishing an academic journal, *Sociology of Sport Journal,* in 1984. Both the society and the journal continue to be crucial, active scholarly spaces.

Since the mid-1970s the number of scholars in sociology of physical activity has increased, as has the range of theories and research methods used by these scholars. The hottest topics have focused on social inequities—especially those connected with gender, race, ethnicity, wealth, sexual orientation, and different cultures around the world. Research interests also include globalization and regional and national differences; exercise and societal conceptions of the human body; disability and ability; and obesity.

Research Methods for Sociology of Physical Activity

Now that you know something about the subdiscipline and its historical roots, you may be wondering how sociologists of physical activity produce their knowledge. What methods do researchers use to answer important questions about the social side of physical activity? They gather both quantitative data (e.g., things you can count or measure in a standardized and often mathematical fashion) and qualitative data (e.g., texts and direct observations of social life that are measured in a deeply contextualized manner). We'll look at six methods: survey research, interviewing, thematic analysis, ethnography, societal analysis, and historical analysis. Of course, sometimes two or more of these are used in the same study. Some studies even use both quantitative and qualitative methods (**mixed methods** research).

▶ Research methods used in sociology of physical activity include survey research, interviewing, thematic analysis, ethnography, societal analysis, and historical analysis.

Survey Research

Doing survey research involves using questionnaires that are completed directly by respondents or filled out by a researcher during brief, highly structured interviews. Questionnaires are used to collect data from a large sample of people. The largest survey project in the country is the U.S. census, conducted every 10 years. Political polls about voting preferences are also surveys. Investigators have conducted surveys dealing with numerous topics related to physical activity, including youngsters' opinions about what led them to become involved in sport, former collegiate athletes' thoughts about leaving competition at the end of their 4 years of eligibility, and college athletic directors' opinions about the characteristics of successful coaches.

▌ Sport sociologists investigate why some activities seem to cater more toward females than males, and vice versa.

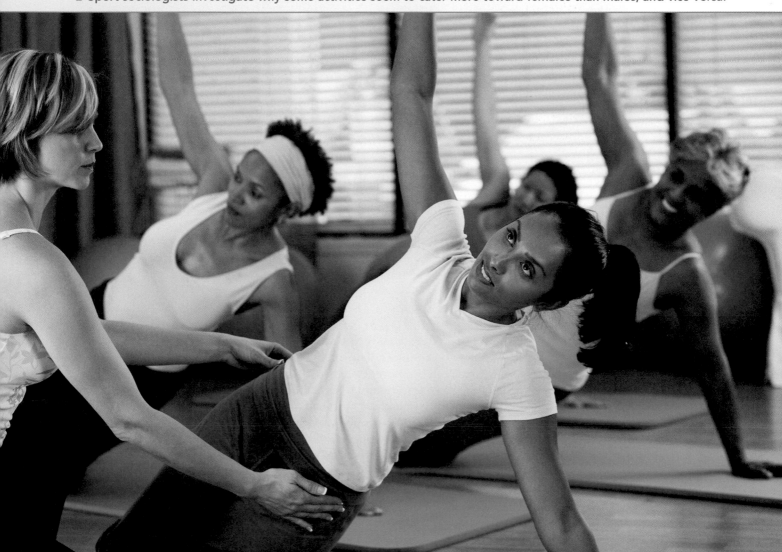

Interviewing

Researchers use interviews when they want broader and deeper information than they can get through a questionnaire or when they want information about activities that would be difficult or impossible to observe themselves. Because interviews are time-consuming, studies often focus on a small number of people. Interviews may be conducted one-on-one or in a group. One form of group interviewing that is increasingly being used in our field is **focus group** research. This method is particularly valuable when the researcher wants to gain insights into people's shared understandings and to see how individuals are influenced by others in a group situation (Gibbs 1997). Typically, interviews are audio recorded, and less often they are also video recorded and later transcribed. Researchers often ask relatively open-ended questions and then probe to get details. One investigator asked top high school, college, and professional athletes to discuss their initial involvement in sport, their participation over the years, and their disengagement from sport. Another asked young people about the characteristics of their athletic heroes.

Thematic Analysis

Researchers use thematic analysis, sometimes called content analysis or textual analysis, to investigate cultural material such as magazine and newspaper articles, photos, the verbal and visual content of television programs, interview data, sporting events, and sport celebrities (Birrell & McDonald 2000). This procedure involves examining the material and then categorizing the content in various ways. For example, magazine photos of female and male Olympic athletes were analyzed using the categories of physical appearance, poses, position of the body, emotional displays, camera angles, and groupings of people. A researcher often ends up with several main categories or themes along with a set of subthemes within each. Sometimes investigators count the number of times each theme and subtheme occur. In other cases the researchers are more interested in qualitative data, such as the richness and complexity of the themes. Identifying themes helps the researcher organize a large mass of data into a manageable number of categories that can be analyzed further in relation to theories of interest.

Ethnography

Researchers using ethnography spend many months or even years observing in a particular social setting. They "hang around" while ordinary day-to-day events take place, often taking part themselves, talking with people about what's happening, and keeping careful field notes so that they can remember details. Although observation is their primary source of information, they may also look at local documents or use interviews and questionnaires. Researchers analyze most of the data that they collect using thematic analyses. Ethnography has been used to study a variety of sport settings, including minor league ice hockey, baseball in the Dominican Republic, women's softball, the Olympic Games, women's professional golf, rodeo, and boys' Little League Baseball.

Societal Analysis

In societal analysis, the researcher's goal is to examine the sweep of social life, usually from the perspective of a broad social theory. This method, of course, isn't the only one tied to theory—all good social research is theoretically based. The theories used in societal analysis, however, are extremely broad—they attempt to explain the most fundamental ways in which societies operate. Examples of these include Marxism (and its many derivatives), modernization, structural-functionalism, figuration theory, various strands of feminism, various cultural studies frameworks, and various forms of postmodernism. These theories are extremely complex; massive collections of scholarly papers and books support, refute, or expand on each

ACTIVITY 7.2
Which Method?
Identify examples of each of the six research methods in this activity in your web study guide.

ACTIVITY 7.3
Practical Plug-In
In this activity in your web study guide, you'll consider what a community youth sport specialist needs to know and do in order to design a successful program.

of them. Illustrations of societal analysis in sociology of physical activity include a study of modernization that focuses on sport in preindustrial and postindustrial societies (Guttmann 1978) and an investigation, focused on social class, of societal constraints and human freedoms in sport (Gruneau 1983).

Historical Analysis

Social scientists interested in large-scale social change frequently incorporate historical research into their work. Historical analyses often are part of a larger study involving theoretical explorations of the kind just noted. Examples of historical research by sociologists of physical activity include an examination of the ways in which development of sport in Canada was related to socioeconomic inequities (Gruneau 1983) and a study of unruliness among soccer spectators in England from the 1890s to the 1980s (Dunning et al. 1988).

> Research in sociology of physical activity involves collecting quantitative and qualitative data using a variety of different methods.

Overview of Knowledge in Sociology of Physical Activity

This section is meant to be a snapshot of the knowledge produced by scholars in the subdiscipline and as such cannot do justice to the richness and variety of the knowledge that scholars have produced. In fact, the body of knowledge is far-reaching and especially complex because it concerns many different societies throughout the world, and social life is not the same everywhere.

Sociologists of physical activity often look at these areas from the standpoint of power relationships or the ways in which external, dominant cultural beliefs about social life, morals, valued behaviors, and so on influence everyday choices for physical activity. Power relations are difficult to see in our everyday lives, but once you learn to recognize them, you may begin to see them everywhere. You'll be able to look more carefully, with a more penetrating gaze, at various fascinating social aspects of physical activity.

For sociologists, **power** is the ability to do what you want without being stopped by others. Money, prestige, body size and strength, and information are major sources of power, hence major sources of inequality. If you look closely at kinesiology contexts—gyms, playgrounds, athletic fields, hiking or biking trails—you will see that the power typically exhibited is not the overt, bullying kind; rather, the forms of power that have the most influence on physical activity involvement are subtle, built-in, and almost hidden. In fact, the most current understandings of power reflect the presence of systems of power rather than focus on power as something individuals either have or don't have. Thus, a dominant power structure that features particular forms of masculinity and femininity exists in the physical education classroom, the exercise facility, and the sports field long before any of us ever step foot in those social spaces. For example, there are no signs saying that the cardio room and group exercise classes are for women, but in most fitness facilities, it is women who populate these spaces. When a male enters these spaces, he may be teased by some or may be celebrated by others who are happy that he broke the gender barrier. Both of these responses reflect a somewhat subtle cultural knowledge that we all share to some degree. This "knowingness," as it is referred to by some scholars, is a form of power in society because it reinforces the way societal members think things *ought to be* organized (e.g., males on the free weights and females in the cardio room).

With specific reference to physical activity, power inequalities sometimes affect our participation as performers as well as our opportunities for involvement in influential leadership roles such as health club owner, coach, athletic director, or cardiac rehabilitation program director. People with more access to power can usually get involved more easily. Those with less access to power often face social conditions that make involvement more difficult. For example, African American players in the NFL have a more difficult time getting jobs in coaching and administrative positions in the league than do whites. Furthermore, movies, television, music, and physical activities frequently display—and thus reinforce—current power differences in American society.

Whether we participate directly in activities such as sport, dance, and exercise or watch others performing them, the activities express subtle information that tends to support and reinforce status quo power differences. For example, telecasts that highlight the physical appearance of female athletes—with special focus on their clothes and hairstyles—contribute to undermining, and thus trivializing, their athletic accomplishments.

An understanding of power in society is crucial for kinesiologists as they identify the best ways to promote meaningful physical activity for all societal members. Jennifer Hargreaves (2001) suggests that "stories of sport are almost exclusively stories of those in power" (p. 8). Thus, kinesiologists might ask, what are the hierarchies of bodies that emerge in physical education classes, rehabilitation settings, or school-based sport programs? What are the social structures that reinforce societal privilege of particular bodies (e.g., laws, policies, school curricula, expert knowledge)? And, most important, how do physical educators, athletic trainers, fitness professionals, and coaches contribute to (or resist) these processes of bodily privilege? In the remainder of this overview, we relate sociological analyses directly to the ability of kinesiologists to promote quality physical activity, including sport and exercise, for all societal members. Power relations underlie social inequalities; they affect people's quality of life, their chances for a better life in the future, and their opportunities for engagement in physical activity. Accordingly, sociologists produce research that examines the influence of systems of power (e.g., gender, race, social class) on the everyday experiences of societal members. We focus on three areas within these systems of power: participation, leadership, and cultural expression.

Gender Relations

Gender is different from sex. As a system of power, gender is not an identity you have (e.g., male or female) but rather a set of norms or expectations about how we should behave, and these are linked to societal understandings of sexuality and procreation. At birth, babies are assigned to a sex category based on the appearance of the genitalia (Lorber 1994). This assignment process, and the subsequent differential treatment of youngsters based on different sex categories, constitute the gendering process. Genders are not natural, biological categories—they are socially defined. You can't inherit a gender—you have to be assigned to it and learn it (Fausto-Sterling 2000).

In American society, men typically have more power in a gender hierarchy than do women, although this relationship changes when race, social class, and sexual orientation are considered (e.g., white, middle-class women often exercise more societal power than will a Mexican working-class male). The situation changed somewhat during the 20th century, but many inequalities remain. In the field of physical activity, inequalities prevail because of beliefs about the appropriateness of certain forms of physical activity for recognized gender groups.

Participation

In the United States, many girls play in organized youth sports, and many of their older sisters play on high school and college teams. This occurrence is relatively recent. The passage of Title IX in 1972, requiring that women be provided with equitable opportunities to participate in sport, is perhaps the single most significant factor in accounting for the huge increase in girls' and women's participation in sport over the last 3 decades. In 1971, only 1 of every 27 high school girls played varsity sports. Since Title IX was enacted in 1972, female high school athletic participation has increased immensely, with 1 of every 2.4 high school girls participating in varsity sports. The average number of women's collegiate teams per school increased from 2.5 in 1970 to 8.65 teams per school, for a total of 9,101 collegiate teams for women in 2008 (Acosta & Carpenter 2008; Women's Sports Foundation 2009). In 2010-2011, the NCAA reported an increase in overall participation rates including women, accounting for 43% of collegiate athletes or 191,131 of the total 444,077 athletes (Brown 2011). In the 2006-2007 school year, girls' participation in high school athletics exceeded 3 million for the first time, and in 2010-2011 this number rose to 3,173,549 along with an increase in boys' high school sports participation rates, according to a survey by the National Federation of State High School Associations (2012).

In North American society, sports are deemed important avenues for exploring and confirming masculinity. They are thought to sharpen a number of qualities traditionally considered appropriate for men and boys, such as toughness, aggressiveness, roughness, working well under pressure, and competitiveness. You probably know girls and women with some of these qualities, a point that shows the indistinct boundaries between the two genders. Nevertheless, in our society people still conceptualize these qualities as masculine.

Although many sports attract large numbers of both males and females, some are more gender specific (table 7.2). For example, girls and women are less visible in weightlifting and cycling, whereas boys and men demonstrate disproportionately lower rates of participation

▶ Although many more girls and women are participating in physical activities today compared with several decades ago, they tend to participate in sports that are considered socially appropriate, involving less body contact, prominent aesthetic dimensions, and less extreme strength development.

TABLE 7.2 American Participation in Sports and Physical Activity by Gender

Sports and physical activities	Total	Male	Female
Exercising walking	95,803	38.7%	61.3%
Exercising with equipment	55,286	46.8%	53.2%
Swimming	51,943	47.4%	52.6%
Bicycle riding	39,789	55.6%	44.4%
Bowling	38,980	52.9%	47.1%
Aerobic exercising	38,541	30.3%	69.7%
Hiking	37,704	51.0%	49.0%
Working out at clubs	36,278	48.2%	51.8%
Running/jogging	35,524	52.5%	47.5%
Weightlifting	31,480	65.5%	34.5%

Top 10 activities listed in thousands (95,803 = 95,803,000).

Data from NSGA, 2010, *Sports participation in 2009* (Mt. Prospect, IL: National Sporting Goods Association).

in exercise walking and aerobic exercise. Yet, as shown in table 7.2, male and female rates of participation have a narrow margin of difference in a significant number of activities (e.g., hiking and working out at a club). If a woman or girl you know started training to be a football player, what would you think? Many Americans would judge this negatively. But what makes it inappropriate? Although many females are now developing muscular strength and endurance through weight training, few take part in sports requiring great strength, physical domination of an opponent through bodily contact, or the manipulation of heavy objects. Furthermore, only in recent decades have women taken up competition in strenuous, long-distance runs, including marathons. An early, cutting-edge analysis showed that the sports we judge more appropriate for females, such as gymnastics or tennis, entail little body contact, the manipulation of light objects, and aesthetically pleasing movements (Metheny 1965).

The continued emphasis on masculinity in sport often means that women's competition is downplayed. Examples include fewer resources for female sports (e.g., less travel money, lower coaching salaries, fewer publicity funds, less media exposure, fewer corporate sponsorships) and sometimes team nicknames that reflect physical ineptitude, such as the Teddy Bears, Blue Chicks, Cotton Blossoms, or Wild Kittens (Coakley 2007, pp. 243-249; Eitzen 2006, pp. 40-43).

Leadership

Our knowledge of gender inequities in physical activity leadership pertains primarily to sport, but let's briefly examine leadership in other physical activities. We know that opportunities for American women in coaching, sport telecasting, and officiating are increasing but are still more limited than they are for men (Acosta & Carpenter 1994, 2008; Coakley 2009). Moreover, from the 1970s into the 1990s, sharp decreases occurred in the percentage of female coaches of female high school and intercollegiate teams and in the proportion of female administrators heading women's intercollegiate programs. For example, in 1977-1978, 79% of the coaches of female basketball teams were women; by 1991-1992 that number had dropped to 64% (Acosta & Carpenter 1994). More generally, the proportion of female collegiate teams led by a female head coach decreased from 90% in 1972 to 43% in 2008. The percentage of female administrators heading women's intercollegiate programs went from 90% in 1972 to a mere 21% in 2008; however, as low as that number was, it did mark the first year since 1972 that the number of women in administrative positions increased, marking the end of a downward trend that began when women's athletics shifted from the AIAW to NCAA structure (Acosta & Carpenter 2008). Also in 2008, 27% of head athletic trainers and 11.3% of head sports information directors were female (Acosta & Carpenter 2008).

> Women occupy a relatively small proportion of coaching and leadership positions in sport.

Why are more men now coaching and administering women's teams and programs? Researchers have postulated a number of reasons. To begin with, the larger number of male players constitutes a larger pool of qualified candidates. In addition, the fact that women's coaching salaries still lag behind those of men may cause some of the top female players to seek careers outside athletics. In addition, men occupy most of the influential athletic leadership positions (e.g., collegiate athletic directors, leaders in the Olympic movement). When they make decisions about hiring coaches or appointing people to important committees, they likely use their network of connections with other men in athletics, usually referred to as the "old-boys network," to find candidates. Finally, men may favor applicants with stereotypical masculine qualities (e.g., those who are aggressive, dominating, physically tough),

ACTIVITY 7.4

Interview with a Practicing Professional
In this interview in your web study guide, you'll hear from Keri Flynn. Keri's educational background is in sociology and psychology, in addition to kinesiology, and she discusses the ways in which her background and interests inform her work as an athletic trainer.

a preference that puts many well-qualified women at a disadvantage (Coakley 2009). More recently, issues such as homophobia in collegiate athletics and incompatibility of parenting roles with coaching roles for both men and women, but especially women, have gained attention as possible barriers to women's involvement in collegiate coaching.

Cultural Expression

Given that gender inequalities exist in American society and that all physical activities have expressive dimensions, it's not surprising that our physical activities frequently demonstrate such inequalities. For example, in some forms of dancing (e.g., the waltz, many country-western dances) men usually move forward and lead women, who usually move backward. This technique communicates a subtle message about men's superiority.

Our most popular sports, the ones we see constantly on television, are celebrations of heterosexual manhood centered on aggressiveness, roughness, and the ability to dominate physically. Girls and women have most often been relegated to marginal positions as spectators of boys' and men's competition or as players in their own less popular versions of these and other sports. One of the messages communicated by this is that females don't matter much in "real sports." Coaches sometimes criticize weak or ineffective plays by boys and men by calling them "sissies" (Coakley 2009, p. 261). Intended to spur better performance, such comments send a message that females are inferior. Coaches communicate a similar message when they tell players that they "throw like a girl." These examples show ways in which sports tend to reinforce ideas about the acceptability of heterosexual male dominance.

> Many sports serve as vehicles for exploring, celebrating, and giving privilege to masculinity; because of this, they express ideas that are problematic for girls and women as well as for boys and men who are not athletically inclined.

Girls and women are not the only ones shortchanged by the gendered attitudes that prevail in sports. Our society often questions boys and young men who aren't interested in athletics or lack ability. They are called "sissies," and their sexual orientation may be challenged with taunts of "gay," "fag," or "queer" (Coakley 2009, p. 268). In high school, boys' popularity is closely connected with being good in athletics. A group of white college graduates who had not played varsity sports in either high school or college said that as boys they experienced social ostracism and name-calling because of their poor athletic ability. One said, "I identified sports as a major aspect of what I was supposed to be like as a male, which oppressed me because I could not do it, no matter how hard I tried" (Stein & Hoffman 1978, p. 148).

Have you ever thought about the degree of coverage of men's and women's sports in the media (e.g., television, newspapers, magazines)? About 90% of sport on television focuses on men's athletics, whereas women's competition constitutes less than 10% of coverage (Coakley 2009). In U.S. newspapers, the space devoted to women's athletics remains less than 15% of the total for all sports (Coakley 2009). This disparity trivializes women's athletic accomplishments and sends an underlying message that women's events aren't worth much.

Masculinity and Femininity Examined

Think about situations you have encountered in physical activities in which questions occurred about a male's masculinity or a female's femininity. What do these examples show us about our shared beliefs concerning appropriate activities for boys and men and for girls and women? What do they show us about gender inequalities in these activities? What do they show us about our society's definitions of masculinity and femininity?

Ethnic and Racial Relations

Race is a historically, culturally, and socially defined category of social difference typically marked by phenotypical variance among people (e.g., skin color, body shape, facial features, hair type). It is crucial to understand that race is socially defined based on the characteristics we select; it is not a natural, biological category. Like gender, race is not a fixed identity that you have but rather a shifting condition of social life that you experience every day. We live in a society that is organized around arbitrary beliefs about gender, race, and social class; thus, kinesiologists must understand how to work through these social conditions and inequalities to make meaningful physical activity available to all societal members.

Ethnicity refers to cultural heritage. People who share important and distinct cultural traditions, often developed over many generations, are classified as an ethnic group. Ethnic markers include such things as language, dialect, religion, music, art, dance, games and sports, and style of dress. Latinos and Latinas, persons of Jewish descent, Inuits, Amish, and Cajuns are examples of ethnic groups in the United States. Sometimes race and ethnicity overlap. For example, African Americans are defined as a racial group, yet they also have distinct cultural traditions. In addition, the African American population itself includes enormous ethnic diversity (e.g., distinct differences from African tribal cultures, heritages from the various African nation-states, influences from different Caribbean cultures, and a variety of European colonial influences).

In the United States, members of racial and ethnic minorities have typically held less power than the white majority. Although society made strides toward equality during the 20th century, many difficulties remain. Physical activity, of course, is not immune to such problems. Our focus here is mostly on sport, and we highlight comparisons between African American and white males because these groups have been most fully researched.

Participation

In elite team sports and in track and field, African American male athletes are overrepresented in relation to their proportion in the general population. For example, in 2010, African Americans made up about 13% of the U.S. population, whereas their numbers were much higher in major league professional sports: 78% in basketball, 67% in football, and 26% in soccer. Baseball and hockey are major league sports in which African American males are underrepresented (8.5% and <1%, respectively) (Lapchick 2011b). The high participation rates probably relate most to American social structure (Eitzen & Sage 2009, pp. 305-312). Whether they are local playground heroes or major league players, African American male athletes receive much public applause from African American men and women alike. Some of the most visible and popular African American role models are top male athletes, and their status may give young African American boys the incentive to spend long hours practicing their sport skills.

But this doesn't account for African Americans' high participation rates in a few sports and their low involvement in most others. Opportunities, and lack of opportunities, probably influence this situation (Eitzen & Sage 2009). African Americans tend to be more visible in sports with easy access to facilities and coaching, primarily in school and community recreation programs. Sports requiring private coaching, expensive equipment, empty land for playing fields, or facilities at private clubs—for example, tennis, golf, swimming, soccer, or gymnastics—are less popular. The extent to which these various factors influence African American athletic involvement remains unclear. The success of African American athletes in historically segregated sports (e.g., Tiger Woods in golf, Venus and Serena Williams in tennis, and Bill Lester in NASCAR) raises new questions about access, identities, and underlying reasons for sport involvement. As well, U.S. higher education is expected to be equitably accessible to all social groups, but a quick review of NCAA Division I representation across leadership and athlete participation reveals disproportionate involvement across race and gender. For example, although African American males account for nearly 25% of male Division I athletes, they make up less than 7% of both athletic directors and head coaches. In comparison, white male student athletes are overrepresented in athletic director and head coaching roles (figure 7.2).

Recreational physical activities in which competition is not essential, as well as exercise programs geared toward improving health and appearance, are probably subject to some of the same factors. For example, surfing, backpacking, exercising at a health club, swimming, and skiing require various combinations of expensive equipment, facilities, lessons, and

Large numbers of African American athletes play major team sports and participate in track and field, but representation remains disproportionately low in most other sports and leadership roles in U.S. society.

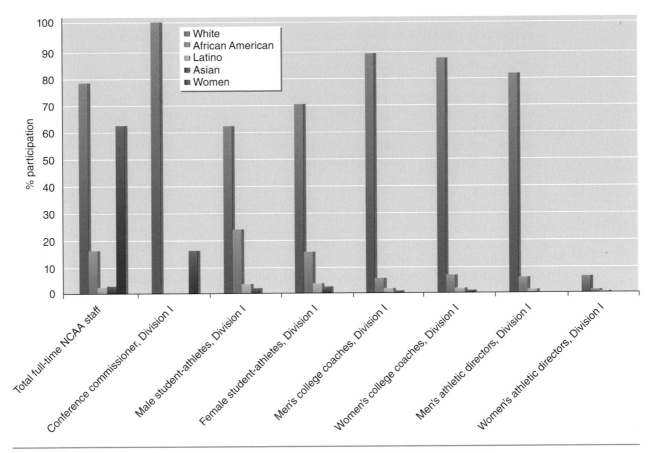

FIGURE 7.2 Race and gender representation in college athletics 2010.

travel, limiting involvement by those in lower-income families. Because a disproportionate number of African Americans live at the poverty level, it is not surprising that they tend to be underrepresented in such activities (U.S. Bureau of the Census 2007).

Leadership

If our society had complete racial equality, then in sports with high proportions of African American players—football, basketball, and baseball—we would expect to find rather high proportions of African American coaches. But this is not the case. In 2011, African Americans represented only 12% of Major League Baseball managers, 22% of NFL head coaches, 33% of National Basketball Association (NBA) head coaches, and 7% of Major League Soccer (MLS) head coaches (Lapchick 2011b, 2011c, 2011d, 2011e). Among NCAA division I-A intercollegiate programs in 2008-2009, African Americans were head coaches of 5% of the football teams, 21% of the men's basketball teams, 15.3% of the women's basketball teams, and 1.1% of division I baseball teams (Lapchick 2011a). Indeed, the 2007 Super Bowl was historic—not only one but both teams were led by African American head coaches, Tony Dungy and Lovie Smith. Some credit has been given to an NFL rule (the Rooney Rule) requiring that at least one minority candidate be interviewed when any team is searching for a new head coach.

In addition, a disproportionately low number of African Americans officiate games, accounting for only 42% of NBA basketball officials in 2011, yet African Americans represent 78% of the players (Lapchick 2011c).

There are two major explanations for these leadership inequities. First, racist stereotypes featuring erroneous conceptions of inferior intellectual capabilities of African Americans are probably operating to keep them from attaining athletic leadership roles in which thinking and working well under pressure are viewed as crucial. When people doing the hiring rely on racialized views and these views are supported by institutional racism, African Americans (and other nonwhite racial groups) are likely to be shortchanged (Eitzen & Sage 2009). Ricky Stokes, an African American assistant basketball coach at Wake Forest University, put it this way: "As an African American assistant coach, sometimes you can be labeled as a recruiter. A lot of times people think that's all you're supposed to do. Sometimes people don't think you can do the job on the floor—coaching, teaching the fundamentals, and, taking another step, speaking to booster clubs." Although Stokes expressed optimism about the likelihood that he would land a head coaching job, he realized that people might perceive hiring him as taking a chance (Ross 1996, pp. C1-C2).

A second reason for leadership inequities is that white athletic administrators in charge of hiring people for entry-level coaching or management positions are likely to learn about candidates from their professional friends in other athletic programs, most of whom are also white and may think in similar ways. Because ethnic and racial minorities are not usually involved in these old-boy networks, athletic administrators can easily overlook them. Relatively few people from marginalized ethnic and racial groups hold leadership positions in programs featuring expensive sport and exercise regimens (Coakley 2009; Eitzen & Sage 2009). A quick tour of facilities such as health clubs, soccer fields, skating rinks, ski areas, gymnastics schools, and tennis courts would probably reveal only a handful of coaches and administrators of color.

> Few African Americans reach important sport leadership positions, even in team sports that boast many African American players.

Cultural Expression

African Americans take pride in the athletic prowess and creativity of African American men and women who have become sport heroes. However, the very sports that make African Americans heroes can also contribute to their racial and ethnic stereotyping.

Expressive behavior by African American male athletes may be one way of illuminating the surveillance that African American males face in our society in terms of performing black masculinities—forms of masculinity that sport organizations and other institutions regularly seek to contain and sanitize in an effort to protect the current racial hierarchy (Majors 1990). Because their access to education, jobs, and power is restricted, African American males often work on proving their masculinity in other ways. One of these is constructing an expressive "cool pose" consisting of "styles of demeanor, speech, gesture, clothing, hairstyle, walk, stance, and hand-shake" (Majors 1990, p. 111). Cool pose sends a message that the person has a potent, interesting lifestyle that deserves respect; it is an embodied recognition of people's right to occupy a space that often excludes them based on racial hierarchies in conjunction with ideas about social class, gender, and, at times, sexuality. Sports are a ready-made arena for cool pose, an affective movement that applies to other social groups as well. Examples include football end-zone dances and spikes; Muhammad Ali's boasting, poetry, and dancing in the ring; and Julius Erving's gravity-defying moves to the basket that often began at the foul line (Majors 1990). Many current African American athletic stars continue to create images rooted in the traditions of cool pose. In addition, cool pose is threaded through a

Assessing the Ethnic and Racial Mix in Sport

Think about the local physical activity programs with which you're familiar in schools and community recreation centers (e.g., physical education classes, high school and college varsity athletic teams, rhythmic aerobics classes at local YMCAs, pickup basketball games on inner-city courts, gymnastics instruction at private centers, youth sport leagues formed by groups such as the American Youth Soccer Organization or Little League Baseball). From your observations, what ethnic and racial minorities are represented among the people in leadership roles in these programs (e.g., supervisors, directors, coaches, officials)? How do the proportions of different ethnic and racial minorities in these positions compare with the proportions of ethnic and racial minorities among the players? What are possible reasons for the similarities or differences?

variety of stylish gestures and "walks" that are embedded in the ordinary activities of African Americans and in distinctive dances such as hip-hop.

Young African American boys and men develop their own renditions of cool pose at playgrounds and school athletic facilities. Their style is a crucial part of their sports—especially basketball (Carlston 1983; Kochman 1981; Majors 1990; Wilson & Sparks 1996). They challenge each other with their creative, flashy moves, assertively daring other players to top them. Their goal is to heighten their reputation by combining effective game skills with a spectacular look—a look that we can even see in a range of media images designed to make profitable a particular form of black masculinity through black athleticism. (See Neal 2006 for further reading.)

Unfortunately, sport also communicates ideas that help maintain racial and ethnic inequities. For example, the fact that many top athletes are African American seems to send an upbeat message to African American boys and young men that they can make it to the pros if they work hard enough. But this message is misleadingly positive. For example, although a number of African American high school varsity basketball and football players believe that they have a chance of playing professionally, only a minuscule number actually make it. And because constant athletic practice usually gets in the way of acquiring other job skills, the combination of racial inequalities and athletic identity may limit one's ability to move beyond athletics for employment opportunities (Coakley 2009; Eitzen & Sage 2009). White youths in higher socioeconomic brackets who dream of playing professional sports also usually end up short of their goal, but because they don't face racial discrimination and are from higher-income families that have more contacts with potential employers, they are often in better positions to pursue other job opportunities.

Racial Stereotyping

Racist stereotypes are sometimes used in television portrayals of white and African American athletes. In two articles published in 1989 and 1996, Derrick Jackson studied the comments of television sportscasters about African American and white athletes. He noticed that sportscasters tended to recognize white basketball and football players for their brains: intelligence, thoughtfulness, and strategy. African Americans, on the other hand, were noted more for their brawn: physical skills, moves, and muscular strength. Such reactions reinforce a prominent racist stereotype. It is heartening to learn, however, that although brains were still mentioned more often in 1996 television portrayals of white athletes, this tendency was not as strong as it was in 1989. Unfortunately, over the same period the "brawn gap" between African Americans and whites doubled. African Americans were twice as likely to receive praise on television for their physicality in 1996. What is crucial to understand from Jackson's research is that our knowledge of racial categories changes over time and place and thus, as consumers of sport, we need to be aware of the racialized rhetoric used to convey sporting stories. The brain versus brawn manner of describing athletic performance is just one way to convey cultural messages through bodily performances. This takes on new formations as we consider how race intersects with gender, social class, sexuality, and ability (Coakley 2007; Douglas & Jamieson 2006; Eitzen 2006).

Michael Jordan, probably the most visible athlete in the world in the 1990s, provides much food for thought in connection with our discussion of expressiveness. A special collection of articles devoted exclusively to this athletic superstar demonstrates the ways in which he was constructed as a media icon who transcends race (Andrews 1996). Part of this involved stripping away some of his "blackness"—his racial identity. For example, advertisements accentuated his physical, athletic black body (a common media portrayal of African American athletes). At the same time, however, the threat of heightened sexuality stereotypically associated with black bodies in American society was defused by frequent references to his wholesomeness and strong family ties. Interestingly, these mediated images of Jordan took hold so well that even following his divorce, he retained a "wholesome" image. Such mixed messages probably contribute to his popularity among whites and African Americans alike (McDonald 1996). Media portrayals of Jordan give us a collage of meanings, some lending

> African American athletes who engage in "cool pose" express creativity, strength, and pride associated with masculinity; however, mass media professionals (e.g., journalists, broadcasters, advertisers) sometimes use racial stereotypes to characterize African American athletes and may refer to them in ways that strip away some of their racial identity.

American Indian Mascot Controversy

One of the ongoing controversies in the sporting world is the use of American Indian team names and mascots for sport teams. A well-known example is the Cleveland Indians and their mascot named Chief Wahoo. If you are a baseball fan, you've probably seen the Cleveland Indians baseball logo. If not, you can find it on the official Cleveland Indians' website (http://indians.mlb.com/index.jsp?c_id=cle). As you will see, the logo features the head of Chief Wahoo, whose toothy grin, hook nose, fire-engine red complexion, and single-feather headdress "identify" him as an American Indian.

Critics of the sport teams that adopt American Indian imagery argue that such representations create false and degrading images of American Indians. These commentators believe that the "rituals" carried out by teams or fans (e.g., the tomahawk chop or the dance done by Chief Illiniwek) show disrespect for Indian culture. Some critics contend that the Indian names and mascots contribute to the miseducation of Americans by producing a kind of cultural illiteracy about indigenous (native) people (Staurowsky 1999). These critics argue that the symbols of Indian life featured in sporting imagery—the war paint, masks, costumes, drumming, chants, and ceremonies—are neither authentic nor accurate. They are what some critics call pseudo-Indian symbols (King 2004) and just feed long-standing stereotypes of American Indians.

A number of institutions and schools—such as Cornell, Dartmouth, Marquette, Stanford, and Syracuse—have seen merit in these arguments and have changed their names and mascots accordingly. Others, including professional sport teams (e.g., the Cleveland Indians, Washington Redskins, and Atlanta Braves), have not.

Those who defend American Indian symbols (and virtually all defenders are white people) assert that their appropriation of Indian imagery is a tribute. To be called a warrior or a brave is a compliment because these words suggest that American Indians are courageous and tough. In addition, defenders argue that it is their right to adopt and subsequently own, for marketing purposes, such American Indian imagery (Staurowsky 2004).

Most American Indians themselves are clear on their position. The violent and aggressive images associated with others' ideas of American Indians offend them. They believe that they own their culture and that non-Indians should not have the right to use it for their particular, often commercial, purposes. They contend that the Indian names, mascots, and logos distort and trivialize their culture. They see Chief Wahoo as a racist caricature of an American Indian. In short, they argue that such American Indian imagery contributes to the long history of oppression that characterizes white dealings with American Indians (King 2004).

 ACTIVITY 7.5

Mascot Controversy
Native American mascots: Research both sides of the issue in this activity in your web study guide.

positive support to a distinctive African American culture and others reinforcing old ideas that help to maintain racial inequalities: Sorting out which are which requires a sociologically informed analysis.

Socioeconomic Relations

Wealth, education, and occupational prestige are all ingredients of socioeconomic status. People with more money, more education, and greater occupational prestige have higher socioeconomic status and usually more access to power. To illustrate, it's not hard to recognize the many advantages of being a corporate chief executive compared with being a custodian at the corporate headquarters. The chief executive can afford a luxurious lifestyle, whereas the custodian might have to scrimp to make ends meet. The chief executive can send his children to the most expensive universities, whereas the children of the custodian might have to work to put themselves through college. The chief executive can make major decisions affecting future directions of the company, whereas the custodian might only be able to recommend better ways to clean the office.

Abundant evidence indicates that the income gap between rich and poor is widening in the United States, and in fact, the pattern of income growth has changed from fast and fair to slow and distorted (Economic Policy Institute 2011). For example, between 1947 and 1973, family income for the bottom 20% of households grew at a rate of 117.4%, whereas the top 20% of income earners experienced family income growth at a rate of 88.7%. The past 30 years, however, ushered in a much less equitable and much less productive rate of family income growth, with the bottom 20% of income earners experiencing a negative growth rate of –7.4% while the top 20% of households had an income growth rate of 49% (Economic Policy Institute 2011).

This vast and growing income disparity in the United States has important implications for many aspects of American life, including education, recreation, health care, and physical activity. For example, among industrialized nations, citizens of countries with wider income gaps generally are less healthy (Siedentop 1996). Much evidence indicates that poor people are more likely to suffer from health problems, largely due to structural inequalities that result from increased privatization of social services and decreased opportunity to access health resources (e.g., reduction in public facilities and increases in fee-for-membership facilities).

In the United States we have seen an enormous growth in private health and exercise clubs that charge membership fees and cater to the upper middle class and the wealthy. Meanwhile, a decline has occurred in publicly funded programs in schools and recreation departments that are more accessible to middle- and working-class people. For example, it is increasingly common for students to pay user fees to participate in varsity athletics in the public schools and for families to pay user fees for youngsters to compete in community youth sport programs (Coakley 2009). Most U.S. citizens could benefit from physical activity, yet barriers are in place that make programs less accessible to low-income people. Table 7.3 shows how participation in certain kinds of physical activity varies by level of income. For example, participation in all physical activities increases at or above the $35,000 annual income marker, and some activities (tennis, salt water fishing, and golf) have quite low participation rates until the $35,000 or $50,000 marker.

Like the United States, Canada has experienced a widening of the gap between the haves and the have-nots. In 2002 a study showed that the net worth of the top 25% of Canadian families rose by 14% between 1994 and 1999 and that their financial wealth increased by 40%. During the same period, however, the bottom 75% of Canadian families reported no gain in either area (Abbate 2002). Not surprisingly, in Canada the health outcomes of poverty follow the pattern described in the United States. Aboriginal peoples in Canada and the poor, compared with the rich, are much more likely to face structural and cultural challenges in gaining access to physical activity and health resources (Reading & Wien 2009). Impoverished Canadians face similar barriers to exercise and physical activity. Community user fees have risen, as have student user fees. Those who are already without resources have less access to sport participation ("User Fees" 2003).

Participation

Research indicates that people's level of education is clearly tied to participation in physical activity. One study showed that highly educated groups were 1.5 to 3.1 times more likely to be active than were the least educated groups (Stephens & Caspersen 1994). Salaried professionals are much more likely to participate in corporate fitness programs than are hourly workers (Eitzen & Sage 2009). To the extent that educational attainment is related to economic resources, these data may also reflect availability of time and resources for participation in sport (Coakley 2009).

TABLE 7.3 Participation in Selected Sports Activities by Household Income: 2009 (in thousands)

Physical activities	Number	Under $15,000	$15,000-$24,999	$25,000-$34,999	$35,000-$49,999	$50,000-$74,999	$75,000-$99,999	$100,000 and over
Total	269,988	25,568	24,659	37,297	39,689	54,549	41,485	56,740
Aerobic exercising	33,138	1,760	2,076	2,443	3,748	7,437	6,287	9,387
Backpacking	12,281	1,325	764	1,218	1,770	2,724	1,613	2,867
Baseball	11,507	573	456	1,076	1,772	2,473	2,366	2,791
Basketball	24,410	1,816	1,078	1,852	3,702	5,069	4,739	6,154
Bicycling riding	38,139	2,433	1,894	2,529	5,266	8,321	6,859	10,837
Billiards	28,172	2,763	1,624	2,300	4,397	5,961	4,863	6,265
Bowling	44,972	3,337	2,414	3,241	6,867	10,415	8,422	10,275
Camping	50,863	4,119	2,685	3,833	7,520	13,219	8,808	10,678
Exercise walking	93,359	6,855	7,061	7,911	12,813	19,961	16,814	21,944
Exercising with equipment	57,206	2,917	2,885	4,423	8,112	12,161	10,707	16,001
Fishing (net)	32,876	2,539	2,152	2,775	5,821	7,500	5,359	6,729
Fishing-fresh water	28,996	2,327	1,869	2,635	5,350	6,552	4,429	5,834
Fishing-salt water	8,195	453	435	453	768	2,056	1,765	2,265
Football (tackle)	8,890	952	1,247	881	1,335	1,640	1,282	1,553
Golf	22,317	606	675	1,078	3,061	4,614	4,589	7,693
Hiking	34,013	2,294	1,858	2,126	4,234	7,587	6,726	9,189
Running/jogging	32,212	1,189	1,784	2,033	4,340	5,787	7,970	9,109
Soccer	13,578	956	539	727	1,644	2,527	2,603	4,583
Softball	11,829	1,055	536	874	2,165	2,658	2,354	2,186
Swimming	50,226	3,171	2,313	4,125	3,442	11,031	8,918	14,227
Tennis	10,818	411	436	509	1,301	2,149	1,955	4,058
Volleyball	10,733	707	651	447	1,917	2,014	1,969	3,027
Weightlifting	34,505	2,029	2,123	2,330	5,224	6,976	7,192	8,631
Yoga	15,738	1,025	1,013	1,254	1,885	3,524	2,574	4,462

Data from U.S. Census Bureau, 2012, *Statistical abstract of the United States*.

Income is also linked to the types of sport and recreational activities in which people take part. Wealthy adults tend to participate in individual activities such as tennis, golf, and skiing, whereas those with lower incomes tend to play team sports such as football, basketball, and baseball (see table 7.3). At least two explanations can account for this difference. Equipment and facility costs are so high in some sports that these sports don't attract people with low incomes. In addition, the working hours of higher-income professionals often fluctuate, making it more difficult to schedule regular competition with a team. Blue-collar workers tend to have standard working hours, a circumstance that makes it easier to schedule regular team sport competition (Eitzen & Sage 2009). Expensive physical activities that lend themselves to flexible scheduling may have more appeal to higher-income people. Examples include skiing, scuba diving, mountain climbing, hang gliding, and white-water kayaking.

We can point to many athletes who grew up in poverty and eventually acquired enormous wealth and fame as professional players. Many people thus think that talented athletes have a sure path to upward social mobility. But only a tiny minority of mostly male athletes successfully travel this route. In 2004, for example, the 1,035 major league professional players in football, men's and women's basketball, and baseball represented only 0.19% (<1%) of high school athletes playing these sports. About 16,000 players from colleges and universities were eligible to be drafted as rookies in the NFL in 2004, but only about 250 ended up on the final team rosters (Coakley 2009, p. 338).

> ► Socioeconomic status influences the types of physical activities to which people have access; physical activities requiring expensive equipment, facilities, and coaching are mostly beyond the reach of people at lower income levels.

ACTIVITY 7.6

Practical Plug-In
Consider what a fitness consultant needs to know and do to market a physical activity program to a diverse population in this activity in your web study guide.

Leadership

The people who control elite sports are generally quite wealthy. At the top are the owners of professional team sport franchises, sport media moguls, and corporate executives. In 2010, *SportsBusiness Journal* reported its list of the 50 most influential people in sports business. The list was heavily populated with media moguls, league commissioners, and corporate giants (Coca Cola, PepsiCo, Verizon). Given the NFL and NBA contract negotiations that year, the list also included player association executives (NFL and NBA), who happened to be the only African American men on the list. From position 41 onward, there were four women listed, two of whom were nonwhite (*SportsBusiness Daily* 2010). Similarly, in 2008, four of the most powerful people in sport as identified by *Bloomberg Businessweek* were George Bodenheimer (president, ESPN), Dick Ebersol (chairman, NBC sports), Phil Knight (chairman, Nike), and Sean McManus (president, CBS news and sports) (Burnsed et al. 2008). Although these two lists are not comprehensive and may reflect mainstream ideas about influential persons in sport, they do illustrate the vast white masculinity that controls sport in the United States and in international settings.

Increasingly, corporate heads are being hired as athletic directors in NCAA Division I college and university programs, and their salaries and other financial compensation have to be high enough to lure them away from private business. Top coaches are also well paid. In 2006, the average NCAA football coach's salary was $950,000 per year not including benefits, perks, and incentives. In 2011 it was $1.47 million, an increase of nearly 55% over the span of six seasons (Brady et al. 2011). Also in 2011, at least 32 coaches in the NCAA's top-tier Football Bowl Subdivision earned more than $2 million. Of those, 9 made more than $3 million, and 3 made more than $4 million. Texas' Mack Brown tops the list, being paid more than $5 million. The top five incomes of NCAA Division I basketball coaches whose teams made the 2011 NCAA tournament were well over $3 million, with Rick Pitino topping the list at $7.5 million (Berkowitz & Upton 2011a). Pat Summitt was the only women's coach on the list with a reported $2 million salary, and the four remaining top-earning Division I basketball coaches all earned $1 million or more (Berkowitz & Upton 2011b). By comparison, the median salary of presidents of public universities in 2010 was $375,442 (Stripling & Fuller 2011). Some people believe that these incomes are justified because many top college coaches are in contention for jobs with professional teams that can offer huge sums of money.

Few opportunities are available for the less affluent to occupy important leadership roles. In the commercial enterprises of our most popular spectator sports, the goal is to make money—either for profit (e.g., professional sport franchises, media companies) or to improve the program (e.g., nonprofit, big-time intercollegiate sports). People who are particularly adept at doing this (e.g., winning coaches; astute collegiate athletic directors; efficient professional franchise managers, television executives, and production personnel; forward-looking team owners) are usually rewarded with higher incomes or more profits. Less affluent people have few opportunities to occupy important leadership roles in this high-stakes business.

People with average incomes have much better chances to assume leadership roles in lower-level collegiate programs, high school athletics, and community recreational programs such as organized youth sports and adult athletic leagues. The competitive structure of these programs often mimics elite athletics, of course. For example, youth sport teams often take the names of major professional squads. Procedures involving cutting players from teams, having playoffs, and naming all-star teams are derived from elite college and professional sports. People at the lowest income levels don't participate in sport as frequently and are therefore not likely to advance to positions of leadership even in lower-level programs and leagues.

Leaders in other types of physical activity programs—for example, noncompetitive recreational activities, instructional programs, and health-related exercise programs—also range from people who are extremely wealthy to those with modest incomes. For example, owners of health club chains, ski resorts, and golf courses are probably among the most affluent. Because these are private businesses, specific financial information is difficult to obtain. As in any industry, their profitability undoubtedly varies widely. People who manage or teach in such facilities are probably from the middle and upper middle classes. Most individuals at the low end of the socioeconomic continuum probably don't participate enough to make it into the leadership ranks in these types of programs.

> Wealthy people occupy influential leadership positions in our popular spectator sports and in some physical activities. Those at the lowest levels of socioeconomic status rarely find themselves in positions of leadership.

Cultural Expression

Sports such as golf and sailing, as well as physical activities that are not essentially competitive such as skiing and scuba diving, are sometimes used to mark economic affluence. Belonging to private country clubs or indulging in vacations at fancy sport and recreational resorts lets others know that you have enough time and money to enjoy luxurious physical activities.

■ Do you perceive losing as a bad thing? How is your answer a mark of your cultural expressions?

In similar fashion, well-exercised, lean, taut bodies serve as status symbols in white North American society. They adorn magazine and television advertisements, model the latest designer clothes, and appear in popular movies and television shows. A thin, well-contoured "hard body" accentuated by spandex tells others that you have the time, money, and self-discipline to shape yourself according to today's difficult-to-attain standards. These ideals are readily observable in almost any magazine or movie. Having a personal trainer sets you off as even more affluent. Most people can't afford all the body work—exercising, dieting, beauty treatments, plastic surgery, and other procedures—needed to achieve this look. The look thus becomes a scarce commodity—something that marks you as wealthy and stylish.

Sport sends other messages that support socioeconomic inequities. For example, winning is the most prevalent organizing theme in newspaper stories and telecasts of sporting events (Kinkema & Harris 1998). Winning is usually attributed to self-discipline, talent, and hard work. If an athlete or a team doesn't win, then we assume that the player or the team was lazy or lacked talent and so didn't deserve to win. Such beliefs underscore the American conception of merit—we often link hard work and talent to financial success. The flip side is that if someone fails financially, it must be because she or he isn't talented or didn't work hard. This reasoning allows us to hold the belief that the rich and poor both deserve whatever money they have. The point here is not that merit is a bad idea. The problem is that this logic often leads us to overlook the societal barriers (e.g., poor nutrition, neighborhood gang violence, poor access to libraries and computers, dysfunctional families, lack of child care) that prevent poor people from developing themselves to the fullest and becoming valuable members of society (Coakley 2009).

Finally, our popular team sports also send messages about the importance of obedience and teamwork. These qualities are valuable for most workers. Although some employers may need employees who can break out of traditional molds and be creative, most simply need people who can work well with others and follow directions. Because sport tends to reinforce obedience and teamwork, it helps maintain our current economic system.

> ➤ A well-sculpted, lean body and participation in expensive sports are often markers that a person is wealthy. In addition, sport expresses other ideas that reinforce the socioeconomic status quo, such as the notion that winners and losers deserve what they get and the value of obedience and teamwork.

Attributions of Failure: Lack of Merit or Lack of Money?

Think about the things coaches typically tell players when a team loses a game. You can recall your direct experiences or talk to other people about theirs. How often did coaches tell players that they didn't work hard enough, and that if they had practiced harder or given more in the game, victory would have been theirs? How often did coaches tell players that they obviously tried hard but lost to a team that was just more talented? What other reasons did coaches offer for why the team lost? Did any (or all) of these reasons make the players feel that they really didn't deserve to win?

Wrap-Up

Sociology of physical activity takes us beyond common, everyday understandings of sport and exercise. It illuminates patterns of change and stability, identifies social problems, and urges modifications aimed at enhancing equality and human well-being. This subdiscipline has so far emphasized sport; but interest in exercise, fitness, and societal conceptions of the human body is increasing. Although researchers have analyzed relationships and trends in social life, for the most part they have not completely nailed down causal factors underlying these trends and relationships. This circumstance is not unusual in the social sciences because it is difficult to bring a slice of social life under the strict laboratory controls needed to demonstrate causation.

If you are interested in what you have learned here, you might enjoy looking through some of the sport sociology textbooks, taking an introductory course in this subdiscipline, and exploring major journals such as *Sociology of Sport Journal, Journal of Sport & Social Issues,* and *International Review for the Sociology of Sport*. You may also wish to review research reports released by major sport studies research centers, such as the Institute for Diversity and Ethics in Sport at Central Florida University, the Centre for Sport Policy Studies at the University of Toronto, the National Research Institute for College Recreational Sports & Wellness at Ohio State University, and the Center for the Study of Sport in Society at Northeastern University. Sociology of physical activity provides information that kinesiologists can use to increase our understanding of our experiences as participants, spectators, and professionals, as well as the experiences of others. It also helps us think more clearly about the changes we would like to make in these physical activity programs as well as the things we would like to keep the same.

Finally, sociology of physical activity is an ongoing reflection of the changing social conditions in which kinesiologists do their work. This leads to a question that is always relevant for kinesiologists: What do I need to know about this social, political, and cultural context for physical activity so that I do not mistakenly reinforce existing social inequalities?

More Information on . . . Sociology of Physical Activity

Organizations

American Alliance for Health, Physical Education, Recreation and Dance

International Sociology of Sport Association

National Association for Kinesiology and Physical Education in Higher Education

North American Society for the Sociology of Sport

Journals

International Review for the Sociology of Sport (Sage)

Journal of Sport & Social Issues (Sage)

Sociology of Sport Journal (Human Kinetics)

 ACTIVITY 7.7

Use the key points review in your web study guide as a study aid for this chapter.

Review Questions

1. List and discuss the goals of sociological study of physical activity.
2. Give an analysis of the expanding research directions in sociology of physical activity from 1970 to the present.
3. List and discuss the six research methods commonly used for sociology of physical activity.
4. Describe the ties between participation in physical activity and power relationships based on gender, race and ethnicity, and socioeconomic status.
5. Describe the ties between leadership in physical activity programs and power relationships based on gender, race and ethnicity, and socioeconomic status.
6. Describe the ties between physical activity expressiveness and power relationships based on gender, race and ethnicity, and socioeconomic status.
7. Identify the contrasting beliefs that underlie the American Indian mascot issue. Which argument would a sociologist of physical activity be most likely to advance? Why?

chapter eight

Motor Behavior

Katherine T. Thomas and Jerry R. Thomas

chapter objectives

In this chapter we will

- explain what a motor behavior researcher does;

- present the goals of motor behavior, including motor learning, motor control, and motor development;

- explain the research process used by scholars in motor behavior; and

- present some of the principles of motor learning, motor control, and motor development

Your roommate is on the college softball team. For the past 6 weeks the coach has had the team warm up with warm-up bats. You remember from high school that a warm-up bat is expensive special equipment that is not used for batting, just for warm ups. Your roommate is excited to be on the softball team and believes in the coach's rationale for this warm-up—that the weighted bats will make the swings at the ball that follow faster. First, you wonder whether this is true. You know that weight training is helpful to most athletes; however, this warm-up is not really weight training. You know that the lifts used in weight training train a specific set of muscles in an optimal way. You wonder whether in fact swinging the weighted bat might have a negative impact because it might change the movement of the bat relative to the movement of the body. Finally, you wonder whether other reasons related to sport techniques explain why none of your classes has presented this technique. What other factors could you consider as you decide whether this type of practice will help the team?

This chapter introduces the subdiscipline of motor behavior and establishes the relationship of motor behavior with other scholarly subdisciplines in the discipline of kinesiology. You will begin to understand why motor behavior is of interest to performers who want to improve movement expertise, to scientists who contribute to scholarly study, and to practitioners in professional practice who seek to assist others in developing motor skills.

Why Use Motor Behavior?

Have you ever wondered why you enjoy certain physical activities? Often, being skilled helps us enjoy an activity. If you pitch a great fastball, run a quick 5K, have a killer volleyball spike, or can hit a terrific backhand in tennis, you are more likely to enjoy these activities. Someone with less skill may find the activities frustrating and less enjoyable. Of course, there are notable exceptions, such as golf, a sport on which some of us spend many hours and a great deal of money demonstrating that we have very little skill!

Goals of Motor Behavior

- To understand how motor skills are learned, how processes such as feedback and practice improve learning and performance of motor skills, and how response selection and response execution become more efficient and effective

- To understand how motor skills are controlled, how the mechanisms in response selection and response execution control the body's movement, and how environmental and individual factors affect the mechanisms of response selection and response execution

- To understand how the learning and control of motor skills change across the life span, how motor learning and control improve during childhood and adolescence, and how motor learning and control deteriorate with aging

---

Think about a skill you like and do well, for example, dribbling a basketball. How did you learn to dribble? What types of practice experiences worked best for you? How did you learn to coordinate the movements of your feet, your hands, and the ball? Remember those early years when you dribbled slowly, could not turn quickly, could dribble only with one hand, and sometimes lost control of the ball? In what ways did your brain and nervous system develop and adjust so that you could improve your control and coordination? Across the life span, practice is an important factor in learning skills. Most people improve motor skills and yet never understand how the nervous system adapts, how it develops or controls movement, or how best to use practice to improve performance. The study of motor behavior focuses on how skills are learned and controlled and how movement changes from birth through the end of life.

People believe that practice improves performance; we see evidence of this in homes, schools, and communities. Parents urge children to practice playing the piano; athletes ask coaches for help practicing a sport skill; and students perform practice exercises to learn keyboard typing. Clearly, some forms of practice are superior to others. *What, how,* and *how much* are important questions about practice. Another important question is, what can be changed with practice? For example, is the expert better because of practice, or was she born with special talent? Research in motor behavior tells us that your parents and teachers were partly right when they said, "Practice makes perfect." More precisely, they might have said, "*Correct* practice makes perfect"; research also tells us that many factors influence how well the practice works.

Is your roommate's coach using good practice technique? The answer is probably no—the practice routine that the coach developed will not improve performance in batting. The athletes are not doing enough repetitions to become stronger, and many training devices such as this may interfere with the timing of their movements, which is critical to most complex skills. (In other words, practice that improves timing would help your roommate and the other athletes on the softball team.) Athletes use many special practice drills hoping to improve their playing performance, but only practice conditions that are similar to actual game performance will benefit performance in athletic contests. This principle, called **specificity of practice**, is one of the most important and easy-to-replicate principles in motor behavior. In addition, other factors such as talent and confidence will influence the skill that the team demonstrates in games. Published research studies show that there is no benefit to swinging a weighted bat prior to batting (Otsuji et al. 2002; Southard & Groomer 2003), in fact, bat speed slows and the pattern is altered. We discuss more about specificity of practice and other principles and theories within motor behavior later in this chapter.

Teachers and coaches should be experts on practice, because much of their training focuses on how to organize and conduct practice. But we all might encounter situations that call for knowledge about practice and skill acquisition. Consider these examples:

- Your golf partner asks you to watch her swing and advise her on how to avoid slicing her drive (of course, you know that the less skilled a golfer is, the more likely it is that she will share her ideas about the golf swing).

- You volunteer for a youth league football program. The topic of discussion is how best to group kids for fair competition.

- You drive to school every day and realize that many days you do not remember a single event during your drive (and you also realize that hundreds of others are driving in the same scary way).

- You work part-time at an assisted living center. You see that many of the residents have difficulty standing up from a sitting position, balancing, and grasping small objects. You wonder why this is so and whether they can regain these skills.

- You and your dad are watching a National Basketball Association game, both of you marvel at the graceful moves and expert skills of the players. How would you explain some players' ability to attain such a high level of expertise?

You have probably faced similar situations that you couldn't solve or explain. Although most of us can talk on a superficial level about motor skill performance—"Did you see how long he was off the ground? He floated through space to dunk that ball!"—few people really understand how a person controls, learns, and develops motor skill over time. The study of motor behavior provides answers to many of these questions and helps us understand the process of the development, control, and learning of motor skills (Thomas 2006; Ulrich & Reeve 2005). Understanding will allow you to answer questions and plan practice.

College students may see themselves merely as students in a class rather than as scholars engaging in scholarly study. But you have accumulated a large body of knowledge that you can use to learn about motor behavior. For example, you may recognize some of the theories and research methods used in motor behavior from your introductory, developmental, or experimental psychology course. The study of motor behavior began as a branch of psychology that used movement or physical activity to understand cognition. Information from biology or zoology is incorporated into motor behavior as heredity, aging, and growth influence physical activity (e.g., see Thomas & Thomas 2008). In addition, researchers apply principles and laws from physics to their study of humans in motion. So, what you know about physics, biology or zoology, and psychology will help you understand motor behavior.

You have some applied experience with motor behavior. Remember your physical education teachers or coaches asking you to practice passing a volleyball or swinging a golf club or dribbling a soccer ball repeatedly? They gave you feedback about your practice: "Bend a little more at the knees." "Follow through with a smooth stroke." "Keep the ball as close to your feet as possible." They were implementing some of the principles of motor behavior—that correct practice and appropriate feedback improve performance. Of course, some of us can recall adults who provided less helpful information, such as "practice harder" or "hit the ball." Motor behavior guides us in providing better situations for learning and practice as well as understanding why some cues and feedback are better than others.

The subdiscipline of motor behavior is part of the behavioral sphere within the study of physical activity, along with psychology of sport and exercise (see figure 8.1). As people practice physical activity, they experience the essence of motor behavior by trying to control and learn movements. Scholars of motor behavior study how motor skills are learned, controlled, and developed to assist people as they practice and experience physical activity. Professionals who try to improve individuals' physical skills use the knowledge from motor behavior research in professional practice. Coaches and physical education teachers are among the many professionals who base their practice on motor behavior knowledge. Other examples include the gerontologist who wants to improve the motor skills of people who are elderly, the physical therapist who rehabilitates injured patients, and the athletic trainer who attempts to prevent injuries and rehabilitate injured players.

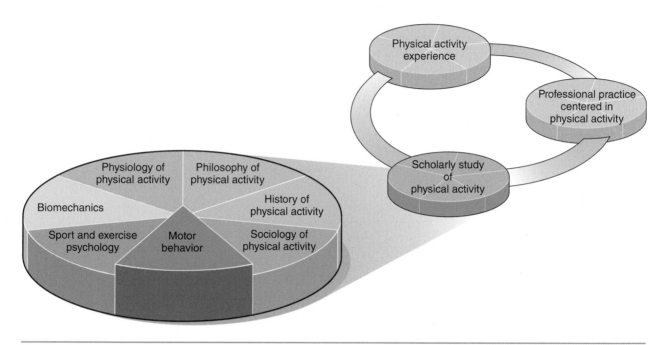

FIGURE 8.1 Motor behavior of physical activity and the other scholarly subdisciplines.

Researchers in motor behavior, exercise and sport psychology, and biomechanics attempt to solve some of the same problems and use some of the same methods and equipment. The pedagogical areas of physical education and physical education for individuals with disabilities (adapted physical education) should be based on what we know about how people learn and control movements. Motor behavior also differs specifically from sport psychology in that sport psychology typically studies elite athletes in competitive settings whereas motor behavior studies people of all skill levels (including elite athletes).

Finally, the pedagogical areas of physical education and adapted physical education use motor behavior knowledge to design instructional programs and evaluate motor skill. For example, a physical education teacher may use motor behavior information to plan a series of instructional classes for 6th graders to improve soccer dribbling techniques. Motor behavior is part of the behavioral sphere of kinesiology and is linked to the other behavioral subdisciplines—psychology of sport and exercise and pedagogy of physical activity. Motor behavior is also closely associated with biomechanics from the biophysical sphere of study.

What Do Motor Behaviorists Do?

Scholars who study motor behavior are often employed at universities, where they teach and do research and service, but some work in research facilities not associated with universities. Motor behavior research can be conducted in a university laboratory, in a clinical setting (e.g., a hospital), or in an industrial or military setting. For example, a researcher might ask, "Does pushing an elevator button multiple times cause the elevator to arrive sooner?"

Motor learning and motor control research has industrial and military applications. Researchers might study, for example, how night vision glasses influence pilots' control of helicopters or what the optimal method is for training workers to assemble a product. Besides conducting research, industrial and military researchers write grants and perform professional services, such as reviewing manuscripts for scholarly journals.

Medical and educational researchers often study motor learning, motor control, and motor development. They might study how the nervous system changes with age in terms of the control of movement, what the best methods are for teaching rehabilitation protocols to patients in physical therapy, how the game performance of youth soccer players can be improved, or what causes movement deterioration with Parkinson's disease.

The duties of a motor behaviorist at a university typically include research in the area of learning, control, or development; teaching; and service. Scholars may write grants to support research or programs related to practice. Teaching duties could include courses in motor behavior (learning, control, and development) or related courses such as biomechanics or sport psychology, research methods, measurement and evaluation, or pedagogy and youth sport. Service for a university faculty member might include evaluating motor disorders, managing a program for individuals with motor disorders, and conducting workshops or clinics about motor disorders.

Motor Learning Theory in Practice

Think about a physical education class in which your teacher introduced a new skill. How did the teacher present the information—demonstration, instructions, an audiovisual aid—and how was practice organized? For example, did you practice a part of the skill and then receive more instruction? Who demonstrated? Compare this class with a class in which you and the other students were performing a well-learned skill. How did the practice differ in the two situations? For a new skill, teachers often explain how you will use the skill and then demonstrate the skill and present instructions so that you will understand what to do first. Next, you will practice the skill several times. After seeing some consistency in your performance, the teacher may present more instructions and demonstrations.

For a well-learned skill, the teacher may begin by having students use the skill in a game or gamelike situation. Instruction, demonstration, and feedback are more likely to focus on strategy and less likely to include information and correction of the skill.

The subdiscipline of motor behavior produces knowledge of how people achieve motor skills across the human life span. The subdiscipline involves three areas: motor learning, motor control, and motor development (a developmental view of motor learning and motor control). One goal of motor behavior is to understand how motor skills are learned. The second goal is to understand how motor skills are controlled. Motor learning research deals with "the acquisition of skilled movements as a result of practice," whereas work in motor control seeks to "understand the neural, physical, and behavioral aspects of movement" (Schmidt 1988, p. 17); both are often evaluated across the life span, resulting in a developmental view. This brings us to the third goal of motor behavior, which is to understand how the learning and control of motor skills change across the life span.

Motor learning and motor control are not distinctly different areas, but they tend to ask slightly different questions. The goals of motor learning can be summarized as understanding the influence of feedback, practice, and individual differences, especially as they relate to the retention and transfer of motor skill. The goals of motor control are to understand how to coordinate the muscles and joints during movement, how to control a sequence of movements, and how to use environmental information to plan and adjust movements. In

Contrasting Motor Control and Learning

Remember when you were learning to drive a car? Name one task that was difficult for you to master (e.g., using the clutch, slowing down to turn without stopping completely while signaling, parallel parking). Now think about the last time you drove somewhere familiar, for example, to school today. Sometimes we cannot remember anything about a familiar drive.

Notice how your skill has changed as a result of learning to drive and then continuing to practice. When you were learning to drive you had to think about every movement—moving your foot from the accelerator to the brake, turning the steering wheel, using the turn signal. Later you thought only about where you were going and the traffic. Now, on an often-traveled route, you only occasionally monitor the traffic. Motor learning helps us understand how we learn skills like driving a car so that the skill becomes this automatic. Can you identify other skills you have learned so well that you do not think about them at all during or after the performance? In contrast, are there situations in which you do have to think about a well-learned skill? For example, when traffic chokes the road or a storm occurs, you must pay more attention to driving (not to mention driving to school while talking on your cell phone, fixing your hair, and eating breakfast). When task demands change, as in this example, we sometimes move from the automatic control of movements to conscious control of movements.

You can see that motor control and learning are related. Motor control is essential for every movement—from poorly skilled to well skilled. Motor learning is responsible for the shift from poorly skilled to highly skilled movements. One way to judge when a movement is well learned or highly skilled is to recognize that the movement is automatic, or executed without conscious control. We do not have as much practice driving in storms and heavy traffic, so that type of driving is not well learned and therefore not automatic.

addition, scholars are often interested in exploring how motor learning and motor control vary with different age groups, from children to senior citizens.

When you think of motor behavior, you may think of sport skills. But consider the many other types of movements that people use in their daily activities:

- Babies learning to use a fork and spoon
- Dentists learning to control the drill while looking in a mirror
- Surgeons controlling a scalpel; microsurgeons using a laser while viewing a magnified video image of the brain
- Children learning to ride a bicycle or roller skate
- Students learning to use a computer
- Teenagers learning to drive
- Dancers performing carefully choreographed movements
- Pilots learning to control an airplane
- Young children learning to control a crayon when coloring

All these activities and many others involve motor behavior and are of interest to researchers and practitioners in this area. Thus, the goals of motor behavior are important not only in sport but also in the total field of physical activity. As we saw in part I, understanding the learning, control, and development of movements plays an essential role in our culture and society. For example, the first clue parents may have that their baby is developing normally may be a reflex or motor milestone, such as reaching. Individuals have probably been asking questions about motor behavior since the beginning of time; scientists have been seeking the answers to those questions for more than a hundred years.

History of Motor Behavior

Research has a long history in all three areas of motor behavior (for reviews, see Thomas 2006 or Ulrich & Reeve 2005), but the reason for the research in each area has changed dramatically. For example, in the late 1800s and early 1900s, researchers were primarily interested in using motor skills as a means to understand the mind; motor skills were a tool to examine cognition (Abernethy & Sparrow 1992). Adams (1987) identified five themes from this early work that have persisted through the years: knowledge of results, distribution of practice, transfer of training, retention, and individual differences. These themes are the foundation of motor learning. Although the themes persist, researchers are now focusing on the motor skills themselves rather than simply using motor skills to understand cognition.

Similarly, motor control research can be traced to research in the late 1800s on the "springlike" qualities of muscle (Blix 1892-1895), and Sherrington's (1906) seminal work on neural control is still useful in explaining how the nervous system controls muscles during movement. In both cases, the purpose was to understand not motor behavior but biology.

The World War II era, from 1939 to 1945, was one of great interest in motor behavior research (Thomas 1997) because the military needed to select and train pilots (Adams 1987). Thus, motor skills were viewed as a necessary component in military efforts, but not because understanding motor skills was an important area of study.

Beginning in the 1960s, and gaining momentum in the early 1970s, motor behavior evolved as a scholarly subdiscipline of kinesiology. Scholars doing motor behavior research were no longer neurophysiologists or psychologists; they were specialists in physical activity. Franklin Henry's (Henry & Rogers 1960) paper on **memory drum theory** was the first major theoretical paper (landmark study; see Ulrich & Reeve 2005, pp. S63-S64) from the discipline of kinesiology (at the time called *physical education*). Henry's theory stated that reaction time was slower for complex movements because those movements took more planning time. For example, movements with several segments—involving moving from one position to a second position and then to a third position—required a longer reaction time than did single-segment movements, because the brain required more time to specify the needed information for the more complex movement. The current work on motor programs (central representations in the brain of the plan for movements) evolved from Henry's memory drum theory.

 ACTIVITY 8.1

Interview with a Practicing Professional
In this interview in your web study guide, Catherine Stevermer, a physical therapist, gives examples of the ways that physical therapists use their knowledge of motor behavior to help their patients.

▶ The subdiscipline continues to evolve because of research, technology, and different theoretical models. Motor learning and motor control are distinct but are linked by a common goal—to understand human movement.

Current research in motor control and learning usually focuses on understanding how the neuromuscular system controls and repeats movements. One purpose of this research is to understand and develop treatments for diseases and injuries such as Parkinson's disease and spinal cord injuries. Another is to improve performance in sport and physical activity.

A developmental approach to motor learning and motor control (also known as *motor development*) originated in developmental psychology and child development. Motor development grew from the "baby biographies," many written before 1900, that described the changes in reflexes and movements of infants. The early work used twins to establish the role of environment and heredity in shaping behavior (Bayley 1935; Dennis 1938; Dennis & Dennis 1940; Galton 1876; Gesell 1928; McGraw 1935, 1939). As with motor control and learning, the initial research in motor development took place not because scholars were interested in motor skills but rather because they were using motor skills to understand other areas of interest.

In the 1940s and 1950s, developmental psychologists lost interest in the developmental aspects of motor skill. Study in this area might have ended if not for three motor development scholars—Ruth Glassow, Larry Rarick, and Anna Espenschade. Their emphasis, which was different than that of the developmental psychologists, was on how children acquire skills—for example, how fundamental movement patterns are formed and how growth affects motor performance. These three scholars maintained the developmental nature of motor learning and motor control in their research through the 1950s and 1960s.

Just as research in motor learning and motor control increased around 1970, developmental research addressing the questions of motor learning and motor control also became popular during that period (Clark & Whitall 1989; Thomas & Thomas 1989). Motor development was considered part of the subdiscipline of motor behavior because the same topics were studied developmentally. But two research themes in motor development continued from the years before 1970—the influence of growth and maturation on motor performance and the developmental patterns of fundamental movements. Growth clearly influences the performance of motor skills, and some of the studies on growth relate growth to performance. Because growth cannot explain all of the improvement in motor performance during childhood, developmental scientists have turned to motor control, motor learning, and biomechanics for more information. Thus, currently we see three lines of motor development research: the study of motor learning in children; the study of motor control in children; and the influence of growth on motor learning, control, and performance.

Motor Expertise

Research on motor expertise often compares novices and experts to determine how they differ within various sports. Motor expertise is a recent and appropriate addition to motor development because age and experience are associated with increases in expertise. The interesting point occurs when a younger person is more expert than an older person—when, for example, your chapter authors find that our grandchildren can fix the DVD player, hit a golf ball out of the sand trap, or ride a skateboard and we, despite our wisdom and age, cannot! (By the way, we play golf in the low 80s; if it is warmer than that, we won't play.)

Research Methods for Motor Behavior

Research begins with a question that is stated as a hypothesis. Hypotheses are undoubtedly familiar to you. The question helps the scientist decide who will participate in the research project, what type of task they will do, and what the researcher will measure to answer the question. This is the same process used in most research in kinesiology. Understanding how movements are learned, how they are controlled, and how they change across the life span is complicated. In fact, to answer one question it is usually necessary to perform many experiments. Motor behavior researchers have concentrated on the use of techniques to measure movement speed and accuracy. Motor control and motor learning researchers use technology similar to that used by researchers in biomechanics. During the past 40 years, technology (e.g., computers, high-speed video, **electromyography [EMG]**) has been invented and adapted to control the testing situation and to record and analyze the movements. Technology has allowed the use of real-world movements instead of movement invented in the laboratory just for research purposes. Motor behavior courses often include laboratory experiences so that students like you can copy the experiments you read about. This type of hands-on learning has demonstrated benefits to your learning.

Types of Studies

There are many ways to answer any particular question in research studies. Three experimental designs, or techniques, are used frequently in motor behavior research. The between-group

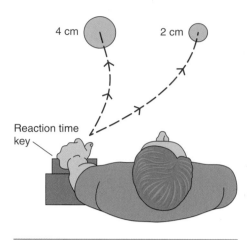

4 cm   2 cm

Reaction time
key

FIGURE 8.2 A reaction time study measures how fast a movement occurs after a signal and whether the size of the target object affects the movement.

design compares two or more groups that were exposed to different treatments (interventions) but tests them using the same task. The researcher could use this design to answer the question, "Does practicing a simple movement increase its speed?" The research design compares two groups (randomly formed) on the same task (movement speed). The treatment in this example is practice: If two groups practice differing amounts, will their performance on the task be different? In the second design, the within-group design, all participants are exposed to two or more different treatments and are tested on the same task.

Suppose you want to study whether a participant's **reaction time**—how quickly the movement begins after a signal—varies because of the size of the target. You could tell the participant that after you say "go," she is to move a stylus (like a pencil) as rapidly as possible to a target 30 centimeters (11.8 inches) away. The target is a circle either 2 or 4 centimeters (0.78 or 1.57 inches) in diameter (see figure 8.2). Does the time between hearing the signal and beginning the movement (reaction time) change depending on target size? In a between-group design, one group could move to the 2-centimeter target and the second to the 4-centimeter target. In a single group (within-group) design, half the participants could move to the 2-centimeter target first and then to the 4-centimeter target, and the other half could move to the 4-centimeter target first and then to the 2-centimeter target. In deciding which design to use, an investigator should consider whether changing target sizes interferes with performance. If so, then a between-group design is preferred; if not, the investigator will use a within-group design because he or she will not have to test as many participants.

The third type of experimental design used in motor behavior research is descriptive research. Here, the investigator measures or observes participants performing a task. Sometimes the investigator observes the same participants several times to trace changes. For example, the investigator may measure reaction time in the same children when they are 4, 6, and 8 years of age. In other cases, the performance of different groups is compared. For example, 4-, 6-, and 8-year-old children are tested for reaction time. Researchers often use this technique to describe age differences or differences between experts and novices. Descriptive research is different from research using the first two techniques because the participants receive no treatment.

Studying the Early Stages of Learning

Motor learning has used **novel learning tasks** to provide certainty that no participant has tried the task. Novel learning tasks were created for use in experiments to eliminate the advantage that some participants might have because they practiced the task before the experiment. Novel learning tasks tend to be simple—so simple that to make the task challenging, the participant is often asked to wear a blindfold (see figure 8.3).

Before the early 1980s, motor behavior research used simple, novel tasks to study early skill learning; this research helps us understand how beginners learn new motor skills.

You might question whether the results of these experiments can be applied to real-world, complex movement situations. This is a good point! These simple tasks allow us to study improvement and have helped us understand a great deal about how movements are learned. However, using simple tasks in research limits what the researcher can learn. One reason is that the outcome of the movement (product) rather than the nature of the movement (process) itself is studied (Christina 1989). Novel learning experiments are not helpful for researchers interested in physical activity or sport tasks in which performers have had thousands of trials (e.g., keyboard typing or baseball batting).

Studying Expert Performers

To address the limitations of studying novel learning tasks, some scholars began to study experts. These investigators asked, "What do expert performers do during practice and competition?" One way to answer the question was to compare expert performers to novice performers and evaluate how they differ on perceptual knowledge—particularly decision-making, skill, and game performance variables. In such studies, a criterion-like national team membership or ranking in the sport establishes expertise.

The knowledge and skills of a sport are often unique to that sport (Thomas & Thomas 1994): Knowing about and being skilled in tennis does not help one play soccer. Videotapes of badminton players have been used to determine how age and expertise influence how well players could predict where a birdie struck by an opponent would land. Participants looked at a video of a badminton player hitting a shot. The video had been altered (e.g., the head, arm, or racket was erased or certain motions were zeroed in on); sometimes information was erased that was needed to predict accurately where the birdie would land. As you might guess, expert players could make better, faster predictions, with less information, about where the birdie would land than could the novice players. Experts also looked at different body parts than novices did and therefore based their decisions on different sources of information than novices did. The advantage seemed to be a result of playing experience rather than age. You can see how this finding could help

FIGURE 8.3 This linear positioning task has been used to study early stages of learning because the task is novel. Learners must rely on what they remember about the movement (e.g., speed, distance, beginning and ending points) because vision is occluded. The task is both novel and simple so that the motor behaviorist can observe learning. The participant is not likely to have done the task before (i.e., it is novel) and will be able to master the task in a short time (it is simple).

a coach, teacher, or player improve badminton play. The information also helps us understand about learning many skills, because researchers have found similar patterns in many sport and motor skills. This type of experiment would be nearly impossible to do in a motor learning class because the students might not represent the range of expertise necessary (novice to expert).

Measuring Movements

The tasks used in motor behavior research provide a number of ways to measure movements and their outcomes. For example, suppose that senior citizens perform a movement that involves reaching 30 centimeters (11.8 inches) and then grasping and lifting containers of different sizes and weights. We could simply count the number of times each senior citizen successfully lifted the containers (reaching and grasping are prerequisites to lifting in this task) and use the number of successful lifts as an outcome measurement. We might find that as objects were smaller or heavier, the number of successful lifts decreased. Although that is an interesting finding, our long-term goal is probably to understand why or how changing the size or weight of the object influences success. So we probably need to examine the process of the movement. To examine process, we could use a high-speed video of the movement, taken with two cameras, and evaluate how this movement differs for 55-, 65-, and 75-year-old participants. From such videos, researchers can measure **kinematics** (location, velocity, acceleration—see chapter 10 on biomechanics). Other measures include muscle activity (EMG) of the arm and hand during the movement, the pinching force between the forefinger and thumb during the grasp, error in the movement, speed of the movement, reaction time before the movement begins, and accuracy of the movement. These process measures might help us understand how the reach, grasp, or lift changed for the various weights and sizes of objects. Using our hypotheses (educated guesses) we could begin to answer our original question. For example, we might hypothesize that the reaching and grasping phase would be the same for objects of the same size, even when weight varied, but that the lift would be different when the weight changed. Do you think reach, grasp, or lift would be different as we changed size but kept the weight the same? If we predict that reach or grasp changes, it is critical to measure the process.

Characteristics of Movement Tasks

In addition to deciding what to measure, motor behavior scientists must consider the characteristics of the movements they study (Gentile 1972). For example, some movements are more continuous (performing a gymnastics routine or riding a bicycle), and some are more discrete, involving a short period with a distinct beginning and end (striking a ball with a bat). Some movements are more open in character, whereas others are more closed. An example of an open movement is hitting a thrown ball—the environmental characteristics change from trial to trial because the batter must respond to the speed and location of the oncoming ball. A closed movement takes place in a more consistent environment, in which the performer tries to do the same thing each time (e.g., in archery or bowling).

Expertise Is More Than Good Stats

Suppose you wanted to research the talent on a women's softball team. You have available standard data that are kept on players—batting averages, RBIs, home runs, and fielding percentages. But you think that these data do not always reflect the contributions that players make to the team. You have videotapes of five games that the women have played, taken with a wide-angle camera that shows the complete field. Identify the kind of data you can get from the videotapes to answer the following questions:

- How many times did each player hit the ball hard while at bat?
- How many times did each player make a special effort that doesn't show up in the scorebook (e.g., backed up a throw, saved a bad throw, moved up a runner)?

Researchers and teachers must identify the characteristics (open or closed, discrete or continuous) of the tasks they study, for two reasons. First, the results of two studies may be different if the task characteristics are different. Second, the characteristics of the task used in the study must match the questions asked. For example, if you want to know why something happens in bowling, you should not use an open continuous task to answer the question. Teachers also must know how to draw these distinctions between task characteristics because skills of one type may require a different instructional strategy than skills of another type.

Measuring Learning and Transfer

The idea of learning—as determined by retention and transfer—is similar to the idea in the previous discussion about experts. That is, a motor skill can be examined at any point from the first attempt to well beyond mastery. The goal of most practice is learning, and learning is defined as retention and transfer. An example that has meaning for most students is related to the three most threatening words in college—"comprehensive final exam." At the end of the semester, how much do you remember when taking a comprehensive final examination? The purpose of the examination is to determine whether you learned what was taught in the course. For those who have really learned something, a comprehensive final examination is not a problem. As you will see in your motor learning class, retention is also used to measure learning in motor skills. Another way to measure learning is with transfer, in which you must do a slightly different version of the task. Experts are better at transferring information or skills than novices, and experts can perform well after periods without practice (called retention intervals). Many coaches are frustrated when the team "learns" a play during practice but has trouble executing during the game. This situation is an example

To study motor skill acquisition, researchers must also study how well skills are retained and how they transfer to other, similar situations.

▌ Before Bradley Wiggins (yellow jersey) won the 2012 Tour de France, he first had to learn the basic skill of riding a bike. This illustrates the concept of learning and transfer.

of transfer, and it suggests that the team did not really learn the play. A great example of a task that is well learned based on both retention and transfer is riding a bike. You can ride successfully after a winter off, and you can negotiate new routes by transferring the skill.

We know from motor learning research what the reason is: Once you learned to ride a bike, you continued to practice until the skill was automatic and "overlearned." Comprehensive final examinations are designed to get students to this same point. By now you should also understand that retention and transfer are the goals of most learning. So the variables that affect retention and transfer are important to motor learning. What is confusing is that everything we see is performance, but some of what we see is also learning. Performance variables can also be interesting to motor learning students and scholars, as you will see later in this chapter. Performance variables are things that influence performance but do not influence learning.

Overview of Knowledge in Motor Behavior

Now let us look at what motor behavior researchers have learned from years of study. Our goal is not to cover all of motor behavior but to present six major principles that are well supported in the motor behavior literature. These principles are based on research using the measurement and methods outlined in the previous section. These speak to the goals of motor behavior presented previously. In motor control, development and learning research and theory were formed around a concept of how the brain and central nervous system work to create movements. So before presenting the principles, we briefly describe that concept.

To help scientists conceptualize the brain as the master controller in planning, organizing, selecting, and controlling movements, a model called *information processing* was adopted. This model represents the motor behavior system as a computer. General commands are sent from the brain (the central processing unit or CPU) through the spinal cord (wiring), which probably reduces the complexity of the information into relatively simple commands, to the muscles or muscle groups (printer, screen). (If your brain were really like a computer, when you accidentally touched a hot burner on a stove, a message would pop up asking "Are you sure?" before you could pull your hand away.) From this perspective, the goal of motor behavior is to explain response selection (how the skill to be used is selected) and response execution (how the skill that is selected is performed).

Motor Learning

> Motor learning is an internal state that is relatively permanent; practice is required in order for it to occur, and it is difficult to observe and measure.

The two principles we have selected to represent motor learning are that (1) correct practice improves performance and supports learning and (2) augmented feedback enhances practice and thereby learning. These principles were selected for several reasons. Numerous motor learning textbooks (e.g., Fischman 2007) include, at a minimum, a chapter on each of these variables. Furthermore, these are likely the two most widely studied variables in motor learning, and scholars agree (Schmidt 1988) that practice and feedback are the two most important variables in motor learning. Two of our favorite historical quotations are, "Practice the results of which are known makes perfect" and "Practice is a necessary but not sufficient condition for learning." Both of these quotes suggest that practice and feedback are independently important variables, and the relationship between them is also important.

Consider why these variables are important for you. Assuming that there is transfer from one learning situation to another, for example, from learning a motor skill to learning the content in anatomy, understanding how to practice for learning could be helpful. As you move toward a career, you will have to help someone learn something. Although you may not be a teacher, understanding how to help someone learn will be to your advantage. For example, you may need to lead an exercise class or teach an employee how to do his or her job, or you may want to learn a new sport or assist your own child as he or she learns a new skill.

The skill acquisition process is an orderly progression. The learner begins by making many large errors while trying to understand the task. Early in learning, the cognitive demands are great; in fact, the task may be more cognitive than motor (Adams 1987; Fitts & Posner 1967; Gentile 1972). With practice the errors become more consistent; rather than making

The adage "You have to crawl before you can walk" is a good way to think of skill acquisition—an orderly process that sometimes leads to falls!

a different error on each trial, performers make the same errors repeatedly. At this point the demands are less cognitive and more motor, and the errors are smaller and less frequent—response execution is improving. When the learner can execute the skill with fewer and smaller errors and doesn't have to think about the skill while performing it, the skill is considered learned or automatic. Now, instead of thinking about what each body part is doing (response execution), the performer can think about the opponent or strategy when deciding which response to select and execute.

Recall that motor learning is an effort to explain and predict conditions that will make skill acquisition easier or faster as well as make learning more permanent. Such conditions include individual differences in the learner, like speed of movement and coordination. Task differences are also important conditions in skill acquisition, because tasks may be more open, as in batting, or closed, as in bowling. Environmental conditions that may affect learning include practice, **feedback (intrinsic and extrinsic)**, and transfer. The idea of automatic or learned skill is especially important in the study of experts. Research suggests that expertise begins to emerge at 10,000 hours of practice (Ericsson 2003). One could guess you are an expert student and that you will begin to become an expert at your job after about 5 years of full-time work!

At this point you might wonder how we know when something is learned. Clearly that is a challenge for motor learning research, theory, and application.

 ACTIVITY 8.2

Practical Plug-In
In this activity in your web study guide, you'll consider the factors involved in learning an activity (golf, in this case) and improving one's performance in that activity.

Although learning reflects the successful acquisition of a skill, performance reflects the degree to which someone can demonstrate that skill at any given time. Performance is the current observable behavior—in other words, what the learner is doing right now. Sometimes performance reflects learning, as when a player is able to demonstrate successfully his or her newly acquired skill, but at other times it does not. For instance, most of us have taken an examination and after turning it in remembered an answer that we knew but could not put on the paper during the examination. We would argue that we had learned the material but just could not produce it for the examination! In this case we are saying that performance does not represent learning.

One way to distinguish between performance and learning variables is to remember that performance variables have a temporary effect whereas learning variables have a relatively permanent effect. Knowing the difference between these effects is a critical part of motor learning. You may have trouble typing a term paper after you have been working and typing late into the night and are tired. Yet the next day, after some rest, you may type rapidly while making few errors. Fatigue depressed your performance, but you had learned the keyboard through practice and could demonstrate this learning later when you were rested.

Because performance does not always reflect learning, researchers prefer to measure learning with retention and transfer tests (Christina 1992; Magill & Hall 1990). In sport, retention refers to how much of a skill a performer can demonstrate after a period of no practice. Examples of retention include being able to swim at the beginning of the summer after having learned the previous summer or typing after a vacation during which you didn't type. Transfer tests require that you use the information in a slightly different way from how the skill was originally practiced. Using a principle from physics to solve a biomechanical problem is an example. You use transfer in several ways—for example, you transfer skills that you learn in practice to a game, or you use your experience throwing a baseball when you throw a javelin for the first time.

▶ Because variables such as fatigue can influence motor performance, tests of retention and transfer provide the best measure of motor learning.

Practice

Practice is repetition, but of course there is more to understanding why practice is such an important factor. More practice is usually associated with better performance and is a requirement for learning. However, practice must be organized differently depending on whether the objective is performance (short term) or learning (long term). Has a teacher, coach, or someone else ever said, "Oh come on, it is easy," when you were struggling to learn a new skill? Hearing that remark can be maddening, because performing the skill is not easy. Practice that leads to learning is hard work. Good teachers and coaches understand this principle of motor learning. In your motor learning class you will study practice in two phases, comprising before-practice variables and during-practice variables, specifically:

Before Practice

- Goal setting (Locke & Latham 1985)
- Instructions
- Demonstrations or modeling (McCullagh 1993; Weiss & Klint 1987)
- Mental practice

During Practice

- Scheduling of practice
- Context of practice (Chamberlin & Lee 1993; Schmidt 1991)

ACTIVITY 8.3

Planning Practice
In this activity, you will determine the type and order of practice variables in two different scenarios.

Once you have successfully completed a motor learning course, you will understand the prepractice variables and how to schedule and organize practice. Some of these variables are presented in the Example of Practice Organization sidebar. For example, constant practice involves doing the same thing over and over, and variable practice involves not doing the same thing on consecutive practice trials. The first step is to consider the learner and the skill, and the next step is to plan the type of practice.

Feedback—Knowledge of Results and Performance

Knowledge of performance is given by an instructor who provides feedback about the nature or process of the movement. Knowledge of results is information about the outcome of the movements that a teacher, coach, or trainer may supply.

As you can see, feedback is an integral part of the practice regimen. Learning a skill correctly cannot occur without feedback. Feedback guides the learner toward performing the task correctly and also reinforces correct performance. Feedback can be intrinsic or extrinsic (also called *augmented*). Intrinsic feedback is information about performance that you obtain for yourself as a result of the movement. Extrinsic feedback is information provided by an outside source such as a teacher or videotape.

Besides distinguishing between intrinsic and extrinsic feedback, motor behaviorists also categorize extrinsic feedback as knowledge of results (KR) and knowledge of performance (KP). Figure 8.4 presents examples of intrinsic and extrinsic KR and KP. Knowledge of results is information about the result (e.g., you missed the target); it helps advanced performers more than it does beginners because advanced performers understand how to make corrections based on KR. That is, they are more likely to understand how to change the movement to influence the outcome. A good typist who is told that he keyed in "Hybe" instead of "June" would know that the right hand was one key away from home. Because he is an advanced typist, he is able to use KR feedback. A beginner typist, however, might need a cue ("Check

Example of Practice Organization

Your sister has asked you to teach your 7-year-old nephew to swim. He remembers choking after inhaling water when he was younger and has avoided swimming lessons since then. Fortunately, he looks up to you and is willing to give swimming a try. Unfortunately, you have not taught swimming before. So you face an important challenge and undoubtedly realize that to be successful you will need to do some planning.

At this point you know that practice and feedback are important to learning. We can consider planning practice in three parts: prepractice, practice, and postpractice. During practice you also need to think about feedback. We know that our participant is a novice. What kind of skill is swimming? *Continuous* and *relatively closed* are good descriptors. As you plan the first practice session, consider what the learner is thinking, what you—the teacher—are thinking, and what you should say. The following table demonstrates this process.

	The nephew thinks:	The teacher thinks:	The teacher says:
Prepractice (goal setting, simple instructions, modeling)	"I am afraid. This is going to be hard. I will probably fail."	"I should set a goal, demonstrate, and keep my instructions simple."	"It is OK to be afraid; you are smart to be afraid. The only way not to be afraid is to learn to swim. So today I want you to see how the water holds me up." (The teacher demonstrates a float.) "To float, you must put your face in the water. Once you get your face wet, then we will work until you keep your face in the water for 5 seconds."
Practice (constant practice, contextual practice, blocked practice, extrinsic feedback)		"So far, so good. He is not choking, and his face is in the water. Short practice trials with rest in between are working. I should show him what I want him to do. I can give him feedback when he comes up for air."	"That was great—you had your face in the water for 5 seconds. I am proud of you. Next time I want you to get your ears wet too, so put your face in far enough to get your ears wet. I know you can do it! Let me show you what I want." (The teacher demonstrates.)
Postpractice (verbal rehearsal, goal setting)	"That wasn't so bad. I can put my face in the water for 5 seconds, and I didn't choke. I could feel the water in my ears."	"Next I need to increase the difficulty, but I must also let him rest so that he doesn't get fatigued and make a mistake."	"You are making progress. Now we will practice walking in the water, and then we'll try putting your face in again. Can you tell me what you've accomplished so far?"
	"I put my face and ears in the water."		"Good. You know, it took a lot of practice before I learned to swim. If we keep working you will learn to swim just like I did."

(continued)

Example of Practice Organization (continued)

A few days later the practice continues, as you can see in this table.

	The nephew thinks:	The teacher thinks:	The teacher says:
Prepractice (goal setting, demonstration, verbal rehearsal)	"So far this is easy. I am pretty good at this. I can put my face in the water, walk in the water, and float."	"He puts his face in the water, walks in the water, and can do each of the parts (float, kick, and arms). I think he is ready to begin to swim. So, I want to provide verbal cues that he can rehearse and use for the progressive part of practice."	"Today you are ready for something new. You can count to three—and there are only three things you need to remember and do to swim: (1) Put your face in the water and float, (2) kick your feet, and (3) move your arms. Let me show you while you count for me." (The teacher demonstrates.) "What do 1, 2, and 3 stand for? Now it's your turn."
Practice (blocked practice, extrinsic feedback after several trials)	"OK, 1 is face in to float, 2 is kick, 3 is arms."	"He did it! Now we need to repeat this several times with just a little rest between. Should I change the practice? Let's see, this is a closed continuous skill, so constant practice is OK. Maybe we should practice in a different pool or at least a different place in this pool next time."	"Good job. You did all three. Now do it again." (Repeating several times.) "You are going about 10 feet before you stand up. That is good. You are working very hard. Let's see if you can go a bit farther. From where you are to where I am standing is about 12 feet. Try to get to me. I am sure you can do it!"
Postpractice (mental and verbal rehearsal, extrinsic feedback)	"I am totally awesome. 1, face in and float; 2, kick; 3, arms."	"He is swimming. We have made a lot of progress, but I am worried that the progress will slow down. I need to keep him motivated."	"You reached the goal—good work! Now say the three steps for me. Good. Can you think of the way it feels as you say the steps? Think about this and say the steps before next time."

your home position") to correct the error. This typist is using KP feedback. Sometimes we tell a beginning driver to "slow down" or "slow up"; why do those phrases mean the same thing? The goal of feedback is ultimately to help performers detect and correct their own errors. The study of feedback has focused on KP and KR with regard to frequency, precision, modality, and processing time. Feedback may also be reinforcing ("That was a good effort") or negative ("You did not try very hard"). However, KR and KP are the forms of feedback discussed here.

 ACTIVITY 8.4

KR and KP

Test your understanding of the two types of knowledge in this activity in your web study guide.

ACTIVITY 8.5

Learning and Performance

Differentiate among practice variables and identify appropriate practice settings in this activity in your web study guide.

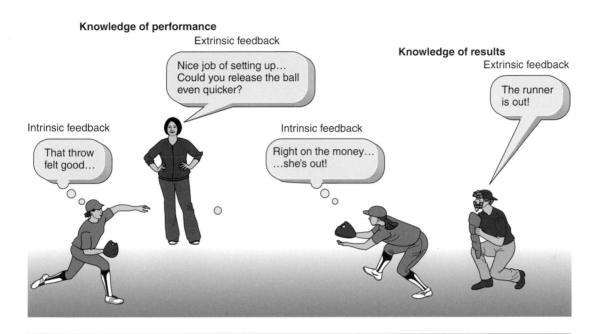

FIGURE 8.4 Four kinds of feedback for a shortstop.

Frequency of feedback refers to the percentage of the practice trials on which feedback is given (e.g., 50% or every other trial vs. 100% or all trials). On the surface, more would seem better, but the performer can become dependent when feedback is too frequent. If the goal is to detect and correct your own errors, too much feedback is counterproductive—you do not learn the detection and correction processes when you receive feedback all the time. Knowledge of performance (or KP) should be given more often at the beginning of learning and then gradually reduced (Salmoni et al. 1984), a process often referred to as *fading* the feedback. Besides considering KP versus KR and the frequency of feedback, you will learn about two other variables in motor learning—precision and modality of feedback. Motor development considers the amount of time that children need to use the feedback, so you will likely learn more about this as well.

In baseball, the batter watches the pitcher for critical information that may give a clue about the pitch; the batter ignores the crowd's yelling and gathers only related advance information

> Learning requires both practice and feedback. Extrinsic feedback (e.g., KR or KP) must be information that the learner could not obtain on his or her own, should be corrective, should be provided on about half of the trials, and must be followed by sufficient time to make corrections before the next attempt.

Feedback in Golf

Here's a fun way to test yourself on how well you can identify types of feedback. Read the joke and try to categorize the caddy's feedback.

Golfer: "Do you think my game is improving?"
Caddy: "Yes, sir, you miss the ball much closer now."
Golfer: "I think I'm going to drown myself in the lake."
Caddy: "Do you think you can keep your head down that long?"
Golfer: "You've got to be the worst caddy in the world."
Caddy: "I don't think so, sir. That would be too much of a coincidence."

about the pitch and where to hit it. Figure 8.5 provides an example of how long the batter has to make decisions about swinging the bat after the pitcher releases the ball. The figure demonstrates that a batter has more decision time (150 milliseconds vs. 130 milliseconds) if he waits longer and speeds up the swing (140 milliseconds vs. 160 milliseconds). The example also demonstrates how critical advance information is to successful performance. Thus, it becomes obvious why batters want to know in advance whether a pitcher is likely to throw a curve or a fastball as well as the potential location of the pitch. You may hear a baseball commentator say, "The count is three balls and one strike; the batter will be looking for a fastball high and inside" (a certain pitch in a specific location). The batter is trying to reduce the decision to a go or no-go situation—he'll swing if the pitch is what he expects—rather than consider all possible options. The batter can do this because in this situation he can decide not to swing even if the pitch is a strike. If the count were three balls and two strikes, the batter would be less likely to look for a specific pitch.

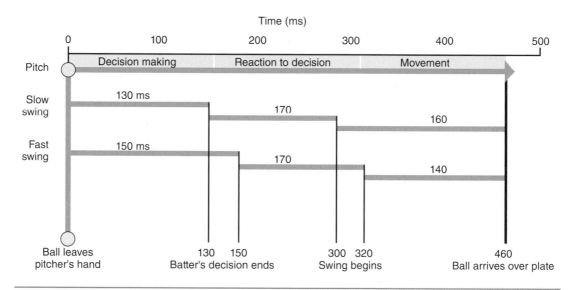

FIGURE 8.5   Time line showing the critical events in hitting a pitched baseball. The movement time is 160 milliseconds for the slow swing and 140 milliseconds for the fast swing.

Reprinted, by permission, from R.A. Schmidt and C.A. Wrisberg, 2007, *Motor learning and performance: A situation-based learning approach*, 4th ed. (Champaign, IL: Human Kinetics), 153.

Motor Control

The first principle of motor control is that the brain uses the central nervous system to initiate and control muscles that make the movements. The second principle of motor control is that a goal of most movements is to rely on the decision-making centers in the brain as little as possible once the movement is initiated. Two theories, and many research studies, led us to select these two principles. First, motor programs (Schmidt 1975) are the theoretical explanation of how we successfully produce and control movements. We can compare a motor program to a computer program that does math problems. First, you select the program to use; this is response selection. The program can add, subtract, multiply, and divide. If you put in the numbers and indicate what math operations to perform, the program outputs the answer; this is response execution. A motor program is like that; you select the program (response selection) and indicate what it should do (operations). Then the program specifies how to do the skill and sends signals through the spinal cord to the muscles that perform the movement (response execution). Schmidt (1991) indicates that motor programs, at a minimum, must do the following five things:

- Specify the muscles involved in the action
- Select the order of muscle involvement

- Determine the forces of muscle contraction
- Specify the relative timing and sequences of contractions
- Determine the duration of contractions

If we had to remember every single movement we have made, our memory would be overloaded. Motor programs explain why we do not have a storage problem. Instead of storing in memory each movement that we have ever done, we store groups of movements with similar characteristics. These groups, called *schema*, are the foundation of motor program theory (Schmidt 1975).

Not all researchers agree with the idea of motor programs from schema theory, so they have proposed another theory called *dynamical systems* (Haken et al. 1985; Kelso 1995; Fischman 2007), which suggests a more direct and less cognitive link between information that the perceptual system picks up and the motor action that occurs. This direct link is called a *coordinated structure*. One characteristic of a coordinated structure is automated movement that relies on very little decision making or central control in the brain.

You may be wondering which theory—motor program or dynamical systems—more accurately represents the process of motor control. A decisive test for theories of motor control is how well they explain motor learning and the development of motor expertise (Abernethy et al. 1993). You will have a chance to examine and compare these theories in your motor learning class.

The study of motor control addresses five areas:

1. Degrees of freedom—coordination (Rosenbaum 1991)
2. Motor equivalency
3. Serial order of movements—coarticulation (Rosenbaum 1991)
4. Perceptual integration during movement (Rosenbaum 1991)
5. Skill acquisition (Rosenbaum 1991)

> ► Motor programs are a proposed memory mechanism that allows movements to be controlled. As motor programs are developed, they become more automatic, allowing the performer to concentrate on the use of the movement in performance situations.

> ► Future research that contrasts predictions and key elements from a dynamical systems view of motor behavior or a motor program view is likely to be both exciting and controversial.

(⚡) ACTIVITY 8.6

Practical Plug-In
In this activity in your web study guide, you'll learn more about motor learning by investigating how the human body responds and adapts after an injury.

The brain initiates the planning of movements, and the nervous system then sends signals through the spinal cord to the muscles, which in turn make the movements. What is not known is how the brain represents the information to be sent to the muscles. The area of motor control is an effort to understand what the brain, nervous system, and muscles are doing to direct movements. Although skill acquisition—the notions of learning and improvement—is the focus of motor learning, it is critical in motor control as well. Skill acquisition accentuates the relationship between motor learning and motor control.

Once again, you may be wondering why understanding motor control is important. The answer is obvious for someone interested in physical therapy, rehabilitative medicine, sports medicine, or athletic training. Anyone who wants to learn or teach a skill uses the principles of motor control by trying to use the simplest movement possible at the early stages of learning. Some of what you will learn about motor control is less intuitive and often misunderstood. For example, consider the baseball batting example from figure 8.5; here motor control tells us that the most rapid adjustments are made when we are not actually thinking about the flight of the pitch or the swing of the bat. That is, we turn the control of the skill over to a more automatic part of the system. The two theories of motor control explain this equally well. However, the idea of automatic responses is not well understood by many sport enthusiasts. When you hear coaches or parents say, "Watch the ball all the way to the

▌ Understanding the theories of motor control enables this therapist to help her elderly patient relearn balance skills.

bat," you know they do not understand how to best use the sensory system to make rapid and accurate adjustments. We hear comments that allude to this shift from central (decision-making) control of movements to peripheral control when athletes say, "It just felt right."

Understanding motor control is important because the range of performance of motor skills from normal to expert depends upon the shift from learning (central control, practice, and feedback) to more automatic movements that are programmed ahead of the movement and controlled peripherally.

Developmental Motor Learning and Control

The two principles of motor development (developmental learning and control) are that (1) children are not miniature adults and (2) children are more alike than different (Thomas & Thomas 2008). After watching a baby and a child perform a motor task, you can see clear differences and similarities related to age. Infants demonstrate more random movements, have a smaller repertoire of movements, and tend to do the same movement repeatedly. Children have a larger repertoire of motor skills, perform more voluntary skills, and exhibit greater skill than infants (Seefeldt & Haubenstricker 1982), although less than adults. Children's movements are not scaled-down versions of adult movements. One explanation for the relative inefficiency of children's movements may be their lack of practice. In addition, adults probably select and plan movements, a cognitive process, better than children do.

Studying children is valuable because children typically change in an orderly fashion so that the order of change is the same across children, whereas the rate varies. Hence the second principle, that children are alike, refers to gender, culture, and developmental age. Understanding how motor skills develop and the factors influencing development allows us to make predictions. Those predictions are important to pediatricians, teachers, and parents. Virtually all the questions in motor learning and motor control can be studied as part of motor development—by examining the same questions across age. In addition, some unique problems arise in studying infants, children, and persons who are elderly. Children differ from adults in several ways: physical growth, information processing, experience, and neurological development (Thomas et al. 2001). Although some topics are unique to developmental motor learning and control, considerable overlap occurs with the previous two goals of motor behavior—motor learning and motor control (Ulrich 2007). In some university programs, motor development is part of the motor learning and control course, and in other universities the class is separate. Motor development includes several topics that are not part of motor learning, for example, physical growth and youth sport.

> If you are planning to study physical activity, you must understand the physiology, biomechanics, and motor behavior underlying the development of movement to address problems such as how children gain control of movement skills, how people achieve expertise, and how these skills deteriorate as people age.

ACTIVITY 8.7

Practical Plug-In

In this activity in your web study guide, you'll think about what a new coach needs to know in order to make sure his beginning soccer players improve and have fun.

Developmental Changes in the Mechanics of Movement

The mechanics of movements are different at different ages partly because the body executing the skill is of a different size and has different proportions. For example, a baby's head is one-quarter of her height, whereas an adult's head is only one-eighth of her height. The legs also change dramatically during childhood. At birth the legs are typically less than 30% of body length; adult leg length is often more than half of body length. Obviously, these physical characteristics will influence balance and locomotion.

Growth is also a factor in motor behavior; children are growing until around 13 to 14 years of age for females and around 18 to 20 years of age for males, whereas growth is not a factor in adult motor learning studies. Children grow in three physical dimensions: overall size, proportions (e.g., leg length, shoulder breadth, chest depth), and body composition (e.g., increases in muscle). Growth influences motor performance, in part because children must contend with changes in their bodies, an issue that adults do not have to deal with. Furthermore, the adult–child differences can be factors explaining child-to-child differences. For example, size and strength have a positive relationship: Larger children are stronger. As children grow, the increase in size produces increases in strength. But strength also increases because of neuromuscular efficiency. Because growth is not under our control, we must understand how growth influences performance to accommodate the challenges of physical growth.

Life Span Development

Besides addressing the effects of growth on motor performance, developmental motor learning and control also deals with the effects of aging on motor performance; thus, this area of study is sometimes referred to as *life span development* (Spirduso & MacRae 1990; Stelmach & Nahom 1992). Changes in strength and motor coordination are most rapid at the extremes of the age continuum. Growth is less important after adolescence, because the changes in physical parameters are less dramatic in adults and elderly people than in infants and young children.

Cognitive processes also change during childhood because of the developing nervous system and because of experience. Children use fewer cognitive processes and use them less effectively than adults do. For example, when asked to remember a movement, a 5-year-old child might "put on his thinking cap"—not a very effective strategy for remembering. A 7-year-old might do the movement repeatedly, whereas a 12-year-old could repeat the movement in a series composed of several movements. Adults and older children know that they must do something to learn and remember, whereas younger children may not recognize the importance of deliberate practice or may select an ineffective practice regimen. One of the reasons for the increasing interest in cognitive processing during motor skill acquisition in children is that the strategies used by adults can be taught to children to improve performance. In other words, this is an area in which research enhances teaching.

Experience

Experience was presented as a factor in children's performance. Experience can help elderly people and can be a negative factor for children. In a situation in which experience is a benefit, those who are elderly have an advantage and children have a disadvantage because children lack experience and elderly people have more experience. Clearly, we can enhance experiences for children in two ways—quality and quantity. Research shows that children with experience can outperform adults with less experience, which means that practice and experience help. This finding has clear implications for pedagogy and curriculum development. In other words, curriculum and instruction need to provide experience to enhance sport performance in children. Furthermore, we know that the quality of the experience can be improved through use of what we know about information processing to help children get more out of the practice. So, if we provide children with adult learning strategies, they will benefit more from a given amount of experience. The key is that children need practice, and that to be effective, practice must help children retain information, skill, and decision making, which they would normally lose. To do this, we teach adult learning strategies to children.

Changing Neuromuscular Systems

The hardware (analogous to the computer itself and its disk drives and monitor), or neurology, of children is different from the hardware of adults. Infants have fewer synapses (connections between neurons in the nervous system) and less myelin (the covering of nerve cells with a fatty sheath that aids nerve impulse transmission). In addition, babies at birth have fewer neurons than adults do. The hardware is nearly complete as children begin school, but it may not function as well as it does in older children and adults.

ACTIVITY 8.8

Practical Plug-In
In this activity in your web study guide, you'll consider what a future practicing professional majoring in kinesiology needs to know to understand and help a family member recover from a stroke.

One of the most intriguing developmental questions in motor behavior is how the brain and nervous system adjust their control to increases in cognitive function, body size, and strength during childhood and to decreases in those variables as people age.

The deteriorating nervous system in persons who are elderly may negatively affect performance. As people age, they lose neurons in the brain and motor neurons in muscles. This change results in slowness and variability in movement control. Older people, however, benefit from experience and use it to compensate for the loss of speed, strength, and control. Understanding the changes in the central nervous system is important because we have little opportunity to accelerate or decelerate these changes.

Motor control issues are important in understanding adult motor behavior, and they are important in developmental learning and control. Critical issues are similar to those listed in the previous section on motor control (for a review, see Clark & Phillips 1991). One additional point is important: How does growth become integrated into the motor control system? As previously mentioned, children may change very rapidly, particularly at puberty (Malina 1984; Malina et al. 2004). How does the motor control system account for changes in size, proportion, and mass? Consider a boy who played baseball from March until July at 12 years of age. By the next March he is 13 years of age, he has grown 4 inches (10.16 centimeters) in height, his arms have grown in length, and he weighs more. He has not practiced batting or throwing much since the last summer, yet he can still bat and throw successfully. How does the nervous system compensate for these changes to produce coordinated movements?

At the other end of the age continuum are persons who are elderly. How does a system of motor control that has functioned for years using one set of parameters account for loss in cognitive function, body mass, flexibility, and strength? No good explanation has been found for these important issues from the developmental aspects of motor learning and control, although some progress is being made (for reviews, see Thelen et al. 1990; Thomas et al. 2001).

Growth and Gender in the Development of Overhand Throwing

We have all seen examples of people of different ages and genders who show markedly different physical abilities. Can we document those differences? What are the reasons for them? Here we look at what researchers have learned about overhand throwing. First, we need to examine physical growth and development. See figure 8.6 for a summary of growth in terms of stature and weight, and figure 8.7 for proportion. Then we will see how these factors affect overhand throwing skills.

Physical growth is rapid during infancy, constant during childhood, and

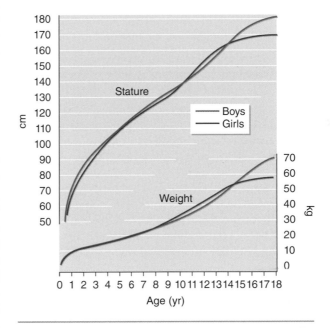

FIGURE 8.6 Average height and weight curves for American boys and girls.

Adapted from R.M. Malina, 1984, Physical growth and maturation. In *Motor development during childhood and adolescence*, edited by J.R. Thomas (Minneapolis, MN: Burgess), 7. By permission of J.R. Thomas.

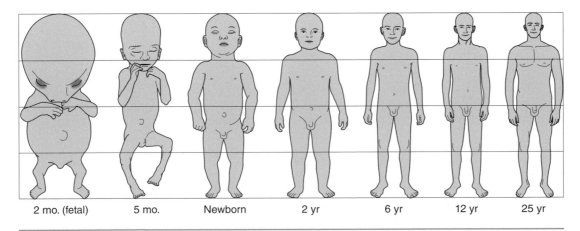

| 2 mo. (fetal) | 5 mo. | Newborn | 2 yr | 6 yr | 12 yr | 25 yr |

FIGURE 8.7 Changes in form and proportion of the human body during fetal and postnatal life.

Adapted from K.M. Newell, 1984, Physical constraints to the development of motor skills. In *Motor development during childhood and adolescence*, edited by J.R. Thomas (Minneapolis, MN: Burgess), 108. By permission of Jerry R. Thomas.

rapid during the growth spurt associated with puberty (Malina 1984; Malina et al. 2004). Children's limbs grow more rapidly than does the torso or head. The increase in proportion of total height attributed to leg length explains some of the improvement in running and jumping performance during childhood. The shoulders and hips become wider, with males' shoulders becoming much broader on average than females' shoulders, giving males an advantage in activities in which shoulder power is helpful, such as throwing. Growth is orderly, with length and then breadth and circumferences increasing. Children in upper elementary school grades often look awkward—their legs and arms appear to be long and skinny. Figure 8.7 shows the change in proportion from birth though adulthood. After most of the length of a limb has been attained, the limb gains thickness. The bone circumference and muscles (and fat!) increase to make the body look more proportional; children fill out. Men's shoulders and chests usually grow (bones and muscle) until about 30 years of age, whereas most women have stopped growing by their late teens.

The changes in physical growth present three challenges: mechanical, adaptive, and absolute. The mechanics of movement change because of different body proportions; the

Development of Body Actions During Throwing

Level	Body actions
Preparatory arm backswing	
Level 1	The child's hand moves directly forward to release from where the child first grasps the ball.
Level 2	The forearm moves backward beside the head and then forward in ball release.
Level 3	The ball moves backward with a circular and upward backswing with elbow flexion.
Level 4	The ball moves backward with a downward swing below the waist and then up behind the head with elbow flexion.
Arm action	
Level 1	The upper arm moves backward and forward at an angle to the body. The forearm is linked to the upper arm movement and does not lag behind it.
Level 2	The upper arm is aligned with the shoulder and moves with it until near release, when the upper arm leads the shoulder. The forearm lags behind the upper arm (i.e., it appears to stay at the same point after the shoulder and upper arm have started forward).
Level 3	The upper arm lags behind the shoulder as the forward movement of the arm begins. The forearm lags even farther behind the upper arm and shoulder, reaching its final point of lag as the body is front facing.
Trunk action	
Level 1	Little or no trunk action occurs; the throw is accomplished by just the forward and backward action of the arm.
Level 2	The trunk rotation is in one block; the spine and hips move together in one rotating action.
Level 3	The hips lead in forward rotation followed by the spine (differentiated rotation).
Foot action	
Level 1	Feet do not move during throwing action.
Level 2	Step occurs but with the foot on the same side as the throwing hand.
Level 3	Step occurs but with the foot on the opposite side of the throwing hand.
Level 4	Same as level 3 except that the step is longer (about one-half of standing height).

Reprinted from M.A. Roberton, 1984, Changing motor patterns during childhood. In *Motor development during childhood and adolescence*, edited by J.R. Thomas (Minneapolis, MN: Burgess Publishing), 75.

individual must adapt to a rapidly changing body. These changes are especially problematic in seasonal sports. Consider the wrestler who experienced a rapid growth spurt from the end of one season to the beginning of the next. His center of gravity has changed, which may influence his balance and the location of optimal points for exerting maximal force. Finally, absolute changes in size may influence performance. Females gain fat at puberty, which adversely influences performance in most physical activities; but males gain muscle, which has a positive influence on performance. During childhood, physical growth interacts with all the other factors that are developing (e.g., motor programs) and therefore must be considered for instruction and research.

A good way to examine structural and functional change and its influence on skill across childhood and adolescence is to consider overhand throwing. Figure 8.8 shows the changes in throwing for distance that occur across childhood and adolescence for girls and boys.

The developmental nature of the overhand throw across childhood and early adolescence is detailed on page 242. This description is separated into various body actions during the throw (e.g., arm, trunk, foot).

Given that prepubescent boys and girls are structurally similar, differences as large as those observed in throwing are unusual. Thomas and French (1985) noted that the gender differences in throwing at 5 years of age were three times as large as those for any of the 20 tasks that they reviewed. Although it seems likely that practice, encouragement, and opportunities account for a large part of these differences, throw-

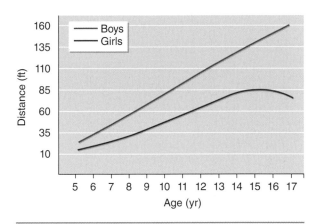

FIGURE 8.8 Age differences in throwing for distance.
Data from Espenschade 1960.

ing may be one of the few motor skills in which biological factors play a significant role in gender performance differences before puberty. Again we see that although we are unique, we (males and females) are more alike than different—both before and after puberty!

Student Learning Demonstrates Teaching Skill

How well did you do in planning swimming practice for your nephew? Compare the information in the sidebar on pages 233 and 234 to the sections on practice, feedback, and transfer. How can you judge whether practice sessions influence the learning of a skill? If your nephew could do what you presented in the first lesson the next time you met, you have evidence of retention. Remember that retention is one measure of learning. What characteristics of practice might disguise whether skills are being learned?

Wrap-Up

Although the study of motor behavior has produced important knowledge for human behavior since the late 1800s, it began to evolve as a significant subdiscipline in the field of physical activity in the 1960s.

Knowledge developed through motor behavior has become increasingly important in all aspects of society. Although we often think of learning and control of sport skills as the main issue, our society depends on human movement in many ways—babies learning to use a spoon, surgeons using a scalpel, pilots learning to control airplanes, children learning to control a pencil in writing, dentists using a drill. Understanding the development, learning, and control of these and other motor skills so that people can use them more effectively is the goal of motor behavior.

Knowledge of motor behavior is essential in several professions, including physical or occupational therapy, physical education, coaching, or working with children in community organizations such as the YMCA or YWCA or Boys & Girls Clubs. Your school probably offers at least one undergraduate course in motor learning, motor control, and motor development.

More Information on . . . Motor Behavior

Organizations

American Geriatrics Society

Canadian Society for Psychomotor Learning and Sport Psychology

International Society of Motor Control

Motor Development and Learning Academy of NASPE

North American Society for the Psychology of Sport and Physical Activity

Society for the Neural Control of Movement

Society for Neuroscience

Journals

Journal of Motor Behavior

Journal of Sport and Exercise Psychology

Kinesiology Review

Motor Control

Pediatric Exercise Science

Research Quarterly for Exercise and Sport

 ACTIVITY 8.9

Use the key points review in your web study guide as a study aid for this chapter.

Review Questions

1. How does motor behavior differ from the psychology of sport?

2. Explain the differences between motor learning and motor control within the field of motor behavior.

3. Why is the change in motor learning and motor control across the life span of interest?

4. Think about the practice issues discussed in this chapter, such as feedback, retention, transfer, goal setting, and scheduling. Choose a sport with which you are familiar and discuss how the practice characteristics would influence your planning if you were a coach. Pick a specific age group or performance level, such as high school, college, or professional coaching.

5. Can you think of an example in which more difficult practice conditions result in better retention and transfer? Why does that happen? Can you plan practices to promote this? How?

6. Discuss when it might be best to provide either knowledge of performance or knowledge of results to a person learning a motor skill.

chapter nine

Sport and Exercise Psychology

Robin S. Vealey

chapter objectives

In this chapter we will

- discuss what sport and exercise psychologists do;

- describe how sport and exercise psychology evolved within the field of physical activity;

- help you understand how professionals in sport and exercise psychology engage in research and practice;

- provide an overview of what research tells us about personality, motivation, arousal and anxiety, and group processes in sport and exercise settings; and

- explain imagery and attentional focus as examples of how mental skills training is used in sport and exercise.

You are an elite swimmer poised on the starting block for the 50-meter freestyle race at the Olympic trials. In the other lanes are the top swimmers in the country who, like you, have trained for years to hone their bodies into finely tuned physical machines—all in preparation for this moment. What are you thinking? Should you even *be* thinking? Where is your attention focused? Do you have negative thoughts and fears about not performing well, or do you feel a confident excitement that you are ready to meet the challenge? Do you feel a relaxed inner calmness or a feverish emotional intensity? You have prepared yourself physically for this moment by swimming hours and hours for miles and miles. But have you prepared yourself mentally?

The psychologist William James wrote that "the greatest discovery . . . is that human beings, by changing the inner attitudes of their minds, can change the outer aspects of their lives." How true that is! Our participation in physical activity is greatly influenced by our "inner attitudes," or how we *think about* what we're doing. This mental aspect of physical activity is the focus of study in sport and exercise psychology. Specifically, sport and exercise psychology focuses on the mental processes of humans as they engage in physical activity.

If you're the Olympic swimmer in the opening scenario, systematic training with a sport psychologist has prepared you to think productively and focus optimally at this crucial moment. Your mental practice has prepared you to focus your attention on productive, energizing thoughts instead of worries and distractions. You have programmed your mind and body to respond automatically at your preferred level of emotional intensity. You have mastered the mental focus and physical readiness that you need for your best performance, and now your body and mind are one, waiting to explode from the platform for the performance of your life!

Most of us will not compete in the Olympic Games, but the mental aspects of physical activity will challenge us repeatedly as we go through our lives. Consider the following situations that you may encounter:

- You take up mountain biking and are amazed at how the activity enhances your self-esteem and confidence along with your physical health.

- You begin an exercise program to enhance your personal fitness, but you lack motivation and quit the program after 2 months.

- You really enjoy your weekly tennis league, but you become very nervous playing doubles because you don't want to let your partner down.

- As the coach of a local youth soccer team, you are deciding on the best ways to provide feedback and structure practices for children.

- Your father is overweight and worried about his health and asks you to help him establish a physical activity routine that he will enjoy.

All these examples illustrate the critical and fascinating link between mind and body inherent in physical activity.

▶ Physical activity always involves the mind as well as the body. The mental aspect of physical activity is the realm of sport and exercise psychologists.

Why Use Sport and Exercise Psychology?

Sport and exercise psychology involves the study of human thought, emotion, and behavior in physical activity. Think of sport and exercise psychology as the study of the ABCs of physical activity. *A* represents the term *affect,* which means emotion. For example, the emotions of anxiety and anger are studied to see how they influence athletes' performances. Emotional moods are influenced by exercise, which has been shown to decrease depression and enhance feelings of well-being.

B is for *behavior.* Examining why people behave differently from one another when engaging in physical activity is fascinating. That is, why are some people so committed to and persistent in their exercise routines? Why do some athletes train exhaustively to achieve excellence, whereas other talented athletes don't work as hard and thus don't reach their potential?

Goals of Sport and Exercise Psychology

- To understand the social–psychological factors that influence people's behavior and performance in physical activity
- To understand the psychological effects derived from participation in physical activity
- To enhance sport and exercise experiences for those who participate in physical activity

 ACTIVITY 9.1

Do You Know Your ABCs?
Test your understanding of affect, behavior, and cognitions in this activity in your web study guide.

Finally, *C* represents *cognitions.* Sport and exercise psychology examines how the thought processes of individuals, that is, their cognitions, influence and are influenced by their participation in physical activity. Attempting to understand why athletes choke under pressure requires an understanding of their thinking processes. Similarly, exercise psychologists are interested in how physical activity can help individuals develop self-esteem and confidence or better ways to think about themselves.

Remember your ABCs to keep in mind that sport and exercise psychology is the study of affect (emotions), behavior (actions), and cognitions (thoughts) related to physical activity participation.

> Sport psychology and exercise psychology, as subdisciplines of kinesiology, focus on the study of human thought, emotion, and behavior in physical activity.

This chapter discusses the psychology of physical activity in relation to the specific areas of sport and exercise (figure 9.1). Sport psychology and exercise psychology have developed into two distinct areas of study in the physical activity field. Exercise psychology is devoted to the psychological aspects of fitness, exercise, health, and wellness. Sport psychology focuses more on the psychological aspects of competitive sport participation. Because the two areas are closely related and share a great deal of theoretical content, however, this chapter covers both.

FIGURE 9.1   Sport and exercise psychology and the other scholarly subdisciplines.

Although sport and exercise psychology are areas of study in kinesiology, they have close ties to the discipline of psychology. Psychology is the science that deals with human thoughts, feelings, and behavior, so sport and exercise psychology deals with the thoughts, feelings, and behaviors of people involved in sport and exercise. For example, researchers attempting to understand why people begin or discontinue participation in fitness programs often apply motivational theories developed in the discipline of psychology.

In summary, sport and exercise psychology studies the ABCs of physical activity to learn how and why people feel (*A* for affect), act (*B* for behavior), and think (*C* for cognitions) in various ways as they participate in sport and exercise. These areas of study focus on the fascinating link between our minds and bodies.

What Do Sport and Exercise Psychologists Do?

Like all kinesiology professionals, sport and exercise psychology professionals attempt to enhance people's experiences in physical activity. As in any subdiscipline of physical activity, sport and exercise psychology professionals work in many career areas.

Kinesiology professors of sport and exercise psychology at universities fulfill the multiple roles of researcher, teacher, and service provider. A kinesiology professor specializing in exercise

---

Exploring the ABCs of Physical Activity

Consider the following questions related to sport and exercise psychology:

1. Recalling that we've defined physical activity as intentional and purposeful movement (chapter 1), think of several forms of physical activity in which you participate. What do you think your motivation is for performing each of these activities?

2. Do you think that participation in physical activity has influenced your character? If so, what was the activity and how do you think it affected you? If not, why do you think that participation has not influenced your character?

3. Have you ever been involved in or watched any form of violence in sport? If so, what happened and why? What do you think caused the violence to occur?

4. Why do you think so few Americans participate in habitual exercise when it has been clearly shown to enhance health and longevity?

---

psychology may develop theory and conduct research on the psychological benefits of aerobic exercise, teach courses in exercise psychology, develop a health promotion program within the university, and conduct workshops for the community on motivation and stress management. A kinesiology professor specializing in sport psychology may conduct research on stress in sport and athlete burnout, teach courses in sport psychology, consult with athletes from the university and surrounding community, and offer regional workshops on coaching effectiveness.

Some sport and exercise psychology–trained kinesiologists focus more exclusively on practitioner or service provider roles. For example, a sport psychology specialist may work on performance enhancement, personal development, and lifestyle management issues of athletes within a university athletic department, a professional sport team, or a private sport training academy. Likewise, an exercise psychology specialist may focus on worksite health promotion for a large corporation, develop fitness and wellness programs for a community recreation department, or coordinate physical activity programs for all age groups through a YMCA. Sometimes sport and exercise psychology professionals become consultants for sports medicine or physical therapy clinics to help injured people through the psychological aspects of injury rehabilitation.

Differentiating between an exercise or sport psychologist trained as a physical activity specialist and one trained as a clinical or counseling psychologist is important. Clinical or counseling psychologists are licensed to provide psychotherapy and consultation for individuals with clinical conditions such as depression, phobias, or anorexia nervosa. The focus of kinesiology-trained practitioners of sport and exercise psychology within the physical activity field is on education or the teaching of skills to enhance the performance or personal fulfillment of individuals involved in sport or exercise. Although some clinical and counseling psychologists provide services for athletes and exercisers, the main focus of their practice is different from that of sport and exercise psychology as a subdiscipline of the physical activity field.

▶ Career opportunities in sport and exercise psychology include positions as university professors, performance enhancement specialists, fitness and health promotion specialists, and sports medicine consultants.

History of Sport and Exercise Psychology

Throughout history, the important connection between mental health and physical activity has been noted. However, sport and exercise psychology was not recognized as a subdiscipline of kinesiology until the 1960s.

Beginnings of Sport and Exercise Psychology

The relationship between exercise and psychological well-being was recognized as early as the 4th century BCE (Buckworth & Dishman 2002). Around the turn of the 20th century, physical educators began to write about the psychological benefits of physical activity.

Researchers began assessing the social influences of the presence of others on motor performance, a research area that later became known as social facilitation. The first person to examine these social influences was Norman Triplett, who in 1898 studied the effects of the presence of other people on bicycling performance. The effects of exercise on depression were reported in 1905 (Franz & Hamilton 1905).

Griffith Era in Sport Psychology

The true beginning of sport psychology in North America dawned with the work of Coleman Griffith, who, as a professor at the University of Illinois, engaged in the first systematic examination of the psychological aspects of sport between 1919 and 1938. Griffith established the Athletic Research Laboratory at Illinois, published numerous research articles, and wrote two classic books—*Psychology of Coaching* in 1926 and *Psychology and Athletics* in 1928. Griffith also interviewed sport celebrities of that time such as Red Grange and Knute Rockne about the mental aspects of their sports. Philip Wrigley hired him in 1938 as a sport psychology consultant for the Chicago Cubs baseball team. During this work, Griffith developed psychological profiles for specific players, such as Dizzy Dean, and researched methods of building confidence and increasing motivation.

▶ Coleman Griffith began systematic research in sport psychology in the 1920s, but because his early work was not extended, the area was not recognized as an academic subdiscipline of physical activity until the 1960s.

Unfortunately, Griffith may be thought of as a prophet without disciples because no one continued his significant work. Although sporadic publications about psychological aspects of sport and exercise appeared after Griffith's time, the subdiscipline largely lay dormant until a reawakening in the 1960s.

▌ Coleman Griffith was a pioneer in the field of sport psychology.

1960s and 1970s

The 1960s saw an upsurge of interest in the social–psychological factors associated with physical activity. The research of this period focused primarily on personality traits related to sport participation and social facilitation or audience effects on motor performance. Bruce Ogilvie, a clinical psychologist at San Jose State University, began early work examining personality in athletes and began applied psychological interventions with athletes. Dorothy Harris, at Pennsylvania State University, is also considered a pioneer of this period because she began a systematic research focus on women in sport. The first meeting of the International Society of Sport Psychology (ISSP) was held in 1965; the first conference of the North American Society for the Psychology of Sport and Physical Activity (NASPSPA) was held in 1967; and the Canadian Society for Psychomotor Learning and Sport Psychology was founded in 1969. All these served as important forums for sharing knowledge in sport psychology.

In the 1970s sport psychology was formally recognized as an established scientific subdiscipline within kinesiology. Systematic research programs were established at leading universities; graduate study became available; and in 1979, the *Journal of Sport Psychology* began publication. Much of the research during this time was experimental research conducted in laboratory settings that involved testing theory from the parent discipline of psychology. Rainer Martens, a leading scholar during this time, pioneered the systematic study of competitive anxiety in sport.

Modern Sport and Exercise Psychology

Three trends marked the decade of the 1980s. First, exercise psychology separated from sport psychology into a distinct subdiscipline within kinesiology. Before this time, pockets of research investigating the psychological components of exercise existed, but it was during the 1980s that systematic research programs and graduate program offerings in exercise psychology were developed. William P. Morgan was a major force in promoting the field of exercise psychology and served as the first president of Division 47, Exercise and Sport Psychology, of the American Psychological Association. Daniel Landers also served as a leader in this area, particularly in pioneering psychophysiological research in the field. In 1988, the *Journal of Sport Psychology* became the *Journal of Sport and Exercise Psychology.*

The second trend of the 1980s was explosive growth in field research in sport and exercise psychology. Much of the research of the 1960s and 1970s occurred in laboratory settings, seemingly to gain academic credibility for the young field. But researchers realized the need to study sport and exercise in the field to gain a better understanding of psychological processes as they operated in competition and various forms of physical activity.

Logically following this move to field research, the third trend of the 1980s was increased interest in applied sport psychology or mental training with athletes. The establishment of two new applied journals, *The Sport Psychologist (TSP)* in 1987 and the *Journal of Applied Sport Psychology (JASP)* in 1989, and a new organization, the Association for the Advancement of Applied Sport Psychology (AAASP) in 1986, indicated the expansion of the subdiscipline to include applied interests. Influential professionals during this period included John Silva, the founding president of AAASP and first editor of *JASP;* Dan Gould, the founding editor of *TSP;* and Terry Orlick, who pioneered applied sport psychology through his mental training books and consulting work with Canadian Olympic athletes.

During the 1990s, sport psychology and exercise psychology grew as separate subdisciplines through the accumulation of research and applied programs. In 1991, AAASP established certification criteria for minimum professional training standards for consultants in sport and exercise psychology. Ethical standards and guidelines for consulting in sport and exercise settings were approved, and the United States Olympic Committee created a registry of certified professionals eligible to consult with Olympic teams and athletes.

In the 2000s, the knowledge base of the field grew rapidly with an increase in published research, providing a sound foundation for sport psychology practice. This decade showed a big increase in qualitative research, which typically involves interview data, as opposed to quantitative research, which presents findings using numbers or statistics. In the 2000s, sport and exercise psychology practice expanded greatly, as consulting services and applied materials (books, videos, websites) proliferated. Online resources included Peak Performance Sports, Virtual Brands sport psychology, and Sport Psychology Today by Mental Edge Athletics. A new journal dedicated to practitioners (coaches, consultants, athletes), the *Journal of Sport Psychology in Action,* was established in 2010.

> ▶ The 1980s saw the emergence of exercise psychology, the growth of field research, and an explosion of applied mental training with athletes.

Research Methods for Sport and Exercise Psychology

By now you should understand the nature and historical development of sport and exercise psychology as a scientific area of study. But how is science conducted in this subdiscipline? What methods do researchers use to ask important questions about the psychological aspects of physical activity participation? Researchers use six methods to systematically assess thoughts, feelings, and behaviors in sport and exercise psychology.

Questionnaires

Questionnaires are widely used in sport and exercise psychology. Questionnaires may be survey instruments that assess demographic variables such as age, sex, or socioeconomic status as well as general information such as the type, frequency, and duration of exercise engaged in during the past week. Most of the questionnaires used, however, are **psychological inventories**, which are standardized measures of specific forms of thoughts, feelings, or behaviors (Anastasi 1988). For example, psychological inventories are used to measure the amount of anxiety, motivation, and confidence an individual feels about exercising or competing in sport.

Have You Got Passion for Physical Activity?

Complete the following questionnaire by marking a response for each item as you think about your favorite physical activity.

Scoring your responses: totally agree = 7; agree = 6; sort of agree = 5; unsure = 4; sort of disagree = 3; don't really agree = 2; don't agree at all = 1.

	Totally agree	Agree	Sort of agree	Unsure	Sort of disagree	Don't really agree	Don't agree at all
1. I am completely taken with my activity.							
2. My activity reflects the qualities that I like about myself.							
3. My activity allows me to live memorable experiences.							
4. My activity is in harmony with other activities in my life.							
5. My activity is a passion that I manage to control.							
6. I cannot live without my activity.							
7. My mood depends on my being able to do my activity.							
8. I have difficulty imagining my life without my activity.							
9. I am emotionally dependent on my activity.							
10. I have an almost obsessive feeling for my activity.							

Add your scores for items 1 through 5 together to get your positive passion score. This represents how much you love to engage in your activity based on your own free will to choose to participate. A high score indicates that you gain great joy from the activity, you don't feel you have to do it, and it doesn't control your life.

Add your scores for items 6 through 10 to get your obsessive passion score. This represents how much you feel you have to engage in your activity, as if your activity controls you and you *have* to do it (for others, for approval, for yourself).

Are you passionate about your physical activity? How much of that passion is positive and how much is obsessive? You want your positive passion to outweigh your obsessive passion, so you feel in control of your life and gain joy from participation, not pressure to perform. Consider ways you can lessen the obsessive passion you feel and increase your positive passion for your physical activity.

Adapted from Vallerand et al. 2003.

Consider for a moment the difficulty of accurately measuring the thoughts, feelings, and behaviors of people. Assessing individuals' levels of self-esteem is much different from measuring the amount of oxygen that they expend during a fitness test or how much weight they lose in the preseason. For this reason, psychological inventories must meet rigorous standards of uniformity of procedures with regard to their development, administration, and scoring so that researchers can gain a valid and reliable assessment of behavior. Therefore, psychological inventories used by researchers are not simply a list of questions thrown together; they are carefully constructed and tested assessment tools that have met specific standards set by experts in the subdiscipline.

Interviews

Interviews are used in sport and exercise psychology when the research question being pursued requires in-depth understanding of individuals' beliefs, experiences, or values. For example, interviews may be useful if one is attempting to understand why children drop out of youth sport programs because interviews allow children to explain things in their own words rather than respond to questionnaires. Like any other scientific method, however, interviews must be structured to be systematic, and researchers must be trained in the use of interviews in order for them to be effective and valid.

Observations

Observation of behavior is the third method used in sport and exercise psychology. For example, researchers often observe the behavior of coaches during practice or competition to assess the frequency of various types of feedback and communication that they provide to athletes. Research examining the motivation of children to engage in physical activity or vigorous play activities benefits from using behavioral observation. Typically, observation studies use some type of behavior checklist or coding system to ensure that observation of the behavior is occurring within a particular set of parameters. Two observers are typically used to code behavior, and their results are checked against each other to ensure a consistent, reliable assessment.

Physiological Measures

Physiological measures of physical, mental, and emotional responses are sometimes referred to as *biofeedback*. For example, blood pressure and heart rate may be measured to assess the effects of psychological stressors on individuals. Exercise psychologists use measures such as these to study the effects of exercise on stress reactivity and existing anxiety levels. Sport psychologists might measure the amount of tension in muscles to assess how well athletes can learn to relax physically through mental training. Researchers can measure brain waves to assess levels of attention or relaxation. Brain-imaging technology, such as fMRI or PET scans, are being used to view how the brain responds when athletes visualize performing their skills. For example, expert athletes activate areas in their brains much differently than beginner athletes, indicating that elite sport performance is "hardwired" in a specific way in the brain.

Biochemical Measures

Biochemical measures are used less frequently in sport and exercise psychology, but brief mention here will be helpful. This type of assessment involves drawing and analyzing blood or urine for chemicals from the body that represent responses to stressors or emotions. Examples are epinephrine and cortisol, which the adrenal gland releases in response to certain types of stressors.

Content Analysis

Researchers use content analysis to analyze written material from various sources such as government documents, newspapers or magazines, and even television programming. For example, a researcher could analyze several popular television shows to assess the levels of physical activity being modeled by the media to viewers. Written or dictated physical activity logs are useful in exercise research because they provide detailed accounting of all or selected types of physical activity performed within a given period.

ACTIVITY 9.2

Interview with a Practicing Professional
In this interview in your web study guide, Michael DeCello, a high school basketball coach, talks about how he employs sport psychology with his players during practice and games.

> Research methods in sport and exercise psychology include questionnaires, interviews, observation, physiological measures, biochemical measures, and content analysis.

Thus, researchers in sport and exercise psychology have various methods available to them. Often, researchers use two or more methods within a single research study. For example, a study examining exercise adherence in an adult fitness program could use observation to assess participation rates, intensity of exercise, and instructor behavior; questionnaires to measure participants' self-confidence and perceived benefits in relation to exercising; and physiological measures of participants' resting heart rate and lung volume. All of these methods are designed to be used in a systematic manner to ensure that accurate and consistent measures are obtained. This systematic approach is the mark of science—and it is very different from the casual observations that we all make in everyday social interactions.

ACTIVITY 9.3

Which Method?
In this activity in your web study guide, you'll review two scenarios and determine which research method to use in each.

All scientific methods in sport and exercise psychology have limitations. Questionnaires typically provide a systematic measure of certain phenomena, but they lack the depth and richness of interviews. However, establishing consistency and a systematic approach is more difficult when one is using interviews compared with questionnaires. Scientists in sport and exercise psychology look at the menu of research methods available and then select the method that best fits the recipe of the particular research study. This menu also includes other choices for researchers, such as whether to conduct the study in a laboratory or field setting and what type of participants to use in the research.

Overview of Knowledge in Sport and Exercise Psychology

This section provides a brief overview of a few selected sport and exercise psychology topics to give you a glimpse of the knowledge that researchers and practitioners have produced. The selected topics presented here include personality, motivation, arousal and anxiety, social and group processes, and mental skills training in physical activity.

Personality

One of the most popular issues in sport and exercise psychology concerns the relationship between personality and physical activity participation. We typically think of **personality** as the unique blend of the characteristics that make individuals different from and similar to each other. Our personalities determine our thoughts, feelings, and behaviors (ABCs) in response to our environment.

Personality and Sport

Despite popular opinion, no distinguishable "athletic personality" has been shown to exist. That is, no consistent research findings show that athletes possess a general personality type distinct from the personality of nonathletes. Also, no research has shown consistent personality differences between athletic subgroups (e.g., team athletes vs. individual sport athletes, contact sport athletes vs. noncontact sport athletes).

Research has identified several differences in personality characteristics between successful and unsuccessful athletes (e.g., Krane & Williams 2010). These differences, however, are not based on innate, deeply ingrained personality traits but rather result from more effective thinking and responding in relation to sport challenges as well as higher levels of motivation. Specifically, successful athletes, compared with less successful athletes, are

- more self-confident,
- better able to cope well with stress and distractions,
- better able to control emotions and remain appropriately activated,
- better at attention focusing and refocusing,

What athletic personality characteristics does Roger Federer possess that have helped make him one of the best tennis players in the world?

- better able to view anxiety as beneficial, and
- more highly determined and committed to excellence in their sport.

Olympic and World champion athletes have defined *mental toughness* as the natural or developed psychological edge that enables you to cope with competitive demands and remain determined, focused, confident, and in control under pressure (Jones et al. 2007). These athletes identify the following as critical personality responses that represent mental toughness: loving the pressure of competition, adapting to and coping with distractions and sudden changes, channeling anxiety, not being fazed by mistakes in the process, being acutely aware of any inappropriate thoughts and feelings and changing them immediately to perform optimally when needed, using failure to drive yourself, learning from failure, and knowing how to rationally handle success—an impressive list of qualities that we all would like to have as part of a mentally tough personality!

Although most sport personality research has focused on the influence of personality on sport behavior, research has also examined the effects of sport participation on personality development and change. A belief commonly held in American society is that sport builds character or that sport participation may develop socially valued personality attributes. Research shows, however, that competition reduces prosocial behaviors such as helping and sharing and that losing magnifies this effect. Sport participation has been shown to increase rivalrous, antisocial behavior and aggression and has been linked to lower levels of moral reasoning.

Nevertheless, the sport story has a positive side. Research in a variety of field settings has demonstrated that children's moral development and prosocial behaviors (cooperation, acceptance, sharing) can be enhanced in sport settings when adult leaders structure situations to foster these positive behaviors (Hellison 2003; Shields & Bredemeier 2007). Interventions with children were successful in building character when naturally occurring conflicts arose and were discussed with the children to enhance their reasoning and values about sport and life events. The moral of the story is this: Sport doesn't build character, people do!

Personality and Exercise

As in sport, researchers have found no "exercise personality" or set of personality characteristics that predict exercise adherence. Exercisers cannot be differentiated from nonexercisers based on an overall personality type. Two personality characteristics, however, are strong predictors of exercise behavior. Individuals who are more confident in their physical abilities tend to exercise more than those who are less physically confident. A second important predictor of exercise behavior, obviously, is self-motivation, with self-motivated individuals beginning and continuing exercise programs and less motivated individuals dropping out or never starting at all.

A personality type termed "obligatory exercisers" describes individuals who participate in exercise at excessive and even harmful levels (Coen & Ogles 1993). For these individuals, exercise becomes the central focus of life, and their behavior becomes pathological in terms of their need to control themselves and their environment. Clinical evidence demonstrates a similar link between anorexia nervosa, a psychopathological eating disorder, and compulsive exercise. Specialists in exercise psychology attempt to help individuals plan and engage in exercise behavior that is healthy and noncontrolling to enhance total well-being.

Echoing the idea that sport builds character, exercise or fitness training has also popularly been associated with positive personality change and mental health (Landers & Arent 2007).

The personality characteristic that researchers have most frequently examined in this area is self-esteem. Self-esteem is our perception of personal worthiness and the emotions associated with that perception. Think of self-esteem as how much we like ourselves. Research has generally confirmed that fitness training improves self-esteem in children, adolescents, and adults. Research has also shown that exercise positively influences perceptions of physical capabilities, or self-confidence. Interestingly, the research indicates that these changes in self-esteem and self-confidence may result from perceived, as opposed to actual, changes in

No set of underlying traits exists for an "athletic personality," but successful athletes possess more positive self-perceptions, have stronger levels of motivation, possess more adaptive perfectionism, and use more productive cognitive coping strategies compared with less successful athletes.

physical fitness. In addition, many aspects of intellectual performance have been related to physical activity, suggesting that cognitive functions respond positively to increased levels of physical activity.

Many people also associate exercise with changes in mood and anxiety. Most individuals say that they "feel better" or "feel good" after vigorous exercise, which emphasizes the important link between physical activity and psychological well-being. In addition, research documents that anxiety and tension decline following acute physical activity. The greatest reductions in anxiety occur in exercise programs that continue for more than 15 weeks. Much research has been conducted to determine whether exercise or fitness reduces people's susceptibility to stress, and the generally accepted conclusion is that aerobically fit individuals demonstrate a reduced psychosocial stress response. A tentative explanation for this finding is that exercise either acts as a coping strategy that reduces the physiological response to stress or serves as an "inoculator" to foster a more effective response to psychosocial stress (Landers & Arent 2007).

Prolonged physical activity is also associated with decreases in depression and a lessening of depressive symptoms in individuals who are clinically depressed at the outset of the exercise treatment. Explanations for these changes range from the distraction hypothesis, which maintains that exercise distracts attention from stress, to other explanations that focus on the physiological and biochemical changes in the body after exercise.

> → Sport has not been found to build socially valued attributes, or character, but exercise has been shown to produce several benefits including enhanced self-concept and psychological well-being and decreased anxiety and depression.

Motivation

Sport and exercise psychologists are interested in understanding what motivates people to engage in physical activity. **Motivation** is a complex set of internal and external forces that influence individuals to behave in certain ways. The behaviors most typically associated with motivation are choice, effort, and persistence. That is, we assume that people are motivated when they make choices to join a fitness program, work intensely during the program, and continue to adhere to their training program when their lives become busy. Motivation directs and energizes our behavior in sport and exercise. Consider how you would answer the following two popular questions about motivation:

1. What is the best way to motivate people?
2. What makes some people motivated and others not?

Intrinsic and Extrinsic Motivation

The first question assumes that motivation is something that you give to others—like a cup of water. You might recall from chapter 4 that this type of motivation is termed **extrinsic motivation**, which means that people engage in a certain behavior to gain some external reward from that participation such as a trophy in sport or weight loss in exercise. Rewards and punishment are often used as incentives to motivate individuals to exercise or exert effort in sport. Gaining popularity in school or eliciting parental approval from sport achievement clearly motivates children to continue this behavior. Adolescent boys gain motivation to lift weights when their bodies become muscular and are admired by others.

Extrinsic motivation and rewards, however, enhance motivation only in the short term and do not fuel long-term commitment to achievement in sport or fitness training. Thus,

---

How's Your Motivation?

What is your favorite physical activity? What motivates you to participate in this activity? Is your motivation more internal or external? Explain your answer by including information about the choices you make in physical activity, the effort you expend, and the persistence with which you pursue it. If your motivation is external, describe the reward you're seeking. If it's internal, explain the pleasure and satisfaction you get from the activity.

Intrinsic motivation is self-fueling over the long term because it is based on controllable feelings of enjoyment and competence compared with extrinsic motivation, which relies on external reinforcers from the social environment.

although extrinsic motivators are always present in society and offer powerful incentives, this type of motivation is short-lived. The enduring motivation necessary for achievement and success is not something that you can give to another person. Rather, enduring motivation comes from within.

Thus, the second question—why are some people motivated and others not?—is more interesting to sport and exercise psychologists. This motivation from within, called **intrinsic motivation**, involves engaging in behavior because you enjoy the process and gain pleasure and satisfaction from that participation. Intrinsically motivated athletes perform because they love the sport. People who exercise regularly do so because they enjoy physical activity. Intrinsic motivation serves as a long-term fueling process for commitment and achievement of important goals in sport and fitness training. Thus, sport and exercise psychologists seek to understand how to develop and enhance intrinsic motivation rather than focus on the quick-fix use of rewards and gimmicks to motivate individuals extrinsically.

The answers to the two motivation questions, then, are the following:

1. The best way to motivate people is to help them develop or increase their intrinsic motivation.

2. People who are more motivated than others typically have higher intrinsic motivation to achieve in a certain activity.

These answers raise another question—how is intrinsic motivation developed?

Developing Intrinsic Motivation

The key to understanding motivation is realizing that all humans, regardless of their individual goals, are motivated to feel competent and self-determining.

Although numerous theories about motivation exist, it is generally accepted that people are motivated to feel competent and self-determining. From the time we are born, we all attempt to be competent in our environment—even toddlers are motivated to gain independence by learning to crawl and walk. As our lives continue, our need to be competent is channeled in various areas through socialization. Some people are motivated to achieve in sport or through fitness training, others in music, others in a career.

Although we are all intrinsically motivated to be competent, competence means different things to different people. Research in sport and exercise psychology has shown that individuals have different goals for achievement in sport and exercise. Therefore, to understand motivation, we must understand how each person defines success or competence for herself. Because people engage in exercise programs for diverse reasons, including social affiliation, personal mastery and fitness, or competitive bodybuilding, exercise programs must assess the goals of participants to fuel individual motivation.

Researchers have found that young children participating in physical activity are motivated to have fun, affiliate with their friends and meet new friends, and develop skills. Although adults often think otherwise, these goals motivate children more than the goal of winning does! For this reason, physical activity programs should provide opportunities for children and youth to attain these goals. Practices and competitions should be structured to be challenging and enjoyable for all children so that they develop their skills, learn important cooperative and social skills, and experience the fun and enjoyment of physical activity. The formula is simple. If we like something, we do it more. If we do it more, we become better at it. When we become better at it, we like it more. Think of this as the positive cycle of motivation (see figure 9.2).

Using Rewards to Enhance Motivation

Rewards, a type of extrinsic motivation, can be used in a positive way to enhance people's feelings of competence, which then increase their intrinsic motivation. Many potential rewards exist in the physical activity environment (e.g., trophies, money, status, weight loss, fitness, popularity). What is important is how individuals perceive rewards that they obtain. If people perceive rewards as controlling, then the rewards only enhance extrinsic motivation and will actually weaken intrinsic motivation. In sport, extrinsic rewards that are used to control athletes such as college scholarships, trophies, and all-star selections often decrease intrinsic motivation. Incentives such as weight loss and prizes for attendance may be motivating to exercisers but could be problematic if used to coerce or manipulate individuals. Behavioral contracts and attendance lotteries are examples of incentives used for exercise adherence, but research indicates that these inducements alone have no lasting effect on exercise behavior. Rewards should be used not to control athletes or exercisers but rather to make them feel that they've earned praise and reward through their effort and competence. Moreover, teachers and coaches must praise and reward the right behaviors. When young children first attempt physical skills, their coaches or teachers should reinforce these mastery attempts. With any luck, children will learn to value individual improvement and mastery in physical activities, which will enhance their self-worth.

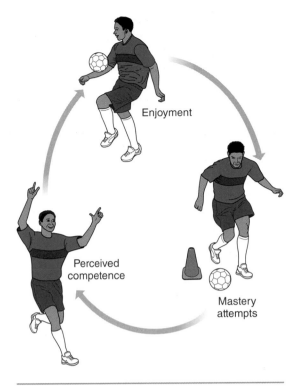

FIGURE 9.2   Positive cycle of motivation.

ACTIVITY 9.4

Practical Plug-In

In this activity in your web study guide, you'll consider a scenario in which a physical education teacher needs to determine how best to help a faltering student, and you'll learn about the concept of "learned helplessness."

From Anxiety to Flow: Understanding Arousal

As just discussed, motivation involves intensity of behavior and the urge to be competent and successful. However, the intensity and urge to be competent sometimes create anxiety for individuals as they participate in sport or exercise. Has your performance ever been disrupted because you were tense or nervous? In contrast, can you remember performing "in the zone" where time seemed to stand still and you experienced an almost spiritual feeling of focus, sense of control, and total immersion in your activity? These examples, gut-wrenching anxiety as well as the exhilaration of peak experience, are two ways in which individuals experience arousal in physical activity. **Arousal** is a state of bodily energy or physical and mental readiness. Arousal results from the ways in which athletes' minds and bodies respond to competition or the ways in which those involved in exercise respond to physical activity. Arousal falls along a continuum ranging from very low (as when in deep sleep) to extremely high (the most "wired" you've ever been). **Anxiety** is simply a negative emotional state in which individuals feel nervous or worried when they experience high levels of arousal. Think of arousal as simply generic energy; when arousal becomes too high and people interpret it negatively ("I feel scared, nervous, anxious."), it is then labeled as anxiety. High levels of arousal labeled positively ("I feel ecstatic!") are often described as joy or happiness. Interestingly, research shows that highly successful elite athletes do not have less anxiety than less successful athletes, but rather that the successful athletes perceive feelings of nervousness and anxiety as positive and helpful (Jones et al. 1994). That is, they know that these feelings represent the increasing readiness of their bodies to perform, so they accept the nervousness, anxiety, and high arousal as normal and helpful. Try this yourself! The next time you feel nervous, instead of labeling it as negative and focusing on how bad it feels, tell yourself that your body is ready and when the time comes to perform, that nervousness will be used as positive energy or arousal to help you succeed. It works!

People often assume that stress is the same thing as anxiety, but stress serves the important purpose of stimulating growth. **Stress** may be defined simply as a demand placed on a person. We stress athletes all the time through weight training and exhaustive physical repetitions of their sport skills, and a careful progression of physical and mental stress builds their stress tolerance so that they can withstand the stress of competitive performance. So stress isn't the same thing as anxiety, but stress often creates anxiety. The anxiety most widely studied in sport psychology is based on the threat of failure or evaluation, as when athletes compete in public. As discussed previously, anxiety is of interest in exercise psychology because research demonstrates that physical activity decreases feelings of anxiety, thus contributing to mental and physical health. Research has also demonstrated that social physique anxiety, or apprehension about one's body and appearance, predicts exercise behavior (Crawford & Eklund 1994). For instance, the less apprehensive people feel about their appearance, the more likely they'll be to participate in activities that require them to reveal more of their bodies (e.g., gymnastics or swimming).

Overall, stress may induce high levels of arousal and even anxiety, but it often has a positive influence on performance. Think back to times that you performed well in sport or academically, because the stress motivated you to prepare or train to be successful in that situation. Stress becomes a problem only when individuals do not allow adequate recovery time from life or competitive demands. Chronic stress without adequate recovery burns out the energy reserves of athletes and others engaged in physical activity.

Revisit Your Flow Experience

Think about a time when you experienced flow. Describe the activity that you were doing and what the experience felt like. What made it different from other experiences? How long did it last, and why do you think it went away when it did?

When individuals' arousal levels are at their optimal zones, the experience of flow occurs (Csikszentmihalyi 1990). Exercisers speak of "runner's high," that elusive state in which physical activity is effortless and immensely enjoyable. Athletes train to achieve the ultimate high of "being in the zone" and enjoying the ultimate thrill of sport characterized by peak performance. Thus, helping physically active individuals identify the special recipe of feeling states that lead to their optimal arousal zones and flow will certainly enhance their performance and overall experience of physical activity.

Many sophisticated models are available to explain how arousal and anxiety influence performance in sport. However, a general rule is that people perform better at moderate levels of arousal. This is called the inverted-U model of arousal (see figure 9.3), in which athletes' performance increases as their arousal levels increase up to an optimal or moderate point, after which athletes go "over the top" and performance decreases as their arousal levels become too high. As shown in figure 9.3, there is an optimal zone of arousal for performance, not too low and not too high.

The optimal intensity of arousal, however, is unique for each person. Individual factors that may affect preferred arousal levels include personality characteristics, coping ability, and skill level. In addition, the optimal intensity of athletes' arousal differs according to situational demands of different activities. A weightlifter can be at a very high arousal level and perform well, whereas a golfer's optimal level for performance is somewhat lower. Arousal influences performance through its effect on muscular tension and the performer's ability to focus his attention properly. The muscles of athletes performing in sport often become tense and

FIGURE 9.3   Inverted-U model of arousal.

lose the smooth coordinated movement needed to perform optimally. Their attention tends to turn inward toward worry and thoughts of inadequacy, or their attention narrows to the point where they miss important competitive cues they need to read to perform optimally.

✔ ACTIVITY 9.5

Your Optimal Energy Profile
In this activity, you'll gain a better understanding of the feelings that lead to good experiences and those that hurt your performance or detract from your enjoyment of an activity, and you'll plan strategies for getting yourself into your zone when you need to.

Sport psychology experts help athletes learn to "find their numbers" (as you did in Activity 9.5) and then to develop focus plans in which they engage in individualized thoughts and activities (e.g., psyching up or calming down) to create the levels of arousal that work best for them in their particular sport activities. It takes some experimentation, but athletes can learn how to find their optimal numbers and how to monitor themselves to then adjust their numbers when needed. Think of it as athletes' ability to optimize the challenge for themselves (stress) to create their desired levels of arousal when they need it to put themselves in the position to experience flow (and avoid the negative effects of extreme anxiety).

Social and Group Processes

In this section, our focus is on social or group processes that influence individuals' behaviors in different ways, including the presence of others, group membership, and aggression.

Presence of Others

How is your behavior different when you are alone as opposed to when you are with other people? Why do some people prefer to exercise at home whereas some like to exercise with others in a public gym? Why do some athletes perform better in front of a big crowd? Does the home advantage really exist? Since the turn of the 20th century, researchers have been fascinated with the effects of an audience on human performance. Research has shown that the presence of other people increases our arousal, which then may hurt or help our performance. Generally, spectators have a negative effect on someone who is learning a skill and a positive effect on someone who is very skilled. Think about a beginning golfer who is learning the game and has to tee off in front of a large group on the first hole. The presence of spectators increases her arousal, and because her skills are not well learned, this arousal causes her to hit a bad shot. But consider a professional tennis player who makes the finals at Wimbledon for the first time. The huge crowd at Centre Court is likely to inspire and elevate her performance.

Although this explanation sounds simple, the effects of others on our behavior are complex. For example, research has shown that it is not the mere presence of others that causes this effect but rather people's perceptions that others are evaluating them. We know from the previous section that every individual has a different optimal arousal zone, so the presence of others could influence performance differently based on individual responses to changes in arousal. One thing we do know from research in this area, however, is that people should avoid situations of excessive evaluation or analysis when learning sport skills; this pressure can hurt the learning process by adversely affecting beginners' arousal levels and quality of attention.

ACTIVITY 9.6

Practical Plug-In

In this activity in your web study guide, you'll consider what strategies a coach can use to help his players overcome their tendency to "psych themselves out" about the home field advantage when playing away games.

> Family support and modeling are highly predictive of the physical activity participation of children, whereas elderly people tend to become less physically active because of social isolation.

The presence of others applies a bit differently to exercise, because the people present are usually socially supportive workout partners or simply other individuals concurrently exercising on their own (e.g., while weight training at a public facility). Family social support is a strong predictor of exercise maintenance for women, and individuals who exercise with their spouses have higher rates of exercise adherence than those who exercise alone. Research has demonstrated that physical activity levels of children are related to the modeling of such activity, shared activities, social support, and encouragement within families. Several studies have demonstrated that active children perceive more parental encouragement than inactive children. The social isolation of many elderly people is problematic because social incentives to engage in physical activity decrease with age. Community-based programs can ensure the important social support for habitual physical activity participation for all ages.

Group Membership

The area of group dynamics focuses on how being a part of a group influences behavior as well as how certain psychosocial factors influence collective group behavior. Are people more likely to adhere to exercise programs if they participate in groups rather than exercise individually? Why do some teams have better chemistry than others? And why do teams with better chemistry often perform better than more talented teams with poorer chemistry?

A group performs better and group members are more satisfied when the group is cohesive. **Cohesion** is the tendency for groups to stick together and remain united in pursuing goals (Eys et al. 2006). Thus, coaches and exercise leaders should strive to develop and nurture cohesion in their teams or groups in several different ways. Emphasizing uniqueness or a positive identity related to group membership facilitates cohesion. Nicknames such as the "Sweat Net" or the "Lunch Bunch" often reflect cohesion and group identification. Athletes sometimes shave their heads or wear team jackets to demonstrate their solidarity and commitment to the group. Cohesion also increases when individual members of teams understand and accept their roles within the group. Studies on adherence to exercise programs indicate that most people prefer to exercise with another person or in groups, an arrangement that enhances enjoyment through affiliation and social support and strengthens commitment to the program. Cohesion and friendship within the group constitute one of the strongest predictors of exercise adherence.

We also know that success breeds cohesion, so some early successes are crucial in the development of group dynamics. When groups experience success initially, cohesion develops accordingly. But cohesion is a dynamic quality that is always changing within a group. For example, conflicts that require resolution may arise, but the group may still be cohesive. The quest for cohesion can go too far, resulting in extreme cohesion that involves conformity and elitism. Although a group needs a strong, positive shared identity, the group should celebrate and respect diversity among its members.

Also of interest in group dynamics research is how group membership influences individual performance. **Social loafing** refers to a decrease in individual performance within groups (Heuze & Brunel 2003). This decline occurs because individuals believe that their performance is not identifiable and that other group members will pick up the slack. Social loafing is not a conscious process—people do not decide to loaf. Rather, it is a psychological tendency when people perform in a group. But increasing the identifiability of individuals by monitoring their performances easily reduces social loafing. Common ways of accomplishing this in sport are using video analysis to assess individual performance and compiling individual statistics that break out a single athlete's performance from the total performance of the team. Social loafing can be monitored in exercise groups if performance totals (e.g., number of sit-ups, amount of weight lifted, distance run) are recorded so that progress can be checked. Research clearly demonstrates that identifying and monitoring individual efforts eliminate social loafing.

> Cohesion and group membership facilitate physical activity performance, but social loafing may occur unless individuals are monitored and their inputs are viewed as important to overall performance.

Aggression

Sport competition in Western society often leads to aggression or violence. Why do fights break out in hockey matches? Why does crowd violence often erupt at soccer matches? **Aggression** is behavior directed toward inflicting harm or injury on another person. A main source of aggression in competitive situations is the inevitable presence of frustration.

Frustration often results when a person's goals are blocked, and in competitive sport, blocking the goal achievement of the opponent is the main objective. Social learning theory views aggression as a learned behavior that develops because of modeling and reinforcement. Ice hockey players are glorified for fighting with opponents, and baseball players are encouraged and even expected to charge the mound and aggress against the pitcher after being hit by a pitched ball. Children learn aggressive behaviors at an early age. Research also links aggression to levels of moral reasoning, and athletes view aggression as more legitimate because of their lower levels of moral reasoning compared with nonathletes (Shields & Bredemeier 2007).

Most people believe that competition or exercise reduces aggressive impulses in humans by providing a release or purging of aggression (called *catharsis*). But research does not support this claim. Instead, it shows that aggressive tendencies increase after we compete, engage in vigorous physical activity, or watch a competitive event. Thus, competitive and physical activity participation and spectatorship do not serve as a catharsis for aggressive responses.

> Aggression, or behavior intended to harm another person, has been shown to be socially learned behavior and is related to lower levels of moral reasoning.

Mental Skills Training in Physical Activity

Mental skills training involves the use of intervention techniques such as goal setting, relaxation training, imagery, and self-talk in conjunction with physical training to enhance important mental skills such as confidence, focus, and arousal management. Mental skills are important, as we discussed in the section on personality earlier in the chapter. For example, in one study the coping skills, confidence, and motivation of professional baseball pitchers were more important in predicting their success in baseball than their physical skills were (Smith et al. 1995). In addition, these mental skills were predictive of all players' survival in professional baseball 2 and 3 years later (Smith & Christensen 1995).

Intervention techniques in sport and exercise psychology can increase exercise adherence, enhance sport performance, develop life skills, aid in injury rehabilitation, and ease career transitions and retirement from sport.

Mental skills training has been used to increase exercise adherence, improve sport performance, develop important life skills for young people participating in physical activity, aid in rehabilitation from injury and disease, and enhance career transition and retirement from sport. For example, the intervention checklist shown in Helping People Get Started With Exercise demonstrates how to set goals and monitor behavior to enhance commitment to physical activity. In this section, to give you an idea about how mental skills training works, we focus on the mental training tool of imagery and then discuss the important mental skill of attentional focus.

ACTIVITY 9.7

Intervention Techniques
Analyze training interventions in this activity in your web study guide.

Helping People Get Started With Exercise

- Introduce people to light and enjoyable exercise (minimal risk of discomfort or injury).
- Help them establish personalized and realistic goals (avoid comparison with others).
- Use the shaping technique by rewarding small achievements along the way to reaching a goal (progressive, graduated exercise goals).
- Avoid situations in which the individual may feel vulnerable or have lapses in confidence (as when activity is viewed as inappropriate or sex typed).
- Use immediate, informational feedback; provide encouragement; and help people regulate their confidence by practicing productive self-talk.
- Physically guide participants through movements in which they initially lack confidence.
- After they have gained skill and knowledge, encourage participants to take gradual control of their exercise behavior through self-monitoring and self-regulation.

Adapted from Smith and Biddle 1995.

Imagery

Imagery, a mental technique that programs the mind to respond as programmed, uses one's senses to create or recreate an experience in the mind (Vealey & Greenleaf 2010). Seeing is believing, as many athletes will tell you. Jack Nicklaus, perhaps the greatest golfer of all time, says that hitting a golf shot is 50% mental picture. Lindsey Vonn, 2010 Olympic downhill skiing champion, mentally rehearsed the downhill course at Whistler while balancing on two pliable bands and visualizing the sensations and shifts of balance needed on the course. Professional baseball player Todd Helton views a video iPod loaded with clips of his hitting technique and visualizes how he will perform against specific upcoming pitchers. Alex Baumann, Olympic double gold medalist in swimming, states:

> "The best way I have learned to prepare mentally for competition is to visualize the race in my mind. . . . In my imagery I concentrate on attaining the splits I have set out to do. About 15 minutes before the race I visualize the race in my mind. I think about my own race and nothing else. I try to get those splits in my mind, and after that I am ready to go. My visualization has been refined more and more

over the years. That is what really got me the world record and Olympic medals." (Orlick 1998, p. 70).

Images guide our beliefs, and beliefs guide our performance (see Believe and Achieve). Much of our behavior is based on deeply held beliefs that we have about ourselves that sometimes are not even accurate. Complete the Believe and Achieve activity to understand the power of imagery and self-belief.

Believe and Achieve

Your self-image serves as a self-fulfilling prophecy. That is, we are what we dwell on or what we think about ourselves. On the left, identify at least one positive and one negative belief or personal image that you have about yourself. It might be that you're good at math, you're not a good conversationalist, or that you're a good putter in golf. On the right side, consider and write down how that image or belief about yourself affects your behavior (e.g., I freeze on exams, avoid meeting new people, and enjoy golf and putt well).

Take time to reflect on the examples you've identified of your personal image or beliefs about yourself and how they impact your life every day. Consider how your life is enriched by positive beliefs and often constrained by negative beliefs or images about yourself. Use imagery daily to visualize yourself in more positive ways to enhance your personal belief system.

Personal Image/Belief About Myself   Impact of That Belief/Image on My Behavior

_____   _____
_____   _____
_____   _____
_____   _____
_____   _____

Research has demonstrated that when used systematically, imagery enhances sport and exercise performance, and although imagery cannot take the place of physical practice, it is better than no practice at all. Many elite athletes use imagery, and they often cite it as an important mental factor in their success. Novice athletes can use imagery to create positive mental blueprints of successful performance, and exercisers can use imagery to visualize their muscles firing and becoming stronger when training for fitness. Exercisers can use imagery to dissociate, or to direct attention away, from the repetitive exercise activity to visualize motivational or pleasant things. Many runners like to use personal audio players when they run to help them create pleasant mental images.

But how does imagery work? How in the world could imagining an experience actually improve your performance? Incredibly, when you vividly imagine performing a well-learned skill (e.g., throwing a ball), your brain fires in very similar ways as when you actually perform the skills. Thus, imagery works by providing a mental blueprint for perfect responses. Each "blueprint" in your brain contains the physical and mental responses that represent perfect performance. When you engage in imagery, these responses are modified and refined, just as when you physically practice them. So, imagery or mental practice works—because it "trains your brain" in a similar way as actually physically practicing the skill. This has been shown to be true with many different sport skills.

Here are some tips about how to use imagery most effectively (Vealey & Greenleaf 2010). A great deal of research has been conducted in sport and exercise psychology to help us understand the most effective ways to practice imagery to have the greatest effects.

- Use as many senses as possible in your imagery (visual, auditory, tactile, smell, kinesthetic). In particular, the kinesthetic sense of feeling your body move in skilled physical activities is important to optimize your imagery.

- Practice imagery using both internal and external perspectives. The internal perspective means you see the image from behind your own eyes, whereas an external perspective is like watching yourself perform in a video from outside your body. Both perspectives have been shown to enhance performance, but more recent work shows that an internal perspective is needed to truly mimic the brain activity used in actually performing the skill.

- Load your images with productive behavioral, physical, and mental responses. Don't just imagine the environment and what you see: Imagery is more productive when you imagine how you will respond to your environment. So don't just see the track, the spectators, and your performance. Imagine running smoothly and easily in perfect rhythm, feeling loose and fluid and strong and fit. Use the energy of the crowd to increase your intensity. See the difference? Be sure to imagine your responses that make your performance successful.

- You don't have to be lying down to use imagery. Research shows it is helpful and more influential on performance to practice imagery while in your specific performance posture and, if applicable, while holding an implement you use in your activity (e.g., hockey stick, golf club). Skaters should stand and move somewhat as they listen to their music and imagine performing their routines perfectly.

- Use simple verbal triggers to prime the image that you want. Thinking "dollar bill" in a sand bunker provides a golfer the image of slicing her wedge through the sand behind and under the ball to splash the sand (and ball) onto the green. Sprinters may think of their legs in the blocks as coiled springs ready to explode at the starting gun.

- Practice imagery daily. Research shows elite athletes create and follow systematic mental training plans that include imagery practice. Don't be discouraged if your imagery skills are not as sharp as you'd like. Elite athletes will tell you that it takes time and practice to perfect your imagery skills (just like it takes time and practice to perfect your physical skills).

Attentional Focus

Maintaining attentional control and focus may be the most important mental skill for any type of physical activity. In sport, performance depends on the cues that athletes process from themselves and from the social and physical environment. In exercise, learning how and when to focus on one's body (called *association*) as well as how to distract oneself (called *dissociation*) with music or visual imagery, for example, can enhance the exercise experience and help performance. Attention, or focus, is trainable. Athletes think that distractions hurt their performance, but in reality what hurts their performance is how they choose to focus when distractions and obstacles arise. Expert athletes are superior to nonexperts in selectively processing cues from the sport environment. Experts are much better at anticipating opponents' movements, predicting the speed and direction of ball or puck movement, and recognizing patterns of play and formations, and are more efficient at visual scanning in the competitive environment.

Attentional focus is affected by one's arousal levels. Remember earlier when we established that moderate levels of arousal were better for sport performance than under- or overarousal? One explanation for this inverted-U model is attentional. Easterbrook's (1959) famous model of attention and arousal (called the "cue utilization" model) is shown in figure 9.4. The model shows that attention varies from very wide to very narrow depending on one's level of arousal. The attentional "funnel" in the middle of the chart shows that attention is too broad at low levels of arousal, and thus irrelevant cues may be accepted that interfere with focused performance. At moderate levels of arousal, attention is predicted to be optimal, because once arousal becomes very high, attention becomes very narrow, in which case athletes fail to process important peripheral cues. An example of this is a quarterback who is so nervous that his narrow focus of attention interferes with his ability to read the defense and widely scan the field for open receivers. So one way to optimize attentional focus is to

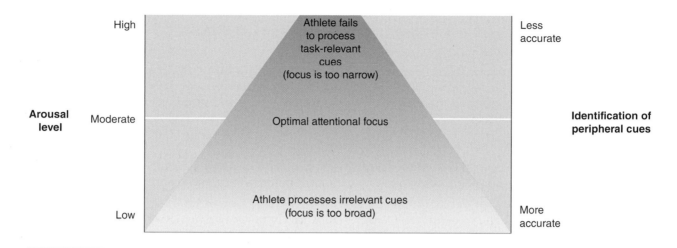

FIGURE 9.4 Relationship between attention and arousal–anxiety.

Reprinted from R. S. Vealey, 2005, *Coaching for the inner edge* (Morgantown, WV: Fitness Information Technology). By permission of R.S. Vealey.

regulate one's arousal levels to avoid the overarousal that causes attention to narrow and impair performance.

Increases in arousal brought on by stress and pressure can cause another attentional focus problem. Arousal and anxiety may make individuals more self-focused, which leads to the dreaded mental phenomenon known as "choking." As the great tennis champion John McEnroe has stated, "Everyone chokes." None of us are immune from this occurrence; we are all faced with achievement situations in which we desperately want to perform well, only to have our strong need to do so unwittingly derail us. **Choking** is the sudden or progressive deterioration of performance below the typical and expected level of expertise for a person performing under pressure, whereby the person is seemingly incapable of regaining control over his performance. The Internet is full of examples of the biggest and most famous all-time chokes in sport. Think about a time when you choked under pressure, and consider why this occurred. What were you thinking and how did that lead you to choke?

The increased arousal and anxiety in pressure situations cause two self-focus problems. The first is that people sometimes focus on their perceived inadequacies or fear they won't be able to perform well. This inward focus, of course, distracts them from the external focus they need to complete the task with full concentration. But the second and biggest culprit of self-focused attention that leads to choking is the attempt to consciously control performance. When people learn skills, they use conscious processing, which means they think through exactly how to do the steps to perform the skill. We also use conscious processing when we attempt to fix a problem in our technique, such as working at the driving range to modify our golf swing. However, high-level performance is not possible using conscious processing. To perform physical skills at a high level, athletes engage in automatic processing during which they are not paying conscious attention to how they perform. Think of this as your autopilot. When athletes experience flow or peak performances, they typically struggle to describe what they were thinking or how they did it. That is because they were fully on autopilot, which is the "performance zone" that you want to be in so that your unconscious.

So, choking, then, occurs when athletes want so badly to perform well that they "pay attention" to what they're doing (conscious processing), which disrupts their autopilot and automatic performance response. And boy, do they pay! Anyone who has choked knows how devastating it is to try so hard (*too* hard, in fact). See the highlighted section on choking for several tips on how you can train to reduce the tendency to choke and learn to respond in situations where choking may occur.

Attentional focus is a key skill for any performer in any physical activity. Other mental skills such as confidence, energy management, self-awareness, and productive thinking are also important for success in physical activity pursuits. Similarly, many mental training tools

Choking: What Can We Do About It?

According to Sian Beilock (2010), who has studied choking in sport and other situations, we can reduce the tendency to choke, but we cannot eliminate it. It is a very difficult mental state to overcome, because you must release conscious control of performance. Here's some suggestions to reduce choking and how to respond when it happens:

- *Get used to it!* Practice under many types of pressure, and get used to performing when others are watching and judging you.

- *Just do it!* The Nike slogan works here, as coaches know when they try to "ice" shooters or place kickers prior to their performances in high-pressure situations. Taking more time can lead to overthinking and thus overanalysis and controlled processing.

- *Distract yourself!* If someone does "ice" you or you're playing a sport in which there are lulls in performance that allow too much thinking, use a preplanned strategy to distract yourself from becoming too focused on how you are performing.

- *Direct you attention out, not in.* Research with college baseball batters found that they performed best when focusing on where they wanted the ball to go off their bats, as opposed to thinking about what their hands were doing or what they wanted the bat to do (Castaneda & Gray 2007). Similarly, elite golfers performed better when they focused on the desired flight of their shots, as opposed to thinking about their wrists or the position of the clubface (Bell & Hardy 2009). Many athletes pick an external focal point to look at (prior to batting or shooting or serving, for example) so they can focus their attention out to a target or focus on the outcome that they want to happen (not on *how* to make it happen).

- *Use holistic trigger words to cue your performance.* Research shows that skilled athletes perform better when they focus on more holistic thoughts about their performance (e.g., smooth, strong, sweep) as opposed to partial descriptions of how to move (quick hands, drive knee, arch back) (Mullen & Hardy 2010).

are used in sport and exercise psychology other than imagery, which was discussed in this chapter. If you are interested in learning more about mental skills training, you can find many books and articles, including testimonials from athletes who have used these methods, in the scholarly and popular literature on sport and exercise psychology.

ACTIVITY 9.8

Practical Plug-In

In this activity, you'll consider a scenario in which a coach needs to help his star basketball player regain the confidence she has lost in her ability to shoot free-throws.

> The goal of physical relaxation training is to teach people how to control the responses of their autonomic functions when faced with stressors in physical activity environments.

People use physical relaxation techniques to control their autonomic functions, including the muscular and hormonal changes that occur during physical activity. These techniques allow participants to engage in physical activity with greater mastery and control over the responses of their bodies to environmental stimuli. Examples of physical relaxation techniques are breathing exercises, muscular tension–relaxation techniques, and various types of meditation. Momentary relaxation before exercising is effective in facilitating smooth, coordinated muscular effort. Individuals can learn how to regulate physiological arousal by reducing their heart and breathing rates to induce a more relaxed state. Obviously, people can use physical relaxation techniques in conjunction with goal setting, imagery, and self-talk to optimize both physical and cognitive readiness to engage in physical activity.

Concluding Your Overview of Knowledge

You've now read about several areas of study in sport and exercise psychology. Although your tour of knowledge didn't cover all the topics in these areas of study, you have had a glimpse of the knowledge that kinesiologists have developed over the years in sport and exercise psychology. You should now have a better idea about such things as how exercise influences our mental health, why intrinsic motivation is critical in fueling exercise and sport behavior, how you can "find your number" and learn to optimize your arousal, why attentional focus is so important in physical activity, and how the mental training technique of imagery may be used to enhance focus.

Burnout

To cap off the chapter, this section focuses on a topic that is typically of great interest in sport and exercise psychology—burnout. The topic has become more important because of the pressure we feel in our society to perform, excel, and juggle many responsibilities at the same time.

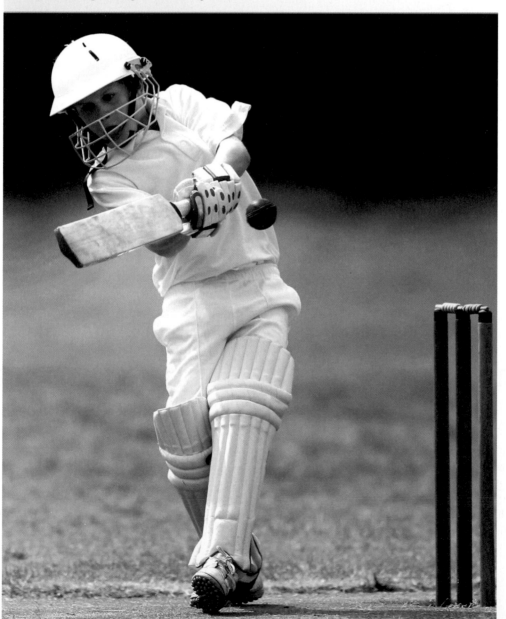

Even youth sport athletes are susceptible to burnout. To help prevent burnout, children should be exposed to many different sports and activities rather than specializing in only one activity.

"I don't know, I just feel burned out." How many times have you heard this statement from people? What is burnout? Is it the same thing as stress? Is it real? Or is it a lack of mental toughness?

Burnout is real, and here's what it looks like. First, it involves feelings of mental, emotional, and physical exhaustion. Second, this exhaustion leads to negative moods and feelings (depression, despair) and a negative change in responses to other people (cynicism, aloofness, lack of empathy). Third, people experiencing burnout feel a lack of accomplishment, which decreases their performance level and feelings of self-esteem. Fourth, burnout causes people to become disillusioned with their involvement in an activity such as sport, their career, or an exercise program. Burnout is a complex condition that occurs when certain personality characteristics of people interact with life stressors.

To help prevent burnout, coaches and trainers should encourage challenge and variety in training. People in fitness programs should engage in different types of activities (cross-training) such as biking, hiking, rowing, and swimming. They could even consider activities like orienteering or Frisbee golf. Coaches should build variety into sport training to break the monotony of repetitive training. They should help individuals distinguish between overload and overtraining. As the social rewards for sport success and fit bodies have become more glamorous, a strong tendency is to think that more is better in training. People should focus on the quality of training and know when to push themselves and when to rest and adequately recover. Most people, especially children, should avoid intense specialization in one activity to the exclusion of all others. We enhance our mental health when we possess multifaceted identities or self-concepts with several areas of interest and expertise. Burnout is not a sign of weakness or a simple response to stress. Athletes and those involved in fitness training programs should watch for early signs of staleness and overtraining and should attempt to stay committed in a positive, passionate way, as opposed to feeling the pressure of *having* to engage in these activities.

Wrap-Up

Sport and exercise psychology, as a young science, has only begun to scratch the surface of understanding the thoughts, feelings, and behaviors related to participation in physical activity. But the knowledge base that has developed over the last 4 decades is impressive as researchers continue to study personality, motivation, arousal and stress, group processes, and mental skills training in physical activity. Kinesiologists are committed to extending and applying this knowledge to enhance participation in physical activity for all people.

If you are especially interested in what you've learned so far about sport and exercise psychology, you might want to take an introductory sport or exercise psychology course, browse through journals, and check out the websites of sport psychology organizations listed in this chapter. You might also want to talk to a professor about career opportunities in either sport or exercise psychology. We cannot all be Olympic athletes, but all of us can engage in meaningful physical activity to derive personal fulfillment and optimal health and well-being.

More Information on . . . Sport and Exercise Psychology

Organizations

American Psychological Association

Association for Applied Sport Psychology

International Society of Sport Psychology

North American Society for the Psychology of Sport and Physical Activity

Journals

International Journal of Sport and Exercise Psychology

Journal of Applied Sport Psychology

Journal of Sport and Exercise Psychology

Journal of Sport Psychology in Action

Sport, Exercise, and Performance Psychology

The Sport Psychologist

 ACTIVITY 9.9

Use the key points review in your web study guide as a study aid for this chapter.

Review Questions

1. What are the ABCs that kinesiologists in sport and exercise psychology study? Identify questions that kinesiologists in this area might study based on these ABCs.

2. What was significant about Coleman Griffith's early work in sport psychology? Why did the subdiscipline not emerge again until the 1960s?

3. Identify the six methods used in sport and exercise psychology and provide one example for how each method is used in the subdiscipline.

4. Does sport build character? Why or why not? Does exercise participation improve mental health? If so, how?

5. Explain why intrinsic motivation is a better source of motivation than extrinsic motivation. How can leaders in both exercise and sport contexts enhance intrinsic motivation in participants?

6. Explain cohesion in groups and discuss how cohesion can facilitate sport performance as well as exercise adherence. What are some ways in which cohesion can be developed and nurtured in groups?

7. Although a distinctive athlete "personality" has not been shown in terms of stable traits, what mental qualities have been shown to separate successful from less successful athletes?

8. What is flow? Describe how mental skills training might be used to enhance flow and peak experiences in both sport and exercise settings.

9. What is imagery, and how can it be used to make the physical activity experience more positive for people?

chapter ten

Biomechanics of Physical Activity

Kathy Simpson

chapter objectives

In this chapter we will

- describe what biomechanics is and what it encompasses;

- explain why biomechanics is important to study and use in physical activity careers;

- explain what biomechanists and related specialists do;

- explore how human movement biomechanics emerged within the field of physical activity;

- observe how professionals in biomechanics and related fields engage in research and analyze movement; and

- discover what is known from biomechanics research about running while wearing minimalist footwear compared with traditional, cushioned running shoes.

Consider this situation: A high school cross-country runner is beginning to develop shin pain. He currently wears a traditional running shoe that has considerable cushioning and foot motion control features. The runner wonders whether he should switch to a "minimalist" shoe (one that uses the least amount of material to protect the foot) and wants your advice. Imagine that you are serving in one of the following professional roles:

- Coach
- Physical therapist, athletic trainer, or other rehabilitation specialist
- Medical personnel (e.g., orthopedist, nurse practitioner)
- Marketing director from the company that manufactures minimalist running shoes who wants the business of this runner
- The runner
- Another role of interest to you

How would you determine recommendations for the runner? What would you want or need to know and why? What resources would you use to get this information?

In your professional role for the situation just described, did you think that observing the runner's technique was important? If so, why? Would you look for atypical movements compared with what is considered the correct technique? Is it important to know about anatomical problems (e.g., flat feet), injury history, training history, current footwear information, running skill level, or something else? Would you ask other runners their opinions about traditional versus minimalist footwear, read footwear reviews in popular running magazines, consult experts, find articles based on scientific evidence, or search the Internet for information about running footwear?

We will use your answers to these questions to demonstrate what human movement biomechanics is about, the goals of biomechanics, and the relevance of biomechanics to you and physical activity professionals. Because the question of whether minimalist versus traditional running shoes are better for injury prevention has been very important to runners, we later explore what is known from biomechanical research. So, what is **biomechanics** and what does it have to do with our runner?

Biomechanics is the application of the mechanical laws of physics and engineering to the motion, structure, and functioning of all living systems, including plants and animals. In this chapter, however, we restrict our interest to human movement biomechanics, as defined in the next section.

Why Use Biomechanics of Physical Activity?

The subdiscipline of kinesiology in which we use mechanical laws to determine how we move during physical activity and how forces act on our bodies to affect the structure and functioning of our bodies is *biomechanics of physical activity* (see figure 10.1), also known as *human movement biomechanics*. The formal definition of **human movement biomechanics**, therefore, is the study of the structure and function of human beings using the principles and methods of mechanics of physics and engineering (see reviews in Atwater 1980; Hatze 1974; Winter 1985).

What do we mean by the "structure and function" of human beings? Structure is our anatomy; function is how we move to accomplish some purpose. For our runner, foot structure, such as the shape of the foot, may affect the choice of running shoes. Physical function for runners means their running technique. It has been suggested that people should be trained to land

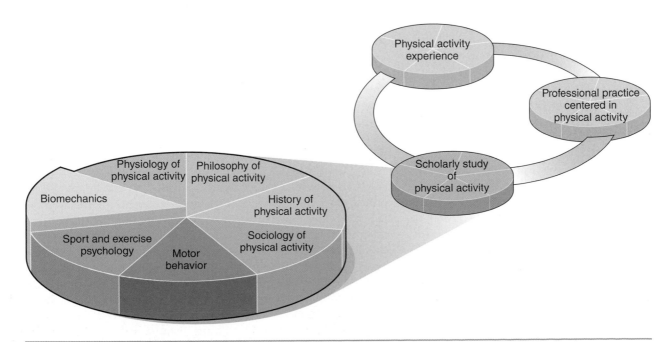

FIGURE 10.1 Biomechanics of physical activity and the other scholarly subdisciplines.

Goals of Biomechanics

- To understand how the basic laws of physics affect and shape the structure and function of the human body
- To apply this understanding in order to (a) improve the outcomes of our movements (i.e., performance effectiveness) or (b) increase or maintain the safety and health of our tissues

differently when wearing a minimalist versus a traditional cushioned shoe. For a client who has undergone a hip replacement, function may mean being able to rise from a chair. Function for bone means being able to support our body weight during standing, protecting our internal tissues (e.g., our brain) from injury, and providing a rigid surface for our muscles to pull against.

Hence, the material in this chapter focuses on two main themes: (1) *function*—how we produce forces to generate, maintain, or slow down movement during physical activity, and (2) *structure*—how forces (such as gravity) affect our body tissues. Therefore, the first main theme, function (how we move), uses mechanical physics, such as Newton's laws, and engineering to understand how we produce movement. Newton's first law of motion says that we won't move (or we won't stop moving if we're already moving) unless a force acts on us. The runner shown at the beginning of this chapter would not be able to move forward without the force of friction to propel him forward. Without gravity, when the runner pushed off the ground, he would never return back to earth.

Forces also must be applied to us to slow us down, whether we are running, riding in a car, or moving in a wheelchair. Another example of a biomechanical principle is Newton's third law: For every action there is an equal and opposite reaction. As shown in figure 10.2, when the runner pushes downward and backward against the ground with some amount of force (the action), the ground will push back onto the runner with equal force in an upward and forward direction, propelling the runner.

How our organs and tissues move and perform their roles also is part of our bodies' functions. Certainly, we need muscle forces to move our bodies. Blood cannot be pumped out of the heart without the force of the heart muscle squeezing blood into the aorta. Bone forces keep our bodies from collapsing onto the ground by resisting the force of gravity. The law of action–reaction is the biomechanical principle that makes this work. When gravity pulls our weight downward (action), our bones push upward (reaction) to keep our body upright.

Forces also are important to our second theme of biomechanics, structure. Whether we are moving or just resting, forces are acting on the outside and inside of our body. These forces, therefore, are affecting the structure, that is, the health of our muscles, bones, connective and other tissues, and organs. For example, astronauts lose bone in space due to a lack of sufficient gravitational force. We can strengthen our muscles, tendons, and ligaments or, conversely, injure them, depending on how much force we apply to them. Hence, certain biomechanical principles have emerged to explain injury mechanisms.

By applying certain engineering principles, we can determine the optimum level of magnitude, rate, and number of repetitions at which forces can be applied to human tissues in order to enhance the tissue's health without immediate or long-term damage. A biomechanical principle (Wolff's law) says that if we apply enough force to bones and do it rapidly, then we can stimulate bone to increase mass. Thus, we encourage people to do weight-bearing exercises that require the performers to hit the ground with more force than otherwise occurs during daily life. One study (Wilks et al. 2009) showed that sprinters, who contact the ground more quickly and with more force compared with longer distance runners, have greater bone mass than long-distance runners. However, Wolff's law also predicts that if we are sedentary and thus do not apply sufficient force to our bones, we will lose bone mass. Hence, Wilks and colleagues also found that sprinters and long-distance runners have more bone mass than sedentary individuals.

Examples of mechanical principles that apply to any movement include Newton's laws of motion (law of inertia, law of acceleration [not discussed], and law of action and reaction).

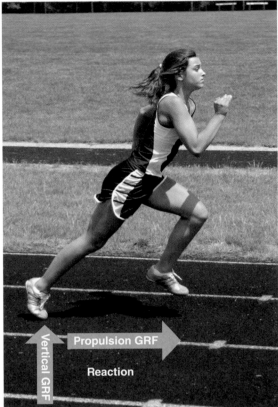

FIGURE 10.2 When a runner wants to move forward and upward into the air, the runner pushes down and back against the ground (the action shown by dashed arrows). The ground simultaneously creates a reaction onto the runner by producing forces of equal magnitude that act in the opposite direction to the action. These forces (solid arrows) produced by the ground are called *ground reaction forces* (GRFs).

Sizing Up Biomechanics

Biomechanics can be applied to biological structures at a variety of sizes. At the smallest end of the spectrum, this area is called **nanobiomechanics**. As an example, scientists who applied forces of about one-trillionth of a newton (approximately 10 trillionths of a kilogram or 0.44 trillionths of a pound) to blood cells infected with malarial parasites found out that malaria causes the cells to become much stiffer, and thus they cannot flex enough to fit and flow through blood capillaries (Fitzgerald 2006). At larger size scales, people have studied the effects of forces on elephants (Hutchinson et al. 2006).

Although biomechanics has been separated here into two themes, they often are interrelated. For example, if you plan to help older adults improve their physical condition, you will learn that at older ages, our muscle structure changes, decreasing leg muscle force (strength). Therefore, older adults who have been sedentary throughout life may have decreased function while moving about. If they start to fall when they stand or walk, they may not have the ability to generate the quick leg power needed to catch themselves (Paterson et al. 2007).

Using the Goals of Biomechanics in Your Career

If you become a physical activity and allied-health specialist or some other professional who is concerned about people's movements or health, you can apply human movement biomechanics to any situation involving human movement. You may ask one or more of the following practical questions of interest in your work:

- How can I improve my client's sport performance?
- What can I recommend to my physical therapy patient to prevent her from sustaining another back injury at work?
- What exercise regimen will improve the bone mass of my clients?
- Which surgical technique during a hip replacement will enhance my patient's ability to walk up stairs?

If you work with clients in your profession, you will always seek answers to many practical questions of interest. However, your best choices of answers are based on our fundamental knowledge of how we produce movement and how the structure and functioning of our body can be changed. Therefore, you may use knowledge from the first major goal of human movement biomechanics: *to understand how people use and are affected by the fundamental principles of mechanical physics and engineering that explain how forces influence our structure and function.* Note that biomechanics is unique in relation to physics and engineering because humans are not just objects that react passively when a force acts on them. People's movements are affected by a variety of factors: sensations (e.g., pain), other performers, the environment, emotions, prior experience, cultural expectations, socioeconomic class, gender, and anatomy.

What Goals of Biomechanics Can Answer Your Question?

Imagine that you could hire a team of biomechanists. Each biomechanist has expertise in one of the goals of biomechanics presented earlier. Think of a question of interest to you, and then determine what each of your biomechanists would be able to contribute to answering your question. Here's an example to get you started: I'm interested in knowing whether collegiate volleyball players should be required to wear ankle braces to prevent ankle injury. A biomechanist whose work is shaped by the general goals of biomechanics would provide information about underlying tissue mechanics that affect the strength of ligaments, such as the orientation of protein fibers that help provide strength to ligaments. Biomechanists whose work centers on the applied goals of biomechanics would analyze the design of the brace in relation to the structure and function of the ankle joint in order to explain how wearing preventive ankle braces would affect the volleyball players' performance.

The following are some examples of fundamental knowledge currently being sought:

- What factors positively influence the composition and strength of damaged ligaments?
- How do the speed and location of impact between the head and ground affect the pattern of loads applied to the skull, neck muscles, and soft tissue?
- What neural, muscular, and mechanical factors explain why humans display particular preparatory movements prior to generating maximal muscle power (e.g., why do we naturally choose to bend our knees just before jumping upward)?

Certainly, if you become a practitioner, the second major goal of biomechanics will be most helpful to you in your career: *to apply to our theoretical understanding to determine how best to (a) improve performance effectiveness (improve function) or (b) increase the safety and health of those body tissues that are affected by forces or those tissues involved in physical activity (improve structure).*

This major goal of biomechanics can be applied to almost any situation that involves physical activity: activities of daily living, repetitive work tasks, exercise, occupational or sport training, sport performance, music, dance, even lying in bed. Thus, if we understand how living organisms can best exploit the mechanical laws that explain how motion is produced (or reduced), then we can intelligently select optimal movement techniques for a performer to use to complete a movement task.

We can use the first major goal of biomechanics, how forces influence the health and strength of tissues and how humans exploit mechanical laws to produce movement, to improve the health and safety of tissues as well. Hence, we can determine how to prevent baseball pitching injuries or develop more effective methods to repair injured tissues. Consider, too, this injury prevention question: Does stretching before engaging in strenuous physical activity prevent injury?

 ACTIVITY 10.1

Practical Plug-In
Consider how a fitness specialist can make best-practice decisions as she designs classes and programs in this activity in your web study guide.

What Do I Need to Know to Use Biomechanics Professionally?

You can see from these examples that all professionals in kinesiology who have training in biomechanics can use their knowledge of how the human body moves, and how it responds to forces acting on the body or within the body, to enhance performance, improve health, and reduce injury potential. These examples also illustrate why biomechanists and other kinesiologists need to understand many areas of kinesiology. Knowledge of biomechanics usually is applied in conjunction with knowledge from a number of the spheres of scholarly study. For example, selecting an appropriate exercise machine requires understanding not only of biomechanics but also of anatomy and exercise physiology. To plan appropriate activities for players, a youth soccer coach needs to understand the interactions between the laws of physics and movement technique, the motor skills that children can learn at their age (motor development), and sport psychology.

Clearly the goals of biomechanics are applicable to many spheres of physical activity. Although biomechanics is a subdiscipline of kinesiology, it is closely related also to mechanical physics and mechanical engineering, which are subdisciplines within the fields of physics and engineering, respectively. What is unique about biomechanics is that it deals with applying the laws of physics, engineering, and biology to human beings (or other living organisms). Branches of engineering closely allied to biomechanics (e.g., biological engineering) investigate problems focused on the design of objects, such as computer furniture, artificial body parts, prostheses, and exercise devices. There is considerable overlap among the goals of biomechanics and the goals of this and other engineering areas.

Tissue biomechanics, for example, is an area of investigation that uses engineering techniques to investigate the properties of human and animal tissues, such as tendons, ligaments, muscles, and cartilage. In conjunction with engineers and orthopedic surgeons, biomechanists can make decisions about the surgical repair materials or implant materials that will best help bone adhere to an artificial joint implant.

▶ Although biomechanics related to physical activity is closely tied to kinesiology, many of the guiding principles and concepts of biomechanics come from the subdisciplines of mechanical physics, mechanical and biological engineering, and biology. For medical applications, much overlap exists between biomechanics and the engineering areas of biological and biomedical engineering and other areas.

What Do Biomechanists Do?

Career opportunities in biomechanics include such positions as researcher, clinical biomechanist, sport performance specialist, ergonomist, forensic biomechanist, and university professor. Related positions are those of a certified orthotist or prosthetist.

Biomechanics researchers work in biomechanics laboratories, performing experiments on problems of interest to various industries or assisting with product development. For example,

| Some biomechanists analyze the gait of patients in clinical settings.

in order to design better footwear, a biomechanist working at a footwear corporation would collaborate with design engineers to understand the interaction between people's anatomy, the way they move, and the forces that act on them. A researcher in a biomechanics laboratory at a hospital may work with physicians and therapists to understand how best to help patients regain normal walking patterns after medical treatment.

 ACTIVITY 10.2

An Ergonomist in Action
Examine an ergonomist's approach and apply it to your own work space in this activity in your web study guide.

A performance enhancement biomechanist might work with elite athletes or professional teams or their coaching staff to improve performance. A biomechanics company or sports medicine clinic could have a facility in which biomechanists analyze the technique of athletes of any skill level and assess performance effectiveness or detect injury-related errors. National-level athletic teams or dance companies may hire this type of biomechanist as a consultant to work with their performers.

Biomechanists may also work as ergonomists. As you saw in chapter 2, ergonomists are experts in improving the efficiency and safety of employees. Some ergonomists work in research and development departments as part of a team of people who design equipment that humans use, such as gardening equipment, commercial airplanes, factory equipment, and sport equipment. Others may work for specialized ergonomics-focused corporations that perform job-site analyses. These analyses involve evaluating how and why employees are currently performing their work tasks. The ergonomist generates data and then recommends appropriate modifications to the tasks or equipment, employee training, or incentives to encourage employees to modify their behaviors to improve their safety or efficiency. Closely entwined with ergonomics is human factors design. Human factors engineers design technology based on how people process information and respond to the information when performing a task. For example, to design an automobile's brakes and steering wheel correctly, an automotive human factors professional would incorporate knowledge of how drivers naturally react when an unexpected object appears in the road. Meister (1999, p. 21) views human factors and ergonomics as one area and defines them as "everything relating the human to technology."

Forensic biomechanists use biomechanical knowledge and principles to answer questions involving civil and criminal law (Schneck 2005). A forensic biomechanist might be hired as an expert witness for a civil lawsuit to testify whether there is a biomechanical basis for an alleged work-related injury. As a simple example, an employee develops a back injury and sues the employer. The employee alleges that the injury was caused because he was required to lift excessively heavy boxes of widgets. The forensic biomechanist would try to determine whether the back muscle forces required when performing the lifting task are high enough to injure this employee and greater than allowed by government standards. The biomechanist also would have to determine whether it is likely that the back problem was caused by other factors unrelated to the lifting tasks, for example, years of spinal disc degeneration that also were made worse by poor health habits.

Clinical biomechanists who work in medical settings, such as research hospitals that include biomechanics laboratories, perform biomechanical analyses of patients or research participants so that physical therapists and physicians can determine treatment. For example, several Shriners hospitals have a biomechanical gait analysis laboratory. The surgeon, physical therapist, and biomechanist work together to assess whether a child with some type of disorder (e.g., cerebral palsy) requires treatment and, if treatment takes place, to determine whether the treatment, such as surgery or physical therapy, sufficiently improved the child's ability to walk. The biomechanist's responsibility is to perform and interpret the gait (walking) analysis and report the findings to the treatment team.

> Ergonomics is the use of a biomechanical approach to design the workplace (or equipment) by fitting the workplace (or equipment) to the worker (Chaffin & Andersson 1991). Human factors, an area closely intertwined with ergonomics, involves designing the workplace based on how people process and respond to the information presented during the performance of a task. Forensic biomechanics is the application of biomechanics to answer questions involving civil and criminal law.

ACTIVITY 10.3

Interview with a Practicing Professional

In this interview in your web study guide, Robin Dorociak, who owns a biomechanics company in Portland, Oregon, talks about the challenges she faces in her work, and she gives advice on how to prepare for a career as a biomechanist.

Although orthotists and prosthetists are not biomechanists, they use many biomechanical concepts and methods. Orthotists assess a client's anatomical structure to understand how the client's anatomy shapes his or her ability to complete tasks effectively. Anatomical deviations may produce injury or pain and in children may be the result of abnormal bone growth, improper muscle functioning, neural defects, or some combination of these. The orthotist then fits the patient with an orthosis, a device that supports a limb, such as a molded plastic brace that holds a child's foot in the proper position. The orthotist then reassesses the effectiveness of the client's movements and watches for potential problems like injury, pain, or abnormal tissue changes such as weakening muscles. Prosthetists are often certified orthotists, too. However, a prosthetist's main focus is to help a person obtain and use a replacement limb such as an artificial hand. By understanding biomechanics, both the

The structure of our bodies, the movements we perform during physical activity, and the ability of our tissues to function effectively are shaped by forces and their effects. The human movement biomechanist uses principles from the mechanical branches of physics and engineering to discover how we can improve the structure and functioning of our bodies.

orthotist and prosthetist can select and adjust the appropriate device and help the client function optimally during daily life and physical activity.

Many biomechanists work in university or college departments of kinesiology, engineering, medicine, or other disciplines. The professor in a kinesiology department teaches biomechanics to a variety of students interested in becoming a physical activity or allied-health professional. The professor also conducts research to examine how the concepts related to generating and absorbing forces help us move and influence the health and structure of our tissues. Some biomechanists may also be clinicians and integrate the care of patients with research and teaching.

Biomechanics is a growing field. Many people who use the knowledge produced by human movement biomechanists are physical activity and allied-health professionals. As shown in the next section, biomechanics can be applied to any movement situation, be it working, exercising, dancing, participating in sport, engaging in physical or occupational therapy, or any other daily activity (even reading this book!).

Does a Career in Biomechanics Interest You?

To determine whether you're interested in a career in biomechanics, answer the following questions:

- Do you like observing and analyzing how people move?
- Do you find yourself trying to figure out a better way to perform a task?
- Do you enjoy trying to solve puzzles?
- Do you like the idea of applying the biology, physics, and math that you've learned to a movement situation of interest?
- Are you interested in investigating the internal structures of the body, such as tendons, to determine how they act and are affected by factors like steroid use?

If you answered yes to some of these questions, consult the websites of the American Society of Biomechanics (www.asbweb.org) and International Society of Biomechanics (http://isbweb.org) for information about biomechanics, jobs, graduate programs, and biomechanists to contact.

History of Biomechanics

Although relatively new to many people, biomechanics has been around for a long time—some say since humans began to make tools (Contini & Drillis 1966b)! The branch of biomechanics that is considered modern human movement biomechanics, however, is relatively young for a subdiscipline. Several historical events and individual scholarly activities led to the development of this field and still influence the methods that biomechanists use to investigate human movement problems. Many people besides those included in this very condensed discussion have contributed to the progress of biomechanics.

Beginnings

One of the earliest known scientists who investigated biomechanical questions was Aristotle (384-322 BCE), who observed animal motion and the walking patterns of humans (Fung 1968). The contributions of Leonardo DaVinci (1452-1519) show the influence of the scientific awakening that occurred during the European Renaissance. DaVinci is credited with developing the first systematic examination of mechanical principles of human and animal movement (Hart 1963). Not until the last quarter of the 19th century and the first quarter of the 20th century, however, when the industrial revolution brought many new inventions, were objective instruments such as cameras developed for measuring movements (Hart 1963). The development of biomechanics and kinesiology in physical education, sport, and dance applications also began in the very late 1800s and early 1900s. Two Swedish men (Posse 1890; Skarstrom 1909) interested in gymnastics were among the first in the United States to apply the term *kinesiology* to the analysis of muscles and movements in a physical education setting (Atwater 1980).

Researchers in the 1920s, including some from areas outside physical activity, had the tools and interest to investigate the mechanics of locomotion (e.g., Fenn 1929), efficiency of human movement in sport and industrial settings (Amar 1920), and energetics of running (Furusawa et al. 1927; Hill 1926). Mechanical analyses of basic movements were not emphasized as part of the formal training of dance and physical educators, however, until 1923, when Ruth Glassow and her students at the University of Illinois began classifying activities into categories such as locomotion, throwing, striking, and balance and applied fundamental principles of mechanics and physics to the skills in each category (Atwater 1980).

Several research leaders in the 1930s supported the use of kinesiology to understand human movement in physical education and orthopedics. Glassow's work continued to expand at the University of Wisconsin (e.g., Glassow 1932). Thomas Cureton, a professor of applied physics and mechanics at the YMCA College in Springfield, Massachusetts, taught the mechanics of sport and physical activities at Springfield College, wrote articles to support the application of mechanical principles to movement (Cureton 1932), and performed research in swimming (e.g., Cureton 1930). Charles McCloy, a leading researcher in many areas of physical activity, identified specific mechanical principles that influence movement (see McCloy 1960 for review of his work). As a professor of orthopedic surgery, Arthur Steindler also profoundly influenced the study of kinesiology. According to Atwater (1980), one of Steindler's legacies was to coin and define the word *biomechanics* (Steindler 1942), although Nelson (1970) believes that Europeans may have used the word first. These early investigators produced a profound effect on the field of physical education and activity through their articles, books, and research because they were among the first to encourage physical education, activity professionals, and medical professionals to identify and use biomechanical principles. Their influence was exponential because many of their students became the core of college professors who would teach future physical activity specialists and college professors (Atwater 1980).

Several societal events prompted the development of other forms of biomechanics. During World War I in the United Kingdom, there was high demand for ergonomic and human factors design analyses to assess the effectiveness of military industries (Meister 1999). Following World War I, researchers in France and Germany applied biomechanics to improve prosthetic devices for war veterans who had lost limbs. This effort stimulated the beginning of physical medicine areas such as physical therapy. Besides helping war victims of World War II, physical medicine research regarding pathological gait (walking) was used to help victims of the polio epidemics (Atwater 1980). Interestingly, the rebuilding of industry in Europe after World War II involved many ergonomic and human factors design analyses of work movements, particularly in Russia and Europe (Contini & Drillis 1966a).

In the 1950s, anthropometry (study of the physical dimensions of people) became even more important, as the new automotive, space, and aviation transport industries (Contini & Drillis 1966a) and the armed services (Thomas 1969) needed to know the measurements of people to design seats, cockpits, and instrument panels that fit these users. For this purpose, sizes, shapes, and weights of various body segments of white males, such as the trunk and limbs, were determined.

The Era of Contemporary Biomechanics

The study of biomechanics of sport and physical activity continued to grow mainly out of college physical education programs. Biomechanical conferences were organized, professional organizations were created, and graduate-level university programs were established in the United States in the 1960s (Atwater 1980). Biomechanists and kinesiologists promoted the

The era of contemporary biomechanics dates to the 1960s, when it was known as kinesiology. In more recent years the term *biomechanics* was adopted to describe this area of inquiry. *Kinesiology* is the term now used to refer to the study of physical activity through experience, scholarly study, and professional practice.

scientific study of human movement within the primary physical activity organization, the American Alliance for Health, Physical Education, Recreation and Dance (AAHPERD). The Kinesiology Section (renamed the Kinesiology Council and now known as the Biomechanics Academy) finally received official recognition in 1965. Its purpose was to provide AAHPERD with a theoretical framework for all foundation areas, such as biomechanics, exercise physiology, and exercise psychology (Atwater 1980). The first international professional association (International Society of Electrophysiology and Kinesiology) was also formed. The 1970s saw a rapid expansion in the number and scope of national (e.g., American Society of Biomechanics) and international professional organizations (e.g., International Society of Biomechanics) and university programs in biomechanics. People in areas outside biomechanics also demonstrated interest in this area. Dance kinesiology evolved during this time as well.

University programs and professional organizations related to biomechanics have continued to expand. Many colleges and universities offer not only courses in biomechanics but also undergraduate- and graduate-level specializations in this area within a kinesiology-based, health allied, or engineering unit. As noted at the end of this chapter, many biomechanics organizations exist across the world, including the International Society of Biomechanics.

Research Methods for Biomechanics

Biomechanics research has greatly benefited from technological advances. Even in the late 1990s it was still common for many biomechanists to spend hours manually tracking thousands of frames of video to analyze the motion of a performer. Now it is possible to use special cameras that can track the motion of the performer in just minutes. What other tools do biomechanists use to accomplish the goals of biomechanics? How, for example, do biomechanists perform gait analyses for patients, research the cushioning properties of newly designed shoes, analyze the performance of Olympic athletes to help them improve, and perform research to answer fundamental questions about how and why we move? Whenever an answer to a biomechanical question is sought, whether by an allied health professional, a coach, or a biomechanist, a common analysis process is used to answer it.

Model for Analysis

Researchers ask questions and then design experiments using systematic methods to answer those questions. Similarly, if you're a practitioner, you will have questions that you want to answer. Although the methods used by practitioners to solve physical activity problems are much different from the methods used in laboratory biomechanics research, they involve a systematic process. To give you a sense of how a teacher or physical therapist might answer a question of interest, we will go through the simplified steps of an analysis process (see figure 10.3). For our example, let us assume that we are working to determine whether a new prosthetic design will improve the gait of individuals who have lost part of a lower limb.

Step 1: Identify Your Question

As shown in the model, the first step is to identify the question of interest: for example, "Does this new shoe model control pronation?" or "Is this rehabilitation protocol effectively strengthening the client's lower back?" Let's assume that the question is "Does prosthetic design XZ-2 or QR-8 allow the best but safest running performance for a long-distance runner who has had most of his leg amputated?"

Step 2: State Performance Goals

Next, identify the overall goals of the movement in conjunction with the participant's goals for performing the movement. A performer may wish to jog 30 minutes for exercise, with the amputated limb performing as naturally and effortlessly as the intact limb. Our performer wants to run a 10,000-meter race as fast as possible without injury.

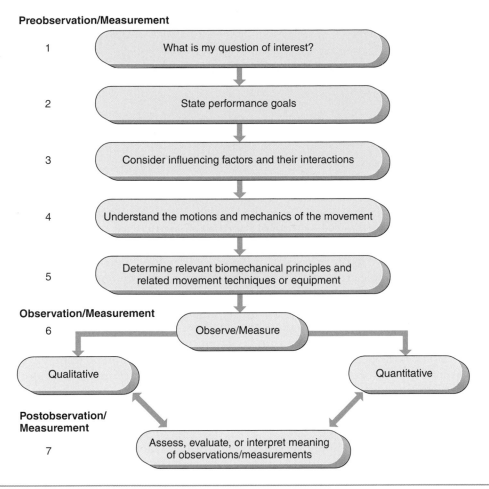

FIGURE 10.3 The movement analysis model used by practitioners or researchers to evaluate a question of interest. Each step of the analysis is labeled by number.

Step 3: Consider Influencing Factors

For this step, we must understand the characteristics that are particular to our runner, that is, physical, emotional, and other characteristics. Our performer may be motivated to run as fast as possible while trying out the new designs or may be hesitant if he perceives that he could be injured while wearing one of the new designs. The strength and length of the amputated limb are important physical characteristics for determining the proper fit of the prosthesis and how well the prosthesis functions. Working with our runner, we also need to understand other factors outside the performer. What type of shoe can he wear while wearing either design? What types of running surfaces are appropriate for these designs?

> Researchers have found that the characteristics of each individual must be considered when attempting to improve a client's performance.

Step 4: Understand Motions and Mechanics

Execution of step 4 requires an understanding of the motions and mechanics involved in the movements used to perform the activity. Running requires performers to support their body weight on one foot and then propel themselves upward and forward through the air. When the leg is not on the ground, runners must pull it through in preparation for placing it back on the ground. Thus, our prosthetic design should allow performers to produce the force necessary to propel themselves forward, should support their body weight without generating excessive pressure on the stump, and should allow them to swing their prosthetic leg forward in a natural manner.

ACTIVITY 10.4

Practical Plug-In

Consider what a tennis coach needs to know and do to help a client in this activity in your web study guide.

Step 5: Determine Relevant Biomechanical Principles and Movement Techniques

The next step involves identifying the mechanical principles most relevant to our question. In addition, we need to determine what movement techniques or equipment designs would help us use these mechanical principles to the best benefit. Newton's laws, principles related to energy, and fluid mechanics are examples of principles that we could choose from. We realize that because the amputated limb is lacking muscles of the lower leg and foot, the prosthetic limb will have to provide some of the propulsive force needed. Thus, we might be interested in the principles related to storing and releasing energy, similar to what happens when a spring is compressed and then allowed to regain its shape. Prostheses that are built specifically for competitive running are designed to take advantage of storing and regaining energy. Newton's third law of action–reaction is evident, too: For every force applied by one object onto another object (say, the performer's foot pushing backward and downward on the ground), there will be an equal and opposite reaction (the ground pushing back forward and upward on the performer's foot). Therefore, the ground can propel our runner forward and upward into the air.

Step 6: Observe or Measure

Finally, we've completed the preobservation (or pre–data collection) stage. Now, we must observe or measure the movements. There are two major methods to gather information. First, kinesiology specialists assess or evaluate the movement of a performer by observing. A movement specialist who observes the movement technique of a performer is doing a **qualitative analysis**. For example, if you were working with the runner who is trying out the

Quantitative data provide numerical values, such as the amount of ankle flexibility. To measure flexibility, the biomechanist is using a special protractor called a goniometer to find the greatest ankle angle (145°) that the client can achieve when extending the foot.

two types of prostheses, you could watch to see whether his prosthetic limb swung excessively outward or listen to hear whether the prosthetic limb was hitting the ground too hard.

For the second method, the biomechanist who is performing research to determine what prosthetic design will best accomplish the client's goals needs to obtain numerical data using biomechanical instrumentation. Clinicians such as physical or occupational therapists or athletic trainers may also wish to obtain numerical data using biomechanical instrumentation to quantify their evaluation that the prosthesis selected is the most appropriate one for the client. The fitness instructor may wish to measure the improvement in running speed of the performer over several months. If a tool is used to measure and evaluate quantities related to space, time, motion, force, or energy during a performance, the process is termed a **quantitative analysis**. (The instrumentation used for biomechanical research and other applications is discussed later in this chapter.)

Step 7: Assess, Evaluate, and Interpret

After obtaining the qualitative or quantitative data for your performer, you must interpret the meaning of the findings in relation to the question of interest, the underlying biomechanical principles, the influencing factors, and the performer's movement techniques. Assume that you were determining which of the two prosthetic designs (XZ-2 and QR-8) was optimal for your performer to enhance running velocity. For each design, the runner ran 50 meters (55 yards) as fast as possible. You could have measured how much the prosthetic knee flexed and how quickly it flexed and extended during various instants in time. Your findings: The knee flexion values and the maximum speed at which the runner could flex the knee of prosthetic design XZ-2 were much closer to the nonprosthetic limb values than were the values for design QR-8. The performer ran at the same speed wearing either design. Your assessment: The XZ-2 design allowed the performer to flex the knee more than QR-8 did during initial contact with the ground, allowing for more effective reduction of impact forces. Performance wasn't affected. Your recommendation: Try the XZ-2 design.

 ACTIVITY 10.5

Movement Analysis Model
Order the steps in the problem-solving process in this activity in your web study guide.

You will need to share this information with a variety of people (e.g., your supervisor or accreditation board reviewers). The professional often is required to document the effectiveness of interventions, be it a teacher assessing a student's progress in a physical education class or a physical therapist submitting evidence to payment providers documenting the client's rehabilitative progress.

The last step in this process can also involve providing the performer qualitative or quantitative feedback. As a physical educator, fitness instructor, athletic trainer, occupational therapist, physiotherapist, physician, or other professional working to enhance performance of your client, you may provide feedback to the performer while the performer is learning or refining skills. Clinicians may choose to provide feedback to a client to enhance motivation as the client improves performance during rehabilitation.

 ACTIVITY 10.6

Practical Plug-In
What knowledge comes into play when a fitness center owner needs to decide what types of equipment to purchase for his facility? Find out in this activity in your web study guide.

Biomechanical Instrumentation and Other Tools

In step 6 of the analysis process, we mentioned that biomechanical instrumentation and other tools could be used during a quantitative analysis. We will now explore some of the tools used to measure biomechanical quantities.

What simple tools besides a tape measure can be used to measure or observe time, movements in space, or force? The following are some of the instruments that may come to mind: stopwatches or metronomes to measure time, barbells and free weights to measure how much force a person can exert, and a video camera to capture the motion of a performer. From the motion images, a movement specialist could measure height, length, and distance that a limb traveled or the length of a dance step. A protractor could be used to measure how many degrees a limb rotated. For example, from a videotape of a fitness walker, we could use a protractor to evaluate whether the walker was landing with the foot at a particular angle. Figure 10.4 shows various parts of instrumentation systems used in a biomechanics laboratory for research, clinical analyses, or ergonomic or other biomechanical analyses.

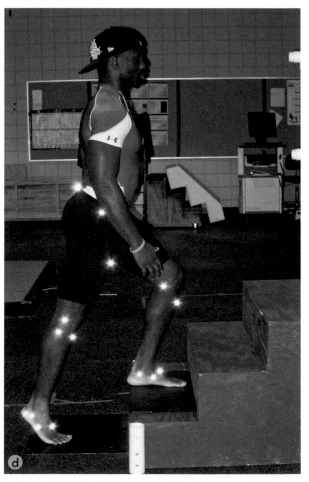

FIGURE 10.4 Biomechanical instrumentation. Components for motion measurement include (a) a digital, high-speed camera that tracks the location of reflective markers; (b) electromagnetic-sensitive markers that are used to track body movements via changes in an electromagnetic field surrounding the participant when the markers move about in space; (c) electrodes and data collection pack used in electromyography; and (d) force platform used to measure ground reaction forces.

© Kathy J. Simpson

Motion Measurement Devices

The use of digital cameras and other motion-detection technology for recording motion in biomechanics is called *motion measurement*. Digital high-speed cameras can be used to trace the motion of reflective markers placed on points on the human body to track the motion of the various body segments. Other types of markers placed on the performer's body (e.g., light emitting diodes [LEDs]) emit signals that are captured by a sensor to track movement. Sensors attached to the performer that establish the position of a body part (e.g., the head) detect magnetic fields. The motions of any body segment (e.g., the jaw) or the entire body can be tracked and used to determine body positions, velocities, and accelerations. Feedback about performance techniques from motion analyses has been helpful to many elite and professional athletes in activities such as pole-vaulting, gymnastics, golf, and professional baseball. Perhaps you would like to hire a private company that uses this technology to analyze your golf stroke. Motion measurement equipment is also used to analyze movements observed in occupational and clinical settings (see www.podiatrytoday.com/article/4433 for an example of a clinical gait analysis case study). Some researchers are also trying to use laser technology similar to that used in commercial games, such as the Kinect.

Force Measurement Devices

Measuring quantities related to force can tell us what is causing motion or injury. To measure how much force is being placed on a joint, ligament, or object, researchers attach tiny force-measuring devices (force transducers) to tissues or place artificial implants inside the body. The latter method is more commonly used in animal experiments because of the practical limitations of implanting devices inside of the performer's body.

Transducers can also be used to measure how much force a patient can exert against a strength-measuring device. To measure ground reaction forces, force platforms have been used. A force platform is a metal plate with force transducers attached to the beams that hold up the plate. Usually a force platform is embedded solidly in the surface so that for gait analyses, a patient can step on and off the platform naturally while walking. Ordinary or specialized force platforms can be used to diagnose postural and balance disorders. Knowledge of these ground reaction forces is helpful to biomechanists for other purposes, too, such as the evaluation of new models of footwear or as a tool to help determine internal force loading on bones and cartilage.

Pressure is the amount of force applied to a given amount of surface area. Areas of high pressure against human tissue can cause health problems. For example, for a person with diabetes who cannot sense discomfort, too much pressure placed on an area of the foot can cause the skin in that area to deteriorate and become infected. Finding the zones of high pressure on the feet of a diabetic patient helps the clinician tailor an insert to be worn inside of the shoe that can spread out the forces more evenly (Ko 2010). Pressure devices can also be placed on wheelchair seats to test padding that will minimize pressure.

Because of the difficulty of directly measuring muscle forces, motion and force platform data are used to calculate estimates of muscle forces via physics and engineering methods. Another method of estimating muscle forces involves measuring the electrical activity of the membranes of the muscle cells (fibers) when nerve cells stimulate them to contract. This method, called electromyography (EMG), requires placing electrodes on the bellies of muscles where the muscle fiber membranes are most likely to be stimulated by nerve cells. When the participant performs the movement of interest, for example, walking, nerve cells send electrical signals to the muscle fibers that tell the fibers to generate force. The measured amount of electrical activity can be converted into the amount of force that the muscle produced during the movement.

Although obtaining accurate muscle force estimates using this method is difficult, EMG is being used for this purpose in complex mathematical models and for simulating movements. Electromyography also has been helpful in identifying the muscles that are active during a particular movement, which indicates what muscles to target for strength, power,

▌ Pressure distributions of the bottom of the feet during walking at one instant in time.

or endurance training. Conversely, EMG studies of exercises can determine what muscles are being trained during an exercise. Electromyography also can be used clinically to determine whether a muscle is functioning correctly. People with movement disorders can learn how to substitute the use of other muscles for a nonfunctioning muscle. Knowledge about muscle activity has led to the development of stimulated walking, in which paralyzed patients wear a device that stimulates various muscles with electricity at the right time to produce the muscle forces necessary for the person to walk.

How Do You Measure Up?

Earlier, you were asked to generate your own question of interest. What instrumentation could you use to help answer your question? What instrumentation could be used to compare the ground reaction forces of the runner when wearing the minimalist versus the traditional cushioned shoe?

All instrumentation described in the previous paragraphs measures some quantity, but mathematical techniques are required to calculate what cannot be directly measured. Computer simulation is based on mathematical modeling and uses the data of various instrumentation devices (EMG and motion measurement, for example) along with database information about the body such as dimensions and locations of muscle attachments. Simulations can help us understand basic mechanics at the cellular, molecular, or whole-body level. One example might be a simulation of knee joint motion, whereby the anatomy and forces involved with various structures, such as ligaments, tendons, cartilage, bones, and muscles, are modeled to

Modeling the movements of a knee joint implanted with a partial knee replacement.

enable understanding of typical knee joint motion. Various properties of the model can be changed. For example, moving an attachment of a particular muscle to a different location to observe whether the model predicts greater friction forces may help determine whether anatomical differences in muscle attachments underlie a mechanism of cartilage degeneration.

Overview of Knowledge in Biomechanics

At the beginning of this chapter, you were given a hypothetical situation: a high school cross-country runner is beginning to develop shin pain and wonders whether switching to a minimalist shoe (one that uses the least amount of material to protect the foot) would be better than the traditional cushion shoe with foot motion control. You then were asked to consider how you would determine recommendations for the runner, including figuring out what information about the runner you'd want and what resources you'd use to answer the question. Now that you know more about biomechanics, would you change your answer? Do you expect that either shoe would be best for all runners? Would the runner's foot–ankle structure (e.g., flat-footed or high-arched) or running technique be important to know?

We hope that one of your answers would be to find the most current recommendations based on biomechanical evidence published in respected, professional journals. That is what Mr. Michael Fedewa, a colleague, and I did to provide you information about this issue. This topic also demonstrates how knowledge in all of the areas of biomechanics can be applied to a specific question of interest. So, here's what we found out.

One reason that barefoot running became popular and minimalist shoes were developed to replace traditional shoes, which provided a high degree of cushioning and foot motion control, was injury rate. Despite technological advances in shoe materials and design, the injury rate of runners remained relatively consistent (Withnall et al. 2006). Additionally, much popular literature had been published that encouraged athletes to run barefoot. According to writers in popular magazines, running shoeless was considered to be the more natural and thus safer running approach, and barefoot runners didn't seem to get hurt (McDougall 2009). Manufacturers used this line of reasoning to promote shoes that mimicked barefoot running while providing a minimal amount of foot protection.

One type of evidence supporting barefoot running was the amount of impact force acting on the runner's foot. Earlier we discussed how the law of action–reaction applies to running—the harder the runner hits the ground, the more force the ground will apply back onto the runner's foot. Thus, we use impact force, that is, the maximum vertical ground reaction force that acts on a runner's foot during the landing phase, as a measure of highest amount of force that the ground applies to a person during landing activities. Impact forces for running can range from 2.0 times body weight (BW) for jogging barefoot to 7.9 BW for sprinting with spikes (see review by Nigg 1983). Compared with runners who wear traditional shoes, barefoot runners tend to protect their feet against high ground reaction forces. To do this, barefoot runners take shorter but more frequent steps and land in a more flexed-knee position (see, for example, De Wit et al. 1996; Divert et al. 2005, 2008; Squadrone et al. 2009).

Reducing impact forces was believed to be an advantage for barefoot running, as some biomechanists believe that high impact forces may be related to some bone and joint injuries (Hreljac et al. 2000). The biomechanical principle to support this notion is that applying high force to bones over many, many repetitions causes bone to deform more, consequently creating microcracks. Perhaps then, the reasoning went, as runners strike the ground with substantial force many times, the shin bones (tibias) begin to bend to a greater degree with time, causing microcracks to occur to some runners' tibias (Donahue et al. 2000).

Lower impact forces during barefoot compared with shod running may also occur if runners self-select a different footstrike pattern. People who are heel strikers when wearing shoes sometimes will exhibit a more flatfooted landing when barefoot. The peak vertical ground reaction force is greater when striking the ground at the heel compared with landing flatfooted or on the forefoot (Goss & Goss 2012; Lieberman et al. 2010), regardless of whether one is running barefoot or with shoes (see nice visual of this at http://barefootrunning.fas.harvard.edu/4BiomechanicsofFootStrike.html). Therefore, two major footwear manufacturers provide instructions with their minimalist footwear that teach runners how to contact the ground on the forefoot.

Injury prevention involves more than reducing impact forces, however. It is unclear whether landing more on the ball versus the heel of the foot is safer or not (Goss & Goss 2012). Another group (Salzer et al. 2012) observed that out of 10 runners who swapped their traditional running shoes to minimalist footwear, 8 exhibited stress fractures of the bones of the forefoot.

You are likely wondering why we are presenting a biomechanical comparison of barefoot to traditional-shoe running rather than providing useful information about minimalist shoes. There simply is not yet much published evidence about the effects of minimalist shoes on people's running biomechanics. The advantages and disadvantages of wearing minimalist running shoes or traditional cushioned shoes or running barefoot, in terms of performance and injury prevention, are being debated currently at professional conferences. Soon you will be able to find sufficient and solid evidence to inform physical activity and rehabilitation specialists when helping their clients select footwear appropriate for performance and injury prevention.

Hit the Ground Running

Select fast walking or running and calculate the maximum amount of vertical impact force that you might generate. To calculate, multiply your weight by 1.5 times your body weight (fast walking) or 2 to 3 times body weight (running). To get a sense of how much force that is, think about whether you could catch that much weight with your arms.

Wrap-Up

The physical laws of nature act on you at all times. Biomechanists as well as other physical activity specialists can apply the principles of these mechanical laws to enhance performance; reduce injury; evaluate the effectiveness of a movement; or select the proper sport equipment, tool, or occupational equipment. The theoretical knowledge regarding how mechanical principles influence our movements and the structure and function of the human body can be applied to occupational work, sports, exercise, dance, daily tasks, and clinical rehabilitation. Furthermore, an understanding of the theoretical knowledge of biomechanics, as well as the use of biomechanical research methods, is helpful to professionals in other subdisciplines of physical activity, ranging from motor control to pedagogy of physical education. If you're fired up to learn more about biomechanics in general, you might start by looking at the International Society of Biomechanics website (www.isbweb.org), which provides information about job opportunities and graduate programs; links to sites focusing on a variety of biomechanical areas such as gait, prostheses, and sport; and abstracts of conferences.

More Information on . . . Biomechanics of Physical Activity

Organizations

International Society of Biomechanics

International Society of Biomechanics in Sports

Journals

Clinical Biomechanics

Clinical Orthopaedics and Related Research

Gait & Posture

International Journal of Experimental and Computational Biomechanics

Journal of Applied Biomechanics

Journal of Biomechanics

Journal of Dance Medicine & Science

Journal of Electromyography and Kinesiology

Journal of Prosthetics and Orthotics

Sports Biomechanics

 ACTIVITY 10.7

Use the key points review in your web study guide as a study aid for this chapter.

Review Questions

1. Define biomechanics and human movement biomechanics.

2. In what types of settings do biomechanists typically work?

3. What are the goals of biomechanics?

4. For each goal of biomechanics, provide an example of how this goal is relevant in an occupational setting. First, describe the profession and the occupational setting; for example, a physical therapist in an orthopedic outpatient clinic. Next, provide a specific situation; for example, a runner is experiencing heel pain thought to be caused by flat feet. Next, describe how you would help the client achieve each goal in this career setting.

5. With a partner, go to a website of a professional biomechanics society (e.g., International Society of Biomechanics: www.isbweb.org) or job-search site and find the listing of biomechanics jobs. Categorize the jobs using a classification system of your choice. For example, one category of jobs might be research jobs in clinical settings; another

might be ergonomic jobs in industry. With your partner, develop a short document or presentation about the job outlook in biomechanics: roughly how many jobs exist, what categories they fall in, what qualifications are necessary for different categories of jobs. Last, do any of these types of employment positions interest you?

chapter eleven

Physiology of Physical Activity

Jennifer L. Caputo

chapter objectives

In this chapter we will

- discover the key features of the subdiscipline of physiology of physical activity and the employment opportunities available to exercise science professionals;

- explain how the physiology of physical activity fits within the discipline of kinesiology;

- review the history and development of physiology of physical activity;

- identify the research methods used by kinesiologists working in exercise science; and

- examine how the body responds to physical activity and how these changes relate to physical performance and health.

The volleyball coach at a local university is curious about what type of warm-up he should use with his athletes. He heard recently that static stretching may be detrimental to the jumping ability of his players. He would like assistance in designing an optimal warm-up for performance.

Is being physically active associated with improved academic performance? A local teacher is curious about this. She wants to know whether it would benefit her students to have additional activity periods during the school day.

An ROTC unit is competing in a Ranger Challenge, which consists of tests of physical fitness and field drills. Some of the necessary skills include marksmanship, muscular strength, navigation, rope climbing, and road marching. Because the exact events remain unknown until the competition, teams must train all aspects of fitness and be ready to perform at a high level in all areas. The captain of the squad is seeking assistance in preparing the team for victory.

Individuals working in the subdiscipline of physiology of physical activity have the opportunity to work with people in a variety of settings and situations. Can you determine how to best assist the people described in the above scenarios?

The questions posed in the chapter opening represent some of the areas of specialization in the kinesiology subdiscipline of physiology of physical activity. If you are curious about how your body and your health status are altered from either the performance of physical activity or a lack of physical activity, the subdiscipline of physiology of physical activity may be appropriate for you. With knowledge of physiology of physical activity, you can help athletes achieve their peak performance, help people safely participate in physical activity, conduct research to help prevent and treat disease through physical activity and exercise, and even help maximize the work capacity of people who work in the extreme heat or cold or at different altitudes.

Understanding how your body responds to meet the immediate demands of being physically active, as well as how your body adapts to repeat bouts of exercise, is a key feature of physiology of physical activity. When you chase down a shot on the tennis court or go for a run, your heart rate and breathing rate increase. Your blood pressure increases and, over time, you begin to sweat. These changes are your body's way of meeting an increased demand for oxygen, energy, and temperature regulation. If you did these physical activities over several weeks, you might notice changes in your body weight and muscle mass. These physiological responses to physical activity and exercise are just a few of the topics that intrigue students of physiology of physical activity.

As shown in figure 11.1, physiology of physical activity, often called **exercise physiology** or **exercise science**, is a subdiscipline of the biophysical sphere, along with biomechanics of physical activity (discussed in chapter 10). Whereas biomechanists apply principles of physics and engineering to physical activity situations, exercise physiologists apply principles of biology and chemistry to understand how the body responds to physical activity.

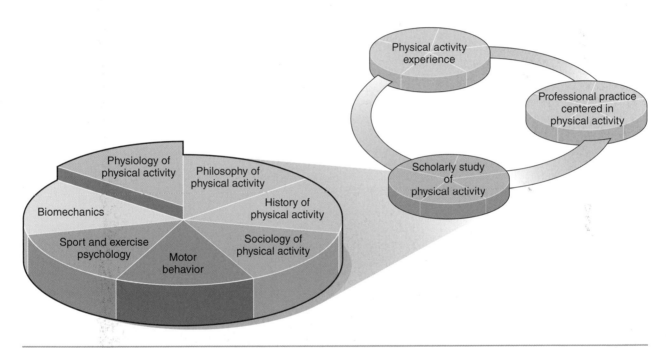

FIGURE 11.1 Physiology of physical activity and the other scholarly subdisciplines.

Exercise scientists study the body's anatomical and physiological responses to physical activity and exercise in order to improve health and physical performance. Exercise physiology is the study of the acute and chronic responses of the body to movement. Exercise science professionals study short-term changes in how the body functions during exercise and also long-term changes due to training. For example, students discover that the heart beats faster as running speed increases and that resting and exercise heart rates decrease following endurance training. Exercise physiologists also study the mechanisms that lead to these changes in heart rate.

Why Use Physiology of Physical Activity?

From chapter 1, we know that exercise is a form of physical activity used to improve or regain performance, health, or physical appearance. Exercise science professionals primarily focus on the exercise components of physical activity including training, health-related

Goals of Exercise Physiology

- To understand how to enhance physical performance
- To understand how to improve physical function in particular environments such as high temperatures and high altitude
- To understand how physical activity and exercise improve health and fitness
- To understand how exercise can be used in treating and preventing disease and alleviating symptoms of disease
- To understand adaptations in anatomy and physiology in response to physical activity

exercise, and therapeutic exercise. Early work in the discipline also helps us to understand the impact of physical activity on the anatomy and physiology of the body. The following sections summarize some of the benefits derived from understanding the subdiscipline of exercise science.

Enhance Sport Performance and Training

The application of physiological techniques to understand and improve human exercise performance has been a major goal of exercise physiology since its inception. Knowledge of the principles of physiology of physical activity serves as the foundation for the development of conditioning programs for various athletes. This area of exercise physiology is often called *sport physiology* and is the application of "the concepts of exercise physiology to enhancing sport performance and optimally training the athlete" (Kenney et al. 2012, p. 3). Think about

The Harvard Fatigue Lab conducted many studies in the early years of the study of sport performance (see the sidebar later in this chapter).

a track coach who needs to develop running programs for sprinters, middle-distance runners, and distance runners. The various speeds and distances covered by these athletes pose special challenges to the coach, who must optimize the training program for each group of runners.

Some of the earliest studies conducted on the physiological responses to strenuous exercise and training came from the Harvard Fatigue Laboratory (Dill et al. 1930; Margaria et al. 1933). Examples of the topics of research currently being conducted by sport physiologists include the impact of stretching and flexibility on performance, the role of genetics in exercise and physical activity performance, barefoot running or using minimalist shoes, and the use of vibration platforms in training. Sport physiologists also use information and techniques from other disciplines and subdisciplines to study ways of enhancing performance. One example is the relationship between carbohydrate intake and performance in endurance events. Sport physiologists may use information gleaned from nutrition research on the carbohydrate content of various foods to work with a sport nutritionist on manipulating the diets of athletes to enhance carbohydrate storage and improve performance. Sport physiologists are also concerned with the effects of the environment on sport performance. Data from sport physiology studies have been used to develop guidelines to avoid heat illness in sport and to prevent health problems brought on by prolonged immersion of scuba divers in water of great depth. Before the 1968 Summer Olympics in Mexico City (7,218 feet [2,200 meters] above sea level), leading exercise physiologists conducted numerous studies on the acute effects of and acclimatization to altitude. Sport governing bodies used the knowledge gained from these studies to better prepare their athletes for Olympic competition.

Strength and Conditioning Specialists

When working with athletes in the weight room, strength coaches need to understand how to alter conditioning programs to meet the needs of athletes from many sport teams. As exercise scientists, strength coaches learn how to produce varying results by altering the weight lifted, the number of repetitions and sets, and the length of rest periods to produce appropriate responses in sprinters, basketball players, softball players, and football players.

 ACTIVITY 11.1

Practical Plug-In
Consider what a coaching director needs to know and do in order to make best-practice decisions for his soccer club in this activity in your web study guide.

Improve Physical Fitness

Understanding the physiological determinants of physical fitness and how training programs improve fitness are also important benefits of studying exercise physiology. Two events in the 1950s—President Dwight D. Eisenhower's heart attack and the publication of a study reporting that American children were less fit than European children—stimulated interest in improving physical fitness in the United States (Berryman 1995). These two events led to the development of numerous training programs for improving fitness. Research conducted over the past 4 decades has resulted in recommendations on the optimal intensity, frequency, and duration of training programs. Individuals working in fitness centers need to understand how to adapt fitness programs to make them safe and appropriate for young and old participants, people who have been sedentary or are already training and want to increase their physical fitness level, and those with special conditions such as pregnancy or diabetes. Pediatric exercise physiologists specialize in studying and working with children and adolescents, and other exercise scientists are interested in gerontology (the study of aging). The influence of heredity on physical fitness components and trainability has attracted considerable research interest among exercise physiologists. The work of Claude Bouchard and his colleagues with identical twins has stimulated much of this interest. The physiology of fitness is also linked with psychology and the kinesiology subdiscipline of exercise and sport psychology. Understanding why some people adhere to training programs and others stop exercising has led to various strategies for motivating people to become physically active.

Promote Health Through the Therapeutic Benefits of Physical Activity

Exercise physiology serves as a foundation for understanding why physical activity and exercise are beneficial in reducing the risk of disease and in treating some forms of disease. In 1996, the Surgeon General of the United States emphasized in a report on physical activity and health that "significant health benefits could be obtained by including a moderate amount of physical activity on most, if not all, days of the week" (United States Department of Health and Human Services [USDHHS] 1996). Many studies have examined the benefits of physical activity and physical fitness in preventing and even reversing coronary heart disease. Researchers have documented the role of physical activity in preventing and treating other diseases such as non-insulin-dependent diabetes and osteoporosis. Some researchers are developing exercise tests to screen for obstructive sleep apnea, and still others are examining the role of exercise in decreasing the risk of dementia and Alzheimer's disease. Sometimes exercise physiologists are interested in the big picture—how physical fitness can prevent disease in large populations. In such cases, exercise physiologists might collaborate with specialists in epidemiology, using two different types of studies—longitudinal (comparing the same group of people over several time periods) and cross-sectional (comparing different groups of people at the same time).

Physiologists who study the role of physical activity in disease management and rehabilitation are known as *clinical exercise physiologists.* Students training in this area become familiar with basic concepts in several medical specialties, such as cardiology and pulmonary medicine, and work under the direction of a clinician. For example, students interested in cardiac rehabilitation need to learn about the various forms of cardiovascular disease and the medical and pharmacological treatments of these diseases. Knowledge of the cardiovascular system and its responses to exercise, and the ability to read an electrocardiogram (ECG, a tracing of the electrical activity of the heart), are also important. Cardiac rehabilitation specialists must understand how to modify exercise mode, intensity, and duration for those who have high blood pressure and for those who have had open heart surgery.

ACTIVITY 11.2

Practical Plug-In

Consider what an exercise physiologist needs to know and do to improve athlete health screening in this activity in your web study guide.

ACTIVITY 11.3

Interview with a Practicing Professional

In this interview in your web study guide you'll hear from Orin Aaker, a health and fitness instructor and exercise physiologist engaged in life-changing work with a special population.

Understand Anatomical and Physiological Changes From Physical Activity

Early work in the exercise sciences focused on the effects of exercise on the functions of organs and body systems. Exercise physiologists have examined the functional responses and structural adaptations of the cardiovascular, respiratory, muscular, and endocrine systems to different types of physical activity. Recently, research interests have focused on the responses of the reproductive, skeletal, and immune systems to acute and chronic exercise. One branch of exercise physiology is closely linked to **biochemistry** (i.e., the chemistry of living things). Investigators looking to examine the fuel sources used by muscles during exercise, for example, have used research techniques developed in biochemistry and physical chemistry. The use of the muscle biopsy needle by Bergstrom and Hultman (Hultman 1967) to sample muscle tissue and examine muscle glycogen (stored carbohydrate) concentration during exercise resulted in one of the most important advances in exercise physiology. Biochemical techniques have been used to develop our understanding of lactate production and use of energy stores during physical activity. Research in molecular biology, one of the newest subdisciplines of biology, has greatly enhanced our understanding of how cells function. Scientists use molecular biological techniques to determine how genes regulate protein synthesis. Exercise physiologists use molecular biological techniques to study how muscle protein synthesis is turned on and off by increased and decreased muscular activity (Booth 1989). This molecular knowledge is important in understanding how muscles increase and decrease in size in response to increased and decreased activity, respectively, and in understanding how muscles are damaged and how muscles recover from injury.

What Do Exercise Physiologists Do?

Many opportunities for employment exist within the subdiscipline of physiology of physical activity. You can choose the setting you most enjoy, such as a fitness or wellness center, hospital, corporation, academic institution, or research laboratory. You can work with different populations ranging from the young to the old and from the apparently healthy to those diagnosed with disease. The variety of jobs offers many opportunities to those who have obtained various levels of education in the field.

Persons with academic knowledge of physiology of physical activity may work as exercise instructors in commercial and corporate fitness centers or as personal trainers and strength and conditioning coaches. There are opportunities for clinical work in cardiac rehabilitation and pulmonary rehabilitation. The U.S. military and the National Aeronautics and Space Administration (NASA) employ exercise physiologists to conduct research (e.g., on preventing heat illnesses during basic training or using physical conditioning while in space to prevent bone loss).

Exercise physiologists usually have a master's degree, whereas teaching or conducting research in academic settings is typically reserved for those who have a doctoral degree in exercise science or exercise physiology. Faculty members teach courses in exercise physiology, anatomy and physiology, exercise testing and prescription, and sport nutrition; conduct laboratory research on topics such as the effects of conditioning programs on sport performance or ways of reducing the risk of chronic diseases; and write research grant proposals that they submit to federal agencies (e.g., the National Institutes of Health [NIH]) or foundations (e.g., the Robert Wood Johnson Foundation) that support research on physical activity.

In addition to obtaining academic degrees in exercise science, physical activity professionals can take certification examinations. Certifications are offered through organizations such as the American College of Sports Medicine (ACSM), the National Strength and Conditioning Association (NSCA), the Aerobics and Fitness Association of America (AFAA), and the American Council on Exercise (ACE).

History of Physiology of Physical Activity

Physiology of physical activity evolved from physiology in the 18th century after Antoine Lavoisier discovered that animals consume oxygen and produce carbon dioxide (USDHHS 1996). As a result, scientists began exploring how humans respond to physical activities and daily tasks (e.g., lifting loads).

Early Beginnings

Two of the early contributors to our understanding of physiology of physical activity were August Krogh of the University of Copenhagen and A.V. Hill of University College, London. Krogh developed one of the first cycle ergometers (exercise bikes), which he used to study the physiological responses to exercise. He received the Nobel Prize in Physiology in 1920 for his research on the regulation of microcirculation (Åstrand 1991). Hill received the Nobel Prize for his work on energy metabolism in 1921. Hill and H. Lupton presented many of the basic concepts of exercise physiology concerning oxygen consumption, lactate production, and oxygen debt (excess oxygen consumption following exercise) in a classic paper in 1923. Two laboratories that developed in physical education departments also had a major influence on research in exercise physiology. Peter V. Karpovich established an exercise physiology laboratory at Springfield College, Springfield, Massachusetts, in 1927 (Kroll 1982). Karpovich became well known for his research on the effects of ergogenic aids on physical performance and as one of the founders of the American College of Sports Medicine. Thomas K. Cureton, Jr., established an exercise physiology laboratory at the University of Illinois in 1944. This laboratory became well known for research on physical fitness. Many of the leading investigators of the physiology of fitness received their training in Dr. Cureton's laboratory.

Significant Events Since 1950

One of the most important events that stimulated research on physiology of physical activity after 1950 occurred in England. In 1953, Jeremy Morris and colleagues published a study on coronary heart disease and physical activity in which they found that drivers of double-decker buses in London had a significantly higher disease risk than conductors on those buses (Morris et al. 1953). This study stimulated interest in epidemiological research on physical activity, physical fitness, and chronic disease (Paffenbarger 1994). Morris' research also stimulated studies on the effects of fitness and endurance training on risk factors for coronary heart disease (e.g., serum cholesterol, blood pressure). Based on the results of many of these studies, U.S. government agencies released two official statements on the role of physical activity in the prevention of chronic diseases in the 1990s. The first followed an NIH Consensus Development Conference on Physical Activity and Cardiovascular Health held in December 1995. Among the consensus development panel's conclusions were that "...physical inactivity is a major risk factor for cardiovascular disease" and that "moderate levels of regular physical activity confer significant health benefits" (NIH Consensus Development Panel on Physical Activity and Cardiovascular Health 1996, p. 245). The other statement was the Surgeon General's first report on physical activity and health by the U.S. Department of Health and Human Services in 1996 (see chapter 2). The report concluded that regular physical activity, in addition to reducing the risk of heart disease, reduces the risk of diabetes, hypertension (high blood pressure), and colon cancer and helps control body weight (USDHHS 1996).

> Both the National Institutes of Health and the Surgeon General concluded that moderate physical activity is beneficial in reducing the risk of chronic diseases (e.g., heart disease, diabetes, hypertension, and colon cancer).

Research Methods for Physiology of Physical Activity

Kinesiologists working in the subdiscipline of exercise science use many methods to measure and determine how the human body responds and adapts to physical activity. Work is conducted both within and outside of the laboratory.

Harvard Fatigue Laboratory

The Harvard Fatigue Laboratory, founded in 1927, was the brainchild of L.J. Henderson, a physical chemist. David Bruce Dill was the director of the laboratory, which included a room containing a treadmill borrowed from the Carnegie Nutrition Laboratory, a large gasometer, a room for basal metabolism studies, an animal room, and a climatic room (Dill 1967). Among the areas of research undertaken by members of this laboratory were environmental studies conducted at high altitude, in the desert, and in steel mills. Although the Harvard Fatigue Laboratory was in existence only until 1946, it had a profound effect on exercise physiology research in the United States and Europe during the second half of the 20th century. Many young investigators received their formative training in the laboratory as postdoctoral or doctoral students, including Ancel Keys, R.E. Johnson, Sid Robinson, and Steve Horvath. Among the many international scientists who spent time in the laboratory were Lucien Brouha, Belgium; Rodolfo Margaria, Italy; and E.H. Christensen, Erling Asmussen, and Marius Nielsen, Denmark (Dill 1967). All of these investigators went on to establish their own laboratories and were responsible for training many of the leading investigators in physiology of physical activity over the past 50 years.

Laboratory Work

Much of the equipment used to monitor and evaluate physiological responses is contained within exercise science laboratories. Working within a laboratory space is beneficial because there is greater opportunity to control factors that can affect responses to exercise such as temperature and humidity variations. It is important to be able to compare work conducted in different laboratories, so the use of standard protocols and techniques is encouraged.

Many techniques are available to assess the health-related components of fitness and the traits associated with lower risk of hypokinetic disease. For example, cardiovascular fitness is assessed either through direct measurement or through estimation of oxygen consumption. This requires measuring the volume of oxygen and carbon dioxide in expired (exhaled) air. Early investigators collected expired air in Douglas bags and analyzed the gas concentrations using chemical analyzers. They then used a spirometer or volume meter to measure how much gas was in the bag. Such analysis is tedious and time-consuming. Now, investigators measure gas concentrations with electronic analyzers and gas volumes with flow meters. When interfaced with computers, these devices provide nearly instantaneous and continuous information on oxygen uptake as people rest or exercise.

By measuring oxygen uptake, researchers can obtain information on how the muscles use oxygen and how much energy is expended during physical activity. The most widely accepted method for determining the cardiovascular fitness of an individual is to increase the intensity of exercise progressively until the person can no longer maintain the required exercise intensity. This method is used to determine **maximal oxygen uptake or $\dot{V}O_2max$**. Physiologists also use submaximal exercise tests to estimate cardiovascular fitness. These tests do not require individuals to exercise at their maximal level; instead, prediction equations are used to estimate maximal oxygen uptake.

Ergometers are used to measure the work performed by muscles during exercise tests. Common ergometers used in exercise science include motorized treadmills and leg or arm cycles. Heart rate, blood pressure, breathing rate, oxygen consumption, and carbon dioxide production are often monitored during exercise. Blood samples may be drawn.

Because body weight is supported during cycling, cycle ergometers are useful for comparing subjects of different body weights. Blood pressure measurements and blood samples during exercise are also easier to obtain when subjects use cycle ergometers because the arm is stationary (figure 11.2). The major disadvantages of the cycle ergometer are that oxygen uptake is generally lower and fatigue occurs earlier because the rider uses only the legs. Motorized treadmills allow people to be tested during walking and running at different speeds. Increasing the slope (grade) of the treadmill can increase exercise intensity at a constant speed. For studies of moderate to heavy exercise, treadmills are preferable to cycle ergometers because running is more strenuous than cycling for most people. But researchers have to account for body weight differences when using the treadmill. A heavy person walking or running at the same speed and grade as a lighter person will work harder because people must lift their body weight with every step. Measuring blood pressure and sampling blood are more difficult when using a treadmill, especially during running, because the arm is moving. Exercise physiologists can use arm ergometers to study upper body exercise. The swimming flume, developed in Sweden, has been used to study the physiological responses to swimming, and rowing ergometers are useful in studying oarsmen.

Body composition is a common measure used by kinesiologists. This assessment is used to determine the percentage of body fat an individual has or the ratio of his or her lean tissue to body fat. The gold standard for determining body composition in humans is hydrostatic

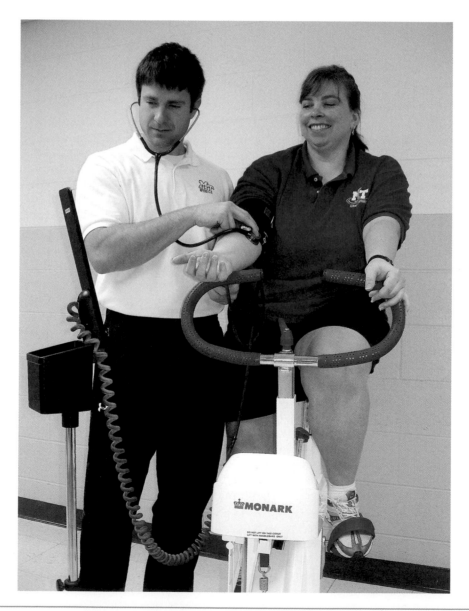

FIGURE 11.2 Blood pressure is easily monitored as a person rides a cycle ergometer.
© Jennifer L. Caputo

weighing, also known as **underwater weighing**, in which the individual is weighed while submerged under water. This technique makes use of Archimedes' principle, which states that the weight of the water displaced by an object is equal to the volume of the submerged object (density = mass ÷ volume). Once body density has been determined, equations are used to estimate the amount of body fat. Other techniques for estimating body density include measuring total body water using isotopes (e.g., deuterium), measuring the thickness of subcutaneous fat (lying beneath the skin) with skinfold calipers (see figure 11.3a), determining tissue impedance with bioelectric impedance analyzers (BIA), and measuring body fat with dual energy X-ray absorptiometry (DEXA), as shown in figure 11.3b.

FIGURE 11.3 (a) Body fat can be assessed through use of the skinfold measurement technique in the laboratory and in the field. (b) Body fat and bone mineral density can be assessed using dual energy X-ray absorptiometry in the laboratory.

a © Jennifer L. Caputo; b Photo courtesy of Middle Tennessee State University

Biochemical Methods

Exercise physiologists use biochemical methods to examine changes at the tissue and cellular levels during and following physical activity. These more invasive techniques include blood sampling and muscle biopsy. Blood samples are obtained from either venipuncture of superficial arm veins or finger pricks. Monitoring changes in the concentration of blood constituents is useful in determining substrate (stored carbohydrate and fat) utilization, acid–base balance, dehydration, immune function, and endocrine responses. A commonly used biochemical technique in exercise physiology is the measurement of blood lactate from fingertip samples taken during exercise as an indicator of the use of anaerobic energy-producing systems. Physiologists use muscle biopsies obtained before and after exercise to examine changes in stored carbohydrate and fat, lactate production, and enzyme activity. Chemical methods used to analyze tissues help determine muscle fiber types in the biopsy samples. Examination of muscle tissue samples under an electron microscope is useful in determining structural changes following different types of training as well as structural damage after exercise.

Animal Models

▶ Advances in technology continue to generate increasing opportunity to conduct research outside of the laboratory. However, it is the combination of thousands of research studies conducted over the past century, both inside and outside of the laboratory and using both humans and animals, which has generated our knowledge of physiology of physical activity.

The effects of physical activity on some organs cannot be studied easily in humans. Examples include the brain, heart, and liver. In some instances, researchers use animals (e.g., rats) to examine both the functional and structural changes that occur in response to single and repeated bouts of exercise on animal treadmills and running wheels. One of the major advantages in using animals in physiological research is that both the subjects and the environment can be more carefully controlled than with humans. The use of animals with genetic abnormalities has also proven helpful for examining the effects of exercise on certain clinical disorders such as obesity and hypertension. Another advantage is that some experimental techniques not approved for human use can be used to study physiological responses. An example is the injection of radioactive isotopes of iron to study changes in tissue iron stores with training. One major criticism of animal research is that not all physiological and chemical mechanisms observed in animal species are identical to those observed in humans. There is some truth to this assertion. For example, growth hormone produced in humans is not identical to that of any other species. Changes in tissues because of training also may differ among species. Investigators must be careful to select the animal, usually a mammal, that most closely reflects the human responses to physical activity in which they are interested.

Field Work

Exercise scientists often work with individuals and conduct research outside of the laboratory. Working in the field can present difficulties with respect to monitoring physiological responses, controlling exercise intensity, and controlling environmental conditions. When scientists work in the field, it is helpful to use tests that require minimal equipment and can be used to screen large numbers of people in a short time period.

Often, kinesiologists conduct research studies in schools. It is not practical to bring a treadmill and a cycle ergometer to a school. However, alternative measures of cardiovascular fitness exist; for example, one can determine how quickly students can run or walk a mile marked off on the school field. The completion time is then used to estimate maximal oxygen consumption. Likewise, the PACER (Progressive Aerobic Cardiovascular Endurance Run), a multistage fitness test adapted from the 20-meter shuttle run, can be performed in school gymnasiums. In addition to allowing multiple students to be tested at a time, these tests require minimal equipment and offer flexibility when space availability differs from one facility to another.

There are alternative means to measure physiological responses while working in the field. Whereas heart rate may be monitored with an ECG machine in the laboratory, low-cost battery-operated heart rate monitors allow values to be monitored outside of the laboratory. Some heart rate monitors allow information to be collected over several days before it is downloaded to a computer. These monitors are also useful in assessing and providing information on exercise intensity in the field.

As an alternative to self-report measures, which can be unreliable, motion sensors are now commonly used to measure physical activity. Pedometers and accelerometers, worn at the hip, ankle, or arm, are useful tools for measuring the quantity or intensity of physical activity (or both) and even estimating caloric expenditure outside of the laboratory. These instruments have become increasingly popular due to their low cost, small size, and ease of use. Whereas pedometers generally measure only step count, accelerometers are able to detect the intensity of movement using an acceleration-versus-time curve. Some pedometers and accelerometers have memory capacity that allows for monitoring of weekly activity and caloric expenditure goals.

Pedometers can be used in the field to monitor step count.

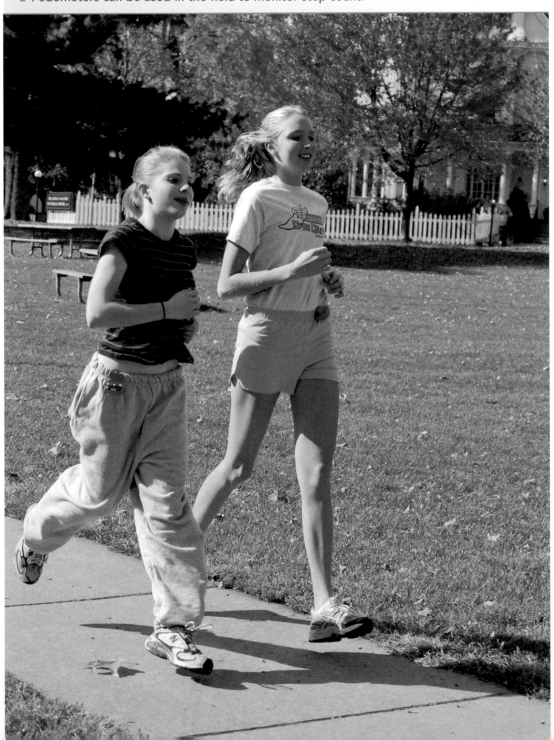

Overview of Knowledge in Physiology of Physical Activity

Exercise physiologists have studied responses to single and repeated bouts of physical activity extensively over the past century. Much of this research has centered on three body systems: the muscular, cardiovascular, and respiratory systems. Physiologists have also examined the influence of environmental factors such as temperature, diet, and altitude on physiological responses and performance. In this overview, we examine how physiological systems respond and adapt to physical activity; review factors that influence physiological responses; and then consider the relationship among fitness, activity, and health.

Skeletal Muscles

The body moves by contracting muscles. The muscles that you use when you are physically active, whether swimming or mowing the lawn, are skeletal muscles, which are under the control of the nervous system. Each muscle is composed of many muscle cells, which are called *muscle fibers*. Inside the muscle cell, many myofibrils run the length of the fiber. The **myofibrils** contain the contractile elements that shorten to generate force and move your bones during physical activity. Human skeletal muscles contain three main types of muscle fibers.

Muscle Fiber Types

Did you ever wonder why some people can run extremely fast for short distances (e.g., the length of a football field) whereas other people are much better at running long distances (e.g., a marathon)? Muscle fiber types in the leg muscles of these individuals are likely different. Muscle fiber types are classified according to the speed at which they contract: fast-twitch (FT) and slow-twitch (ST). Fast-twitch fibers are further subdivided according to how energy is generated. Fast-twitch fibers that use almost exclusively **anaerobic** (without benefit of oxygen) energy systems are called *fast glycolytic* (FG) fibers, whereas fast-twitch fibers that use both **aerobic** (with oxygen) and anaerobic (also called glycolytic) energy systems are called *fast oxidative glycolytic* (FOG) fibers. Slow-twitch fibers primarily use aerobic energy systems. During light- to moderate-intensity exercise such as walking to class or jogging, slow-twitch fibers are recruited first. Because these fibers fatigue slowly, light to moderate physical activity can be sustained for prolonged periods. As the intensity of activity increases, FOG fibers are recruited next, and the FG fibers are recruited at the highest intensities (e.g., in an all-out sprint). Although the FG fibers produce much greater force and power, they fatigue rapidly. Thus, we can sustain the highest intensities of physical activity for only a short time (<1 minute).

Resistance Training

Muscular **strength** is defined as the maximal amount of force exerted by a muscle group. Muscular power is the product of the force times the speed of movement. The ability of a muscle to repeatedly exert force over a prolonged period is known as **muscular endurance**. Resistance training programs to improve muscular strength may use **isometric** (tension is produced without change in the muscle length), **isotonic** (muscle changes length without changing tension), or **isokinetic** (muscle changes length at a constant rate of velocity) exercises. An example of an isometric contraction is putting the palms of your hands against a wall and pushing as hard as you can. Lifting a free weight (e.g., barbell) is an example of an isotonic contraction. Resistance training equipment that allows you to set the speed of movement is isokinetic (see figure 11.4). Regardless of the type of exercise used, the exerciser must overload the muscle group in order for strength to increase. This is the training principle of progressive overload. There must be a gradual increase in the stress placed on the physiological system or tissue in order for improvements to be observed. Likewise, when you stop overloading the system, the training adaptation is lost. This is the training principle of reversibility.

> Muscles increase in strength due to progressive overload.

FIGURE 11.4   Isokinetic machines are often used in the rehabilitation of injuries.

Photo courtesy of Middle Tennessee State University

The principle of specificity applies to resistance training. If you are interested in using resistance training to improve your performance in a given sport, the type of training that you select should match the sport techniques as closely as possible. Strength gains will be greatest over the range of movement and velocity of the resistance exercises used (Morrissey et al. 1995). In some physical activities, the ability to develop force rapidly is critical to performance. For example, a volleyball player must rapidly extend her legs in jumping to block a spike at the net. If your sport requires you to move at high speed, higher-velocity isokinetic or isotonic training will be more beneficial than low velocities or isometric training.

Increases in muscular strength are believed to result from two factors: an increase in muscle size and neural adaptations. An increase in muscle size because of resistance training is due to the increase in size of individual muscle fibers (**hypertrophy**). Although evidence from studies of resistance training in cats suggests that muscle fibers split and increase (**hyperplasia**) (Gonyea 1980), researchers have not observed increased numbers of fibers in other animal species (Gollnick et al. 1981; Timson et al. 1985). Neural adaptations that may occur include recruitment of additional muscle fibers (motor units), better synchronization of muscle fiber contraction, and reduction in neural inhibition.

Physical Activities and Resistance Training

Think about the physical activities that you participate in most often. Explain which activities require muscular strength and which activities require muscular endurance. Select one of these activities and describe the type of movement involved. Given the speed and duration of the movement required, what type of resistance training would be most beneficial in improving your performance in this activity?

Cardiovascular System

The transport of oxygen to all tissues is a primary function of the cardiovascular system, which is composed of the heart, blood vessels, and blood. To increase the supply of oxygen to skeletal muscles, the cardiovascular system responds immediately to physical activity in several ways. Both the volume and distribution of blood flow change during physical activity. In addition, chronic physical activity results in several important physiological adaptations in cardiac function that improve endurance.

Cardiac Output

Cardiac output, defined as the amount of blood pumped out of the heart each minute, is a function of both the heart rate (number of heartbeats per minute) and **stroke volume** (amount of blood pumped per beat). Resting cardiac output is remarkably constant in adult humans at approximately 5 liters (approximately 5 quarts) per minute. During physical activity, cardiac output increases as the muscles use more oxygen. Oxygen uptake (written as $\dot{V}O_2$), or the amount of oxygen used by muscle tissues, increases in direct proportion to the intensity of exercise until maximal oxygen uptake ($\dot{V}O_2$max) is reached. The amount of oxygen delivered to the tissues depends on how much oxygen is in the blood and how much blood the heart is pumping. As your muscles take up more oxygen, your cardiac output increases. At lower exercise intensities, this increase is due to increases in the amount of blood pumped per beat (stroke volume) and the number of beats per minute (heart rate). However, as your exercise intensity increases above 40% of $\dot{V}O_2$max, your stroke volume plateaus because you have reached the limit of how much blood your heart can pump per beat. Your heart rate continues to increase along with oxygen uptake, however, until you achieve your maximal heart rate (see figure 11.5). At this point, you have also reached your limit for oxygen uptake. So, as you can see, the capacity of the heart to pump blood, the maximal cardiac output, appears to be one of the primary factors limiting maximal oxygen uptake. Because $\dot{V}O_2$max and maximal heart rate are reached simultaneously, the fitness instructor can use submaximal exercise tests to estimate $\dot{V}O_2$max. This method is useful in exercise testing and prescription because it does not require the client to exercise to fatigue.

Blood Flow Distribution

When you are at rest, most of your blood flow (cardiac output) goes to your brain and internal organs (e.g., liver, kidneys), with only 15% going to your skeletal muscles. When you become physically active, your muscles need more blood to supply oxygen and nutrients and to remove waste products. At the onset of activity, your blood vessels constrict (vasoconstrict) in regions of your body that need less blood flow and dilate (vasodilate) in your skeletal muscles and heart. During heavy exercise, two-thirds of your blood flow goes to your skeletal muscles through a shift in blood flow away from your kidneys and digestive organs.

Cardiovascular Adaptations to Training

Your ability to exercise at moderate to heavy intensities for prolonged periods is referred to as your *aerobic* or *cardiovascular* endurance. You have learned that one of the best indicators of aerobic endurance is your $\dot{V}O_2$max, also known as maximal aerobic power. You can increase your maximal aerobic power with endurance training. Much of the improvement in your $\dot{V}O_2$max is due to an increase in stroke volume. The cardiac muscle of your heart actually increases in size and contracts more forcefully in response to endurance training. As a result of training, you will have increased stroke volume both at rest and during exercise, and your heart rate will decrease. In other words, you will be able to pump more blood with less work! Of course, you will want to know whether your training regimen was successful. Exercise testing before and after an endurance training program is important for two reasons: It establishes your baseline fitness level, and it allows you to determine the effectiveness of the training program. A graded exercise test, in which intensity progressively increases, is one way to test the efficacy of your training program. At each stage of the test you should notice that your heart rate is lower than it was before you started training (see figure 11.6).

Your endurance training regimen will also result in an increase in the number of capillaries in your skeletal muscles. This change means that you can receive more blood and therefore more oxygen in your working muscle fibers during activity. Researchers have discovered that following training, the difference in the amount of oxygen in the arteries (which move blood from the heart) and veins (which move blood back to the heart) is greater, suggesting that the muscles are extracting more oxygen from the blood. The increase in muscle capillaries may be responsible for the increase in oxygen extraction. Despite your best efforts at fitness training, you may find that others remain far ahead of you in gains, some seemingly without

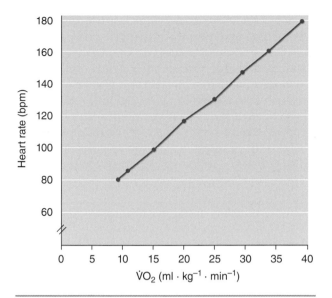

FIGURE 11.5 Relationship between heart rate and $\dot{V}O_2$ during a graded exercise test.

FIGURE 11.6 Comparison of the heart rate responses of fit and unfit people during a graded exercise test.

even trying. This point brings us to a discussion of the other factors that affect maximal aerobic power. One is your genetic makeup. Studies of the $\dot{V}O_2$max of identical and fraternal twins have shown greater variation between fraternal twin pairs than identical twin pairs (Bouchard et al. 1986). In other words, individuals with identical genes (e.g., identical twins) are more alike in maximal aerobic power, showing the strong genetic contribution to maximal aerobic power. Another factor that influences maximal aerobic power is age. $\dot{V}O_2$max begins to decrease after age 30 years because of a decrease in maximal heart rate and therefore maximal cardiac output. The decline in $\dot{V}O_2$max with aging may also be due in part to a decrease in physical activity, as researchers have observed that athletes who maintain their training levels experience a slower decline in $\dot{V}O_2$max (Hagberg 1987).

Determining Your Resting Heart Rate

Determine your resting heart rate by counting your pulse for 15 seconds and multiplying by 4. The average heart rate is 70 beats per minute. Is your resting heart rate less than 70 beats per minute? On the basis of the amount and types of physical activities that you participate in daily, can you explain why your resting heart rate is above or below 70 beats per minute?

Respiratory System

▶ Ventilation increases rapidly at the onset of physical activity and increases as a function of exercise intensity.

As you now know, the cardiovascular system transports oxygen to skeletal muscles. But where does that oxygen come from? The respiratory system regulates the exchange of gases (including oxygen) between the external environment (air) and the internal environment (inside the body). For this exchange to occur, air must move from the nasal cavity or mouth through the respiratory passages to the alveoli within the lungs. After the fresh air enters the alveoli, oxygen can diffuse into the blood in the pulmonary capillaries, and carbon dioxide can leave the blood and enter the lungs for exhalation into the environment. The process of moving air in and out of the lungs is known as **ventilation**. The amount of air exhaled per minute is known as the minute volume ($\dot{V}E$). It is the product of the amount of air exhaled per breath (**tidal volume**) and the number of breaths per minute (respiratory frequency). The minute volume of a person at rest is approximately 6 liters (6.3 quarts) per minute.

At the beginning of physical activity, you may notice that your breathing rate increases rapidly during the first minute until it reaches a plateau. Researchers believe that the stimulus to increase ventilation comes from sensory **receptors** in the moving limbs (e.g., muscle spindles, joint receptors) as well as the motor cortex (i.e., the part of the brain that stimulates muscles to contract). At low exercise intensities, the increase in minute volume (air exhaled) is due primarily to an increase in the amount of air exhaled with each breath; the number of breaths remains constant. At higher exercise intensities, however, both respiratory frequency and tidal volume increase—you take more frequent and bigger breaths.

As your level of exercise intensity increases, your breathing also increases—steadily at first and then much more rapidly at higher intensities (see figure 11.7). The point at which your breathing begins to increase rapidly is known as the **ventilatory threshold**. The ventilatory threshold occurs at exercise intensities between 50% and 75% of $\dot{V}O_2$max. After you have been training for a while, your maximal ventilation—the amount of air entering and leaving your lungs—increases when you exercise. Moreover, your ventilatory threshold occurs at a higher exercise intensity after training.

Temperature Effects

In addition to studying the three primary body systems affected by physical activity, exercise physiologists have studied the influence of environmental factors on our experience of sport and exercise. One factor is temperature variation. Our bodies have complex and effective ways of dealing with temperature changes in the outside environment. Humans are able to regulate their internal (core) temperature so that it remains relatively constant over a wide range of environmental temperatures (Haymes & Wells 1986). When you begin to exercise, muscular contractions produce heat. Blood flowing through these regions of the body warms up and distributes heat to other regions of your body. As your body temperature rises, your skin blood vessels begin to vasodilate and you begin to sweat. In comfortable ambient temperatures, your body core temperature reaches a plateau in 20 to 30 minutes (see figure 11.8). The higher the intensity of your physical activity, the higher your core temperature will be when it reaches a plateau.

Have you ever wondered why you sweat more profusely during physical activity in warm environments? When air temperature approaches or exceeds skin temperature, the body's mechanism of bringing warm blood to the surface to cool is not as effective. In this case, the evaporation of sweat becomes the major avenue of heat loss. In hot, humid environments, heat loss through evaporation is also limited. This results in a higher body temperature and increased risk of heat illness (e.g., heat exhaustion, heatstroke). We tolerate physical activity in cool environments better because our bodies lose heat from blood close to the skin as well as from the evaporation of sweat. In cold water, however, heat loss from the skin increases dramatically because water is an excellent conductor of heat. Submersion in cold water stimulates shivering and an increase in **metabolic rate** (the rate at which the body uses energy). Swimming in cold water also elevates the metabolic rate, which may lead to an earlier onset of fatigue. Similarly, individuals whose clothing becomes wet during physical activity in cold environments lose heat more rapidly, begin to shiver, and are at risk of **hypothermia** (below-normal body temperatures).

FIGURE 11.7 Relationship of ventilation to $\dot{V}O_2$ as exercise intensity increases progressively during a graded exercise test.

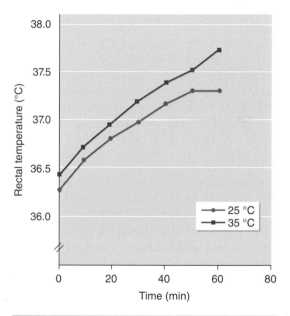

FIGURE 11.8 Increases in body core (rectal) temperature during exercise in comfortable (77 °F; 25 °C) and warm (95 °F; 35 °C) ambient temperatures.

> During physical activity, increased heat production by skeletal muscles stimulates vasodilation of skin blood vessels and sweating.

Nutritional Intake and Physical Activity

Nutritional intake is another factor studied by physiologists of physical activity. Although most of us have enough stored fat to sustain low-intensity activities for many days, we need shorter-term energy sources, such as carbohydrates, for moderately heavy endurance activities. Because the amount of carbohydrate stored in the body is less than 1 pound

(0.45 kilograms), daily carbohydrate intake is crucial to athletic performance. We explore the role of carbohydrates as well as that of two other nutrients, water and iron, in the following sections.

Carbohydrates

Carbohydrate is stored as glycogen primarily in skeletal muscles and the liver. During high-intensity physical activities, muscle glycogen stores are the primary source of energy. Normal muscle glycogen stores are depleted in approximately 90 minutes of continuous exercise at 75% $\dot{V}O_2$max. The amount of glycogen stored is directly related to the carbohydrate content of the diet. Coaches and athletes know that a diet high in carbohydrate (70% or more of the calories), consumed for 2 or 3 days, will increase muscle glycogen storage. Some endurance athletes use this technique, known as *carbohydrate loading*, before competition. Carbohydrate loading is most effective in activities lasting more than 2 hours (e.g., marathons). There is no apparent advantage to carbohydrate loading for activities lasting less than 1 hour.

Maintaining carbohydrate levels during prolonged exercise is important because low blood glucose is often associated with fatigue. Consumption of carbohydrate drinks containing 6% to 8% carbohydrate by weight during prolonged exercise has been shown to improve performance and delay the onset of fatigue (Coggan & Coyle 1991). Drinks containing a higher concentration of carbohydrate (>10% by weight), however, delay gastric emptying and can lead to gastric distress (Davis et al. 1988). Intermittent physical activity and sports can also deplete glycogen stores. For example, soccer players often deplete their muscle glycogen stores during the second half of a game. Saltin (1973) showed that players with depleted glycogen stores covered less distance and spent more time walking and less time running during the second half of a game. Also, some athletes with low carbohydrate intakes (i.e., <50% of the total calories) may progressively deplete their glycogen stores through daily training. To maintain adequate glycogen stores, physically active individuals should consume a diet containing 55% to 60% of the total calories as carbohydrate. Consumption of foods high in carbohydrate immediately following exercise is also beneficial because this increases the rate of muscle glycogen storage (Ivy et al. 1989).

Fluid Intake

Water makes up 55% to 60% of the human body. During physical activity, you lose some of this water through sweating. When you exercise on a warm day, even just walking to class, you will notice that you sweat even more. Because the volume of sweat lost can be substantial, fluid replacement is important. Sweat loss decreases body fluids within and between cells and decreases **plasma volume** (the fluid portion of blood) as well. When plasma volume decreases, less blood returns to the heart, which reduces the amount of blood in each heartbeat (stroke volume). To compensate for the reduced stroke volume, heart rate increases. If fluid losses are not replaced, the sweat rate will decrease, causing an increase in body temperature. This puts people at greater risk of developing heat illnesses, especially heat exhaustion.

Failure to adequately replace fluids can decrease performance in some physical activities. For example, Armstrong and colleagues (1985) found that running velocities at distances ranging from 1,500 to 10,000 meters (approximately 1-6 miles) decreased 3% to 7% following a 2% decrease in body weight. Dehydration is less likely to affect short, high-intensity activities such as sprinting than more prolonged events. To ensure adequate fluid replacement during physical activity, you should drink fluids (e.g., water, carbohydrate–electrolyte drinks) at regular intervals (every 15-20 minutes). Fluid replacement is important during many types of physical activities, including hiking, bicycling, aerobics classes, and working outdoors. In a sport such as soccer or field hockey, fluid replacement during play is difficult because the rules do not allow time-outs during competition except for injury. Drink additional fluid before a match and during halftime in these sports to reduce the fluid deficit. Following activity, you should continue to drink fluids even though you may not feel thirsty. Unfortunately, thirst is not an accurate indicator of the need for fluid. During activities lasting less than 1

Body heat generated during exercise is lost through evaporation of sweat.

Inadequate fluid intake during physical activity results in elevated body temperature, greater risk of developing heat illness, and possible decrements in performance.

hour, consuming carbohydrate–electrolyte drinks offers no advantage. In more prolonged activities, fluids containing a small amount of carbohydrate are more likely to be beneficial in enhancing fluid absorption and maintaining blood glucose.

Iron Intake

As you know, physical activity cannot take place without the transport of oxygen-carrying blood to the muscles. The blood is able to transport oxygen with the help of iron. Oxygen essentially rides on the iron atoms in the **hemoglobin** found in red blood cells. Each hemoglobin contains four iron atoms that bond with four oxygen molecules. A person with low iron cannot adequately synthesize hemoglobin. As hemoglobin concentrations decrease, the amount of oxygen transported to the tissues decreases. Eventually, the person becomes anemic. **Anemia** is defined as a hemoglobin concentration below 12 grams per deciliter of blood in women and 13 grams per deciliter in men. Both maximal oxygen uptake and endurance are reduced in people who have anemia (Celsing et al. 1986).

> ► Depletion of iron stores can lead to anemia, which reduces maximal oxygen uptake and endurance.

✔ ACTIVITY 11.4

Track Your Foods and Physical Activity
Use this activity in your web study guide to start tracking your eating and physical activity habits.

Iron deficiency is one of the most common nutritional deficiencies in the United States, especially among adolescent girls and women. The most likely causes of iron depletion in physically active women are inadequate iron intake and excessive blood loss through the menses. Low iron intake in female athletes most commonly occurs in those who restrict their caloric intake and those who consume diets that are low in meat (Clarkson & Haymes 1995). Heme iron found in meat, fish, and poultry is more highly absorbed from the gastrointestinal tract than is the non–heme iron found in other food sources. People can typically avoid iron deficiency by consuming iron-rich foods daily.

⊡ ACTIVITY 11.5

Practical Plug-In
Consider how an athletic trainer should go about advising an athlete on post-exercise nutrition in this activity in your web study guide.

Physical Activity, Fitness, and Health

Strenuous exercise programs are not required to gain health benefits from physical activity. On the contrary, scientific evidence suggests that engaging in moderate-intensity physical activities on a regular basis confers health benefits (Pate et al. 1995). Moderate-intensity physical activities include brisk walking (3-4 miles per hour, or 4.8-6.4 kilometers per hour), playing table tennis, cycling less than 10 miles per hour (16 kilometers per hour), swimming at moderate speed, playing badminton, golfing (pulling a cart), climbing stairs, and mowing the lawn. The 1996 report titled *Physical Activity and Health: A Report of the Surgeon General* recommended accumulating 30 minutes or more of moderate-intensity physical activity on most, preferably all, days of the week (USDHHS 1996). The activity need not be continuous for 30 minutes; it may include intermittent activities as well. The goal should be to accumulate activities that use approximately 150 to 200 calories over a 30-minute period. The 2008 *Physical Activity Guidelines for Americans* affirmed the health benefits of physical activity with the recommendation to accumulate 150 minutes per week of moderate-intensity physical activity (Physical Activity Guidelines Advisory Committee 2008). With this recommendation, individuals can accumulate 150 minutes in various ways, including 30 minutes 5 days per week or 50 minutes 3 days per week. By emphasizing moderate instead of strenuous physical activities, experts hope that a higher percentage of the adult population, and women and older adults in particular, will regularly engage in physical activity.

Greater volumes and intensity of exercise will result in additional health and fitness benefits. To improve cardiovascular endurance, the American College of Sports Medicine (2010) recommends that adults take part in physical activity 20 to 60 minutes per day, 3 to 5 days per week, at 70% to 94% of maximal heart rate. Exercise intensities that are 64% to 76% of maximal heart rate are usually perceived as moderate exercise, and exercise intensities of 77% to 93% of maximal heart rate are considered hard or vigorous exercise. Activities that are aerobic and can be maintained for prolonged periods are best for improving cardiovascular endurance. These activities include running, hiking, walking, swimming, cross-country skiing, bicycling, aerobic dancing, stair climbing, and rowing. Sports such as soccer, field hockey, and tennis also help develop cardiovascular fitness because they are high-intensity, intermittent activities carried out over prolonged periods.

Accumulating 30 minutes of moderate-intensity exercise each day improves health.

One of the most common reasons for stopping an exercise training program is injury (Pate et al. 1995). For this reason, exercise programs designed to improve fitness and health should begin with moderate-intensity activities and gradually increase the intensity and duration of the activities as fitness improves. For example, an exercise program for sedentary adults may begin with walking at a speed of 3 to 4 miles per hour (4.8-6.4 kilometers per hour) (closer to the moderate intensity of 64% of maximal heart rate) and gradually increase the duration from 20 to 60 minutes. People can begin by walking, progress to alternating between walking and jogging, and advance to jogging.

Effects of Age on Fitness

Although research has shown that maximal oxygen uptake ($\dot{V}O_2$max) declines with age at the rate of approximately 10% per decade, the decrease is smaller (5% per decade) in individuals who remain physically active (Hagberg 1987). The decline in cardiovascular endurance in older individuals may be due in part to a reduction in the intensity and duration of physical activity. Master athletes who maintain their training intensity experience little change in maximal oxygen uptake as they age. $\dot{V}O_2$max, the primary indicator of cardiovascular endurance in adults, increases in absolute terms (liters per minute) with growth in children. Because body mass increases with growth, more oxygen is required to supply the active tissue. When $\dot{V}O_2$max is expressed per kilogram of body weight, however, it remains relatively constant or decreases during growth (Bar-Or 1983). Results of several studies of children suggest that $\dot{V}O_2$max per kilogram does not increase with training in prepubescent children until the peak of the growth spurt occurs (Zwiren 1989). This finding does not mean, however, that children's endurance does not improve before puberty. Improvements in children's endurance with training may be related to improvements in other factors such as running economy and anaerobic capacity (Rowland 1989).

ACTIVITY 11.6

Practical Plug-In
What factors come into play for a personal trainer working with a special needs client? Find out in this activity in your web study guide.

Physical Activity, Fitness, and Coronary Heart Disease

In general, higher levels of fitness are associated with improved health status. This generally held assumption is based on research conducted by exercise physiologists. Blair and colleagues (1989) found that individuals who are the most fit have a lower relative risk of developing or dying from cardiovascular disease or cancer than those who are less fit. The investigators based their determination of physical fitness on a cardiovascular endurance test. Participation in physical activities, as opposed to levels of fitness, also reduces the risk of cardiovascular disease and colon cancer. Paffenbarger (1994) estimated that engaging in moderate-intensity physical activities added approximately 1.5 years to life. Using the number of calories expended weekly in walking, stair climbing, and leisure-time sport as an index of physical activity, his study showed that Harvard alumni had a 46% lower risk of dying from cardiovascular disease if they expended 2,500 calories or more per week in physical activities (Paffenbarger 1994). Participants in moderate-intensity activities (e.g., brisk walking, swimming, cycling) had a 37% lower risk of cardiovascular disease compared with those who did not participate in leisure-time sports.

High levels of physical fitness lower the risk of developing and dying from cardiovascular disease. Regular participation in moderate-intensity physical activity also reduces the risk of cardiovascular disease and colon cancer.

Physical activity and fitness also play a role in reducing the risk of hypertension, or high blood pressure. Over time, endurance training lowers blood pressure in hypertensive individuals by approximately 10 mmHg (millimeters of mercury). Lower-intensity exercise programs (40-70% of $\dot{V}O_2$max) are as effective in reducing blood pressure as are those of higher intensity (Fagard 2001). Among people with normal blood pressure, those who are more physically active have lower blood pressures than those who are less active.

ACTIVITY 11.7

Your Target Heart Rate
Work through this activity in your web study guide to calculate your optimal heart rate range for exercise.

Physical Activity and Weight Control

Participation in daily physical activity helps control body weight. Weight gain and weight loss are determined by the interplay between consumption of calories and expenditure of calories. An excess caloric consumption results in weight gain. Exercise physiologists have added to our understanding of caloric expenditure by investigating the process by which the body burns calories and uses energy. The metabolic rate—the rate at which the body uses energy—increases in direct proportion to the intensity of activity. Lower-intensity activities such as walking at 3 miles per hour (4.8 kilometers per hour) increase the metabolic rate threefold, a caloric expenditure of approximately 4 calories per minute. Higher-intensity activities such running at 7 miles per hour (11 kilometers per hour) increase the metabolic rate 10-fold, which is the equivalent of expending 12.5 calories per minute. Total energy expenditure also depends on the duration of physical activity. Moreover, energy expenditure does not return to resting levels immediately at the end of an exercise bout. Researchers have shown that the metabolic rate remains elevated during recovery from physical activity. Thus, the total energy expended from a single bout of exercise is somewhat greater than the energy cost of the activity.

> Energy expenditure during physical activity is directly proportional to the intensity and duration of the activity.

Although some seek to maintain body weight through exercise, others are looking to lose body weight and body fat. The recommended exercise dose for weight loss is approximately 50 to 60 minutes of daily activity (American College of Sports Medicine 2001). When caloric intake exceeds caloric expenditure, the excess calories are converted to fat and stored in adipose tissue. Sedentary individuals are more likely to be overweight or obese, a condition that puts them at increased risk of developing coronary heart disease (Willett et al. 1995). There is also a higher risk of non-insulin-dependent diabetes mellitus (NIDDM) in those participating in low levels of physical activity (Paffenbarger 1994). Reducing body weight is effective in lowering serum triglycerides and reducing blood pressure in individuals with hypertension. Furthermore, participation in daily physical activities is beneficial not only in reducing the risk of NIDDM but also in helping individuals with NIDDM regulate their blood glucose levels.

Now that you have some basic knowledge about the physiological systems and some of their responses during physical activity, we will examine one of the most pressing problems facing physiologists of physical activity today: how to decrease obesity rates.

Large-scale population studies have shown that obesity has increased dramatically in the United States over the past 25 years. Obesity, defined as a body mass index (BMI) greater than 30 kg/m^2, is calculated using a person's weight and height: BMI = weight in kilograms divided by height in meters squared. The percentage of adults who were classified as obese went from 15% in 1975 to 30% in 2000, a twofold increase. In 2009-2010, the rate reached 35.7% of adults (Ogden et al. 2012a). The problem of obesity is also apparent for younger segments of the population. The prevalence of obesity in children aged 2 to 5 years in 2009-2010 was 12.1% (Ogden et al. 2012b). The rates are even higher for children aged 6 to 11 years (18.0%) and adolescents from 12 to 19 years of age (18.4%). These statistics are alarming, especially given that the risk of developing many chronic diseases is greater in individuals who are obese or overweight.

Obese men and women have a higher risk of developing heart disease, hypertension, and NIDDM. Recall that physical inactivity is also associated with these diseases. Impaired glucose tolerance, a preliminary stage in the development of NIDDM, was found in 25% of obese children and 21% of obese adolescents (Sinha et al. 2002). Overweight and obesity are also related to an elevated risk of some forms of cancer in adults, including endometrial (inner lining of the uterus) and breast cancer in women and colon cancer in both women and men.

People can lose body weight by reducing caloric intake, increasing physical activity, or doing both. Exercise physiologists are particularly interested in studying the effects of increased physical activity or combined diet and exercise programs on reducing body weight. Researchers have found that creating a caloric deficit through a combined physical activity and diet program is as effective as dieting alone in reducing body weight. More important, however, individuals who combined physical activity and diet were more likely to maintain the weight loss after 18 months, whereas those who only dieted regained most or all of the weight lost (Pavlou et al. 1989). Also, preschool children, ages 4 to 6 years, who participated in the highest amount of vigorous physical activity had less body fat than did children who were least active (Janz et al. 2002).

The challenge is how to get more people to participate regularly in physical activity. Fewer than half of adults in the United States regularly participate in 30 minutes per day of moderate physical activity (U.S. Centers for Disease Control and Prevention 2007), and approximately one in every four adults participates in no leisure-time physical activity (U.S. Centers for Disease Control and Prevention 2008). Women are less likely to participate regularly in physical activity than men, and older adults are more likely to be sedentary than young adults. Several strategies have been proposed for increasing daily physical activity, including increasing the amount of physical activity in physical education classes; providing more opportunities for children and adolescents to participate in physical activity before, during, and after school; building walking and bicycle paths that are separated from roadways; and increasing opportunities for adult participation in physical activity at the worksite, in shopping malls, and in community facilities.

How Far Do You Need to Run to Lose One Pound?

An excess caloric expenditure of 3,500 calories is needed to lose 1 pound (0.45 kilograms). If a person burns 100 calories per mile when running (jogging), how many miles would this individual need to run to lose 1 pound? How many days per week do you run? How many miles do you usually run each time? Assuming that your diet is constant, calculate how long it will take you to lose 1 pound from running.

Wrap-Up

Physiology of physical activity is a subdiscipline of kinesiology that centers on the acute and chronic changes to the physiology of the body in response to physical activity. Exercise scientists have studied these changes in an effort to help people be more physically active, healthy, and physically fit. Exercise science professionals work in many arenas including clinical, corporate, and commercial fitness programs and universities. There are also research opportunities in the military and other government organizations.

This chapter provided a summary of the research methods used in physiology of physical activity and an overview of the knowledge base of the subdiscipline. We discussed typical measures that are used to assess health and fitness and learned that many systems of the body are affected by physical activity. If you are interested in learning more about how the body responds and adapts to physical activity and exercise, you can begin taking courses in exercise physiology, anatomy, and physiology. You are also encouraged to visit the Internet sources listed in this chapter to learn more about the educational and certification opportunities available to you in exercise science.

More Information on . . . Physiology of Physical Activity

Organizations

American College of Sports Medicine

American Council on Exercise

American Society of Exercise Physiologists

Canadian Society for Exercise Physiology

Exercise & Sports Science Australia

National Strength and Conditioning Association

Journals

ACSM's Health & Fitness Journal

American Journal of Sports Medicine

Clinics in Sports Medicine

Exercise and Sport Sciences Reviews

International Journal of Exercise Science

Journal of Exercise Physiology Online

Journal of Science and Medicine in Sport

Journal of Sports Medicine and Physical Fitness

Journal of Sports Sciences

Journal of Strength & Conditioning Research

Measurement in Physical Education and Exercise Science

Medicine & Science in Sports & Exercise

Pediatric Exercise Science

The Physician and Sportsmedicine

Research Quarterly for Exercise and Sport

Sports Medicine

Strength and Conditioning Journal

 ACTIVITY 11.8

Use the key points review in your web study guide as a study aid for this chapter.

Review Questions

1. How does physiology of physical activity fit within the discipline of kinesiology?

2. What career path in exercise science is of most interest to you?

3. Describe the contributions that A.V. Hill and David Bruce Dill made to the physiology of physical activity.

4. Give an example of how maximal oxygen uptake can be measured in the laboratory and estimated in the field.

5. Explain how knowledge of exercise physiology can be used to help each of the following: a college athlete, a cardiac patient, and a person trying to lose weight.

6. Calculate your maximum heart rate and then measure your heart rate while you are performing each of three physical activities. Which of the activities meets the intensity guideline to help improve your cardiovascular endurance? Based on your goals, how many days per week and for how long should you perform the activity?

7. Get up and move at a high intensity for a few minutes. Make a list of all of the physiological changes that you notice. Can you explain why each of these changes occurred?

8. What improvements in health result from being physically active on a regular basis?

Practicing a Profession in Physical Activity

For some people, experiencing physical activity and studying about it are ends in themselves. Like psychology, English, or history majors, some kinesiology majors may have chosen their major not because they plan to carve out a career in the field but because they simply are interested in acquiring as much knowledge about the topic as they can. As the discipline of kinesiology becomes more organized, its popularity as a liberal arts subject has increased. This trend is understandable, given that the goal of a liberal education is to understand humanity in its totality and that physical activity plays a profound role in helping us express our humanity.

For most kinesiology majors, however, experiencing and studying physical activity are means toward specific ends: preparing for careers in the physical activity professions. This preparation requires mastery of a different type of knowledge than described in parts I and II. Professional practice—the application of skills to help clients, students, athletes, or patients achieve specific goals—rests on professional practice knowledge. Remember that knowledge derived from professional practice often becomes incorporated into the discipline of kinesiology. When systematic knowledge—usually developed by careful observations of the effect of various manipulations of physical activity on clients (e.g., students, health club members, patients)—is included in kinesiology curriculums, it becomes part of the discipline. Because it is part of the curriculum, you will be required to master this knowledge of professional practice by demonstrating both your understanding of it and your ability to perform it.

If you already have chosen your career path, you probably did so with an eye toward working in some particular aspect of physical activity. The array of careers open to kinesiology majors is vast: sport coaches; sport administrators; athletic trainers; physical education teachers; fitness trainers, consultants, and programmers; rehabilitation specialists in the fields of cardiology, physical therapy, and occupational therapy; and many others. Each of these careers may require a different area of concentration in undergraduate course work, but all require a core understanding of the spheres of physical activity experience and the spheres of scholarly study of physical activity.

All those who work in the field of physical activity assume the responsibilities and obligations that belong to all professionals. Chapter 12 will stimulate your thinking about what it means to be a professional, the obligations and opportunities that belong to professionals, and the steps that you can take now to guarantee your success as a professional. Although experts in the discipline of kinesiology generally recognize seven spheres of professional practice, we will explore only five in this text: careers in health and fitness, careers in therapeutic exercise, careers in physical education, careers in coaching and sport instruction, and careers in sport management. Chapters 13 through 17, written by experts in each of these spheres, focus on specific careers in detail. These chapters may be your first exposure to careers in physical activity. If not, they may help you consolidate your thinking and affirm your commitment to a long-anticipated career. To help you organize your thoughts about the types of work and worksites available to kinesiology majors, we have grouped the many different types of careers according to general objectives and goals, working environments, and qualifications for professionals.

The chapters in part III do not cover all careers that are open to graduates of kinesiology departments. You might be surprised to learn, for example, that scholarly study can be an area of professional practice. Think about the people who are responsible for compiling the spheres of scholarly study described in part II. Who discovered all this knowledge of physical activity? Most of these people were professionals who devoted their careers to studying, researching, and teaching about physical activity at an advanced level, usually at a college or university (as a college professor) but possibly at a research or scientific laboratory (as a research scientist). Your professors are examples of people who have embarked on such careers. These positions normally require a PhD in a specialized area in kinesiology. Because students usually make decisions about this type of career much later in their educational programs, we chose not to address them in this text. If, as you study about physical activity, you become interested in pursuing such work, don't hesitate to consult with your faculty advisor about steps that you might take to prepare yourself for this career.

If you're like some of your peers, you may be somewhat uncertain about which physical activity career you will pursue—if you will pursue one at all. This uncertainty is normal, and it shouldn't cause you any anxiety at this point. In fact, one of the objectives of this section is to help you determine whether you are suited to a career in kinesiology and, if you decide that you are, to help you chart a course for your professional future. Part III should move you closer to selecting a career. As you do this, keep in mind that this choice will be only the first of several career choices that you are likely to face over the course of your working years. Two, three, and even four changes in careers are likely for those currently entering the workforce. Nevertheless, you should make the most intelligent career choice possible, and this section will help you accomplish that important goal.

 PART III PORTFOLIO

Practicing in the Real World: A Case Study

When you've completed the part III chapters, go to the Part III Portfolio activity in your web study guide. In the activity, you will respond to a scenario involving a person who has diabetes and consider the ways in which his condition might be addressed by various practicing professionals.

chapter twelve

Becoming a Physical Activity Professional

Shirl J. Hoffman

chapter objectives

In this chapter we will

- acquaint you with the characteristics of a profession,

- differentiate professional from nonprofessional work,

- explain the types of knowledge and skills essential for performing professional work,

- discuss what you need to do during your undergraduate years to gain entry into and succeed in your chosen professional field, and

- help you determine whether you are suited for a career in a physical activity profession.

Larry grew up playing sports. He was a better than average high school football and basketball player and had above-average grades. He didn't give much thought to a career until he began applying to colleges. Because of his background in sports, he naturally thought about a career as a coach or physical education teacher. But he was far from certain whether he wanted to spend the rest of his life in one of those jobs. He had been injured several times during his high school athletic experience and was always impressed with the athletic trainers who treated him, so that career track had some attraction as well. Through a friend, he met the media and communications director for a local college athletic department, and that job interested him as well. And then there was the possibility of becoming a physical or occupational therapist, even though he knew that would entail several more years of education and considerable expense.

By the time he began his freshman year in college, Larry was paralyzed with uncertainty. He had entered the "big black void," the never-never land where students who haven't planned well for a career often reside during their first few years of college.

One of the problems that college students like Larry face in choosing a career, of course, is that the decision must be made fairly early in their college years, often long before they have acquired the experience and knowledge essential for making the best possible choice. How can you know whether you will like being a coach or a personal trainer or an athletic trainer if you haven't had a chance to experience it? This is why many college students forestall choosing a career while in college. They bounce around from course to course, major to major, without committing to preparing themselves for the workforce. This chapter and those that follow are designed to help all of the Larrys out there make a career choice early in their career. And for those who have already made the choice, the text will give you important information about how you can best prepare yourself to be successful in your chosen career.

Committing to a career early in your college experience can be a scary proposition. If this is your fear, consider the fact that more than five million people between the ages of 44 and 70 are embarking on "encore" (second or third) careers. In all likelihood, your first career choice will not be your last. However, there are distinct advantages to choosing a career early. First, midstream career changes can disrupt your personal and family life and inevitably set you back in your progress toward an ultimate career goal. Second, students who identify with a career track early in their college experience are much better able to relate their course work to their anticipated career. Also, knowing the type of work they want to do, these students can gain experience in professional roles and settings and begin to develop a network of associates in the profession while still in college. So, if you have already committed to a specific career in the field of physical activity, you are on track. You're likely to enjoy your study of kinesiology a bit more and derive more from it than students who have yet to make that commitment. If you are still trying to decide on a career, this part of the text should help you narrow your career choices. In fact, even if you have already decided on a career, the chapters in this section should interest you as a budding kinesiologist who shares with his or her colleagues a special interest in physical activity.

Getting a respectable job is at the top of most students' lists of why they went to college. You probably are expecting your college degree to help you find a well-paying job that is respected by the community, offers good working conditions, and provides a reasonable amount of time off for leisure pursuits. No doubt you also want your job to count for something, to contribute to society in some substantial way. Ideally, other people should benefit directly from your work. Most people want a job that is so enjoyable that they willingly invest more time than the job requires. And the ideal job should not be one that anybody could do; it should require some thinking and skills that set you apart from others in the workforce. You may also want to be in a position to make most of the decisions concerning what you do and how you go about your work. (You surely remember previous jobs in which you didn't have much freedom to make decisions.) And last, but certainly not least, you want a job that centers on physical activity.

Having the freedom to make decisions is one of the advantages of being a professional. But with freedom comes responsibility, because the quality of your decisions will determine how successful you will be in your chosen profession. Will you be a good decision maker? Will the decisions you make increase the health, well-being, and enjoyment of the students, clients, patients, or customers whom you serve, or will your decisions result in wasted time and effort and perhaps jeopardize the safety of those who depend on you? Will your decisions be grounded in the knowledge and skills that you learned in your kinesiology program, or will you make decisions entirely by the "seat of your pants"? An undergraduate education in kinesiology teaches you to make correct decisions based on available knowledge.

If this describes the position you hope to secure following graduation, then you have already narrowed your choice down to a profession, but there are many professions. This chapter is intended to help you decide whether a career in a physical activity profession (figure 12.1) is right for you. The chapter also explores the general nature of the curricula used by colleges and universities to prepare you for your chosen field.

All workers are not professionals. A profession is a particular line of work or occupation, and a professional is a particular type of worker. Teachers, doctors, lawyers, and physical therapists are professionals. The work of janitors, housepainters, plumbers, electricians, and carpenters, although they may describe their expertise as "professional," lacks certain characteristics normally associated with the professions. Knowing why this is the case will help you better understand the distinctive roles and responsibilities of the type of work to which you aspire.

FIGURE 12.1 This chapter introduces the physical activity professions, one of the three dimensions of kinesiology.

A degree in kinesiology prepares you for a career in the physical activity professions, a career category that includes specialized work in which physical activity plays a central role in accomplishing specific goals. For example, physical activity can be a medium for therapy, education, health, recreation, or entertainment. Teaching physical education, coaching, being a fitness consultant or fitness director for a corporation, and serving as an athletic trainer or conditioning specialist for an athletic team are some examples of positions in the physical activity professions. They can be grouped into relatively distinct **spheres of professional practice** in physical activity. Figure 12.2 depicts the major spheres of the physical activity professions. Keep in mind that these spheres represent general categories; each is made up of several different professional occupations.

FIGURE 12.2   The spheres of professional practice centered on physical activity.

The chapters that follow in this part of the book describe some of the most popular physical activity professions available to students graduating from kinesiology programs. Since all of these fit the general category of "profession," it will be important for you to know exactly what is meant by the term. By joining the ranks of the professionals, you will share some of the attributes, expectations, and obligations of those working in all professional fields, whether they are lawyers, counselors, clergy, physicians, or kinesiologists.

Perhaps you already know what it means to be a professional. But do you know the benefits and obligations of being a professional? Do you know the type of college preparation required of professionals? Are you able to recognize the traits that college professors and future employers look for in students to determine who are most likely to become respected, qualified professionals? These are important questions for anyone seeking admission to the professions; and, if you're like most of your peers, you probably haven't given them a great deal of thought.

Learning about the skills and knowledge required for specific occupations such as teaching or coaching, health and fitness promotion or consulting, community sport leadership, athletic training, cardiac rehabilitation, and other careers will come later. Right now it is time to pause and take stock of your own interests, talents, and commitments to find out whether you have what it takes to enter the workforce as a professional in the physical activity field.

What Is a Profession?

Often we use the term *professional* in an informal way to describe the quality of a person's performance. The work of a cabinetmaker may be described as "very professional," or a person who is an excellent plumber may be described as "a real pro." Sometimes we use the term to describe those who play or teach a sport as a full-time source of income (e.g., professional baseball players, or tennis or golf professionals who teach at clubs) to distinguish them from those who play only part-time and receive no remuneration (amateurs). These usages, however, do not convey the strict, formal meaning of the term. Normally, we do not think of cabinetmakers or athletes as members of a profession in the same way that we think of lawyers, physicians, teachers, counselors, physical therapists, rehabilitation specialists, or clergy as members of a profession. Surely, members of a profession are known for doing their work well and for doing it full-time and for pay, but that is far from the whole story.

A cluster of attributes have been used to describe professional types of occupations (Greenwood 1957; Jackson 1970). Although some characterizations of a profession may seem a bit idealistic, they provide a useful model for helping us understand the distinctions and responsibilities of this type of work. You will discover that most types of work aren't easily classified as being completely professional or completely nonprofessional. Although physicians may be located at the extreme professional pole and day laborers may be located at the extreme nonprofessional pole, most types of work probably fall at some midpoint on the continuum depicted in figure 12.3. To underscore this point, some of the physical activity professions are termed *minor professions* or *semiprofessions* (Lawson 1984). Workers in the jobs positioned closest to the professional pole of the continuum

- master complex skills that are grounded in and guided by systematic theory and research;
- perform services for others, known as clients or patients;
- are granted a monopoly by the community to supply certain services to its members;
- are guided by formal and informal ethical codes intended to preserve the health and well-being of their clients; and
- meet the expectations and standards prescribed by their professional subcultures.

Let's look more carefully at each of these characteristics.

> Those who are preparing for a position in the physical activity professions should know what a profession is, the type of work professionals do, how one gains entry to and acceptance in a profession, the obligations of professionals, and the most important factors to consider in preparing for a career in the physical activity professions.

FIGURE 12.3 Occupations can be located along a continuum from professions to nonprofessions.

Professionals Have Mastered Complex Skills Grounded in Theory

The overriding hallmark of professional work is that it draws on a complex body of knowledge and theory that is developed through systematic research. Strictly manual work (day laborer, some restaurant work, some lawn work) does not require this. Because the workplace in general is becoming more technologically oriented, advanced specialized knowledge is becoming more important, even in occupations that once were not considered professional. For example, welding is generally considered to be a fabrication craft; but now, because it increasingly involves complex technology and sophisticated evaluation of the structure of materials, it is being touted as a profession (Albright & Smith 2006). This point deserves emphasis: Determining what is and what isn't a profession can be a difficult challenge.

Professionals are sometimes referred to as **practitioners** (derived from *practice*), a term signifying that they are not merely gifted thinkers but also gifted *doers*. Professionals establish their reputations by bringing about predetermined outcomes efficiently and effectively, usually on behalf of others. Doing this requires skill; in fact, the expert skill of professionals is what sets them apart from other types of workers. We usually associate the concept of skill with moving—the speed of a world-class sprinter or the agility of a pro basketball player. Surely some professions do require deft motor skills—dentistry and surgery, for example— but motor skills are merely one category of skill. The **cognitive skills** of analysis, deduction, diagnosis, prescription, and high-level reasoning; the **perceptual skills** that enable you to identify and recognize problems that have occurred or anticipate what is likely to occur; and **interpersonal skills** such as listening, communicating, and motivating all are critically important in professional work. Development of these skills is essential if professionals are to make expert decisions regarding their professional practices.

For example, workplace biomechanists must have a keen eye for movement (perceptual skill), must be able to analyze workers' movements to determine the reason for a high injury rate (cognitive skill), must be able to devise strategies to correct the movements and reduce injury (cognitive skill), and must have the ability to explain their findings to workers and supervisors (interpersonal skill). Athletic trainers must possess perceptual skills that enable them to spot hazards and potential sources of injuries, cognitive skills of analysis and deduction through which they determine the correct emergency procedures to be used, and communication skills essential for relating their findings to physicians or for motivating athletes to adhere to their rehabilitation schedules. Becoming a professional, then, involves recognizing what factors must be taken into consideration in any situation; knowing the steps that must be taken to bring about the desired result; possessing the skill required to achieve the result; and communicating a plan of action to a client, student, or patient.

How are these important skills acquired? In many cases, professional expertise is developed from experience performing and watching others perform physical activity. This was discussed as physical activity experience in part I (see figure 12.4). For example, although not all soccer, basketball, or gymnastics coaches were elite performers of the sports they coach, most have spent time performing and watching the activity and are familiar with and comfortable in the contexts in which the sports take place. (Later in this chapter we talk more about the contribution of physical activity experience to your professional development.)

The skills of the physical activity professional also derive from knowledge of theory about physical activity. As we saw in part II, **kinesiology theory** is embedded in concepts and principles related to physical activity that are taught in the subdisciplines.

FIGURE 12.4 Many kinds of knowledge form the foundation of professional expertise.

Obviously, not all kinesiology theory is directly applicable to every task that the professional performs, just as not all theoretical knowledge in medicine is applicable to every case a physician treats. For example, learning how to calculate the theoretically optimal angle of the projection of the shot put in biomechanics class may not translate into useful prescriptions for a track coach, just as learning the details of pulmonary function in exercise physiology class may not be directly applicable to the athletic trainer's work.

A Simple Rule of Thumb

Here is an important practical tip for professionals from someone who has succeeded in the business world. The sentiment he expresses applies to most professional settings (Satell 2010):

"Over the years, I've come up with a simple litmus test that has been extremely useful in making personnel decisions and choosing partners: *Professionals solve more problems than they create.* Whenever I need to make a decision about entering or extending a relationship, I apply that standard. Problems are rarely caused for want of competence but usually stem from ill-placed pride and a lack of common decency. Many feel that because they work for a big company, have a big title, and know a lot of things (especially acronyms), they are entitled to make it difficult for others. How else can they advertise their stature? True professionals, however, look to solve problems. They want to help people, seek out 'win-win' situations, and are willing to shoulder the burden when things go wrong. Their self-worth is defined by what they do for others, not what they can do to them and get away with."

Which one are you?

Nevertheless, because kinesiology theory focuses on broad principles about physical activity, it may have a potent, albeit indirect, effect on the way professionals think and the skills that they use to solve problems. Theory offers a broad framework of knowledge, some of which can be generalized to specific problems. Having a clear understanding of the broad picture can help professionals adapt knowledge to new situations, integrate knowledge about the principles of physical activity, and think in highly flexible ways. Contrast this, for example, with bricklaying and carpentry, truly respectable forms of work known as *trades* or *crafts*. These jobs certainly require use of cognitive and perceptual skills, but the skills are almost always applied to a stable set of problems and contexts. Because one brick wall is pretty much like another, even laypersons can buy books on home repair and, simply by reading, serve as their own bricklayers. The work of the physical activity professional entails solving a wide variety of complex problems with complex human beings. People cannot learn to do this by reading manuals.

For example, physical education teachers may observe errors in students' motor performance that they have never seen before. By systematically applying their knowledge of movement mechanics, motor learning, and pedagogy, as well as by using a diagnostic eye developed through practice, these teachers will be able to devise practice experiences to correct the errors. Likewise, an athletic trainer confronted in an emergency with a rare injury may rely on her in-depth knowledge of biomechanics, exercise physiology, and other kinesiology concepts to determine the best course of emergency action. Mastering theoretical knowledge of kinesiology enables professionals to understand the fundamental bases of problems (e.g., physiological, social), ask intelligent questions about their work, and devise flexible strategies for solving problems, even those they have never before confronted.

Physical activity experience and theoretical knowledge by themselves are not sufficient to guide the skillful behavior of professionals. Professionals' reputations are determined largely by their wealth of professional practice knowledge. This is not knowledge that is acquired

▶ Professionals develop a range of cognitive, perceptual, and motor skills, anchored in theoretical, workplace, and practice knowledge, that enable them to achieve predetermined outcomes efficiently and effectively.

so much by reading, nor is it knowledge that can be tested on written examinations. Professional practice knowledge is what some have called "knowledge in action" (Schon 1995). Professional practice knowledge can be seen in the physician's competency in diagnosing diseases or in the skill with which a lawyer assembles a convincing case for a court trial. Professional practice knowledge in kinesiology may be reflected in an athletic director's expertise in recruiting and retaining effective coaches, in a fitness leader's skill in prescribing effective exercises for a client who is recovering from a heart attack, or in an athletic trainer's skill in treating an on-the-field injury. Here it is in a nutshell: You demonstrate your command of professional practice knowledge not by answering questions on a true–false or multiple-choice test but by showing how well you can perform a series of professional skills. In many cases, highly competent professionals are unable to explain in any detailed fashion how they manage to bring about such effective results.

What Does Being a Professional Really Mean?

The term *professional* is used in many ways. Do you think the term is being used appropriately or inappropriately in each of the following examples?

- A physical therapist has a habit of talking to her patients about her own personal problems. Her supervisor reprimands her for not conducting herself as a professional.
- An award is bestowed on a physical education professor for having provided 25 years of distinguished professional service to the university.
- A personal trainer is heard repeatedly giving clients misinformation about nutrition and exercise. The fitness director tells him that he is not behaving in a professional manner.
- The athletic director praises an athletic trainer for having the highest professional standards.
- An employee in the ticket office at a local football stadium prides himself on the quality of his work, noting that he is regarded as a valuable member of the ticket-taking profession.
- A physical education tells her colleague that he should join the American Alliance for Health, Physical Education, Recreation and Dance because it is the professional thing to do.

Notice the widely varying definitions used by those in health care management, sociology, and the insurance industry:

- "Professionalism . . . is the ability to align personal and organizational conduct with ethical and professional standards that include a responsibility to the patient and community, a service orientation, and a commitment to lifelong learning and improvement" (Garman et al. 2006, p. 219).
- Professionalism "exists when an organized occupation gains the power to determine who is qualified to perform a defined set of tasks, to prevent all others from performing that work, and to control the criteria by which to evaluate performance" (Friedson 2001, p. 12).
- "The attributes of professionalism include reliance on a high personal standard of competence in providing professional services; the means by which a person promotes or maintains the image of the profession; willingness to pursue development opportunities that improve skills; the pursuit of quality, competence, and ideals within the profession; a sense of pride about the profession" (McGuigan 2007).

What are the similarities and differences (either literal or implied) in these definitions of professionalism?

Finally, professionals also rely on **workplace knowledge**, usually to perform relatively mundane tasks associated with their jobs. This type of knowledge requires little formal education or training. For example, cardiac rehabilitation technologists must know how to maintain the electronic equipment they use, how to schedule patients for examinations, or the proper procedures for opening and closing the facility. Such tasks are not professional in their scope and do not require highly sophisticated knowledge, but they are important. A cardiac rehabilitation technician who does not maintain a clean testing facility or ensure that the equipment is kept in proper working condition may soon find herself unemployed, regardless of the quality of her professional skills.

Professionals Perform Services for Clients or Patients

Most professional work takes the form of services that are performed for clients. **Service** connotes giving assistance or advantage to others, usually in a spirit of helpfulness and concern. Most professionals are conscious of their roles as service providers. Those in nursing, social work, family counseling, public health, recreation, and kinesiology increasingly refer to their professions as helping fields or **helping professions** (Lawson 1998b). The beneficiaries of a professional's services are known as *clients*. (In the allied medical fields, clients are patients; in teaching, they are students.) Professionals are in the "people business"; they like meeting and working with people, are comfortable in social settings, and are confident in their social skills. They derive great enjoyment from helping people meet their needs and accomplish important goals.

You always should keep three important aspects of service in mind:

- *Professionals are committed to helping others.* One often hears professionals describe their career as a calling, suggesting that their primary motivation is to enhance the health, education, enjoyment, and general well-being of those they serve. In fact, concern for the well-being of clients often supersedes professionals' concern for their own comfort and satisfaction. You may have known physicians or lawyers who sacrificed lucrative practices to work for poor and disadvantaged people in inner cities or developing countries. The same spirit appears in the teacher who volunteers her time to give remedial instruction to a student on the weekend or the geriatric fitness leader who stays late following each session to give special attention to an elderly man too embarrassed to exercise with others present. Obviously, becoming an effective professional need not entail such extreme sacrifices, but wanting to help others is a prerequisite for professionals.

- *Professionals render expert service.* Being a professional means more than simply being willing to help. Professionals know how to determine a client's needs and how to initiate and evaluate an appropriate course of action to meet those needs. Because of this, professionals are granted a certain amount of autonomy to carry out their work. Their opinions are respected, and they usually have a great deal of freedom to make decisions about how they will help clients. Some professionals have more autonomy than others. The personal trainer in private practice may enjoy much more autonomy in decision making than the fitness leader who works for a gym. The physical education teacher, whose actions may be limited by school board policies, the dictates of a school principal, and the curriculum guide, will enjoy less autonomy than the professional sport instructor who is self-employed.

> ► The touchstone of professional work is the delivery of expert services to improve the quality of life for others, always with priority given to the client's welfare.

- *Clients should not become totally dependent on the professional.* That clients, patients, and students become dependent to a certain degree upon authority figures is inevitable but not always desirable. The dependence of clients can be undesirable when it is carried to extremes and results in professionals' adopting a paternalistic attitude toward those whom they serve rather than a collaborative attitude in which the professional and the client work together to achieve common goals. A dance student depends on his teacher to improve his technique just as a patient depends on her exercise therapist to design a program to help her recover from an injury. Unlike customers at a department store, who usually have a clear idea of what they want and can evaluate the products that they buy, clients usually do not have the knowledge essential for judging the quality of the professional's conduct. Consequently, they can only have faith that the professional will keep their best interests in mind when making decisions. One of the most egregious violations of professional conduct is exploiting the vulnerability of clients. For that reason, the coach who orders an injured player back into the game will quickly lose the respect of colleagues, as will a physical therapist who, not wanting to suffer financial loss, fails to refer a patient to a competitor even though the competitor may be better equipped to meet the patient's needs.

Professionalism In the Workplace

Certain key values, attitudes, and behaviors are important in any type of professional work. Those listed below are the recommendations from a leader in the communication industry (Feignbaum 2012).

Personal Appearance

At work, perception matters. Someone with talent, skill, and ability may not be taken seriously if he or she is ungroomed, disheveled, or dressed inappropriately.

Emotional Baggage

Checking your personal issues at the door when at work is part of being professional. Doing your job properly and professionally means not letting what happens at home affect your work performance. It also means not letting people trigger your emotional hot buttons.

Respect

Respect is the foundation of workplace success with superiors, coworkers, employees, customers, and clients. Whatever your company culture, everyone responds to respect. Keep in mind your workplace conduct doesn't represent just your organization—it represents you.

Communication

Savvy professionals know how to communicate. Knowing who to keep informed, when to use discretion, and when to clearly disseminate information is what separates the good and the great. Listening and really understanding people will not go unnoticed.

Integrity

Along with respect, integrity is one of the highest professional values. In fact, some professions such as law, medicine, and journalism have professional codes of ethics. Even when the industry doesn't mandate them, maintaining your own code of ethics makes you an invaluable and respected professional.

Professionals Possess a Monopoly on the Delivery of Services

> Professionals are granted monopolies because only they possess the knowledge and skills to meet particular needs of the community.

Society values professionals because they provide a needed service unavailable through any other occupational group. If you need a cavity filled, you go to a dentist. If you need spiritual guidance, you may go to a member of the clergy. If you need to rehabilitate an injured knee to play basketball again, you go to a sport physical therapist.

The physical activity professions came into existence and continue to flourish because society recognizes that people in such positions make a unique and valuable contribution to the health and well-being of society. Society needs specialists who can help people regain or maintain their health, become physically fit, increase their proficiency in motor and sport skills, recover from injuries, or participate in sport programs. No discipline prepares people with the background knowledge and skills needed to meet this diverse range of needs except kinesiology. In this sense, kinesiology and the physical activity professions it prepares possess a monopoly on the delivery of physical activity services. Various certifications and licensing arrangements often protect the monopoly. Teachers, for example, usually are required to possess a certificate granted by the state or province in which they operate. The purpose of this certificate is to ensure that only professionals who have met the minimum standards of educational preparation will be granted a monopoly to practice the profession of teaching.

Professionals Collaborate With Colleagues to Ensure High Standards and Ethical Practices

Professionals recognize that society has granted them authority to regulate their own conduct and expects them to do it. Violation of accepted professional standards by a single professional threatens all members of the profession by bringing into question the legitimacy of the monopoly they have been granted. Thus, professionals have a huge stake in maintaining or improving not only the quality of their own practices but also the quality of service delivered by all members of their profession. In this sense, although professionals may work at clinics, schools, hospitals, or other agencies, they view themselves primarily as representatives of the profession as a whole.

Because the knowledge base for the physical activity professions is expanding continually, exercise leaders, teachers, trainers, sport leaders, and other professionals can easily become outdated in their techniques. Professionals whose practices are not in accord with the latest information become a threat to all professionals in the field. A clinical exercise physiologist who fails to keep abreast of her field is in danger of recommending activities to clients that recent research has shown to be inferior or dangerous. Such a person may face accusations of unprofessional practice.

Keeping updated will require you to participate in professional organizations; regularly read professional books and articles; and attend conferences, workshops, and other meetings sponsored by professional societies. Unlike salespersons, business executives, or others in the commercial marketplace, who tend to protect information from their competitors, members of a profession relate to each other as colleagues and willingly share information to serve the best interests of their clients. If you want to be a top-notch professional, you must commit to educating yourself continually. It is not too early to join a professional organization while still in your undergraduate years.

Operating on a less than adequate knowledge base is not the only way in which professionals breach standards of conduct. Professionals also can be sanctioned for unethical behavior in the workplace. Unethical conduct occurs when the behavior of a professional at the worksite comes into conflict with generally accepted standards of practices in the field. To ensure that members clearly understand ethical expectations, professional organizations often publish a **code of ethical principles and standards**. The primary objective of such codes is to protect the rights of students, clients, and patients. Codes may instruct professionals to describe and advertise their competencies accurately, may describe the nature of the relationship that should exist between clients and professionals and between subordinates and professionals, and may clarify matters involving financial transactions with clients.

The excerpt in the sidebar from the code of ethics of the National Strength and Conditioning Association is representative of most ethical guidelines. The code is intended to guide the practice of member strength and conditioning specialists as well as the actions of member scientists who conduct research in strength and conditioning. Every professional association has a code of ethics which, in most cases, is available on its website.

One way in which professions attempt to ensure that members understand the prevailing code of conduct and are competent to implement safe and effective practice is to license their members. A license is formal authority granted by law to practice a profession. Professions that require licenses can suspend a professional's license for any number of infractions, effectively terminating the professional's authority to practice.

Although the medical and legal professions have made effective use of licensing provisions to control the quality of professional services, relatively few physical activity professions are

> Professionals have a stake in maintaining high standards of conduct for all practitioners in their profession. To accomplish this, professional organizations make opportunities available for continued education by their members and publish guidelines specifying acceptable and unacceptable conduct.

Excerpt From the National Strength and Conditioning Association Code of Ethics

Principle 1

Members shall respect the rights, welfare, and dignity of all individuals.

1.1. Members shall not discriminate on the basis of race, color, sex, age, religion, or national origin.

1.2. Members shall provide competent, fair, and equal treatment to all individuals.

1.3. Members shall preserve the confidentiality of personal and privileged information of the athlete, client or the NSCA.

1.4. Members shall not release any information to a third party not involved with the athlete's or client's care without a written release unless required by law.

Principle 2

Members shall comply with all applicable state, local, and federal laws.

2.1. Members shall comply with all institutional guidelines.

2.2. Members shall comply with all copyright laws.

2.3. Members shall be familiar with and follow the NSCA bylaws and all applicable policies, procedures, rules, standards, and guidelines.

2.4. Members shall not condone or engage in any illegal behavior.

Principle 3

Members shall maintain and promote high standards.

3.1. Members shall not misrepresent, either directly or indirectly, their skills, training, professional credentials, identity, or services.

3.2. Members shall only provide services that they are qualified to provide through education or experience and which are allowed by practice acts and other pertinent regulations.

3.3. Members shall refer athletes or clients to more qualified fitness, medical, or health care professional when appropriate.

3.4. Members who are researchers or educators shall maintain and promote ethical conduct in research and educational activities.

3.5. Members should strive to continuously improve knowledge, skills, and techniques to protect the athlete or client from injury.

Note: This is not a complete explication of the NSCA Code of Ethics. The full code can be found on the NSCA website.

Reprinted from National Strength and Conditioning Association 2008.

licensed. An exception is physical therapy and physical education teaching. Some states are now licensing athletic trainers, a trend that is moving across the United States. The lack of licensing for physical activity professions doesn't mean that unethical or improper conduct by physical activity professionals goes unpunished. Usually, local entities administer the appropriate sanction at the local workplace, often in consultation with a professional organization. For example, the local school board may terminate a schoolteacher whose conduct violates ethical codes. A university may fire a conditioning coach who administers illegal anabolic steroids to athletes. Although the American College of Sports Medicine (ACSM) cannot suspend a cardiac rehabilitation therapist whose conduct puts a client at risk, the clinic director or other supervisor at the worksite may decide to terminate the worker after consulting ACSM guidelines.

Professionals Adhere to Standards of Their Professional Subculture

Although professionals work in a variety of different locales and contexts, they tend to hold high expectations concerning the way members present themselves and relate to others. For example, professionals emphasize respect and politeness not only to those whom they serve but to colleagues as well. Professionals tend to be well organized in their work, are not clock watchers, and don't hesitate to work extra hours when required. They dress according to accepted occupational standards and are attentive to matters of personal grooming and hygiene. A sport program director who comes to a board meeting in dirty and disheveled clothes, a cardiac rehabilitation technician who forgets to wear her lab coat, and a physical education instructor who teaches classes in street clothes are not dressed professionally. The physical therapist who never combs or brushes his hair or the personal trainer who smells of body odor are guilty of presenting themselves in an unprofessional manner. These general expectations apply to professionals in any field.

Physical activity professionals often face a more specific set of expectations. For example, fitness leaders or physical education teachers might be expected to model the physically active life, to refrain from smoking, and to maintain a level of fitness appropriate for their age.

Different physical activity professions are likely to require unique types of knowledge, employ different types of skills, operate according to different standards and codes of conduct, require different languages and terminology, require different certification and continuing education procedures, and often involve different types of dress and on-the-job conduct. When an individual learns to adapt to these profession-specific expectations, he or she is said to have been socialized into the profession. **Socialization** into a profession means learning how to adhere to the roles and responsibilities of the workplace. The ease, confidence, and efficiency with which professionals perform their daily activities and communicate with their clients and coworkers are good indications of how well professionals have been socialized into their professional roles.

Obviously, the sooner you become socialized into your profession, the more valuable you will be to an employer. For that reason, among others, increasing numbers of students

Are All Fitness Instructors Professionals?

Researchers who examined the fitness knowledge of a random sample of 115 health fitness "professionals" in the areas of nutrition, health screening, testing protocols, exercise prescription, and general training knowledge regarding special populations found that 70% of them of did not have a degree in a field related to exercise science (Malek et al. 2002). Those who did not have the degree scored 30% lower on the test than those who had a degree. A bachelor's degree in the field of exercise science and possession of either ACSM or NSCA certification as opposed to other certifications were strong predictors of knowledge, whereas years of experience working as a fitness instructor was not related to knowledge. The researchers argue that "these findings suggest that personal fitness trainers should have licensing requirements, such as a bachelor's degree in exercise science and certification by an organization whose criteria are extensive and widely accepted, before being allowed to practice their craft" (p.19).

The study was conducted in 2002. While there is reason to believe that the preparation of fitness instructors has improved in recent years, the certification process for fitness workers continues to be an issue. There are reportedly more than 100 organizations that offer certification in fitness instruction. Many of these require little more than a short on-line experience or a weekend workshop. As mentioned, the most respected certifications are those awarded by ACSM and NSCA.

in professional programs are seeking opportunities to volunteer in clinics, YMCAs, schools, and community agencies. Although usually not counted as course work, such extracurricular experiences can give you a head start in learning how to assume professional roles. Similar growth and leadership opportunities are available in student major clubs, athletics, student government organizations, and other on-campus activities.

How Do Our Values Shape Our Professional Conduct?

> Your personal values will influence how you act out your professional role. Mechanical, market-driven professionals value the profession, profit, personal prestige, and status over the rights and needs of clients. Social trustee, civic professionals value clients and the social good more than themselves or their profession.

Hal Lawson (1998a) has described two types of extreme value orientations that may characterize professionals in any field—**mechanical, market-driven professionalism** and **social trustee, civic professionalism**. Although these orientations are "ideal types," which is to say that most professionals don't fit completely into one category or the other, they can serve as references in helping you determine how you want to operate as a professional.

Mechanical, Market-Driven Professionalism

Mechanical, market-driven professionals tend to be so fascinated with their professional aspirations and methods that they lose sight of their clients' needs and desires. In many cases, the work of these professionals becomes its own justification, valued by them and their colleagues regardless of whether it has any real benefits for society. The medical, nursing, and law professions sometimes are accused of this type of professionalism, but some of the physical activity professions can be guilty of it too.

Mechanical, market-driven professionals are likely to serve their clients in a fragmented, compartmentalized fashion. That is, these professionals view their professional contributions within the strict limits of their specialization and ignore the variety of relevant forces acting on the client, such as family disintegration, illness, or drug or alcohol dependency. These professionals may dismiss these other problems as "somebody else's responsibility" and never go out of their way to see that proper referrals are made. Such professionals may treat clients humanely but with little feeling, instead believing that an objective, arm's-length attitude is critical to professional success.

Mechanical, market-driven professionals also are likely to devote much time to enhancing their status and competing with other professionals, even if they must sacrifice the quality of service that they give to their clients. They pride themselves on their expertise, view themselves as superior to those they serve, and often encourage clients to develop a dependency on them and their services.

Social Trustee, Civic Professionalism

Social trustee, civic professionals hold to a much different set of values. They believe that the worth of a profession is measured by its effectiveness in promoting the social welfare, enhancing social and economic development and democracy, and ensuring social responsibility and social justice. The actions of a social trustee, civic professional are guided not by a fascination with technology and technique or by a concern for status and prestige but by a vision of the good and just society. They operate according to this rule: "Healthy people and a good society first, me and my profession second in service of this greater good" (Lawson 1998a, p. 7). They recognize that clients live in a multiplicity of worlds (e.g., work, school, family, church), and they understand the ways in which worlds can interact to affect their clients' lives. Because of this, these professionals often work in teams with other specialists to achieve desirable goals. They make no pretense about being objective in their relationships with clients: Their professional practices reflect their personal values. They don't view clients as dependent on them, believing that the professional–client interaction can be a mutual growth experience in which each benefits from the other's knowledge and skill.

How Does Your Personality Match Up With Your Career?

Career counselors often use occupational attitude inventories to determine the best matches between an individual's abilities and personality and possible career choices. Many of these tests are based on researcher John Holland's six vocational personalities and work environments.

- **Realistic personality:** Values practical and mechanical experiences such as operating machines or farming. Does not enjoy social interaction and would not enjoy teaching. Possible careers include firefighting, electrical work, or carpentry. Opportunities are limited in the field of kinesiology.

- **Investigative personality:** Skilled at understanding and analyzing science and math problems. Not usually leaders or salespersons; more comfortable in laboratory environments. Those who pursue careers in kinesiology as researchers in exercise physiology, biomechanics, sport psychology, or sport sociology might have investigative personalities.

- **Artistic personality:** Enjoys creative activities such as art, drama, crafts, dance, music, or creative writing. Best suited to work as writer, actor, artist, graphic designer, musician, or dancer.

- **Social personality:** Enjoys helping people; values friendliness and trustworthiness in others. Tends to excel in careers such as nursing, counseling, physical therapy, and medicine. Socially-oriented individuals often pursue careers in athletic training and sport medicine.

- **Enterprising personality:** Ambitious, energetic, and sociable. Tends to be a self-starter who enjoys being in charge. Less interested in analyzing or conducting scientific studies than in getting something practical accomplished. Often pursues a career in business or politics. Managers of fitness facilities, coaches, and school principals often have enterprising personalities.

- **Conventional personality:** Likes to work with numbers, machines, facts, and records. Prefers highly structured, orderly work that follows a set plan. Often is a bookkeeper, bank teller, or clerk. Most careers in kinesiology are not compatible with this type.

Which personality type do you think best describes you? What careers in kinesiology do you believe your personality best matches?

Reproduced by special permission of the publisher, Psychological Assessment Resources, Inc. 16204 North Florida Avenue, Lutz, FL 33549, from *Making Vocational Choices*, Third Edition, Copyright 1973, 1985, 1992, 1997 by Psychological Assessment Resources, Inc. All rights reserved.

How Are Physical Activity Professionals Educated for the Workforce?

Now that you've examined some of the general characteristics of the professions, let's consider what formal education is required. By now you are aware that professional work requires advanced knowledge that you can obtain only through formal education at the undergraduate and graduate levels. If you're like most college students, you may not always understand why certain courses are required or what the purposes are of various phases of the curriculum. You might be interested to learn that kinesiologists themselves are engaged in a debate concerning the structure of the kinesiology curriculum. Although the curriculum at your institution may not offer an exact replica of the curriculum described here, it probably is similar in terms of its general features. The kinesiology curriculum model incorporates five general types of academic experiences essential for preparing kinesiologists for their professional roles:

- Course work in liberal arts and sciences
- Course work in physical activity
- Course work in theoretical and applied theoretical knowledge in kinesiology
- Course work in professional practice knowledge and skills for particular professions
- Apprenticeship or internship experience at the worksite

The best professional preparation programs immerse students in all academic experiences.

Course Work in Liberal Arts and Sciences

All college graduates, regardless of their career aspirations, are assumed to be educated people. Used in this sense, the word *educated* describes someone who has struggled with the great works of literature, has learned about the historical development of human civilization, has developed critical skills essential for analyzing arguments logically, is able to determine the truthfulness of ideas and to identify the values embedded in them, has developed sufficient knowledge about the arts to appreciate the richness that they can add to one's life, and has developed insights into the problems and prospects of their own culture and the culture of others. Accomplishing this goal requires courses in science, philosophy, history, literature, sociology, mathematics, art, music, and the dramatic arts, among others. Usually, these courses are part of the "general education" or **liberal studies** requirements in college and are taken during the first 2 years of undergraduate study. They form the foundation for your professional education, which usually comes later.

Liberal education derives its name from freedom; only by immersing yourself in the arts and sciences can you liberate, or free, yourself from dependence on the thinking of others.

▌ A liberal education expands your approach to physical activity by incorporating the totality of your personal experiences.

Liberally educated people have developed a capacity for seeking the truth and for making intelligent choices when they confront life's problems and opportunities. They also are aware that nobody is ever completely educated; everyone is always in the process of becoming educated. These people are aware of what they do not know and continually seek opportunities to expand their intellectual and cultural horizons by reading books and newspapers, attending concerts, patronizing the arts, and engaging in other lifelong educational activities.

Thus, the liberal education you receive in college is just the beginning. Whether this initial immersion in the liberal arts and sciences has accomplished its purpose will be reflected in your level of interest in continuing to educate yourself after graduation. What types of books will you read? What kind of music will you listen to? How will you speak and write? Will your analysis of the claims of politicians and other public figures be logical and based in fact? Your approach to these and hundreds of other life experiences should bear witness to the fact that you are a liberally educated person. Of course, a liberal education should have short-term effects too. The most important is to provide you with the broad-based knowledge and intellectual skills essential for undertaking more advanced study in your professional curriculum. A liberal education frees you from depending on others to do your thinking for you. It is the foundation on which all other educational experiences are constructed.

Course Work in Physical Activity

Most kinesiology departments require their majors to enroll in physical activity courses, either as part of their liberal education requirements or in conjunction with their professional programs. As noted in part I, you entered college with a wealth of experiential knowledge about physical activity and no doubt found it to be a vehicle for learning much about yourself, others, and the world around you. But the physical activity experiences offered within the framework of a liberal or professional education should result in qualitatively richer experiences than the experiences you had before you came to college. Let's take a moment to consider how physical activity courses at the college level might affect you.

If you've already taken a physical activity course at your college, you may have discovered that the course was more intense and taught at a more advanced level than classes in high school. In contrast, those who had an excellent physical education background in high school may have discovered that the college-level physical activity course wasn't much different from the courses they took in high school. In both cases the courses focused on sport, outdoor activities, or fitness. But even if the course of instruction seemed to be the same, you are different. Everything we learn occurs in a context of the totality of our personal experiences. As your liberal education continues to expand your horizons, your approach to physical activity experiences will change as well. You should be a more intelligent analyst of your performances and a more persuasive critic of the performances of others, and you should be able to play a more astute role in the planning and preparation for your own physical activity experiences. You should also be able to draw connections between your physical activity experiences and the world around you.

A course in martial arts should attune you to the nuances of pace, tempo, and rhythm; a course in fencing should foster an appreciation of beauty, force, and form in movement. Your course work in physical activity may sensitize you to ethical and moral questions that you might not have noticed before you were exposed to a liberal education.

Physical activity courses also are required as part of your kinesiology major. If you plan a career in teaching, coaching, or fitness leadership, the requirement may seem reasonable to you, but why does it exist for students in sport administration, prephysical therapy, or other professions in which physical activity performance is not part of the professional responsibilities? If you're struggling with this question, think about the general model of physical activity described in chapter 1, which recognizes that knowledge gained through physical activity provides one of the three legs of the stool on which the discipline of kinesiology rests. Whatever career you enter, performance of physical activity will be at center stage, and you should have acquired a wealth of firsthand physical activity experiences,

College courses in physical activities are taken in conjunction with liberal studies and are an essential part of your professional preparation. As a result, you should develop the ability to examine your physical activity experiences from an informed, educated perspective and to integrate the knowledge learned from them into a coherent professional perspective.

not only to help you identify with physical activity problems confronted by your clients, but also to broaden and enrich your comprehensive understanding of the phenomenon of physical activity.

Course Work in Theoretical Kinesiology

The chapters in part II gave you a general overview of the subdisciplines of kinesiology. Courses in these subdisciplines teach theoretical knowledge not only about sport and fitness but also about physical activity in general. Obviously, much of it is applicable to professional practice. For example, theoretical knowledge about how the body responds to exercise (drawn from the subdiscipline of physiology of physical activity) seems directly applicable to the work of a swimming coach or a cardiac rehabilitation specialist. Other theoretical knowledge, in sport history, for example, or biomechanics, may not seem applicable immediately, but over time, the knowledge may have a profound effect on how a professional thinks and thus acts.

Some professors would argue that students should not expect all theoretical knowledge in kinesiology to apply to specific careers, just as theoretical knowledge in literature, philosophy, biology, and physics should not be expected to prepare students with occupational skills. These professors would claim that the value of theoretical knowledge is its power to sharpen students' perspectives on themselves and the world around them and help them develop new and creative ways of thinking. Thus some would say a "discipline" of kinesiology (theoretical kinesiology) should not directly apply to professional work more than should other liberal studies such as poetry or medieval history. (All kinesiology professors do not agree with this; some believe that kinesiology is strictly a professional field and that its only justification is that it prepares people for jobs in the physical activity field. A detailed discussion of this debate, however, is not within the scope of this textbook.)

ACTIVITY 12.1

Theoretical Kinesiology
Connect parent disciplines with ways of studying physical activity in this matching exercise in your web study guide.

Figure 12.5 shows how theoretical kinesiology is knowledge about physical activity that has its roots in the traditional parent disciplines in the humanities; biological, behavioral, and sociological studies; and the arts. Theoretical knowledge in kinesiology builds on these foundational areas by relating each to the emerging body of knowledge about physical activity. The result is a stimulating body of knowledge for students planning to enter the physical activity professions. At present, isolated courses in exercise physiology, motor behavior, and so on represent this body of knowledge. The important challenge for kinesiologists is to integrate this knowledge the way practitioners must do to make decisions in their work.

The evolution of theoretical knowledge of kinesiology has been an exciting development in the field of physical activity, but questions about how much of this knowledge should be required of students embarking on careers in the physical activity professions continue to surface. Should future athletic trainers, physical education teachers, or aquatic therapists be expected to master the entire body of knowledge of kinesiology? Sport management majors may well ask why they must take courses in exercise physiology or biomechanics, and the student planning to be a conditioning coach may ask why she should be expected to know about the history and philosophy of sport, especially when the concepts may not apply to the professional problems that she will face in the workplace.

But a good case can be made for requiring all undergraduate kinesiology majors, regardless of their career aspirations, to learn the body of knowledge of kinesiology (Newell 1990).

Theoretical kinesiology

FIGURE 12.5   Much theoretical knowledge in kinesiology is derived from existing parent disciplines.

It seems logical to expect students earning a degree in a field to know about the field in its entirety rather than isolated parts.

Another argument for requiring students to master theoretical kinesiology is to ensure that all professionals in the field are united around a common core of physical activity theory, without which kinesiology might fragment into a number of completely different fields. Those who believe that all physical activity professionals should be required to study theoretical kinesiology argue that it might benefit future professionals in ways that we do not fully understand. Just as a liberal education alters students' ways of approaching problems and opportunities in their lives, so an immersion in the theoretical knowledge of the discipline may equip physical activity professionals with general concepts and ways of thinking about physical activity that will assist them in solving the myriad problems they will face in professional practice.

Because some disagreement still exists concerning the relationship of kinesiology theory to professional practice, course requirements in theoretical kinesiology tend to vary from institution to institution. However, most undergraduate programs in kinesiology require students to take some core courses in kinesiology theory regardless of their career aspirations.

Becoming Familiar With Your Academic Program

Do you know what core courses in the kinesiology department are required of all kinesiology or exercise science majors? Is the core requirement different for different career tracks? Why are these courses required? Do some courses have prerequisites? Are students in your department required to enroll in physical activity classes? Why are these courses required? Are you required to complete an internship (or student teaching) as part of your degree requirements? Why is this required? How do you apply for an internship placement? Does the internship have prerequisites? Must you maintain a minimum grade point average to enroll in higher-level courses? What is it? One barometer of students' depth of engagement in their field of study is their familiarity with the requirements of the academic program in which they are enrolled. (If you aren't familiar with the requirements of your program, consult the bulletin or calendar for your institution that lists courses required in your major.)

These core courses also may vary from institution to institution. The American Kinesiology Association has recommended that all programs offer core coursework in four areas (American Kinesiology Association 2003):

- Physical activity in health, wellness, and quality of life
- Scientific foundations of physical activity
- Cultural, historical, and philosophical dimensions of physical activity
- The practice of physical activity

Do the core requirements in your department fit within this general scheme?

Course Work in Professional Practice Knowledge and Professional Skills

Knowledge of the discipline itself will not equip you to perform the tasks required in your chosen profession. You develop the competencies required in a profession through course work that provides you with practice in performing these skills.

Your preparation as a professional wouldn't be complete if you didn't acquire the professional practice knowledge essential for success in your career. The point made earlier bears repeating: Knowledge of kinesiology theory is not sufficient to prepare you for professional practice. Physical education teachers need to master the difficult skills of planning and organizing lessons, communicating with and motivating students, and evaluating students' performances. Teachers must be able to operate efficiently and effectively in the social context of the schools. They learn this professional practice knowledge in courses that are substantially different from courses in kinesiology theory.

Whereas theories of kinesiology tend to focus on physical activity alone, mastering professional practices requires knowledge that will help you identify clients' needs and desires, analyze impediments to the realization of these needs and desires, and implement actions appropriate to the context and other relevant factors. Professional practice knowledge is incorporated into **practice theories**, which are developed through applied research and professional experience. A practice theory may guide a professional in selecting the best way to perform cardiovascular tests on elderly persons at a nursing home with limited equipment, determining how to motivate a soccer player who is having academic difficulties and misses off-season training sessions, or selecting the types of balls and bats to use to speed the learning of children with autism.

Unlike kinesiology theory, practice theories and the skills of professional practice cannot be mastered simply through studying them; they are learned through action. Students acquire these theories and skills by hands-on practical experiences centered on solving real-life professional problems. You can learn about administering a stress test by reading a book, but you learn *how* to administer a stress test only by following up your reading with practical experience. By committing errors and receiving instructive feedback from your professors, you will gradually develop and sharpen your command of professional practice knowledge.

These initial attempts at putting practice theory into practice probably will be in relatively small classes on campus in which your performance will be evaluated more intensely than it might have been in kinesiology theory courses. Your classmates may serve as simulated clients, and your initial attempts to perform professional tasks will be in the safe environment of the classroom, where your mistakes will not lead to serious consequences. If you perform up to departmental standards in these series of courses, you will be permitted to enter the final stage of your professional education: the apprenticeship or internship.

Internships

An apprentice is a novice who learns how to perform a job by working closely with an experienced worker. Apprentices usually assist a veteran for little or no pay in return for the opportunity to learn the skills required to enter the full-time workforce. The term is used most commonly in connection with crafts or trades (e.g., plumbing, electrical contracting,

The internship is the culminating educational and evaluative experience for preprofessionals. As an educational experience, the internship provides you with an opportunity to apply the knowledge and skills that you learned in the undergraduate curriculum while working under the supervision of a trained professional at an off-campus site. The internship is also an evaluative experience in that it tests your level of preparedness to enter your chosen profession.

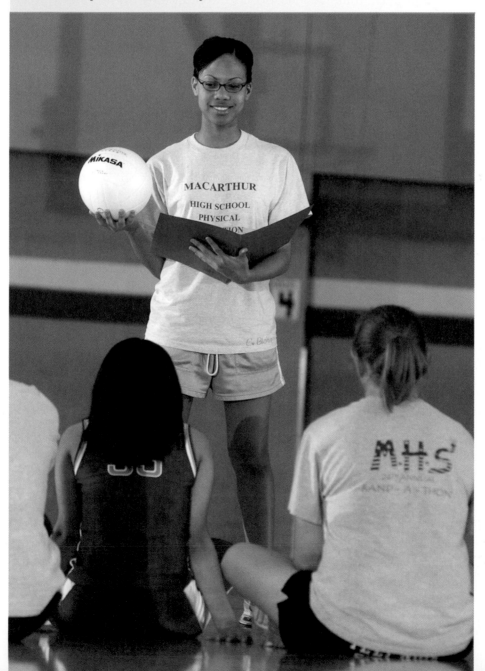

An internship provides real-world experience, which can help you decide to continue your course of study.

carpentry) in which the apprenticeship is the only type of training offered for the job. In professional education programs, the apprenticeship phase is called an **internship**, and it is the culminating experience of 4 years of study. Unlike apprenticeships for the crafts or trades, internships for professional careers are preceded by a rigorous course of study.

In the physical activity professions, the internship is known by a variety of names. It may be referred to as an internship in sport management and fitness instructor programs, as student teaching in physical education teacher education programs, and as clinical training in athletic training programs. Regardless of what it is called, the experience has the following two purposes:

- To teach you how to apply the knowledge and skills that you have learned in your professional program to a real-life situation
- To test your level of preparedness to enter professional practice

For most students, this is the first opportunity to assume the role of a member of their chosen profession. The internship usually is the most challenging experience of the undergraduate years and the most enjoyable. For these reasons, most students feel that it is the most valuable experience of their undergraduate education.

The internship differs from traditional course work in three ways. First, it occurs at the worksite rather than at the college or university. This arrangement means that the subculture of the worksite rather than the subculture of your undergraduate department will dictate your conduct. You will be evaluated not only on the knowledge you display but also on how well you conform to the rules of the workplace. If you were engaged in an athletic training internship for a Division I athletic department, for example, you would need to pay attention to the local rules that govern behavior in the training room. Are there dress codes, and are they strictly enforced? Does the supervisor of the training room expect you to arrive 15 minutes before your scheduled starting time? What local terms are used to describe interventions, locations, or procedures? What safety measures are you expected to follow? What are the established emergency procedures? Does the supervisor like interns to show initiative, or does she prefer that you ask for advice or permission before undertaking an action? You must consider these and other such issues to ensure success in the internship as well as in the profession.

 ACTIVITY 12.2

Build the Pyramid

Test your understanding of the path to becoming a physical activity professional in this activity in your web study guide.

A second way in which an internship differs from your regular course work is that an experienced professional who probably is not a member of the faculty in your department will supervise your performance on a day-to-day basis. A university supervisor may visit the intern site occasionally, but the primary overseer and evaluator of the quality of your work will be the professional to whom you are assigned. Faculty have chosen these on-site supervisors because they are considered to be good role models for future professionals. However, you shouldn't be surprised if their opinions, philosophies, and methods of operation differ from those that you learned in your professional course work. Exposure to a variety of philosophies and methods will usually expand your knowledge of the profession.

The third way in which the internship differs from typical course work is that you will be serving real, not simulated, clients. Your actions will have important consequences, and they can be serious. Serving real clients also means that you must learn the social graces expected of a professional. You will need to demonstrate poise, courtesy, alertness, and initiative. You will be expected to demonstrate a level of confidence that makes clients feel comfortable and encourages them to place their trust in you.

Internship requirements vary across institutions and professional specializations. If you are studying sport management, your internship may involve a semester of 20-hour weeks working in an athletic administration office where you might be required (like most professionals) to perform a range of duties, including mundane tasks such as handling mailings in the ticket office or answering the phone in the information office. You also have measured amounts of responsibility in performing the types of tasks that will be required of you as a sport management professional. If you are undertaking a student teaching experience in physical education, you may be required to spend an entire semester of full days in which you assume major responsibility for teaching classes. If you are a physical education intern (i.e., a student teacher), you will be responsible for logging a preestablished number of hours specified by your department and the state education department.

Are You Suited for a Career in the Physical Activity Professions?

Now that you know something about the type of work that professionals do and the educational experiences required to prepare you for such work, it's time to ask yourself how well suited you are for a career in the physical activity professions. The five-step approach demonstrated in activity 12.3 in your web study guide will help you undertake this self-examination process.

 ACTIVITY 12.3

Selecting a Career
Work through the decision tree in this activity to clarify whether a career in the physical activity professions is right for you.

Do My Attitudes, Values, and Goals Match Those of Professionals?

The first step is to consider how compatible you are with professional work generally, regardless of the particular field. Approach this task honestly and objectively. If you decide that you're not well suited to a professional life, you should consider alternative careers.

Taking stock of your strengths, weaknesses, and potential in relation to the personal attributes and preferences likely to be required in most types of professional work is an important first step. Nobody is in a better position to judge our attitudes, attributes, and preferences than we are, but sometimes our perceptions of ourselves can be a bit clouded. One way to bring our own attitudes into sharper focus is to compare them with those of our friends and peers. The results of a survey by the Higher Education Research Institute at UCLA, involving more than 200,000 freshmen entering 4-year institutions in the United States in 2011, offer a convenient template against which you can assess your own attributes and preferences (The American Freshman: National Norms 2011).

If you are a typical undergraduate student, chances are that you have already identified a career track (14% say they haven't), and you have enrolled in college to gain a general education and appreciation of ideas (72%) and to "get a better job" (86%). This is a good start for future professionals. There are some encouraging signs that incoming college students are becoming more serious about their academic performance. For example, 39% of college freshman reported spending 6 or more hours a week studying or doing homework in high school, a trend that has been on the rise since 2005. (Expect to do a lot more than this in college!) An upward trend also was seen in the percentage of college freshmen (70%) who said they frequently took notes during class as high school seniors, whereas the percentage

> ► If you decide that your attitudes, values, and life goals match those normally associated with a professional career, you are ready to proceed to the next question in the decision model (see activity 12.3). If you conclude that you are not a good match for a professional career, seek advice from a faculty member or the career counseling center at your institution.

of students (36%) who reported being frequently bored during class decreased. Forty-nine percent said there was a good chance they would discuss course content outside of class, another good sign.

The survey also raised some red flags. Only 34% of your peers plan to communicate regularly with their professors, and only 46% plan to become involved in clubs and on campus activities. Most college faculty view such forms of engagement as being very important. Another red flag is the percentage (60%) of freshmen who plan to pursue graduate education. As a rule, higher paying jobs usually come with some administrative and supervisory responsibilities, and advancement in the professions is usually linked closely to educational credentials. Generally, graduate degrees open doors to such career paths. Because educational standards for professional practice usually increase over time, increasing numbers of undergraduate students will discover that a master's degree has become the minimal educational qualification for entering many of the physical activity professions.

The fact that a relatively large percentage (88%) of your peers indicated that they had performed volunteer work in the previous year is a very good sign, but the fact that only 34% plan to continue to do volunteer work or community service work in college is troubling. A student who is not interested in volunteer work may not be well suited for a career in the professions, given that a spirit of service is the lifeblood of professional work. On a more pragmatic note, employers and graduate school admissions officers usually interpret a continuing commitment to volunteer work throughout college, even if only for 2 or 3 hours per week, as an indication that an applicant understands the serious service responsibilities that come with a career in the helping professions.

Almost a third of students (approximately 29%) reported feeling overwhelmed by all of their obligations during their high school years, but professionals in the workforce sometimes feel this way too. Thus, stamina and robust health are especially important personal attributes for physical activity professionals, yet the survey results indicated that only 47% of respondents frequently or occasionally exercise and play sports. On a positive note, though, it is quite likely that the percentage of students who plan to engage in sports and exercise is much higher among the subset of students who plan to major in kinesiology.

Is That Jolt from Red Bull Worth It?

Advertisements for the popular energy drink Red Bull advise students that "Nobody ever wishes they'd slept more during college." Although research has shown that getting adequate amounts of sleep is very important for high cognitive functioning, students continue to rely on caffeine-laced drinks to help them pull "all-nighters." Are these drinks healthful, and do they help or hurt your academic performance? A study of 136 undergraduate students that explored relationships regarding perceived stress, energy drink consumption, and academic performance suggests that the jolts provided by energy drinks come at a price students often are not aware of (Pettit & Debarr 2011). Researchers found that approximately two-thirds of the students tested had consumed at least one energy drink each day for the past 30 days. Not surprisingly, the study found positive relationships between students' perceived stress and the amount of energy drinks they consumed. (The study's design did not permit researchers to determine whether high perceived stress levels were the result of the energy drinks or whether students drank energy drinks because they felt stressed.) The study also found negative correlations between energy drink consumption (largest number of drinks consumed on any one occasion) and academic performance: The higher the consumption, the lower the academic performance. Other researchers have discovered links between energy drink consumption and undesirable behaviors such as risky sexual practices and the use of alcohol, cigarettes, and prescription drugs, findings that suggest students should exercise care in consuming these products (Miller 2008). Clearly more research on this topic is needed, but until conclusive evidence is in hand, the best strategy may be good time management so all-nighters aren't necessary.

After giving the matter some thought, you may have concluded that you really aren't cut out for a professional career. Perhaps you don't find working with people attractive, you don't particularly like helping to solve other people's problems, and you would rather work for a supervisor than take responsibility for acting on your own authority. Maybe the prospect of continually educating yourself to keep abreast of the professional literature or attending to the ethical ramifications of your work behavior is not part of the vision you have for your career. If so, it seems unlikely that a professional career will interest you, regardless of the area of specialization. We suggest that you continue to collect more information by reading the other chapters in this section of the book, talking to faculty members, and consulting with the career counseling center on your campus.

If you are convinced that you would like a career in the physical activity professions but you recognize some glaring deficiencies (fear of speaking before large groups or a lack of organizational skills, for example), consult with your faculty advisor about including academic experiences during your undergraduate years to bolster these weaknesses.

Am I Interested—Really Interested—in Physical Activity?

If what you have read about the physical activity professions excites you and matches your interests and attitudes, the second step is to consider how much you like physical activity. Those who enter the physical activity professions should be more than merely interested in physical activity; they should be fascinated by it. If you're not intrigued by performing, studying, and watching physical activity; if you aren't curious about the mysteries of body movement in all of its manifestations (biophysical, behavioral, and sociocultural); and if you're not prepared to apply your intellect and skills to exploring these mysteries, you may be better off selecting another profession.

Do My Attitudes, Interests, and Talents Lend Themselves to a Specific Physical Activity Profession?

The third step in this personal evaluation is to consider whether your attitudes, interests, and talents match the specific characteristics of the occupation that you are considering. For example, if you envision a career in therapeutic exercise or physical therapy, consider whether you enjoy interacting with sick and injured people. Obviously, most people—even medical students—require some time before they feel completely comfortable in medical environments, but you need to understand, fairly early on, your degree of comfort with the environments in which allied health professionals work.

If you like being around vigorous, healthy, active people, a career in physical fitness counseling, personal training, coaching, or athletic training probably appeals to you. Do you have a special place in your heart for people with disabilities such as mental retardation, autism, blindness, or deafness? Would you find it challenging to help such people maximize their physical and mental potential? If so, a career in adapted physical education seems a reasonable goal. If you like being the center of attention; organizing and speaking to large groups; planning, monitoring, and evaluating activities that others perform; and accepting responsibility for the actions of people younger than you, a career in teaching fitness, coaching, or physical education may be a good choice.

Does working at a desk dressed in business attire and adhering to a 9-to-5 workday appeal to you? If so, you might be well suited to a career in sport management. Are you more attracted to fluid and informal working environments in which your work schedule might be more flexible, even though work might intrude on your weekends and evenings? If your answer is yes, then you might be suited to more entrepreneurial types of professional positions such as personal trainer, professional sport instructor, or university professor. Chapters 13 through 17 will give you more complete descriptions of various physical activity professions to help you assess which careers may best suit you.

If you become convinced that your needs, interests, and attitudes do not closely match the specific professional occupation you had in mind, don't hesitate to explore other career

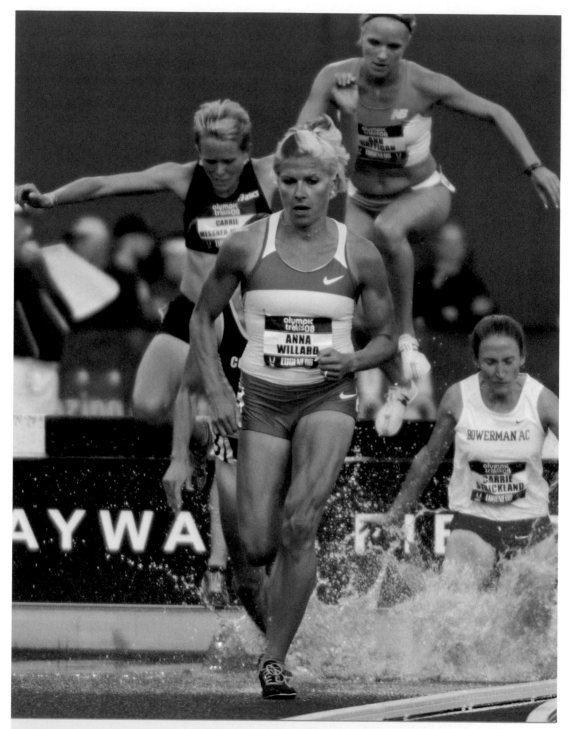

▌ It is important to identify your likes, dislikes, and skills before making a career decision. If you don't like working with athletes, being a collegiate track coach probably isn't for you!

possibilities within the physical activity professions. If you're not sure, consider taking advantage of opportunities to volunteer at clinics, commercial or governmental agencies, schools, or other physical activity settings. Volunteering will give you firsthand experiences with various working environments and will increase the probability that the career you choose will match your interests, talents, and preferences. If possible, coordinate these volunteer experiences with your faculty advisor.

Will My College or University Program Prepare Me Well?

The fourth step is to determine whether the program in which you are presently enrolled will prepare you well for your chosen career. If, for example, you are most interested in a career in teaching and coaching but your department offers only concentrations in exercise physiology, sport administration, and health promotion, the department's curriculum would seem a poor match for your career objectives. The same would be true if the program slants heavily toward the preparation of physical education teachers or professionals in the fitness industry and you are most interested in a career in sport management. If you have doubts about the fit between the course requirements of the program you are in and your career aspirations, don't hesitate to talk to a faculty member.

@ ACTIVITY 12.4

Are You Qualified?
In this activity in your web study guide, you'll discover the qualifications needed for an entry-level position in your field. Follow up by talking with your instructor, an advisor, or practicing professional to ensure you're on the right track.

Perhaps your career goals have changed since you enrolled in your undergraduate program. Although this would occur less often if students took the five-step approach recommended here, changing career goals during the undergraduate years is common. Remember that switching majors or concentrations within the department (e.g., from physical education to exercise therapy) can delay your graduation for a year or more. Changing also can be inconvenient. If you decide to switch majors to an entirely new field, you may push your graduation even farther into the future. Making such a drastic change also will require you to become socialized into another department, become accustomed to new professors, and form new alliances with classmates. However inconvenient switching your major may be, it is better to endure delays and inconvenience than to continue training for a career in which you have little interest. If you are deep into your undergraduate program and find that your career interests have changed to another physical activity profession, the best action might be to complete your present program and follow it up with a master's degree in your area of interest. Faculty members in your department will be ready and willing to advise you, even if doing so means referring you to other departments or other institutions.

Asking questions about the pertinence of your program of study to your life goals is risky business. You may not like how you answer your own questions! But this vital exercise is one that students too often ignore. Remember that those who reach career decisions early in their undergraduate years are more likely to enjoy their undergraduate studies and appreciate their relevance, particularly in courses that they take in their major.

How Committed Am I to Preparing to Be the Best Professional Possible?

At the first practice of the season, coaches often rhetorically ask their teams, "How badly do you want it?" They refer, of course, to the team's willingness to invest the time, energy, and hard work essential for winning the league championship. Now the question for you is, how badly do *you* want it? How much energy are you willing to invest to achieve success in your chosen profession? This fifth and last step in the process may be the most important because it supersedes all other questions in the process. You may have had little difficulty answering the questions up to this point, but if you aren't committed to success, there's little point in continuing to pay tuition only to have your career plans end up on the scrap heap. Are you willing to commit to preparing yourself to be the best professional possible? Let's examine some of the ways in which such a commitment might manifest.

What Should You Look For in a Department of Kinesiology?

Deciding which college you will attend often revolves around a variety of practical issues such as the school's location, tuition charges, and family traditions. But the most important consideration is whether the school can offer you a first-rate preparation for the career you wish to pursue. Here are some recommendations to help you decide which school to attend and whether the program in which you are currently enrolled is compatible with your career interests:

- Select a college or university that suits your general tastes. Are you afraid of getting lost in the shuffle of a large state university? If so, there are scores of smaller liberal arts institutions that offer programs in kinesiology. Although such schools usually do not sponsor high-profile intercollegiate sports programs, which many students find attractive, these schools usually offer smaller classes and the opportunity to establish closer relationships with faculty than do larger schools. However, if you think you would be happier in a larger school, limit your search to those institutions.

- Examine the academic catalog; is ample coursework offered in the kinesiology department to support your study? Does the program offer a concentration or a full-fledged degree program in kinesiology?

- Review the departmental website. Does the department seem alive? Are exciting things going on in the department?

- Examine the faculty directory of the department (on the website). Take time to review their curriculum vitae (resumes) if listed on the web site. Does it appear that some faculty share your interests? Are they fully engaged in their profession? Are they active in academic or professional societies? Are they sponsoring interesting projects? Are they publishing in academic and professional journals?

- Plan to visit the department. Set up an appointment with the department head or with faculty who are working in an area that interests you. If possible, send them an e-mail telling them something about yourself before meeting with them.

- If possible, talk to students during your visit. Generally, students can offer a unique perspective on the department that others cannot. Caution here: It is dangerous to accept one person's opinion; if a student has had a bad experience with a professor, it can taint his feelings about the department. It always is best to talk to a number of different students.

Adapted, by permission, from S.J. Hoffman, 2011, Examining the big picture. In *Careers in sport, fitness, and exercise*, edited by American Kinesiology Association (Champaign, IL; Human Kinetics), 12-13.

Excellence in Academic Work

Graduating with a superb academic record is regarded by most employers as a baseline indicator of your level of commitment to becoming an outstanding professional. Obviously, some students are more academically gifted than others, but often students can overcome slight shortcomings in academic ability by special application of effort. Few things make a more indelible impression on professors than a student's willingness to work hard to achieve academic success.

What types of behaviors suggest a willingness to work hard? Attending class regularly and on time, visiting the library regularly, and reading unassigned journal articles and books about the topics that you are studying in class all are signs that you are taking your academic work seriously. Most professors view students who approach them after class with questions about the material in the day's lesson as more committed than those students who avoid them at all costs.

Perhaps the best predictor of your success in the physical activity professions is the level of commitment you make to preparing yourself to be the most knowledgeable and highly skilled practitioner possible.

Early Identification With the Professional Field

How early in the undergraduate program students identify with their chosen career is also a good indicator of commitment. In the chapters that follow, you will find many references to professional associations. Joining a professional association while still an undergraduate may be the most reliable indicator of commitment. Professional associations usually admit preprofessionals at reduced rates and offer reduced registration rates at conferences. Members receive the organization's publication and other information on a regular basis. Attending professional conferences is another indicator. Professional associations usually hold annual regional and national meetings that feature lectures, workshops, and exhibits of equipment used in professional practice. Attending these can be expensive, but then again, taking a spring break trip to the Florida beaches is costly. Remember, keep your eye on your long-term goals and be willing to sacrifice to attain them.

Students who identify with a career early tend to approach their studies with special excitement and vigor. Their orientation changes from that of a student who views courses as obstacles to be overcome in obtaining a degree to that of a professional who tries to get out of each course the knowledge and skills that will help her develop into the best qualified professional possible. When you identify with your chosen profession, you will seek the advice of veteran professionals and observe them in practice. You will begin to establish a communications network with practicing professionals, and you will establish the habit of reading the journals in your field. You will learn the language of the field and begin to feel comfortable around experienced professionals, although you are still in the preparation phase of your career.

Another way to identify early with a physical activity profession is to obtain certification in an area in which you plan to work. Certifications are available from such groups as the American College of Sports Medicine, the National Strength and Conditioning Association, the American Council on Exercise, and the Aerobics and Fitness Association of America.

> ► Identifying early with the profession you plan to enter by joining appropriate professional organizations, attending conferences, establishing alliances with veteran professionals, and obtaining professional certifications will give you a head start on developing a successful career.

Engagement in College or University Life

Being engaged means being connected to what is happening around you. Next to academic performance, your level of engagement in your department, college or university, and community is one of the best indicators of your level of commitment to becoming a successful professional. Engaged students are those who take responsibility for their academic experiences and their professional futures. They are curious, and they constantly test their personal limits by seeking new ways to become involved in life experiences. They are active, energetic leaders who view the department as their department and seek ways to participate in its operational life. They are likely to be leaders in their major clubs or student government on campus or coordinators of charity events or other community activities outside the academic environment.

> ► Becoming involved in activities within and outside your department is one indication of commitment to developing a successful career in the professions.

Who Are the Engaged Students?

Research has begun to shed light on the populations of students who are most likely to show high levels of engagement in college (Kuh 2003, p. 27). These groups include the following:

Women

Full-time students

Students living on campus

Native students (who start and graduate from the same school)

Learning community students

International students

Students with diversity experiences

Besides becoming engaged in social and organizational extracurricular activities, you should become involved in learning as well. Beyond doing well on course assignments and exams, ask yourself, "How can I take on even more responsibility for learning what I need to know to succeed? Does a professor need students to assist on a research project? Does your institution sponsor undergraduate research assistantships? Does your department sponsor community-based projects on which you can work? Will a professor sponsor you for work on an independent study of a topic of interest to you? Don't be afraid to register interest in such projects. Independent study not only reflects a high level of commitment on your part but also offers an excellent opportunity to develop the type of leadership skills and knowledge essential for success in the world beyond college.

Participating in Volunteer Services

Volunteering regularly for service to a community agency indicates that you share your profession's commitment to service. Seeking in-depth, advanced education in graduate school may also be a sign that you are investing your interest and personal resources in a career in the physical activity professions.

We have seen that the driving force of professionals, especially those in the helping professions, is a spirit of wanting to improve the quality of life for those in the community. Although it is easy for preprofessionals to say that they want to serve others, nothing speaks more loudly than actions. If two students have approximately equal academic records, the advantage always will go to the student whose resume displays clear evidence of volunteer service to community agencies. These might include such programs as Meals on Wheels, Boys & Girls Clubs, local food banks, Boy Scouts and Girl Scouts, shelters for the homeless, church-based programs, walk-a-thons for various charities, American Red Cross activities, and other community-based programs. When your volunteer experience is in an activity that bears some relation to the profession for which you are preparing, it has the added benefit of socializing you into your future profession. Physical therapy and sports medicine clinics, YMCAs and YWCAs, youth sport organizations, schools, Special Olympics, and other agencies associated with the physical activity professions are some potential volunteer opportunities that you might want to explore.

Attending Graduate School

Evidence of commitment to a profession may be reflected in plans to pursue advanced graduate work, and approximately 60% of entering freshmen say they plan to do so. In professions such as physical therapy, a master's degree is a minimal entry degree, and the athletic training profession may require a graduate-level degree in the near future. Permanent certification for teaching physical education in many states requires a master's degree or its equivalent. You may not be prepared to think about graduate school at this early point in your undergraduate preparation, or you may have decided to delay that decision until a few years after you have graduated. For many, this may be a sensible plan. Overall, however, making an early decision to continue with advanced graduate education in the physical activity professions is another indication of commitment.

Generally, master's programs in kinesiology offer advanced education in a number of subdisciplines such as exercise physiology, motor learning, and sport history and in professional areas such as teaching physical education, fitness leadership, athletic training, and sport management. A faculty member at your institution may be an excellent source concerning which institutions you might consider attending. In addition, you may find posters or brochures advertising graduate study at various institutions posted in your departmental office. Normally, master's degrees require 30 to 36 credit hours of work depending on the specialization. In most cases, full-time master's students are able to complete the requirements of the degree in 1½ to 2½ years. Master's programs that require a thesis have slightly fewer hours of course work but include an in-depth research project that may take up to 1 year to complete. Some master's programs do not require a thesis but may include additional course work. Sometimes programs that do not require a thesis do require an internship. Doctoral education in kinesiology should be considered by those who envision a career as a professor in a college or university. If you delay attending graduate school until after you have secured a position and plan to do your graduate work by attending classes in the evenings, you may need as long as 5 years to complete a master's degree.

Thumbnail Sketch of College Freshmen 2010

Educational

- 73% believe that the chief benefit of a college education is that it increases earning power.
- 78% believe that gaining training for a specific career is a very important reason to go to college.
- 60% believe that preparing for graduate school is a very important reason to go to college.
- 65% expect to make at least a B average.
- 34% expect to communicate regularly with their professors.
- 46% expect to become involved in clubs on campus.
- 34% expect to engage in volunteer or community service work.

Personal

- 41% frequently or occasionally drank wine or liquor and 35% drank beer as high school seniors.
- 73% frequently or occasionally attended religious service during the last year.
- 31% frequently or occasionally discussed politics in the past year.
- 29% frequently felt overwhelmed by all they had to do.
- 88% performed volunteer work.
- 70% socialized with someone of another racial or ethnic group.
- 47% exercised or played sports 6 hours or less each week.
- 53% spent more than 3 hours in social networking during their senior year in high school.
- 82% used the Internet for homework.
- 25% frequently looked up scientific research articles and resources.

The following percentages of students rated themselves in the highest 10% or as "above average" in these qualities compared with the average person their age:

- 71% in academic ability
- 55% in physical health
- 61% in leadership ability
- 45% in mathematical ability
- 52% in emotional health
- 77% in drive to achieve

Data from CIRP, 2011, *The American freshman national norms*, 2011. Available: http://heri.ucla.edu/PDFs/pubs/TFS/Norms/Monographs/TheAmericanFreshman2011.pdf

The Decision-Making Process

By going through the decision-making steps in activity 12.3, you have embarked on an important process. If you answer yes to all five questions, you are well on your way to a successful career in a physical activity profession. The next five chapters will give you an in-depth look at some of the professions you have to choose from.

If you hesitated at some questions or were not sure of your answers, the next five chapters may help you make a decision. Use this as an opportunity to assess honestly and objectively whether you are suited to professional work, especially to professional work in one of the physical activity professions.

Wrap-Up

Career decisions are among the most important choices that people make during their lifetime. Such decisions are not irrevocable; in fact, you likely will change your career several times over your working years. Nevertheless, you will gain nothing by delaying your commitment to a career, and you have much to lose if you delay. Students who make career commitments early are in a better position to benefit from their undergraduate education than are students who have only tentative plans about the type of work that they will do when they graduate. If you have decided to attain a degree in kinesiology by default (nothing else seemed interesting) and life beyond graduation looms like "the great black void," you should take steps to assess your compatibility with a career in the physical activity professions.

As we have explained, this necessarily involves learning what it means to be a professional and learning about the types of work and work environments associated with each of the physical activity professions. It also involves making a realistic assessment of your level of excitement and commitment to kinesiology and the physical activity professions. If this chapter achieved its goal, you will already have begun this self-examination process. Now it is time to learn more about the physical activity professions by studying the next five chapters. These chapters, written by experts in the professions being described, provide specific information about what it is like to work in each of these areas. When you have finished, you should be in a good position to make this important decision about your future career.

 ACTIVITY 12.5

Use the key points review in your web study guide as a study aid for this chapter.

Review Questions

1. List three ways in which professional work differs from nonprofessional work.
2. Why is it important for professionals to attend professional conferences and read the professional literature?
3. List three differences that you might observe between a community sport program leader who adheres to a mechanical, market-driven professionalism and one who adheres to a social trustee, civic professionalism.
4. What is the value of the liberal arts and sciences and why are the liberal arts courses required of kinesiology majors important?
5. Why is the internship an important experience in preparing kinesiology students for professional practice?
6. What are five questions that all kinesiology students should ask themselves before deciding to major in kinesiology?
7. What evidence would you—as an employer—use to determine an applicant's general suitability for a professional position?

8. What would be the ideal work history of a kinesiology graduate seeking to enter the athletic training field? The health-fitness field? The sport management field?

9. If, during your second year of the kinesiology major, you decide to pursue a career as a physical education teacher after having entered as a sport management major, what is the best course of action to take?

chapter thirteen

Careers in Health and Fitness

Sandra Minor Bulmer and Jeremy Howell

chapter objectives

In this chapter we will

- acquaint you with the wide range of professional opportunities in the sphere of health and fitness;

- familiarize you with the purpose and types of work done by health and fitness professionals;

- explain how the sphere of health and fitness is evolving and progressing, especially because of marketplace trends;

- inform you about the educational requirements and experiences necessary for becoming an active, competent health and fitness professional; and

- help you identify whether one of these professions fits your skills, aptitudes, and professional desires.

An office worker plans to participate in a cycling event to raise money for a nonprofit organization. The office worker has hired a personal trainer from his local fitness center to guide him in gaining strength for cycling and improving his cardiovascular endurance safely.

A group exercise instructor meets with the class members of a new session and sees that one of the participants has brought along her teenage daughter, who is very overweight. The mother tells the instructor that her daughter was just diagnosed with type 2 diabetes.

A health and fitness director realizes that her facility should expand its programs next year to reach less physically fit populations more effectively. What additional staff will she need to hire? What should their qualifications be? What new equipment and facility renovations will be required? Will her budget allow for this expansion?

In an interview with a health and fitness counselor, a working mother mentions that she is having trouble fitting her exercise program into her busy lifestyle. She feels overwhelmed, irritable, and constantly exhausted. How will the health and fitness counselor begin to develop an exercise program that improves this client's quality of life?

Health and fitness professionals face these types of situations every day in their careers. In each case, they must make decisions that will improve the health and fitness of participants and clients safely while keeping within the goals and objectives of the program or facility where they work.

Health and fitness, one of the spheres of professional practice (see figure 13.1), is made up of many dynamic professions. The sphere of health and fitness professions includes such positions as group fitness instructor; health and fitness specialist; wellness coach; personal trainer; health and fitness director; and specialty positions such as clinical exercise physiologist, registered dietitian, and public health educator.

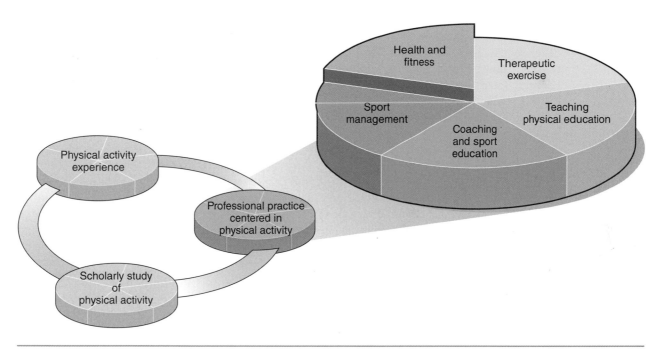

FIGURE 13.1   Health and fitness is one of the spheres of professional practice.

Although the scope of the professional work in this sphere is expanding to include a far more multifaceted view of health and fitness, the core goal of these professions is to improve physical functioning and physical health of individuals and communities. Health and fitness professions have traditionally been conducted in worksite, clinical, commercial, and community settings, but in recent years the lines between these settings have become blurred as each type of program or facility has expanded its offerings to reach more people and affect their lives in more ways.

This expanding scope has created tremendous opportunities for kinesiology graduates and is making health and fitness positions some of the more exciting and dynamic careers people can enter. Thus, our purpose in this chapter is threefold. First, we describe the opportunities that exist for professionals in the health and fitness field. Exactly what do health and fitness professionals do and where do they do it? Second, we analyze why these opportunities are appearing at this time. What social forces are creating new possibilities for current kinesiology graduates? Third, we discuss how you can best take advantage of these opportunities as you complete your education and move into your chosen career. What are the educational requirements and experiences necessary for becoming an active, competent professional in this area?

 ACTIVITY 13.1

Job Skills

What jobs come to mind when you think "health and fitness"? What do those jobs entail, and what skills and qualifications do they require? Explore the answers to these questions in your web study guide.

What Is the Health and Fitness Profession?

This is an exciting time for those involved in the health and fitness professions. Despite the many studies demonstrating that the American public is not nearly as active as it should be (see chapter 2), the marketplace for physical activity professionals has never been better. The health and fitness settings in which physical activities take place and the jobs that physical activity professionals perform have been expanding for nearly two decades.

Today, scientific research and popular culture have popularized the notion that regular exercise can increase lean body mass, heart and lung function, immune function, blood vessels, blood flow, muscular strength, flexibility, bone density, balance, and self-esteem. Perhaps most important is the research that links physical activity to lowering the risk of death because of coronary heart disease, the leading cause of mortality in the United States.

Indeed, numerous government reports extol the undisputable virtues of exercise for good health. In 1996, the U.S. Department of Health and Human Services (USDHHS) published a document titled *Physical Activity and Health: A Report of the Surgeon General*. This was significant because for more than a century, the surgeon general has focused the nation's attention on the public health issues of the day. Previous reports on issues such as smoking, nutrition, violence, and HIV/AIDS heightened America's awareness and led to some innovative public health initiatives. This document concluded that Americans should engage in a minimum of 30 minutes of moderate activity most days of the week. In 2008, the USDHHS published *Physical Activity Guidelines for Americans (PAG)*, which described the department's updated recommendations for physical activity. These new guidelines recommend that children and adolescents engage in 60 minutes or more of physical activity daily and adults engage in at least 150 minutes a week of moderate intensity or 75 minutes a week of vigorous intensity aerobic activity, or an equivalent combination, along with 2 or more days of muscle strengthening exercises per week. Most recently, the USDHHS published *Healthy People 2020*, a document that sets out specific overall health objectives for the nation. The objectives are based on a set of 26 leading health indicators (LHIs), of which physical activity is one. *Healthy People 2020* proposes to reduce the proportion of adults who engage in no leisure-time physical activity and increase the proportion of both children and adults who meet the *PAG* recommendations. The publication also proposes increasing the proportion of the nation's schools that require daily physical education, provide regularly scheduled recess, and allow access to their physical activity spaces and facilities outside of normal school hours. Another objective is increasing the proportion of physician office visits that include counseling or education related to physical activity (USDHHS 2010). A list of select physical activity–related objectives and baseline measures is provided in table 13.1. The full list of physical activity objectives and other resources can be found at the Healthy People 2020 website, www.healthypeople.gov.

What all of these government documents show is that no longer do people perceive physical activity only as a recreational endeavor; it is firmly embedded in the national preventive health consciousness. However, as the Healthy People 2020 baseline measures show, finding creative solutions to healthy living is not easy. Rather than just being asked to train, measure, and test people for improved athletic performance, health and fitness professionals have to develop individual and group exercise programs that connect to other physical, intellectual, emotional, social, and spiritual dimensions. Similarly, much better health and fitness strategies need to be in place in community, commercial, corporate, and clinical settings. This means there are enormous opportunities for kinesiology students at this time.

In this chapter, we first explain the traditional settings within which health and fitness professionals operate as well as the most common professions. Then we look closely at how the changing trends in society and culture are reshaping the health and fitness sphere of professional practice.

Settings for Health and Fitness

The health and fitness field is relatively young and continues to evolve with expanded job descriptions and a greater number of positions for entry-level and experienced professionals.

TABLE 13.1 Selected Healthy People 2020 Physical Activity Objectives

Objective	Baseline	Target
PA-1 Reduce the proportion of adults who engage in no leisure-time physical activity	36.2%	32.6%
PA-2.1: Increase the proportion of adults who engage in aerobic physical activity of at least moderate intensity for at least 150 minutes/week, vigorous intensity for at least 75 minutes/week, or an equivalent combination.	43.5%	47.9%
PA-2.2: Increase the proportion of adults who engage in aerobic physical activity of at least moderate intensity for more than 300 minutes/week, vigorous intensity for more than 150 minutes/week, or an equivalent combination.	28.4%	31.3%
PA-2.3: Increase the proportion of adults who perform muscle-strengthening activities on 2 or more days of the week.	21.9%	24.1%
PA-3: Increase the proportion of adolescents who meet current federal physical activity guidelines for aerobic physical activity and for muscle-strengthening activity.	18.4%	20.2%
PA-4: Increase the proportion of the nation's public and private schools that require daily physical education for all students.		
Elementary schools	3.8%	4.2%
Middle and junior high schools	7.9%	8.6%
High schools	2.1%	2.3%
PA-11: Increase the proportion of physician office visits that include counseling or education related to physical activity.	13.0%	14.3%

Adapted from Healthy People 2020, *Physical activity objectives.*

Traditionally, health and fitness professionals operate in four primary settings—worksite, commercial, clinical, and community. At one time, each of these settings had a distinct mission, target market, and set of objectives for programming. More recently, however, the various health and fitness settings have become more similar than dissimilar. Examples include commercial settings that offer worksite health promotion programs, clinical settings that operate for-profit fitness centers, and community settings such as local health departments that provide medically based fitness programs in partnership with area hospitals. Thus, the lines between the various settings are no longer distinct.

This section describes each type of health and fitness setting and provides an overview of the objectives, types of programs, and target markets for each setting. It also describes the changes that have occurred over time and the trends emerging in the future.

> Traditionally, health and fitness professions have operated in four settings: worksite, commercial, clinical, and community. Recently, the lines between these settings have blurred, resulting in expanded job descriptions and a greater number of positions.

Worksite Settings

Worksite health and fitness programs (often referred to as worksite health promotion programs) have experienced tremendous growth during the past several decades. In the early 1980s, fewer than 5% of employers offered worksite health and fitness programs (O'Donnell & Harris 1994). According to the 2004 National Worksite Health Promotion Survey (Linnan et al. 2008), 66.1% of employers with 750 or more employees offered programs or services to promote physical activity. One of the *Healthy People 2020* objectives is to increase the proportion of employed adults who have access to and participate in employer-based exercise facilities and exercise programs (www.healthypeople.gov). Even the U.S. military is onboard. The U.S. Army Public Health Command, Provisional Army Wellness Center Program, is focused on improving the health and wellness of soldiers, retirees, and their families and Department of Defense civilians.

Early worksite programs focused primarily on physical fitness, nutrition, weight control, stress management, and smoking cessation and were available exclusively to employees of the sponsoring company. Pioneering corporations such as Johnson & Johnson, Coors Brewing

Company, and Tenneco believed that health and fitness programs would help to reduce health care costs, increase productivity and morale, decrease absenteeism, and improve their corporate image. Since these early days, worksite programs have continued to evolve, and many research studies have documented positive outcomes. Specifically, the 2007 U.S. Centers for Disease Control and Prevention (CDC) Community Task Force reviewed more than 50 research studies and concluded that programs to improve a range of health behaviors at the population level have substantial positive outcomes (Goetzel et al. 2008).

The Health Enhancement Research Organization (HERO) has provided strong evidence to support the economic value of worksite health promotion programs. Analysis of the HERO database, which includes 47,500 employees who have completed a health risk appraisal, has revealed that there is a strong association between modifiable risk factors and health care expenditures (Goetzel et al. 1998). For example, employees with self-reported persistent depression had health care expenses 70% greater than did those who reported not being depressed. In addition, costs were 46% higher for those with uncontrolled stress, 35% higher for those with high blood glucose, 21% higher for those who were obese, and 10% higher for those with poor exercise habits. Additional analysis with this database in 2000 revealed that risk factors were associated with 25% of total health care expenditures (Anderson et al. 2000). Bungum and colleagues (2003) found that people who had healthier body weights, calculated by body mass index (BMI), reported fewer absent work days and decreased health care costs.

As worksite programs have continued to develop, additional services have been added to increase employee awareness and help employees make healthy lifestyle changes. Examples of these services are health education classes, company health fairs, the distribution of educational materials, and the use of health risk appraisals to identify employee risk factors. More

Some companies have invested in active workstations that allow employees to reduce the amount of time spent sitting during a typical workday.

recently, worksite programs have expanded offerings even further to provide a work environment that further *supports* healthy lifestyles. In this expanded role, worksite programs have developed and implemented on-site work policies regarding campus-wide smoking behavior, availability of nutritious food, and access to active work stations such as treadmills with desktops. Work–life balance initiatives have become popular as ways to further promote employee wellness in all dimensions of health. These initiatives include flexible work schedules, on-site child care, and leave-of-absence policies for caregiving and other life events. Other popular worksite programs include recreation activities, volunteer service opportunities, financial planning, and **employee assistance programs (EAPs)**. In an effort to reduce the medical costs of all individuals covered by the company health plan, some worksite programs now promote health and fitness services to employees' family members and company retirees.

In economically prosperous times, many companies have built elaborate health and fitness facilities as a way to recruit and retain new employees. Other companies have contracted with local health clubs to offer access to exercise facilities for free or at a reduced rate. In either case, the trend has been for companies to establish partnerships with outside organizations that have expertise in health and fitness to manage their worksite health promotion facilities and programs. For instance, Club One Inc., a San Francisco–based corporation, manages the fitness and health programs and facilities for more than 60 corporations. Another example is MediFit Corporate Services, based in New Jersey. This company has an extensive list of corporate clients in the Atlantic region of the country and provides a variety of contracted services including facility management, on-site programming, and web-based programming for employees who work from home or in smaller satellite branch offices.

Perhaps the most innovative change in workplace health promotion has come about as many large corporations have become self-insured. The relationship between personal health practices and soaring health care costs for both the employer and employee cannot be ignored. According to a 2012 large-scale employer survey by the Kaiser Family Foundation, the average annual health insurance premium for family coverage is increasing. In 2012, the figure was $15,745. Of that amount, $11,429 was employer paid and $4,316 was contributed by the employee. That total represents a 102% increase in premiums over the 2002 figure of $8,003 (Kaiser Family Foundation 2012). Many large companies have now decided to operate their own health insurance plan as opposed to purchasing coverage from an insurance company. Usually the employer still pays a third party (such as an insurance company) to administer the plan it has designed, but the employer does so at a much lower cost. But although the company saves money on costs, it becomes responsible for all claims that are made. This raises the exposure of the company to greater risks. For self-insured companies, improvement in employees' health can save companies money; thus companies are interested in investing in employees' health.

According to a 2010 research brief by the National Institute for Health Care Reform, self-insured employers' interest in health promotion and wellness programs has increased dramatically in recent years. This may be due in part to passage of the recent health care reform called the Patient Protection and Affordable Care Act, which includes support for businesses to expand wellness programs and services to employees. The report indicates that many employers implement wellness programs to reduce direct medical costs or to boost productivity and reduce indirect medical costs such as disability claims. Increasingly, employers view health promotion programs as an effective business strategy to improve the reputation of their company and enhance their ability to attract and retain the best employees (Tu & Mayrell 2010).

As an example, the grocery store chain Safeway Inc. has created a solution to providing its employees with health insurance. In a *Wall Street Journal* OpEd piece, CEO Steven Burd (2009) describes Safeway's Healthy Measures program as a completely voluntary insurance plan that covers 74% of the insured nonunion work force. As with most employers, Safeway's employees pay a portion of their own health care through premiums. However, in the Healthy Measures program, the amount of that premium is reduced based on the results of tests associated with their tobacco use, weight, blood pressure, and cholesterol levels. To further encourage healthy behaviors, the company has financial incentive programs in place to promote cost-conscious health care decisions. Additional components of the program include a

> ► Employers have found that health and fitness programs reduce health care costs, increase productivity and morale, decrease absenteeism, and improve their corporate image.

24-hour information line and health advisor, wellness and lifestyle management programs, healthier options for food, large employee discounts for healthier food, and state-of-the-art fitness facilities at the corporate office.

Commercial Settings

▶ Because commercial health and fitness facilities need to make a profit, they compete for customers, and this requirement encourages innovative programming. Commercial health and fitness facilities are either sales based or retention based.

Any description of the commercial setting must begin with an acknowledgment that many types of businesses fit into this category and not all of them offer quality health and fitness activities delivered by qualified personnel. Fortunately, one of the most significant trends in the commercial setting during the past decade has been an increase in the number of high-quality health and fitness services being offered to members. Commercial health and fitness facilities typically operate with the objective of generating a profit. The manner in which the facility generates this profit places it in one of two distinct categories: **sales-based facilities** and **retention-based facilities**. The following sections offer a comparison of these two business models.

Commercial facilities offer health and fitness programs similar to those offered in the workplace. Popular programs include fitness assessment and health risk screening, nutrition counseling, personal fitness training, massage therapy, group exercise classes, spa services, wellness education classes, and worksite facility and program management. In addition, specialized exercise programs for children, prenatal and postnatal women, older adults, people needing physical therapy, and those with chronic disease conditions have increased in popularity. Commercial facilities also cater to niche markets such as women's fitness, athletic training, private training, dance, and yoga. The nature of a for-profit business encourages innovation as a way to compete for customers in the marketplace. As a result, the commercial health and fitness business is dynamic and offers many opportunities for health and fitness professionals to become involved with a variety of programs.

Sales-Based Facilities

Membership sales activities are the major focus in a sales-based health and fitness facility. These facilities rarely place limits on the number of memberships sold and often offer highly discounted rates as a way of attracting a large number of new customers. Historically, another characteristic of a sales-based facility has been the sale of long-term membership contracts. Within the industry, the philosophy under which sales-based facilities operate is often referred to as a "future service contract." To receive the best possible rate, new members must make a long-term commitment in advance rather than pay for their membership monthly. Because of this approach, sales-based businesses may not be highly motivated to meet members' long-term needs or focus on retention activities such as offering high-quality programs run by qualified personnel. Instead, businesses often allocate extensive resources to advertising campaigns that offer very low joining rates. Despite this, sales-based facilities might suit people who can regularly exercise at hours when the club is not full or who do not need extensive assistance from expert personnel. Many of these facilities occupy a useful niche in the marketplace by providing low-cost access to fitness equipment and indoor facilities.

Sales-based fitness facilities present two major barriers to improving the health and fitness of the general population: their marketing practices and sales strategies. These types of facilities may use billboards, TV, internet, and print advertisements featuring slender, muscular young models, suggesting that exercise is only for those who are already fit and healthy. The result of such advertising is that people who would benefit most from health and fitness programs—those who are older, have special health concerns, or need to lose weight—may feel discouraged from joining. Another challenge is that sales-based facilities may lure prospective members into facilities through the use of manipulative and deceptive sales strategies that offer substantial membership discounts in exchange for long-term commitments. Customers who make a long-term purchase and whose needs are not then met can have a negative first experience with fitness that discourages them from getting involved

■ What barriers does this fitness facility present to potential clients?

with other facilities or programs in the future. The U.S. economic recession that began in 2008 has increased competition between fitness clubs, and some sales-based facilities have been forced to stop requiring long-term contracts.

Retention-Based Facilities

Selling memberships is an important aspect of any commercial facility, but retention-based facilities have a financial stake in making a sincere effort to meet the long-term needs of current members. These types of commercial facilities typically focus on delivering high-quality programs and services so that current members will continue their membership month in and month out and will be more likely to recommend the facility to others. Retention-based facilities use what many in the industry refer to as a "voluntary dues" philosophy—that is, people pay dues voluntarily from month to month rather than pay one large membership fee up front.

Because retention-based facilities may not invest in extensive advertising, referrals from current members are essential to achieving the goal of profitability. Retention-based facilities place maximum limits on the number of memberships sold, usually based on a ratio of members to available space. If the membership base is at maximum, staff engage in sales activities only to replace members who have left the club for reasons such as dissatisfaction, relocation, or other changes in life circumstances. The greatest emphasis in a retention-based business is on high-quality facilities and programming that meet members' needs. Because of the additional cost of offering quality programs and hiring qualified personnel, monthly dues in retention-based facilities tend to be higher than those in sales-based facilities.

▶ The credibility of the commercial fitness industry has improved during the past decade as advertising has shifted to provide a more balanced image of health and fitness that includes all people regardless of age, race, ethnicity, body type, or current fitness level.

Unfortunately, retention-based facilities typically engage in less advertising than sales-based facilities and therefore limit the public's exposure to a credible and professional image of the fitness industry. Such facilities generally depend on word-of-mouth referrals to market their services, leaving the advertising market to the sales-based facilities. As a result, many people have an incomplete image of the commercial health and fitness industry.

Many exciting and higher-paying job opportunities available in the health and fitness professions are in the commercial sector. Recently, retention-based facilities have expended more effort getting the word out to the public regarding the high-quality health and fitness programs they offer, which has increased employment opportunities in this sector.

Clinical Settings

The **clinical setting**, which includes hospitals, outpatient medical facilities, and physical therapy clinics, has been a large growth area for health and fitness programs in the past decade. Today, administrators in clinical settings recognize the value of providing health and fitness programs. The objective of these programs is to help each patient manage medical conditions and keep the subscribers of their health insurance partners and programs healthy in order to avoid expensive medical procedures in the future. Increasingly, many of these facilities also offer memberships to the public similar to those offered by commercial clubs.

Examples of popular clinical programs are preventive screenings, physical therapy, cardiac rehabilitation, chronic disease management, water exercise therapy, childbirth and parenting education, weight management, and nutrition counseling. By definition, fitness professionals working within clinical settings will have more contact and work in partnership with both medically based professionals and clients who have diagnosed medical conditions.

Current trends in the clinical setting are not unlike those in the worksite and commercial settings. Organizations in clinical settings are also building fitness facilities and offering a wide range of health and fitness programs delivered by qualified staff. In addition, these services and programs extend into other dimensions of health besides physical fitness. For example, Eastern Connecticut Health Network provides health and wellness services reflecting the philosophy that the hospital exists not just to treat illness or injury but to help people stay well and engage in lifestyle behaviors that maintain health throughout life.

ACTIVITY 13.2

Settings
Test your understanding of the four types of health and fitness settings in this activity in your web study guide.

Much like worksite settings, some clinical settings have developed partnerships with commercial facilities or agencies that provide expertise in the management of health and fitness facilities. This combination of commercial settings with expertise in facility and program management and clinical settings with expertise in specialized health services has become a popular business model that presents qualified college graduates with new job opportunities. One company that has been very active in partnering with hospitals to design and manage health and fitness facilities is MedFit Partners, based in Evanston, Illinois. This company's many hospital partnerships include the Wilfred R. Cameron Wellness Center of the Washington Hospital in Washington, Pennsylvania. Another example is the Dedham Health and Athletic Club in Dedham, Massachusetts. This 240,000-square-foot facility has developed successful partnerships with several local medical providers, including a diabetes management partnership with the Joslin Clinic and a physician referral program for individuals with diagnosed medical conditions such as hypertension, obesity, and depression.

Community Setting

The **community setting** includes many types of health and fitness facilities. Local branches of nonprofit organizations such as the YMCA, the YWCA, the Jewish Community Center (JCC), and city parks and recreation departments make up one of the largest segments of the community setting. Another large segment of this setting is local health departments, state agencies, and voluntary health agencies such as the American Heart Association, the American Lung Association, and the American Diabetes Association. Community settings also include health and fitness facilities and programs that operate in churches, universities, and apartment complexes. Health and fitness programs offered in community settings may be funded by government grants or other philanthropic donations and may be connected to specific research projects.

Typically, health and fitness programs in community settings seek to fill a specific community need. Most target a specific community group and attempt to reach as many individuals in that group as resources will allow. Many of these facilities receive outside funding and are classified as nonprofit organizations for tax purposes, allowing them to offer health and fitness programs at low cost and to provide financial assistance to individuals who could not otherwise afford to participate.

Commercial Fitness With a Medical Mission

In March 2005, Augie Nieto, the cofounder of Life Fitness, the world's largest commercial manufacturer of fitness equipment, received a diagnosis of ALS (amyotrophic lateral sclerosis, or Lou Gehrig's disease). ALS is a progressive disease that destroys the nerve cells controlling muscles, ultimately causing complete paralysis. Under the banner Augie's Quest, Augie has since dedicated his time and energy to fundraising and research in order to find more effective treatments and, ultimately, a cure for ALS. The International Health, Racquet & Sportsclub Association (IHRSA), the trade association of high-quality commercial clubs, has assisted Augie by hosting an annual fundraising gala at its annual conference and trade show. Many IHRSA member clubs have also participated via their own fundraising programs. Since 2006, Augie's Quest has raised more than $30 million. Through this funding and a partnership with the ALS Therapy Development Institute, scientists at the ALS Therapy Development Institute are searching for the elusive treatment and cure. To learn more about Augie's Quest and the way in which a health and fitness community can come together in innovative ways, visit www.augiesquest.org.

The types of programs offered in community settings are as diverse as those offered in worksite, commercial, and clinical settings and might include parks and recreation programs for seniors, community cardiac rehabilitation programs, strength training, group exercise, and yoga classes. Resources are typically more limited in the community setting, leading to greater reliance on volunteers or minimally trained employees to meet many staffing needs. In addition, some programs are offered on a for-profit basis as a way of generating revenue to deliver other community service programs at little or no cost to participants. For example, many YMCAs sell health club memberships to the public as a method of generating revenue to offset the cost of nonprofit programs such as youth summer camps or programs for individuals with special needs.

Community-based nonprofit organizations are not the only entities that offer community health and fitness programs. The mission to improve the health of the community has crept into worksite, commercial, and clinical settings in recent years. Although they do not typically operate as nonprofit entities, an increasing number of organizations in these settings have expanded their missions to include contributing to the health of surrounding communities. The motives behind these activities are varied and can include a desire to help, improve public relations in the community, offer volunteer opportunities to employees, reduce tax exposure, or improve long-term business strength. Each year, the International Health, Racquet & Sportsclub Association (IHRSA) recognizes the importance of community service by awarding an outstanding community service award to a member who has made a longstanding commitment to making a difference in his or her community.

In summary, each of the four settings for health promotion programs has specific objectives, targets a specific population, and offers a particular menu of health and fitness programs. As a professional who may one day seek employment in one of these physical activity settings, you should look at each organization on its merits and not place too much emphasis on the distinctions between worksite, commercial, clinical, and community settings. As we have shown, these settings overlap considerably, and all are rapidly expanding the scope of the health and fitness programs they offer. According to IHRSA's *Profiles of Success* (2011), more than 50 million people are members of facilities that IHRSA categorizes as commercial, not-for-profit (JCCs, YMCAs, hospital-based clubs, residential, municipal, university, and military), and miscellaneous (corporate fitness centers, resorts, spas, hotels, and country clubs). Many exciting professional jobs are available in each setting. Your ideal job will be the one that best matches your goals as a professional.

Roles for Health and Fitness Professionals

The types of jobs available in a health and fitness facility depend more on the specific programs being offered than whether the facility is in a worksite, commercial, clinical, or community setting. The number of organizations offering high-quality health and fitness programming in each type of setting continues to increase annually, and many types of facilities now offer similar programs.

In this section we describe the types of jobs that are available across the four different health and fitness settings. More important, we explain how these jobs have evolved and continue to evolve in response to social forces and market demands.

Group Exercise Instructor

Once called "aerobics instructor," the position of group exercise instructor has changed greatly over the past decade. Group exercise instructors continue to lead aerobic exercise classes, but now they serve a broad population and provide instruction on a variety of activities. Group exercise instructors teach outdoor activities, aquatic fitness, and exercise classes for specific populations such as prenatal and postnatal women, seniors, children, and clients who need medically supervised exercise. The aerobics studio is no longer an empty room with mirrors. Group exercise instructors use a variety of functional equipment such as steps, slides, spinning bikes, exercise bands, dumbbells, fitness and exercise balls, and barbells. In most places, this position is a part-time job that is paid on an hourly basis.

The aging of the American population has brought an increasing number of older and higher-risk clients into the group exercise setting. What was once an aerobics studio filled exclusively with young adults performing complicated dance routines is now a studio filled with a diverse population. Group exercise instructors need to be knowledgeable and dynamic, have excellent leadership skills, and enjoy working with people in a group setting.

A trend in many health and fitness settings has been the emergence of yoga, tai chi, and Pilates. These activities are often combined to form a broader mind–body approach to physical activity. The Pilates programs available in most settings are vastly different from the form taught by originator Joseph Pilates. Although many instructors have formal training, other practitioners now have backgrounds in a variety of physical therapeutic, massage, and chiropractic techniques. Much Pilates training occurs in a one-on-one or very small group setting. Training usually involves use of a highly specialized apparatus and dedicated studio space. But given the culture of group exercise programming in many health and fitness settings, larger-group Pilates-inspired activities that use mats, core boards, rotator disks, and resistance rings rather than large specialized equipment are now common. Whereas Pilates is a relatively modern movement form, yoga has been around for centuries. Physically demanding, yoga is an inward practice meant to unite the body, mind, and spirit. According to Jason Crandell, editorial advisory board member of *Yoga Journal*, most American practitioners gravitate toward hatha yoga, the physical practice of postures (asanas), breathing exercises (pranayama), and relaxation techniques. However, because yoga is an ancient movement form, there are many practices throughout the world. Although there are no national standardized educational requirements for teaching yoga, training with established and respected gurus, often through generational lineage and affiliation, is common. A review of the popular *Yoga Journal* will lead you to numerous organizations offering workshops for beginning, intermediate, and advanced practitioners.

Exploring New Careers

As you read this section, compare the job descriptions to the list that you created in activity 13.1. What positions can you add to your list? What job duties are different from those that you imagined? We hope that you will become as excited as we are about the numerous possibilities that exist for health and fitness professionals.

Fitness Instructor

The fitness instructor position is predominantly an entry-level position for the fitness professional who may have practical experience with fitness or who is in the process of earning a bachelor's degree in kinesiology or other discipline related to health and fitness. Fitness instructors work one on one or in group settings with primarily healthy adults, although as the population ages, it has become more common for people with some functional limitations and chronic disease conditions to seek assistance in physical activity programs. Job duties typically include orienting clients to the fitness equipment, designing basic exercise programs, and leading groups through basic exercise routines. Some fitness instructors stay in touch with participants as they progress with their programs. In other facilities, professionals such as health and fitness specialists, wellness coaches, or personal trainers provide follow-up appointments. The fitness instructor position can be either a part-time or a full-time job, and compensation is usually an hourly wage. Many organizations offer certifications for entry-level fitness instructors including the American Council on Exercise (ACE) and the American College of Sports Medicine (ACSM). Some facilities do not employ fitness instructors and instead hire only individuals who have completed their degree in a health and fitness related discipline or have advanced training and specialized skills.

Health and Fitness Specialist

The health and fitness specialist position is typically a professional position for individuals who have completed a bachelor of science degree in kinesiology or exercise science and additional training and certifications. Job duties include conducting fitness and functional assessments and designing comprehensive individualized exercise programs that incorporate strength, flexibility, and aerobic fitness components and maximize safety and long-term results. This job provides opportunities to introduce diverse individuals to the benefits of exercise. For many sedentary adults who are joining their first health and fitness facility, the health and fitness specialist is the initial contact person who can either excite or discourage clients in their endeavor to become more healthy and fit. Most reputable health and fitness facilities provide a health screening or fitness assessment for new participants. The health and fitness specialist is responsible for assessing such things as blood pressure, body composition, aerobic capacity, muscular strength, muscular endurance, and flexibility. The results of these assessments are then used to design individualized programs that meet the client's specific health and fitness needs.

▌ Assessment results are used by a health and fitness specialist to design individualized programs.

As highly qualified professionals, health and fitness specialists also design exercise programs for individuals with specialized needs. The American College of Sports Medicine offers specialty certifications for individuals who have a bachelor's degree, health- and fitness-related certification, and experience. These specialty certifications include the ACSM/ACS Certified Cancer Exercise Trainer, who administers assessments and develops exercise programs for individuals with cancer; ACSM/NCPAD Certified Inclusive Fitness Trainer, who provides adapted exercise training to individuals with disabilities; and Physical Activity in Public Health Specialist, who works in partnership with government agencies and other organizations to plan and coordinate physical activity interventions at the community level.

With the increase in childhood obesity rates, health and fitness specialists are called upon to develop programs targeted specifically at children and adolescents. These programs center on the unique needs of children, focusing on play while emphasizing a level of movement and activity that allows for conditioning, caloric expenditure, and skill development. Fitness professionals who have training in elementary school physical education should find an increasing number of job opportunities across all types of health and fitness settings.

Wellness Coach

The wellness coach position represents an exciting growth area in the health and fitness professions. In some facilities this position has replaced that of the fitness instructor or health and fitness specialist. In others, it has led to the creation of a new category of employee. The wellness coach provides guidance to a diverse population on a broad range of health and fitness topics. Many of these jobs are full-time, salaried positions and include benefits.

Wellness coaches have job descriptions that are consistent with O'Donnell's (1989) definition of health promotion: "the art and science of helping people change their lifestyle to move toward a state of optimal health" (p. 5). As a result, job duties are broad and include working with clients on behavior change, stress management, relaxation techniques, time management, smoking cessation, and weight management, besides prescription of exercise programs.

Health screening and health and fitness assessment are important skills for a wellness coach, but coaches view these only as tools to assist them in working with clients. Coaches do not apply these techniques universally to all clients during an initial appointment, instead, wellness coaches assess clients' readiness for change (Prochaska & DiClemente 1986), help them set short- and long-term goals, and assist them in moving at an appropriate pace toward a state of optimal health. In addition to having job opportunities in commercial fitness facilities, wellness coaches may be hired by health insurance providers or worksites as a strategy to improve employee health and decrease medical expenditures.

ACTIVITY 13.3

What's My Profession?
Guess which profession is being described in this game in your web study guide.

Sutter Health Partners, a division of the Sutter Health network, runs an extensive Live Well for Life program in which wellness coaches develop lifestyle solutions for clients, patients, and employees who have specific risk factors such as diabetes, heart disease, high cholesterol, hypertension, and obesity. Sutter Health's wellness coaches have a myriad of undergraduate and graduate backgrounds in fields such as applied behavioral science, exercise and sport science, leisure and recreation, nutrition, gerontology, health education, business, sports medicine, health science, industrial and organizational psychology, and human development. Many have career training as nurse practitioners, registered dietitians, certified personal trainers, and group exercise instructors. With an emphasis on optimal health, the customized programs they design promote physical, emotional, intellectual, social, and spiritual health. As such, wellness coaches have an array of counseling skills to inform clients about their personal behavior and health risk factors, discuss obstacles, review lifelong habits, and provide personal guidance and unwavering encouragement to make lifelong changes.

Although educational credentials and certifications, along with constant continuing education, are essential for the wellness coach, equally important is the ability to provide counseling on a broad range of health topics. Wellness coaches need to be organized, understand principles of behavior change, have excellent communication skills, and be effective in marketing and promoting programs and services. Most importantly, they must know how to effectively engage people. Several training programs and certifications are available for wellness coaching. The American College of Sports Medicine has endorsed the Wellcoaches Certified Health and Wellness Coach certification. Additionally, the National Wellness Institute offers the Certified Wellness Practitioner certification.

Personal Trainer

Personal trainers provide many of the same services offered by both health and fitness specialists and wellness coaches but typically provide these services at a much higher fee-for-service rate. Some personal trainers work independently, traveling to individual clients' homes, conducting training sessions outdoors, or subcontracting with a health and fitness facility or private gym. Trainers may bill for their services hourly, weekly, or monthly. More typically, a facility employs personal trainers, and trainers market their services directly to members. The qualifications that a personal trainer needs vary dramatically between different types of health and fitness facilities. In some facilities, personal trainers are not required to have degrees or certifications. In other facilities, personal trainers are required to have not only a degree in their field and certification from a nationally recognized association but also specialized training and many years of experience. An increasing trend in commercial health and fitness facilities is the classification of personal trainers (e.g., from trainer to master trainer) based on skill, experience, sales volume, education, and certifications held. Usually, experienced trainers are able to charge higher rates for their services and therefore earn higher rates of pay from their facilities. Personal trainers are often compensated substantially in comparison with other health and fitness professionals, in part because of their unique job demands. Personal trainers work when their clients are available, which often means early morning, late evening, and weekends. Additionally, some facilities hire personal trainers as independent contractors, therefore obligating those trainers to operate their own businesses, pay their own taxes, obtain their own health insurance, and forego paid vacation and sick days.

Working as a personal trainer provides a unique opportunity for the health and fitness professional to develop ongoing relationships with clients. This relationship allows the trainer to more accurately assess individual clients' needs and quickly implement new strategies for behavior change. Personal trainers work with clients on a number of health issues such as weight management, chronic disease management, general fitness, and sport conditioning. Current trends in this position include specialization in working with specific populations, such as medically based clients, older adults, and children, as well as working toward specific client goals such as improving appearance, injury rehabilitation, running a 10K race, or completing a triathlon. Increasingly, personal trainers cater to these market segments by offering small-group specialty fitness programs as opposed to only individual services. In this regard, specialty fitness programs bridge the gap between one-on-one personal training and group exercise. Programs are for small groups (two to six people) for a defined time and usually focus on a specific goal (e.g., weight loss). These programs have gained in popularity because they allow people access to personalized training within a small group, making this service more appealing and affordable to a wider population.

Personal trainers need a variety of skills to be successful. Besides having a strong knowledge base in fitness assessment and exercise prescription, personal trainers must be excellent communicators and counselors; have strong business skills in sales, marketing, administration, and time management; and be effective in working with a diverse client population. Personal trainers are the artists of the health and fitness professions. Besides evaluating individual client needs and designing individualized programs, personal trainers must continually present new and exciting ways to keep clients on track, interested, motivated, and excited about making changes and maintaining new health behaviors as part of their lifestyle.

Over the past decade, the public has become more knowledgeable about the types of credentials and certifications that are appropriate for the personal trainer position. Simply being fit is no longer an acceptable qualification. Personal trainers need to pursue an interdisciplinary education, just as wellness coaches do.

Personal trainers are rarely hired directly from bachelor's degree programs unless they also have extensive experience working with clients on a variety of health and fitness issues. For this reason, an aspiring personal trainer likely will begin a career as a fitness instructor or health and fitness specialist and move into a personal trainer position after gaining considerable experience and a specialized certification. Personal trainers keep skills current by attending conferences, workshops, and other continuing education programs.

Case Study in Transitioning from School to a Work Setting

Jim was hired 4 weeks ago as a personal trainer at a large gym that specializes in training athletes. He earned his bachelor of science degree in kinesiology from an accredited program that emphasized the importance of fitness assessments, health screening, and safe exercise techniques. During his first few weeks on the job he observed that many of his fellow trainers were using techniques that appeared to be risky for their clients. He also noticed that few trainers were using the fitness assessments and health screening methods he learned in his degree program. He knew from reading the trainer biographies posted on the wall that many of the trainers in this facility did not have formal degrees in kinesiology. Most of their biographies described their athletic accomplishments and credentials and certificates for different types of training techniques. Jim is worried that his fellow trainers are being negligent with their clients and putting them at risk for injury. Furthermore, he worries that without doing fitness assessments the trainers are not able to provide tailored programs to meet the specific needs of each of their clients. Jim is thinking about approaching his fellow trainers and offering to teach them about proper weight lifting techniques and fitness assessment methods.

Your Decision

What do you think Jim should do in this situation?

Resolution

Jim approached his manager and expressed his concerns. His manager listened very carefully then explained that many of the trainers at the facility were not using standard fitness assessments because their clients were athletes who required specialized performance assessments in order to develop fitness improvements in areas that were specific to their sport. He also acknowledged that many exercises that were used in this facility would not be appropriate for average clients but were actually essential for elite athletes in order for them to prevent injuries from the unique demands in their sport. Jim's manager realized that it was important to provide more hands-on experience for Jim, so he assigned one of the senior trainers in the facility to serve as his mentor. He also signed Jim up to attend a workshop on functional training and assessments. Jim's manager appreciated some of Jim's suggestions and agreed that the facility needed to conduct more health screening for all athletes in the facility prior to their participation and asked Jim to help him with that task. Jim's manager also said that he would arrange a time for Jim to provide an in-service training for the other trainers to update them on body composition assessment techniques he had learned in his degree program. He told Jim that he really appreciated how Jim had come directly to him with his concerns rather than criticize the other trainers directly.

Specialty Positions

Many health and fitness settings are now incorporating specialty positions into their programs, including the following (see also chapter 14):

- Clinical exercise physiologist
- Physical therapist
- Athletic trainer
- Sport coach
- Registered dietitian
- Population specialist (e.g., working with older adults, children, people with chronic disease)
- Public health educator

Many specialty positions require graduate-level studies, certifications, or licensure along with a bachelor's degree in kinesiology or another health- and fitness-related discipline. Specialty positions provide services that go beyond those provided by other health and fitness professionals. These jobs can be exciting and challenging and allow for increased earning potential.

What are the differences between health and fitness professionals and specialty positions? Consider the provision of nutrition information to clients in a health and fitness facility. Health and fitness specialists, wellness coaches, or personal trainers are often called on to provide basic nutrition advice to healthy adult clients. These professionals would not be qualified, however, to prescribe a specific type of diet or supplement or make recommendations to individuals who have specific health conditions. Only registered dietitians have the appropriate qualifications to make such recommendations. Attaining that status requires additional course work beyond what a kinesiology department typically offers.

Injury rehabilitation provides another example of the boundaries between various professionals and specialty positions in a health and fitness facility. Personal trainers are required to obtain a full injury history on each client and to consider this information when designing an exercise program. An appropriately designed exercise program can help people prevent injuries in the future. A personal trainer, however, should not design a program for specific injury rehabilitation. Physical therapists and athletic trainers are the specialists with the appropriate knowledge and qualifications to provide injury rehabilitation services to clients. In an ideal scenario, a personal trainer works in partnership with a physical therapist or registered dietitian. The trainer may supervise specific aspects of a program and communicate with the specialist regarding the client's progress.

Many different specialty positions are focused on particular populations, some of which are listed in this chapter as additional certification options under the health and fitness specialist job title. With the recent addition of specialized certification programs in cancer exercise, inclusive training, and public health, the American College of Sports Medicine has recognized the importance of requiring training beyond the bachelor's degree for professionals who work with specific populations who may have special needs related to exercise training.

▌ There are many careers in the health and fitness field.

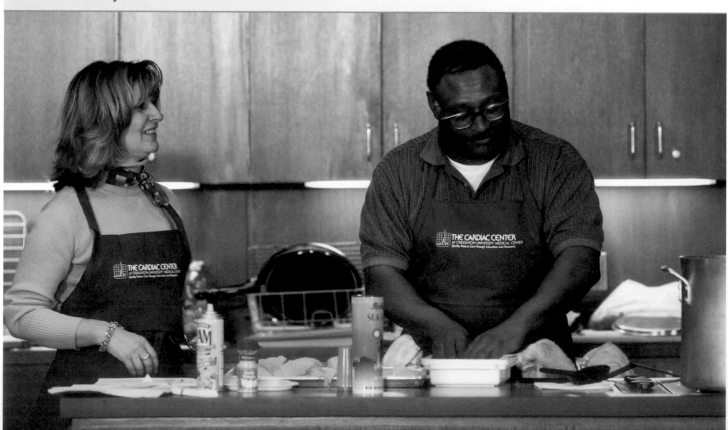

There are also new opportunities for professionals who can navigate the legislative process; partner with local, state, and federal agencies; design, implement, and evaluate programs; and work to effectively promote physical activity as an essential public health component. Public health educators have the training necessary to fill this role and design programs for implementation at the community level. Job settings for public health educators include local health departments, nonprofit organizations, schools, colleges, hospitals, and health insurance companies. The National Commission for Health Education Credentialing provides the Certified Health Education Specialist (CHES) credential for those wishing to pursue the public health educator position.

As mentioned earlier, many specialty positions require graduate-level education in addition to a bachelor's degree in kinesiology or another health- and fitness-related discipline. If you are interested in pursuing a specialty position, check with your academic advisor to find out how you can best prepare for an appropriate graduate program.

Health and Fitness Director

The health and fitness director is a key position in most health and fitness settings. This individual is typically responsible for managing the health and fitness services and programs in a facility. In larger facilities, the health and fitness director may be part of the general management team and supervise a team of managers who handle specific responsibilities within departments. Examples of these management positions include directors for group exercise, yoga, Pilates, personal training, aquatics, athletics and recreation, and youth activities. You should consider all these management positions as job opportunities for health and fitness professionals.

> Health and fitness professionals must understand the ways in which social and cultural forces affect their industry.

The job description for a health and fitness director position includes hiring, training, and providing support for a diverse staff of group exercise instructors, fitness instructors, health and fitness specialists, wellness coaches, personal trainers, specialty positions, and equipment maintenance personnel. In addition, health and fitness directors are responsible for business planning, managing budgets, planning facility renovations, selecting equipment, designing and marketing programs, and forecasting future trends in health and fitness programs. The health and fitness director must have a broad, interdisciplinary education that includes business planning, budgeting, marketing, and staff management as well as a strong foundation in kinesiology.

Educational requirements for the health and fitness director position emphasize academic degrees more than certifications. A degree in kinesiology is especially appropriate for students aspiring to be health and fitness directors. Many facilities require health and fitness directors to have a graduate degree in kinesiology or another health- and fitness-related discipline. Other requirements include experience in a position such as health and fitness specialist or personal trainer. Through experience in such positions, directors gain a full understanding of individualized program design and strategies for successful behavior change. In addition, these job experiences provide valuable insight into the management responsibilities of hiring, training, and supporting staff.

Health and fitness directors need to be effective leaders. Besides managing day-to-day operations, these individuals must be capable of articulating a vision and motivating other individuals to work toward specific goals. The health and fitness director is both a visionary who keeps current with the field of health and fitness and a mentor who supports staff members in moving toward individual goals and aspirations.

Certification and Continuing Education

To be hired for a position in the health and fitness professions, you will need to fulfill specific requirements in your education and training. Once you are hired, you will need to continue seeking information and developing new skills, because the health and fitness field is dynamic, and new information is continually becoming available.

Typically, corporate, commercial, and community organizations prefer to hire health and fitness instructors, health and fitness specialists, and personal trainers with a minimum of a bachelor's degree in kinesiology or a health- and fitness-related discipline. Indeed, many organizations require this degree. This is a positive move because club members, employees, and patients are demanding high-caliber educated professionals who have extensive knowledge of health and fitness. However, many talented professionals remain in the field who do not have kinesiology degrees but do have extensive work experience and a great deal of continuing education and practical training. In either case, most employers now also require individuals to have a certification from a nationally recognized organization.

The most widely recognized and respected certifications are from the American College of Sports Medicine (ACSM), American Council on Exercise (ACE), National Academy of Sports Medicine (NASM), National Strength and Conditioning Association (NSCA), National Federation of Professional Trainers (NFPT), Cooper Institute (CI), and National Council on Strength & Fitness (NCSF). Each of these organizations has received third-party accreditation of its certification procedures and practices from the National Commission for Certifying Agencies (NCCA).

Several educational certificates are recognized by health and fitness organizations when hiring and promoting employees. The Gray Institute for Functional Transformation (GIFT) fellowship is a unique 10-month program in applied functional science at the Gray Institute in Adrian, Michigan (www.grayinstitute.com). In a strategic partnership with the American College of Sports Medicine (ACSM), the Wellcoaches corporation is attempting to set the gold standard for the newer profession of wellness and lifestyle coach (www.wellcoaches.com).

You must do your research when applying for health and fitness positions so that you will know what educational certifications and programs specific organizations require, recommend, and support. In addition, certification organizations require you to continue your education by taking recognized continuing education units (CEUs) in order to maintain that certification. It is professionally essential to keep up to date with new scientific material long after the completion of your undergraduate degree.

Joining a professional association is an important first step. Many associations offer discounted membership rates for students; provide journal subscriptions, educational materials, and training; and sponsor regional and national conferences. Important organizations in this profession are listed in this chapter. New Internet-based companies also offer valuable networking opportunities for professionals in the field (e.g., www.ptonthenet.com). Finally, various governmental agencies within the U.S. Department of Health and Human Services (www.hhs.gov) provide a wealth of health data, materials, and resources for health and fitness. To find out about membership, publications, workshops, and regional and national conferences, contact one of the organizations listed on page 385.

Trends and Opportunities in Health and Fitness

Across every industry, the best leaders are those who read extensively about many fields and keep up to date on social and cultural trends and the latest research in their discipline. They continually ask questions about how the world around them affects their particular organization. Now that you have an understanding of the various positions available in the health and fitness professions, you should broaden your view by adding an understanding of the trends that affect these professions.

As a future health and fitness professional, you should be aware of the ever-changing context in which these professions are immersed. This is a dynamic time in the health and fitness field; personal, social, cultural, and political issues have a powerful influence on the way we view health and fitness—and on the way we work in the profession. Although it is beyond the scope of this chapter to outline all these forces, we deal with four major issues currently influencing these professions: increasing interest in a multidimensional model of wellness, health care reform, emerging demographic trends, and the rise of noncommunicable diseases.

Checking Skills and Competencies Required in the Health and Fitness Profession

Take one final look at the list of job titles, job duties, and required skills and competencies that you created in activity 13.1. Now compare your list with that provided in the table of job descriptions. How did your list of knowledge and skills compare? Although this list is by no means complete, it illustrates that health and fitness professionals must have a broad range of skills and competencies.

Job title	Evolution	Job duties	Skills and competencies
Group exercise instructor	The group exercise instructor position no longer exclusively involves teaching aerobic dance classes. The group instructor now teaches a broad range of classes to a diverse population of individuals.	Lead a wide variety of group exercise classes for various population groups.	Musts: Certification by a nationally recognized organization Strong teaching skills Helpful: Additional certifications or training in specialized areas
Fitness instructor	The fitness instructor is an entry-level position in many health and fitness facilities.	Orient clients to exercise equipment and design basic programs.	Musts: Certification by a nationally recognized organization Strong teaching skills Preferred: Bachelor's degree in a health- and fitness-related field (e.g., kinesiology, physical education, or health education) Helpful: Additional certifications or training in specialized areas
Health and fitness specialist	To meet the needs of a changing demographic of clients who may have health concerns, many facilities now require their fitness professionals to have formal degrees in their field and nationally recognized certifications.	Design customized exercise programs to meet the needs of diverse participants. Conduct fitness assessments for the purpose of informing program design and monitoring progress. Provide expert guidance to assist clients in meeting their personal exercise related goals. Design and implement programs.	Musts: Bachelor's or graduate degree in a health- and fitness-related field Certification by a nationally recognized organization Preferred: Skills and experience with fitness testing , exercise prescription, and program planning Helpful: Additional certifications or training in specialized areas

(continued)

Checking Skills and Competencies Required in the Health and Fitness Profession *(continued)*

Job title	Evolution	Job duties	Skills and competencies
Wellness coach	Having evolved from the more traditional fitness instructor position, the wellness coach provides counseling on a broad range of health topics in addition to conducting fitness assessments and designing exercise programs.	Provide guidance to a diverse population of individuals in areas such as behavior change, stress management, smoking cessation, weight management, and exercise programming.	Musts: Bachelor's or graduate degree in a health- and fitness-related field; Certification by a nationally recognized organization; Skills in counseling, behavior change, cultural diversity, teaching, and exercise prescription; Preferred: Specialized wellness coach certification; Marketing and promotional skills
Personal trainer	This position has changed from exclusively providing individualized exercise programs to providing services on a broad range of health topics to both individuals and small groups.	Provide ongoing support and guidance to a diverse population of clients on topics such as physical fitness, weight management, stress management, and sport conditioning.	Musts: Certification by a nationally accredited organization; Counseling and teaching skills; Business, marketing, sales, and promotion training; Skills and experience with fitness testing and exercise prescription; Preferred: Bachelor's or graduate degree in a health- and fitness-related field; Helpful: Additional certifications or training in specialized areas
Specialty positions	Number and types of positions have expanded in recent years and moved from exclusively clinical settings to a variety of health and fitness settings.	Provide specialized health and fitness services to clients with special needs. Examples include physical therapist and public health educator.	Musts: Graduate degree may be required. Other specific experiences and skills required for each type of specialist position; Preferred: Additional certifications and licensure—may be required to practice in specific states
Health and fitness director	This position has evolved from a fitness instructor who has additional administrative responsibilities to a full-time manager who actively participates in all aspects of personnel and facility management.	Manage all aspects of a health and fitness department. Responsibilities include departmental leadership, staff management, program planning, and all aspects of business administration.	Musts: Bachelor's or graduate degree in a health- and fitness-related field; Additional skills in business administration, management, marketing, and promotion; Preferred: Previous experience as a health and fitness specialist or similar position

ACTIVITY 13.4

Dimensions of Wellness
See if you can identify the five dimensions of wellness in this activity in your web study guide.

Multidimensional Model of Wellness

A trend across all health and fitness settings is the expansion of program offerings to address other dimensions of wellness in addition to physical fitness. The multidimensional model of wellness has five dimensions: (1) physical—the ability to conduct daily tasks with vigor; (2) emotional—the ability to control emotions and express them appropriately and comfortably; (3) intellectual—the ability to learn, grow from experience, and use intellectual capabilities; (4) spiritual—a guiding sense of meaning or value in life that may involve belief in some unifying or universal force; and (5) social—the ability to have satisfying interpersonal relationships and interactions with others (Dintiman & Greenberg 1986) (see figure 13.2). Our previous description of clinical health and fitness settings included an example of a hospital group that offers a wide variety of wellness programs and services. Besides offering exercise programs, this hospital provides classes on topics such as meditation, parenting, and self-esteem. In our description of worksite settings, we discussed ways in which worksite programs have evolved to provide services that not only assist employees in making changes but also support them in maintaining healthy lifestyle behaviors. Besides offering employees state-of-the-art fitness facilities, many worksites now offer family activities, volunteer service opportunities, and programs for facilitating personal growth and improving communication skills.

As facilities expand their program offerings, they will also expand their job descriptions. If you elect to focus exclusively on the physical dimension of wellness, you will continue to find many exciting job opportunities. But if you embrace the challenge of delivering services that address other dimensions of wellness, you are likely to find even more job opportunities.

As we have shown in our description of various job categories, many health and fitness positions have already expanded to involve a broader range of responsibilities than in the past. Wellness coaches and personal trainers continue to design exercise programs but also assist clients with stress management, personal growth, and involvement in social activities. In short, industry leaders in all settings are looking for bright people who can deliver a broad range of services and integrate concepts of exercise and fitness into a broader definition of health.

> Health and fitness programs increasingly address other dimensions of wellness besides physical fitness, drawing from a multidimensional model that includes physical, intellectual, emotional, social, and spiritual well-being.

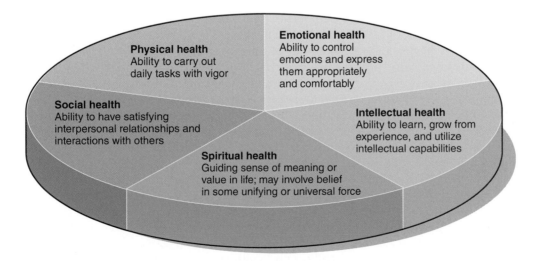

FIGURE 13.2 Dimensions of wellness.

ACTIVITY 13.5

Day in the Life

What does it take to become a personal trainer? What does a typical day look like? Find out in this activity in your web study guide.

ACTIVITY 13.6

Day in the Life

In this activity in your web study guide, you'll round out your understanding of what's involved in becoming a personal trainer and what a typical work day looks like. Compare Brian Bergerson's education, training, and typical day with that of Nick Walker, the physical therapist you met earlier.

Multidimensional Approach to Fitness and Healing

University of California, San Francisco (UCSF) cancer patients have the opportunity to meet with cancer exercise trainers certified by the American College of Sports Medicine. The trainers demonstrate simple exercises that can be done at home, design personalized exercise programs, and refer patients to community cancer-specific yoga and group exercise classes. The UCSF Exercise Counseling program is operated by the Ida & Joseph Friend Cancer Resource Center with a grant from the Mount Zion Health Fund. The exercise trainers, in association with UCSF's department of physical therapy and rehabilitation services, also have written a manual titled "Moving Through Cancer: A Guide to Exercise for Cancer Survivors." The manual is located on the patient education online library (www.ucsfhealth.org). Such cancer exercise programs continue to increase in popularity as new research studies confirm the relationship between physical activity and improved quality of life for individuals diagnosed with cancer (Bulmer et al. 2012).

Health Care Reform

As mentioned earlier in the chapter, health care costs have skyrocketed over the past decade. Here is an example on how the payment system for insured patients works: Mike, a long-time employee at his company, feels a pain in his right knee. Because Mike is an employee, his company pays the bulk of his monthly family health care insurance. He (the consumer) pays the difference by having money taken out of his paycheck before it is taxed every month. Because of his knee pain, Mike takes a sick day from work and is treated by a doctor at the local hospital. The provider (the doctor) then submits a claim to a third-party payer (a national insurance plan such as Aetna or a regional plan such as Blue Cross Blue Shield). The payer processes the claim based on the patient's insurance coverage and the payer's contract with the provider. Once the claim is approved, the payer sends its portion of the claim to the provider. Based on the amount covered by the payer, the provider then determines whether additional money is owed by the patient. Subsequently, Mike may receive a bill from the provider (doctor) that is the total cost of the treatment minus the amount covered by the payer (insurance plan). It is then Mike's (the consumer's) responsibility to pay the provider.

The total cost of health care in the United States is considerably higher than that of other industrialized nations, and experts warn that these costs are unsustainable. In their publication *Health Care Costs: A Primer*, the Kaiser Family Foundation (2012) provides the following facts: "In 2010 the U.S. spent $2.6 trillion on health care, an average of $8,402 per person; the share of economic activity (GDP) devoted to health care has increased from 7.2% in

1970 to 17.9% in 2009 and 2010; and half of health care spending is used to treat just 5% of the population" (p. 1). Despite the large dollar amount spent annually, many in the United States do not receive the health care they need. The U.S. Census estimates that there were 48.6 million uninsured Americans in 2011 representing 15.7% of the population (DeNavas-Walt et al. 2012). Clearly, health care reform is needed to deal with these enormous financial transactions in the system, and such reform must be significant and address health system redesign, coverage expansion, reimbursement reform, and insurance market reform.

Consequently, and with much fanfare and debate, President Obama signed into law the Affordable Care Act on March 23, 2010. The controversial law is set to put in place comprehensive health insurance reforms, with most changes taking place by 2014. For a more detailed timeline of the Affordable Care Act, visit www.healthcare.gov/law/timeline/.

Even with reform, it might be said that the best way to control health care costs is to stay healthy, and a good way to stay healthy is to adopt and maintain healthy lifestyle behaviors, including daily physical activity. But both the rising cost of medical care and the manner in which health care is paid for in the United States have slowed the progress toward offering more preventive services. Health care organizations are so busy developing cost-effective ways to treat people who are sick that they seem unable to focus on the development of better strategies for prevention. In addition, when employees leave their jobs, they tend to change health insurance providers. Health insurance providers will be motivated to invest money in preventive services only when they are confident that they will reap the benefits in future cost savings.

However, there is clearly a move in the right direction. Many health plans now offer an extensive menu of health and fitness programs free of charge to their members. In some cases, this effort has the goal of improving the overall health of potential subscribers in a geographic area. In other cases, it is a marketing strategy aimed at increasing the number of people who enroll in the health plan. Blue Shield of California was one of the first health plans to actively promote healthy lifestyles as part of its business model. Its online Healthy Lifestyle Rewards program helps people adopt healthy lifestyle choices related to diet, stress, and physical activity.

Our previous discussion covers the role of wellness coaches in the Sutter Health Partners program. Sutter Health, one of the nation's leading not-for-profit networks of community-based physicians, hospitals, and other health care providers, delivers high-quality care in more than 100 northern California communities.

As another example, American Specialty Health (the largest network of its kind in the United States) provides wellness, fitness, and specialty managed care programs to health plans, insurance carriers, and employers. This network provides lifestyle coaching, incentive programs, online wellness information, personal health risk assessments, exercise and walking programs, home fitness programs, worksite programs, Internet-based programs, and discounted access to alternative health care providers (e.g., chiropractors, acupuncturists, massage therapists, dietitians, and naturopathic doctors). This direct intervention by health networks and health plan providers certainly seems to be the wave of the future. For the first time, physical activity is increasingly seen as an integral component of the nation's health care delivery system. Quite possibly the model hospital of the future will have on-site fitness facilities and a staff of health and fitness counselors and personal trainers in addition to traditional medical personnel.

Health and fitness professionals must recognize that the addition of insurance-sponsored prevention programs is not the only answer. Creative solutions mean not just providing individuals with access to preventive programs and services but also ensuring participation. Encouraging people to adhere to good diets and other health habits and to adopt physically active lifestyles is just the beginning. Developing and implementing effective outcome-driven programs will be the key.

Demographics

Demographic trends are radically redefining the meaning of healthy lifestyles and healthy aging. According to the U.S. Census Bureau's 2010 data, the median age of the total U.S. population increased from 23 years in 1900 to 30 years in 1950, then to 35.3 years in 2000 and to 37.2 in 2010. In the 2010 census, the number of people under the age of 18 was 74.2

million (24% of the total population). People aged 18 to 44 represented 112.8 million (37%). Those aged 45 to 64 made up 81.5 million persons (26%). Finally, the 65+ population was 40.3 million persons (13%). Perhaps most striking is the rate of growth within each demographic group between 2000 and 2010 (figure 13.3). Whereas the growth rate for those younger than 18 years was 2.6% and only 0.6% for those between 18 and 44, the population aged 45 to 64 increased at a rate of 31.5% and at 15.1% for those over the age of 65 (Howden & Meyer 2011). Driving this increase is the aging of the baby boom generation, the 77 million people born in the United States between 1946 and 1964. For the past quarter century, this generation has had a huge effect on the marketplace because of its sheer numbers.

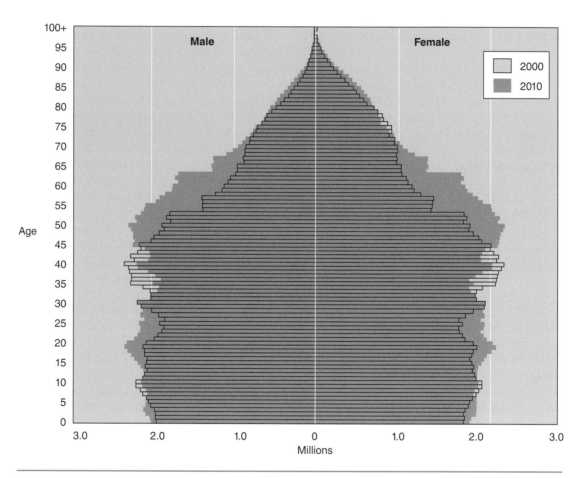

FIGURE 13.3   Population by age and sex, 2000-2010.
Reprinted from Howden and Meyer 2011.

What might this "age wave" (Dychtwald 1990) mean for the health and fitness fields? What does it mean when people say that baby boomers not only will grow older chronologically but will age youthfully? Will middle age be reinvented? And what will happen when future customers, clients, and patients retire at 65 years of age and then have another 25 years of living to do? Consider the fact that the number of people over age 85 is expected to increase to 9 million in 2030 and to 19 million in 2050. These demographic shifts will surely challenge our definition of physical activity as it relates to sport, competition, adventure, movement, pleasure, and health.

Many job opportunities are likely to arise because of the demographic patterns shaping the United States. We will all witness a far more diverse nation in the coming years. For instance, according to the 2010 census report, more than half of the 27.7 million increase in the U.S. population between 2000 and 2010 was due to an increase in the Hispanic population. But the fastest rate of growth during that time was in the Asian population, a remarkable growth of 43% (Humes et al. 2011).

Despite the increasing methodological, empirical, and theoretical difficulty of defining and categorizing the U.S. population into specific racial and ethnic groupings, such trends are having an enormous influence on the way many American health and fitness businesses are developing their strategic plans for the next decade. Many settings are already taking advantage of this demographic shift in their hiring, programming, and marketing practices. New kinesiology graduates will likely be working in multigenerational and multiracial work teams as the need for new allied health professionals continues.

Rise of Noncommunicable Diseases

Perhaps the most alarming trend that will drive future fitness and health industry growth is the incidence of noncommunicable diseases (NCDs). This is a global problem. In September 2011, the United Nations held a high-level summit on noncommunicable disease prevention and control. (This was only the second time in its history that the UN held such a summit; the first concerned AIDS.) At the 2011 summit, experts spoke on the alarming incidence of chronic diseases, or NCDs, throughout the world. Evidence shows that cardiovascular diseases, cancers, chronic respiratory diseases, and diabetes are the world's biggest killers, causing an estimated 36 million deaths each year—63% of all deaths globally. Every year, NCDs kill 9 million people under the age of 60. Apart from the tragic personal, familial, and public loss, the socioeconomic impact is staggering. Brazil, Russia, India, and China together lose more than 20 million productive life-years annually to chronic diseases; the number is expected to increase 65% by 2030. U.S. estimates are that nearly half of Americans are sufferers; $277 billion is spent on treatment and there is a $1.1 trillion impact on the economy through lost productivity. Canada, Australia, the United Kingdom, and other countries with an increasing older population, a disease-based medical model, and a shrinking taxation base are not immune (www.who.int/nmh/events/un_ncd_summit2011/en).

This problem is certainly too big to be solved by any one profession or industry. But it is one that may well come to define the field of kinesiology. And it will be a challenge because although interest in personal wellness, behavior change, and health and lifestyle improvement is high, government lifestyle indicators (such as Healthy People 2020) show that actual effort to improve and maintain health is low. Thirty years of wellness programs and education initiatives have been unsuccessful: Delivery has been based on risk management and clinical compliance. Behavior change either has not been achieved or has not been sustained. The missing component over the last few decades has been "sustainable engagement," and solving this challenge is the key goal for any employer, funder (insurer or government), or educational initiative. It will be one of the great challenges of our time, and you, as current kinesiology students and future allied health professionals, have an enormous role to play.

New Technology and Innovations in Health and Fitness Products

With the advent of cloud technology, there are myriad products and apps (largely emerging from the gamification industries [e.g., video games]) that measure our sleep, food intake, and exercise data. Consider how these products may affect the fitness and health industries now and in the future.

Advice for Health and Fitness Students

This is an exciting time for the health and fitness professions. An incredible transformation is currently under way as we move toward a broader understanding of health and fitness services. We are beginning to view physical activity as a key component in the health care delivery system. We are also experiencing a considerable demographic shift in population that challenges us to work with increasingly diverse groups of people.

Today's health and fitness professional needs to obtain a broad education across several scientific, behavioral, and humanity-based disciplines (see the job descriptions in the Checking Skills and Competencies Required sidebar). To develop the necessary skills and competencies, you should seek a combined degree in kinesiology and health with additional courses from a wide range of areas such as teaching; behavior change psychology; neuroscience; social and personal media; computer technology; gerontology; marketing; and the sociology of race, gender, and ethnicity. Course work in theoretical kinesiology (part II of this text) will be extremely valuable, and other specialized health- and fitness-related courses will round out your professional practice knowledge. You must also gain practical experience counseling clients on a variety of health and fitness topics and be able to relate to a diverse population of clients from different age groups and ethnic backgrounds. Involve yourself in a health and fitness job setting through an internship, cooperative education experience, or work-study program to keep up to date on what is happening in the field.

When you graduate from college, you will compete for jobs with people who have extensive practical experiences and insightful life experiences. For this reason you should obtain practical skills as part of your education. In addition, read the relevant health and fitness journals and industry publications. Most of these publications are available through the associations listed in this chapter. Many are available to students at substantially reduced rates, and some are even provided free of charge to health and fitness professionals. A number of online resources are available for health and fitness professionals who wish to share knowledge and ideas. Many have a membership subscription fee but offer continuing education workshops and provide workshops in the community (e.g., www.ptonthenet.com).

Given the competitive nature of the job market, you must maximize your educational experience while in college. Work with your advisor to create a degree plan that will prepare you for the current and future job market. Besides taking the courses required by your major, select elective courses that will prepare you for the type of job you want. Take advantage of as many internship and community service learning experiences as possible. In many cases, a degree in a health and fitness discipline alone will not be sufficient to get you the job you want. Besides having the credentials on paper, you must be able to demonstrate that you can apply the skills you have learned.

> Maximize your educational experience while in college, selecting a broad spectrum of elective courses in your degree program.

Newly graduated students often show up at interviews, excited, with degrees in hand, but unprepared for the types of jobs available. In some cases, students interviewing for health and fitness specialist positions report that they do not have any teaching experience because they chose an area of emphasis that deals strictly with the biophysical sphere of physical activity. Because teaching courses fall within another area of emphasis (i.e., the behavioral sphere of physical activity), these students didn't believe that they needed them. Don't limit yourself to one area of emphasis at the expense of gaining skills that will help you be successful in the job you want.

Take special care in choosing elective courses in your degree program. If your department does not offer a course in teaching or counseling skills, then inquire what departments on campus might. Speak to your advisor; perhaps a faculty member with teaching or coaching expertise can supervise you in an independent study. We can make a similar argument for classes based on business and marketing. Clearly, in our media-saturated culture, marketing skills are important for health and fitness professionals. After all, a big part of your job as a health and fitness professional will be to motivate people to take part in the programs that you offer and to promote the idea of making healthy lifestyle changes. Again, if your department does not offer a specific class in health and fitness marketing, look for another course on campus that provides general marketing information that you can apply to your discipline. What will matter most when you interview for a job is that you have knowledge and skills to do the job successfully. Do your homework early and make sure that you obtain the appropriate skills along the way.

 ACTIVITY 13.7

Interview with a Practicing Professional

Jessica Scibek is the manager of a university fitness center. She talks about her training and qualifications, responsibilities, what she finds most rewarding about her work, and more in this interview in your web study guide.

 ACTIVITY 13.8

Interview with a Practicing Professional

Brian Christy is a strength and conditioning coach. He discusses his path to success, what he finds most fulfilling about his job, the advice he'd give to students interested in pursuing a similar career, and more in this interview in your web study guide.

Wrap-Up

In this chapter we have described the health and fitness field and summarized the types of settings and jobs that are available for health and fitness professionals. In addition, we have provided some insight into current and future trends in health and fitness and described ways in which these trends affect the field. Return to the situations described in the chapter opening. Consider how those in health and fitness careers are directly responsible and able to assist each of the individuals described there. The world of health and fitness is ever changing, which is partly what makes this profession so dynamic and exciting. If you are considering a career in the health and fitness professions, learn as much as you can about them. It is never too early to begin to tailor your education to your personal passion. Consider industry trends as you complete this phase of what should be an ongoing educational process. This discussion about the health and fitness professions should serve only as the point from which you begin your process of discovery. Enjoy the journey.

More Information on . . . Careers in Health and Fitness

Organizations

American Alliance for Health, Physical Education, Recreation and Dance

American College of Sports Medicine

American Council on Exercise

American Public Health Association

IDEA Health & Fitness Association

International Health, Racquet & Sportsclub Association

Medical Fitness Association

Medical Wellness Association

National Commission for Health Education Credentialing

National Strength and Conditioning Association

National Wellness Institute

Society for Public Health Education

Wellness Coaches USA

Journals

ACSM's Health & Fitness Journal
American Journal of Health Behavior
American Journal of Health Education
American Journal of Health Promotion
American Journal of Public Health
Exercise and Sport Sciences Reviews
Health Education & Behavior
The Health Educator
Health Promotion Practice
Journal of Nutrition
Journal of Physical Education, Recreation and Dance
Journal of Strength and Conditioning Research
Medicine & Science in Sports & Exercise
NSCA's Performance Training Journal
Nutrition Journal
Research Quarterly for Exercise and Sport
Strength and Conditioning Journal
Women in Sport and Physical Activity Journal

 ACTIVITY 13.9

Use the key points review in your web study guide as a study aid for this chapter.

Review Questions

1. Identify program objectives, and target markets, for each of the following health and fitness settings: (a) worksite, (b) commercial, (c) community, and (d) clinical.

2. Compare and contrast sales-based and retention-based commercial health and fitness business models.

3. Describe typical job duties for the following health and fitness positions: (a) group exercise instructor, (b) fitness instructor, (c) health and fitness specialist, (d) wellness coach, (e) personal trainer, and (f) health and fitness director.

4. How will demographic trends in the U.S. population (e.g., aging baby boomers) affect the health and fitness professions?

5. List the knowledge, skills, and abilities that you need to obtain to be competitive in the health and fitness job market.

6. List specific physical activity-related companies, programs, equipment, resources, and services now available that were created in the past 10 years. Forecast what new companies, programs, equipment, resources, and services might be developed in the next 10 years.

chapter fourteen

Careers in Therapeutic Exercise

Chad Starkey

chapter objectives

In this chapter we will

- acquaint you with the wide range of professional opportunities in the sphere of therapeutic exercise;

- familiarize you with the purpose and types of work done by professionals in therapeutic exercise;

- inform you about the educational requirements and experiences necessary to become an active, competent professional in the area; and

- help you identify whether one of these professions fits your skills, aptitudes, and professional desires.

While going in for a layup, a basketball player has her legs taken out from under her. The athletic trainer arrives on the scene and finds that the player has a fractured right wrist. The athletic trainer will guide the immediate care of this athlete, refer the athlete to a physician, and develop a treatment program to safely return her to sport.

Following a stroke, a 54-year-old chief financial officer is recovering from surgery. Along his road to recovery he will meet physical therapists, occupational therapists, and cardio-vascular rehabilitation specialists. Each of these professionals will assist in restoring his function, allowing him return to his career.

Surgeons successfully repair an infant's clubbed feet, but the child is showing a delay in walking upright. The physician refers the child to a physical therapist for the rehabilitation necessary to achieve normal development.

A 30-year-old man is planning a career in law enforcement. To help him meet the physical requirements such as a timed run, strength, and other measures, the candidate seeks the services of a strength and conditioning specialist.

Each of these scenarios depicts situations that therapeutic exercise professionals face daily in work settings such as hospitals, laboratories, clinics, and sports medicine facilities. Although their patients and clients may differ, athletic trainers, clinical exercise physiologists, occupational therapists, physical therapists, therapeutic recreation specialists, and strength and conditioning specialists all apply exercise and movement experiences to improve a person's physical functioning.

Figure 14.1 illustrates the spheres of professional practice, including the sphere of therapeutic exercise. Physical activity professionals in this sphere help people restore lost function (rehabilitative therapeutic exercise) or acquire skills and functions considered normal or expected for function (habilitative therapeutic exercise).

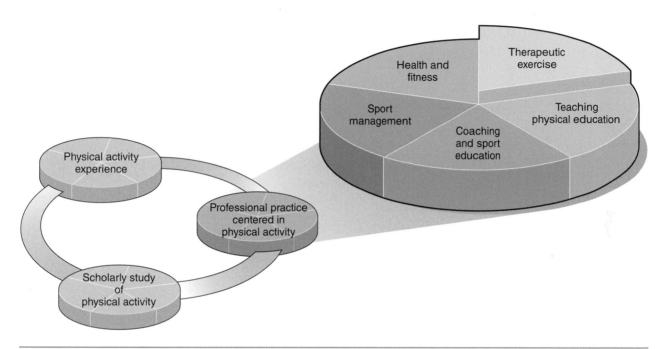

FIGURE 14.1 Therapeutic exercise is one of the spheres of professional practice.

What Is Therapeutic Exercise?

Exercise therapists design and implement movement experiences to restore or improve motor function to a level that enables people to reach personal or career goals unencumbered by physical limitations. Reaching these goals requires knowledge of human anatomy, human physiology, exercise physiology, and kinesiology in structured activity programs. To develop therapeutic goals, clinicians must call on their knowledge of the effects of exercise on the muscular, nervous, skeletal, and cardiovascular systems and relate those effects to the patient's needs. Depending on the patient, workplace, and conditions being treated, therapeutic goals may include restoring muscular function and strength, joint range of motion, proprioception, cardiovascular and pulmonary function, and metabolic function.

Reassessing "Disability"

Historically, health care and medicine tended to focus on what patients and clients were unable to do as the result of injury, illness, or inactivity. This state, disability, carries a stigma that emphasizes the negative aspects of the person's state of well-being. In 2002, the World Health Organization (WHO) developed a system that places the emphasis on what the patient *can* do, the **International Classification of Functioning, Disability and Health** (ICF) (World Health Organization 2002). The ICF system incorporates the medical aspects of resolving pathology and the social model of reducing the negative impact the condition has on the person's life (figure 14.2). This model illustrates the relationship of function and disability as the result of the interactions between health conditions (injury and disease) and contextual factors. Contextual factors consist of external environmental factors such as

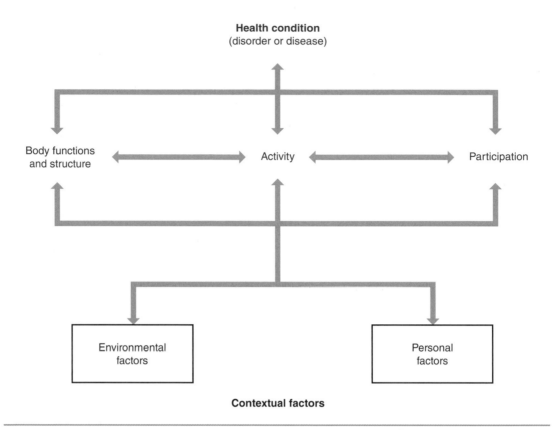

FIGURE 14.2 International Classification of Functioning, Disability and Health.

Reprinted, by permission, from World Health Organization, 2002, *Towards a common language for functioning, disability and health: The international classification of functioning, disability and health* (Geneva, Switzerland: WHO), 9. Available: http://www.who.int/classifications/icf/training/icfbeginnersguide.pdf

social attitudes, the physical characteristics of the place where the person lives and works, and personal factors including gender, profession, and life experiences.

The ICF focuses on the person's level of function and identifies interventions that maximize function that lead to a more individualized course of care. Activity (function) is the core of the ICF model. Therapeutic exercise specialists either help otherwise healthy people reach the level of activity they desire or need to function (habilitative therapeutic exercise) or return injured or ill individuals to their prior level of function (rehabilitative therapeutic exercise).

Therapeutic Exercise, Rehabilitation, and Habilitation Defined

Therapeutic exercise can be classified as rehabilitative (restoring lost function) or habilitative (helping to acquire normal function).

What is therapeutic exercise? In technical terms, therapeutic exercise is an intervention that uses the systematic and scientific application of exercise and movement experiences to *develop* or *restore* muscular strength, endurance, or flexibility; neuromuscular coordination; cardiovascular efficiency; and other health and performance factors. In practice, therapeutic exercise is programmed physical activity aimed at improving or restoring the quality of life. Thus, therapeutic exercise is classified as being rehabilitative or habilitative. In this chapter we draw a distinction between these two aspects of therapeutic exercise to help you understand the different types of professional work associated with this area.

Broadly speaking, **rehabilitation** describes processes and treatments that restore skills or functions that were previously acquired but have been lost because of injury, disease, or behavioral traits such as voluntary inactivity. If you have ever strained a muscle or broken a bone, then you know the value of rehabilitation in helping you regain lost functions. Although interventions often include ice, heat, electricity, ultrasound, manual therapy, or psychological counseling, we focus in this chapter on interventions that rely heavily on physical activity.

Rehabilitation specialists require a thorough knowledge of the pathological aspects of injury and disease, the limitations that they impose on human performance, and the types of treatments required to meet the patient's functional needs. Because people are more than just muscles and bones, rehabilitation specialists must take into account the psychological effect of the injury, the patient's personal goals and expectations, and the subsequent rehabilitation program as well. The regimen used by an athletic trainer working to restore a running back's knee so that he can return to competition is an example of rehabilitative therapeutic exercise.

Generally, the concept of rehabilitation includes the notion of **habilitation**, the processes and treatments leading to the acquisition of skills and functions that are normal and expected for an individual of a particular age and status (Olivares 2011). The standards or expectations that signal a need for habilitation may be vastly different for individuals of the same age. For example, we would expect different physical performance standards of a lawyer than we would of an athlete. The lawyer who is physically fit according to the definition in chapter 3 is probably not in need of habilitation—his state of physical fitness is normal for his age, and he does not need special physical abilities for his occupation. In contrast, an athlete who is fit but lacks the cardiovascular endurance or strength to perform the tasks expected of her in field hockey is a candidate for habilitation involving intensive conditioning, training, and other skill-specific exercise.

In both rehabilitative and habilitative therapeutic exercise, the physical status of the individual regarding any permanent disabilities or impairments, such as blindness, amputation, or paralysis, must be considered. For example, a paraplegic person's functional loss of use of the lower extremities would not be a cause for rehabilitation, but lack of upper body strength relative to that considered normal for a spinal cord injury would require care. Moreover, a physical therapist who is attempting to correct congenital postural problems is practicing **habilitative therapeutic exercise** because it involves bringing the client to a level of functioning not previously attained.

Obviously, health and fitness programs targeted at unfit populations may involve habilitative therapeutic exercise because the populations served are usually performing at levels below that considered normal and may not have previously attained the level of health or fitness desired. Because we addressed professional opportunities in fitness and health promotion in chapter 13, we do not consider them here. We do, however, consider other types of professional services that include habilitative and rehabilitative therapeutic exercise.

Each of the professions described in this chapter emphasizes the importance of preventing injury and disease. An athletic trainer, for example, helps to identify musculoskeletal problems such as lax joints, weak muscles, and family medical conditions that could predispose an athlete to injury or even death. Likewise, a physical therapist may assess a person's posture or evaluate a worker's biomechanics when lifting heavy objects to prevent job-related injuries. A clinical exercise physiologist who performs electrocardiograms and exercise stress tests is also practicing preventive medicine by identifying people who are at risk of experiencing a heart attack. By developing appropriate exercise programs to remedy various problems, these specialists help reduce the likelihood of specific injuries and diseases.

> "An ounce of prevention is worth a pound of cure." Both rehabilitative and habilitative therapeutic exercises focus on developing the body's systems so that injury or disease is less likely.

Although preventive therapeutic exercise programs are usually habilitative, at times they may be considered rehabilitative. In habilitative therapeutic exercise, bringing a person's physical state up to expected standards is preventive because it reduces the risk of injury

Considering a Career in Therapeutic Exercise?

All therapeutic exercise professions discussed in this chapter involve working closely with people. These professionals work with a number of clients or patients daily, routinely discussing protocols and treatment plans with colleagues or other physical activity professionals. Do you enjoy contact with people? Do you have good communication skills? Do you have good problem-solving skills? If you are considering a career in therapeutic exercise, in what areas might you need to seek out further instruction and education?

or illness. **Rehabilitative therapeutic exercise** may also be preventive; by restoring the patient's physical skills and functions, rehabilitation reduces the risk of reinjury and of suffering different types of injuries.

Rehabilitative Therapeutic Exercise

Restoring physical function often involves the combination of passive treatments such as heat, cold, and electrical stimulation and active exercises such as those used to improve strength, range of motion, and balance. Passive treatments are often used prior to or after active treatments. Collectively, active and passive therapies are called **interventions**.

Physical dysfunction is characterized by the inability to use one or more limbs or the torso. If you have ever broken an ankle or sprained your wrist, you have experienced physical dysfunction. Physical rehabilitation specialists help individuals who are experiencing physical dysfunction stemming from traumatic injury, congenital defects, or disease to regain the use of the affected body part or compensate for its disability. Physical therapists, occupational therapists, and athletic trainers specialize in rehabilitating these conditions. They work closely with other health care professionals such as physicians, nurses, and dietitians.

Exercise Therapy for the Rehabilitation of Musculoskeletal Injuries

Strokes, spinal cord injuries, back problems, and other injuries from automobile, industrial, sport, and home accidents can prevent the nervous system or muscles from functioning properly. Physical activity also carries the risk of trauma, as when repetitive motions lead to neuromuscular trauma (see chapters 2 and 10). Not only does injury to muscles, joints, and bones directly affect the traumatized tissues, but the accompanying immobilization and associated decreased physical activity can cause more widespread problems.

Following a significant injury that inhibits the ability of a limb to function, the resulting disuse may affect the limb and the activity of the entire body. Consequently, over time the efficiency of the heart and lungs may degrade. Perhaps you have experienced this effect after injuring a bone or joint; after regaining full motion you may have noticed that running up a flight of stairs or playing basketball left you more winded than such activities would have before your injury.

The goal of orthopedic rehabilitation programs is to restore symptom-free movement and restore the function of the cardiopulmonary system. Restoring the function of a limb consists of increasing joint range of motion, increasing muscular strength and endurance, and building cardiovascular endurance.

The heart is a muscle, and respiration is driven by the contraction and relaxation of muscles. Based on this, the goal of cardiorespiratory rehabilitation parallels that of orthopedic rehabilitation: Increasing the strength and efficiency of muscular contractions.

Although **neuromuscular injuries** directly affect the extremities, the resulting loss of function and inactivity decrease cardiovascular function. Most neuromuscular rehabilitation programs also emphasize restoring the function of the heart and lungs. Clinicians involved in this type of rehabilitation use passive and active exercise to restore limb function. Strength and muscular endurance training follows, which protects the limb. Various forms of heat and cold, electrical stimulation, therapeutic ultrasound, and manual therapy techniques augment these exercises. Throughout this process, clinicians attempt to maintain the patient's level of cardiovascular endurance. Most types of neuromuscular rehabilitation occur in outpatient physical therapy clinics, hospitals, and athletic training clinics.

Exercise Therapy for the Rehabilitation of Athletic Injuries
You may have experienced or witnessed an athletic injury at one time or another—whether you broke a finger playing sports or anxiously waited during an injury time-out to find out how seriously your favorite football player was injured. Almost immediately following a sport injury, the athlete's rehabilitation begins. Restoring the body part to its prior level of function is only part of the rehabilitation process. Once the limb begins to regain strength, range of motion, and neuromuscular coordination, the athlete must be reintegrated into athletic activity. A unique aspect of athletic rehabilitation is the sport-specific functional progression. Here, traditional rehabilitation protocol merges with the skills and tasks needed to compete in a particular sport.

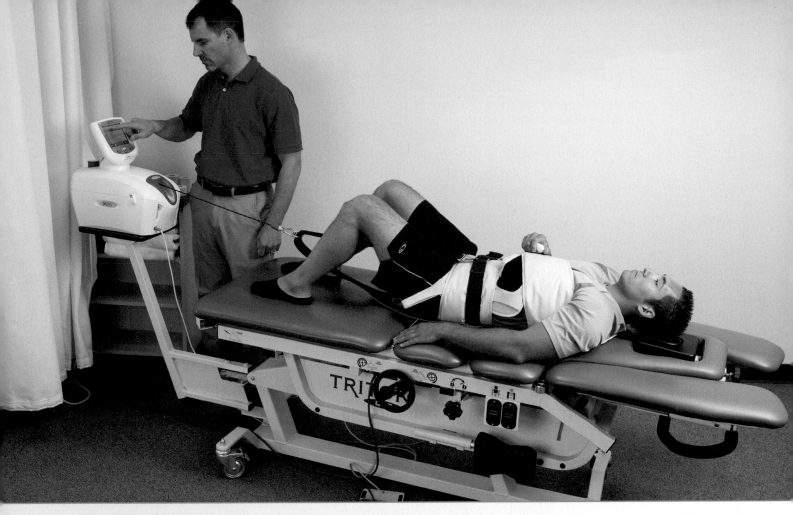

Therapeutic interventions can take place in many settings and with varied individuals, including in an athletic training room with an injured athlete.

Sports medicine is an aspect of therapeutic exercise that is exclusively dedicated to the prevention, treatment, and rehabilitation of athletic injuries. But sports medicine is distinct from other aspects of therapeutic exercise in that the rehabilitation program used is more aggressive than that used for the general population. The spectrum of sports medicine is broad and encompasses a range of activities that are beyond the scope of this text. Students who are interested in more information regarding sports medicine should explore sections on athletic training and physical therapy that appear later in this chapter.

No longer unique to the sports medicine setting, a new form of therapeutic exercise has emerged from this area: **prehabilitation**. Whereas rehabilitation occurs following an injury or after surgery, prehabilitation occurs following the injury but before surgery. When a physician deems that surgery is necessary for an orthopedic injury such as an anterior cruciate ligament (ACL) tear, part of the rehabilitation process must focus on restoring lost function. Prehabilitation involves developing as much strength and range of motion as possible before surgery. Then, following surgery, the patient will have an increased functional level on which to build. This concept is sometimes also applied to strengthen the heart and improve circulation of people who will undergo cardiac surgery.

> *Sports medicine* is a generic term that refers to the practice of medicine, the art of rehabilitation, and the science of research as they relate to athletic participation and improve performance (Prentice 2011).

Exercise Therapy for the Rehabilitation of Postsurgical Trauma

Although surgery is performed to restore a person's health, the process of surgery itself has detrimental effects on the body. The incision affects the involved muscles, nerves, and tissues, and the decreased activity associated with recovery leads to decreased muscle mass and function and decreased cardiovascular efficiency. Although this effect is most evident when surgery involves the limbs, the heart and lungs are also affected. Just as skeletal muscle wastes away when it is not used, cardiac muscle responds similarly. In most cases, people recovering from long-term disability must participate in neuromuscular rehabilitation to restore normal cardiopulmonary function.

Exercise Therapy in Cardiopulmonary Rehabilitation

Diseases of the cardiopulmonary system include coronary artery disease, arrhythmia, hypertension, heart attacks, and emphysema. Collectively, these diseases are the leading causes of death and long-term disability among adults in the United States. Undiagnosed heart disease is a leading cause of death among young athletes (Sheikh and Sharma 2011). Early identification of people at risk for these conditions, and subsequent intervention, will reduce the likelihood of their occurrence and increase the chance of survival following an episode.

Working closely with a physician, the exercise therapist participates in planning, implementing, and supervising physical activity programs designed to help restore individuals to normal function. Therapeutic exercise specialists must be aware of certain conditions that can make rehabilitative exercise a potentially deadly activity. Asthma, for example, can be dangerous because the lungs cannot exchange oxygen and carbon dioxide at the rate needed to support the body's metabolism during exercise. Because the risk is greatest when exercise occurs in a hot, dry environment, therapeutic exercise specialists treating asthmatic clients will prescribe regimens that the client can more easily tolerate, such as aquatic exercise routines.

> Because of the nature of the diseases being treated, a physician supervises most cardiopulmonary rehabilitation programs.

Cardiovascular exercises to improve the efficiency of the heart and lungs include breathing exercises, walking, running, swimming, and resistance exercises such as weight training. These are prescribed at progressive levels of intensity and duration. Diagnostic tests such as stress testing, $\dot{V}O_2$max testing, electrocardiograms, and echocardiograms are used to monitor the patient's progress and adapt the level of activity accordingly. Stress tests and other diagnostic tests are usually conducted in hospitals or other medical environments. Cardiovascular exercise programs normally take place in clinics, health clubs, or wellness centers, and the patient also is prescribed independent programs for home use. A licensed physician must supervise all diagnostic tests and exercise prescribed to high-risk patients in **cardiopulmonary rehabilitation** programs.

Exercise Therapy for the Rehabilitation of Older Populations

Humans are living longer, and the percentage of the population aged 65 and over continues to increase. The natural decrease in strength and flexibility and the increase in age-related skeletal diseases such as arthritis hinder the quality of life for many people. As we saw in chapter 2, loss of ability to perform activities of daily living can have profound physical and emotional effects on people's lives. In addition, as a person enters the later years, the aerobic ability of the body decreases and the percentage of body fat increases, decreasing one's capacity to do physical work.

Various forms of therapeutic exercise designed to increase strength, slow the loss of flexibility, and improve cardiovascular condition can delay the onset of related problems and minimize their effects, thereby helping seniors regain or maintain independent, active living (Olivares et al. 2011). Programs that involve physical activity can decrease blood pressure, control cholesterol, and reduce the risks of stroke and heart attack, and physical activity programs may lower the risk of certain types of cancer as well. Exercise in the older population, however, is often a double-edged sword. Although physical activity is required to maintain an appropriate level of fitness, the body is less capable of withstanding the forces of exercise. Clinicians working with clients who have arthritis, for example, whose joints cannot tolerate the physical stresses associated with extensive walking, jogging, or running, must modify the exercise program to minimize the amount of force exerted on the joints. In this case, water aerobics can maintain cardiovascular fitness without further injuring the joints.

Exercise Therapy for the Rehabilitation of Psychological Disorders

The physical benefits of exercise should now be apparent to you. But one area is often overlooked: the psychological and emotional benefits of physical activity. Exercise is a wonderful way to reduce stress and unwind from the toils of life. People who are involved in regular physical activity tend to sleep better and suffer fewer emotional disorders than those who are not. Twenty to 30 minutes of aerobic exercise can produce changes that are on a par with standard forms of psychotherapy (Sims et al. 2006). In the healthy population, aerobic exercise can prevent the onset of some types of psychological maladies; for people with these types of conditions, exercise can often be considered a form of treatment.

Mental health care providers often prescribe exercise regimens as part of their patients' therapy programs. The older population referred to in the previous section can also gain psychological benefits from exercise. Elderly people who engage in physical activity have decreased signs of depression, increased self-image, and improved morale, all leading to better quality of life (Conn et al. 2003). Smoking and alcohol cessation programs often include exercise, although scientific data have yet to show that this is effective (Ussher et al. 2012).

> Exercise is as important for the mind as it is for the body. Improving one's physical health through exercise also improves one's mental state.

Habilitative Therapeutic Exercise

Habilitative exercises help bring people in line with established physical standards such as those presented in chapter 3. The goal of this type of therapeutic exercise is to help people reach the expected level of physical fitness for their demographic classification. Throughout this text we have described the relationship of physical activity to health and the kinesiologist's role in improving health through physical activity. Many of the professions described in chapter 13 also meet the definition of habilitative therapeutic exercise. The following sections describe some of the areas in which therapeutic exercise is used for habilitative purposes.

Exercise Therapy for Specialized Habilitation

Although not recognized as such, specialized habilitation is probably the single largest role of habilitative therapeutic exercise. Specialized habilitation involves bringing specific groups of people in line with standards that exceed rather than merely meet those of the general population. These may include running a 5-minute mile, having body mass index within a certain range, or being able to lift a certain amount of weight. Preseason sport training camps, military boot camps, and police and firefighter academies are examples of settings in which specialized habilitative therapeutic exercise takes place. Exercise specialists in this area must fully understand the muscular and cardiovascular capabilities needed by people in these special groups. Often, the exercise specialist has a professional background in the specialty area. For example, a former baseball player with a degree in kinesiology may specialize in organizing habilitative exercise regimens for spring baseball training camps. Such experience can help keep the exercise program in line with physical expectations of the baseball players involved.

Exercise Therapy for the Habilitation of Obese Populations

Physical fitness, exercising, and losing weight are controversial American obsessions. Many people overemphasize weight loss, and others ignore all aspects of healthy eating or exercise habits. Both approaches are hazardous and potentially deadly.

Fad dieting, exercise addiction, and distorted body image have reached almost epidemic proportions among females of high school and college age. Medical conditions associated with such obsessions, especially anorexia and bulimia, are increasing at alarming rates. At the other end of the spectrum is the overweight, sedentary portion of our population, whose lifestyles predispose them to cardiovascular disease and diabetes. The mass media continually bombards us with conflicting and confusing information on which diets do—and do not—work. These contradictions may leave most Americans unclear about what is meant by healthy eating.

Because of its role in bringing people in line with established standards, therapeutic exercise used to control weight and body mass is considered habilitative. Proper nutritional and exercise counseling are essential in promoting a healthy lifestyle and decreasing the incidence of disease. Specialists such as exercise physiologists, strength and conditioning

As obesity rates continue to rise, more professionals will be needed who can work with individuals to create attainable fitness and nutrition goals.

specialists, and personal trainers work in cooperation with physicians, nutritionists, counselors, and other professionals to develop and implement programs that help people return to normal, functional, healthy lifestyles.

Habilitative therapeutic exercise helps people reach minimal physical fitness levels based on their age, gender, and desired activity. This form of therapeutic exercise can begin shortly after birth and continue throughout the life span.

Exercise Therapy for the Habilitation of Children With Developmental Problems

In the not-too-distant past, children born with certain physical conditions faced life with functional limitations and participation restrictions. Early identification of specific conditions followed by appropriate intervention has provided countless people with the opportunity to have a more active life.

The goal of this type of habilitative exercise is to help the child adapt to, or compensate for, functional anatomical and physiological deficits. In certain cases the underlying condition may have been surgically corrected. Other cases may involve teaching the child how to use a

 ACTIVITY 14.1

Hab or Rehab?

Use this web study guide activity to distinguish between the two types of therapy.

prosthetic device or how to perform basic skills such as rolling over, walking, or eating. For example, exercise therapy may be used to strengthen the muscles of children with cerebral palsy, thereby improving their walking performance.

Although working with children has a unique set of rewards, it likewise presents a unique set of challenges to therapists. Issues such as communication between the therapist and child, home care problems, and schooling are unique to this population. Regardless of the challenges facing the therapist, the ultimate goal of habilitative regimens for this population is to promote physical, social, and cognitive development at a rate near the norm for the child's age group.

Exercise Therapy for Habilitation for General Fitness

With all the benefits of exercise described in this text, the logical question is, why doesn't everyone participate in exercise? Although more people are now regularly exercising than ever before, most of our population live sedentary lives. Although no one factor is singularly to blame for this lack of exercise, the longtime popularity of television and the recent growth in computer fixation both contribute.

Research has shown that exercise can help prevent disease and enable people to reach their maximum life expectancy. Therapeutic exercise professionals play an important role in helping the general population achieve healthy standards of physical fitness by introducing them to exercise through hospital-based fitness centers, commercial gymnasiums and fitness centers, employer-based wellness centers, and community-based nonprofit agencies.

> During the industrial age, a larger proportion of the population was engaged daily in strenuous physical activity. The information age has created a relatively sedentary group of people. Even casual exercise can offer a more balanced lifestyle.

Settings for Therapeutic Exercise

Therapeutic exercise professionals work in a variety of settings, including inpatient facilities (such as hospitals and residential facilities), outpatient clinics (such as physical therapy offices and athletic training clinics), and even clients' homes. Some are employed by others, and some work in their own private practices. Although the decision to pursue a given profession in the sphere of therapeutic exercise may dictate the setting in which you work, you may still have opportunities to work with professionals from a wide range of work settings. Most health care facilities use the team approach in which representatives from several different professions collaborate in the planning and delivery of patient care.

Inpatient Facilities

Rehabilitation hospitals provide specialized care that strives to return individuals to their maximum level of function. People with severe conditions such as brain or spinal cord trauma, or those recuperating from severe disease, may require a long-term stay in a rehabilitation facility. **Custodial care facilities** such as nursing homes provide services to assist in activities of daily living as well as to meet people's specific medical needs. In both types of **inpatient facilities**, the level of patient disability usually dictates the degree of coordinated effort that physicians, rehabilitation specialists, and social services personnel must arrange.

Besides having the skills described in this chapter, therapeutic exercise professionals who work at inpatient facilities often must have knowledge of ambulation and patient transfer techniques, prevention of bedsores, and other skills unique to long-term care.

Outpatient Clinic Settings

Outpatient clinic settings are characterized by short-term patient visits (patients do not stay overnight). These types of facilities comprise the most diverse settings for therapeutic exercise professionals and include physical therapy clinics, sports medicine clinics, and cardiac rehabilitation facilities. Outpatient facilities may have a wide range of specialized equipment to meet the needs of their clientele. The more diverse the range of people being treated, the broader the range of equipment. Most outpatient clinics consist of examination areas,

▌ Therapeutic exercise professionals provide care for people recovering from illness or dealing with chronic diseases.

specialized treatment and rehabilitation areas, hydrotherapy pools, and open-space exercise areas. Some outpatient clinics are specialized and accept only certain patients—for example, people with cardiovascular disease, hand injuries, or spinal conditions.

Social, economic, or other factors can often prohibit people from traveling regularly to an **outpatient facility** for treatment. As a result, many outpatient clinics send therapists to patients' homes to provide both rehabilitative and habilitative home care.

Sport Team Settings

Athletic training clinics are a specific type of outpatient facility. Located in high schools, colleges, and professional team facilities, the first athletic "training rooms" were simply a table located in the corner of the locker room. Although the size and complexity may vary, many athletic training facilities in some high schools and at major colleges and universities are state-of-the-art health care clinics.

Facilities typically include treatment and rehabilitation areas, examination rooms, hydrotherapy areas, and space dedicated to injury prevention methods such as taping, wrapping, and bracing. Most professional sports medicine facilities, and many of those in colleges and universities, have X-ray rooms and space where minor medical procedures can be performed.

Many sport team settings also offer weight rooms, cardiovascular fitness centers, swimming pools, and exercise physiology laboratories. These types of facilities expand the scope and breadth of therapeutic exercise programs that can be offered to habilitate and rehabilitate athletes.

Private Practice

The thought of being one's own boss is intriguing to many people. **Private practice** is an entrepreneurial venture in which the professional establishes his or her own place of work. Physical therapy and occupational therapy are capital-intensive professions requiring expensive equipment and a lease on the physical facility, so a business loan, personal wealth, or investors are required for start-up costs. People must also consider other expenses such as utilities and support staff before venturing into private practice.

 ACTIVITY 14.2

Therapeutic Exercise Settings
In this web study guide activity, you'll identify the characteristics of the different therapeutic exercise settings.

Most professionals do not begin their careers in their own practice; they usually make the transition into private practice following employment in an established clinical practice. By working first in an established practice, beginning professionals can learn about the nuances of the business world and build a patient base and professional reputation. Before venturing into private practice, beginning therapeutic exercise professionals should first gain work experience, amass capital, and establish a strong patient base.

Roles for Therapeutic Exercise Professionals

Because of their background in the anatomical and physical principles of human movement, kinesiology students are well prepared to enter health care professions that use therapeutic exercise as a part of their habilitation or rehabilitation treatment regimens, including athletic training, cardiac rehabilitation, occupational therapy, physical therapy, and strength and conditioning. This section describes these professions, the settings in which they take place, and the educational requirements and credentials needed to practice them.

You will notice that the professional roles discussed here often overlap. Indeed, many people pursue multiple credentials such as athletic training and physical therapy, or physical therapy and strength and conditioning (table 14.1). As competition for health care expenditures increases, and with the health care job market tightening, the need for multiskilled and multicredentialed individuals will increase in the future.

Many therapeutic exercise professions such as physical therapy and occupational therapy require post baccalaureate education to enter the profession. Physical therapists currently must graduate from a master's degree program in physical therapy. Almost all physical therapy programs award the clinical doctorate in physical therapy (DPT) degree. Occupational therapy required an entry-level post baccalaureate degree as of 2007. Currently, the entry-level education requirement for physical therapy assistants and occupational therapy assistants is an associate's degree.

TABLE 14.1 Skill Matrix

	Athletic trainer	Clinical exercise physiologist	Occupational therapist	Physical therapist	Therapeutic recreation specialist	Strength and conditioning specialist
HABILITATION						
Fitness assessment		X				X
Analysis of function	X	X	X	X	X	X
Neuromuscular conditioning	X	X	X	X		X
Diagnostic	X		X	X		
REHABILITATION						
Develop intervention program	X	X	X	X	X	
Restore range of motion	X		X	X		
Restore strength	X	X	X	X		X
Improve cardiovascular function	X	X	X	X		X
Restore activities of daily living	X	X	X	X	X	X
Restore work or sport function	X	X	X	X	X	X

An undergraduate education in kinesiology will provide you with comprehensive knowledge of human movement, the effects of exercise, and normal anatomical and physiological function. Many graduate-level therapeutic exercise programs view an undergraduate degree in kinesiology as an ideal background for advanced studies. You should always check the current educational requirements for any profession you intend to pursue and use elective courses to fulfill any specific programmatic requirements.

Identifying Whom You Would Like to Work With

People exploring a career in therapeutic exercise often first identify the population with whom they wish to work. Take a moment to think about your favorite populations. Put a check mark by the following groups that seem appealing to you.

____ Infants
____ Children
____ Adults
____ Senior citizens
____ People with injuries
____ People with limited function

____ Athletes or other physically active people
____ Nonathletes
____ Healthy people
____ Recovering patients
____ People requiring long-term care
____ Other: _____

The educational requirements of these professions vary depending on their focus, but most have the common traits of a strong science base and an active clinical education component. Course work in theoretical kinesiology (part II of this text) will be valuable, and other specialized courses and clinical practicums will round out your professional knowledge. Your background in kinesiology will provide you with a well-rounded education including many of the skills that will be expected of tomorrow's health care professional.

Differing levels of preparation and regulation are found among therapeutic exercise professions. Health care professionals must pay close attention to state licensure requirements, especially when changing location. The professions discussed in this chapter and the professional organizations and journals that serve them are listed at the end of this chapter. These additional resources will allow you to explore the professions in detail.

Athletic Trainer

High school, college, and professional athletes become injured at staggering rates, with as many as one-third of the 1 million high school football players suffering an injury each year (Fernandez et al. 2007). Combining the excitement of athletics with the demands of health care, athletic training is the only profession that falls entirely within the realm of sports medicine. The title *athletic trainer* can be misleading. These professionals are not coaches, personal trainers, or other performance-improving personnel. Athletic training is a health care profession that addresses the prevention and risk management, clinical diagnosis, immediate care, treatment, and rehabilitation of injuries and other conditions experienced by athletes and other physically active individuals. An athletic trainer (AT) works under the direction of a licensed medical or osteopathic physician. Besides attending to the direct health care of athletes and others engaged in physical activity, the athletic trainer coordinates referrals to appropriate specialists.

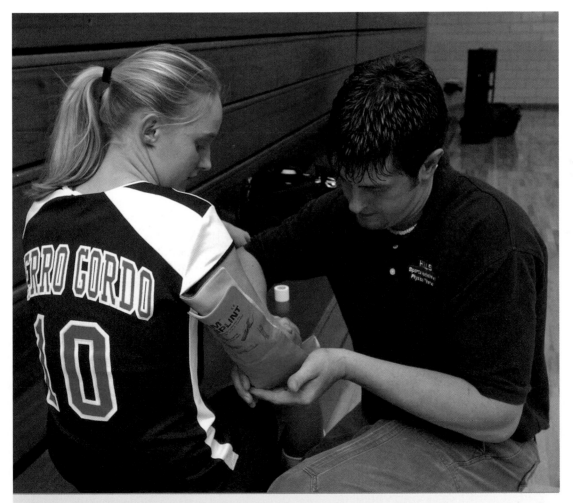

❙ Athletic trainers have many employment options, from high school through professional athletics.

Employment Settings

Athletic trainers traditionally have been employed by high schools, colleges and universities, and professional sport teams. New roles are emerging for athletic trainers in hospitals, sports medicine clinics, industrial rehabilitation clinics, and other allied medical environments. The athletic trainer's roles and responsibilities vary depending on the work setting.

States are gradually beginning to mandate athletic training coverage of high-risk high school sports. The trend toward hiring full-time athletic trainers is increasing, although many secondary schools that receive the services of athletic trainers do so by contracting services from local sports medicine clinics and hospitals. Colleges and professional teams hire their own full-time athletic trainers.

- *Athletic setting.* Athletic trainers are in a unique position to implement procedures to prevent or decrease the occurrence of sport injuries. They have the opportunity to work with the same athletes before an injury, examine the injury to form a clinical diagnosis, and then work with the patient during the rehabilitation process.

- *Clinical setting.* Community recreation leagues, road races, and personal workout regimens have exposed millions of Americans to the possibility of incurring athletic-related injuries. Add the fact that most high school athletes do not receive the direct services of an athletic trainer, and you can understand the increase in the number of sports medicine and physical therapy clinics. The emergence of athletic trainers in these settings has spurred growth in other rehabilitation settings, including corporate and industrial clinics, hospitals, and private practice. Athletic trainers' roles in these emerging settings may vary from those of athletic trainers in more traditional settings. These roles may also vary among states. Athletic trainers are responsible for the prevention, evaluation, management, treatment, and rehabilitation of musculoskeletal injuries.

Education and Credentials

Athletic training is unique in the health care field; it is one of the few professions that still has an undergraduate entry for professional practice (nursing also has an undergraduate option). However, many athletic training programs are transitioning to an entry-level master's degree program. Regardless of the route selected, an athletic training degree from a program accredited by the Commission on Accreditation of Athletic Training Education (CAATE) is required (www.caate.net). Academic course work is supplemented by clinical education that permits students to affiliate with high school, college, and professional sport team athletic training clinics and provides the opportunity to learn in private clinics and hospitals. Many athletic trainers go on to pursue a postprofessional degree in athletic training.

The Board of Certification (BOC) conducts national certification testing of athletic trainers. An athletic trainer needs BOC certification to work as an athletic trainer in major colleges and universities, with professional sport teams, and with the U.S. Olympic Committee. Additionally, 48 U.S. states require BOC certification as a prerequisite for licensure, which is needed to practice athletic training (BOC 2012).

In Canada, athletic trainers (referred to as athletic therapists) must complete an academic program accredited by the Canadian Athletic Therapists Association (CATA) and complete 1,200 hours of clinical experience under the supervision of a certified athletic therapist (CAT(C)). BOC and CATA have a reciprocity agreement for certification between the United States and Canada.

ACTIVITY 14.3

Day in the Life
What does it take to become an athletic trainer, and what does a typical work day look like? Go to your web study guide to discover how closely your ideas about what this career entails match the reality as reported by one practicing professional.

Employment Opportunities

The U.S. Bureau of Labor Statistics (BLS) projects that jobs for athletic trainers will grow at a much faster rate than for other professions through 2018 (U.S. BLS 2009). School districts are replacing part-time athletic training positions with full-time positions and adding more athletic trainers to their existing sports medicine department to follow a collegiate athletic health care model. Employment opportunities in hospitals, physical therapy clinics, and sports medicine clinics, where athletic trainers work as part of a rehabilitation team, may also increase. Additional job growth is predicted in industrial settings, where athletic trainers care for line employees, and in physicians' practices and hospitals, where athletic trainers extend the functions of a physician. The average salary in 2011 for athletic trainers who held a master's degree was $51,000 (Lowe 2011).

Clinical Exercise Physiologist

When supervised by a clinical exercise physiologist (CEP), exercise is an effective treatment for people with chronic cardiovascular, pulmonary, and metabolic diseases. Clinical exercise physiologists administer cardiopulmonary exercise tests in hospital exercise testing laboratories, administer fitness testing, and implement and deliver cardiovascular conditioning programs for individuals who are apparently disease free.

Exercise testing such as graded exercise tests and exercise prescription are the responsibility of the clinical exercise specialist (CES), who also provides patient education and counseling. Depending on the specialist's educational background and place of employment, job responsibilities range from taking vital signs (blood pressure, pulse, temperature) to performing echocardiograms and electrocardiograms before, during, and after exercise. Because the clinical exercise specialist works closely with patients, excellent interpersonal communication skills are essential. The registered clinical exercise physiologist (RCEP) is responsible for developing and directing clinical exercise programs and working in conjunction with a physician who has overall control of the patient's rehabilitation program. The registered clinical exercise physiologist is also responsible for the administration of the rehabilitation center and the education of the cardiac rehabilitation staff and is often engaged in research.

> Credentials similar to the American College of Sports Medicine's ES and RCEP certifications are available to people who want to work in the fitness setting.

A hospital may choose to offer a cardiac rehabilitation exercise program for people recovering from heart surgery. The registered clinical exercise physiologist and physician, working as a team, collect the client's medical history and evaluate his cardiopulmonary function. Using these data, the two clinicians will develop a structured exercise program. An exercise specialist (ES) will then work with the client in progressing through the exercise program. At regular intervals the program director, physician, and exercise specialist reevaluate the patient's progress.

Employment Settings

Registered clinical exercise physiologists and exercise specialists are employed in hospitals, specialty clinics, health and fitness centers, and urgent care centers. Most clinical exercise physiologists work in rehabilitation centers where patients are recovering from injury, disease, or surgery. Many work in stand-alone cardiac rehabilitation centers where specialized exercise equipment and rooms are available in an outpatient facility. Some hold exercise classes at YMCAs, schools, or other external facilities. Generally, clinical exercise physiologists work closely with nurses, nurse practitioners, physicians, and physician assistants. Besides serving an important role in rehabilitation programs, exercise specialists may work in prevention programs as discussed in chapter 13. In 2011, the average salary for an exercise physiologist was $45,081 (www1.salary.com/Exercise-Physiologist-Salary.html).

Education and Credentials

Typical course work required of people pursuing a career in clinical exercise physiology includes courses in anatomy, physiology, nutrition, biomechanics, exercise testing and

Clinical exercise physiologists work with clients who have chronic cardiovascular, pulmonary, and metabolic diseases.

prescription, and cardiac testing and prescription. Besides taking courses in these areas, students should pursue clinical rotations and fieldwork to supplement their learning experiences.

A bachelor's degree is required for certification as an exercise specialist, although a master's degree is preferred. The registered clinical exercise physiologist requires a master's degree and 1,200 hours of experience working with a variety of people who have chronic diseases.

The American College of Sports Medicine also offers certifications for similar professions in fitness settings, such as health and fitness instructors.

Employment Opportunities

The U.S. Bureau of Labor Statistics does not collect specific data on exercise physiologists, but the aging of the U.S. population will continue to create a job demand for those involved in cardiovascular rehabilitation. The U.S. Bureau of Labor Statistics (2011a) projects a 24% increase in the demand for cardiac rehabilitation specialists through 2018. Employment opportunities are increased when registered clinical exercise physiologist is combined with another degree or credential such as nursing.

ACTIVITY 14.4

Interview with a Practicing Professional
In this activity in your web study guide, Abby Smith, a program coordinator at a medical center, talks about her educational background, how she got started in her career, and what, if anything, she would do differently.

Occupational Therapist

An occupational therapist (OT) assists people with physical, emotional, or mental disabilities to restore or maintain as much independence as possible in daily living and work throughout their lives (note that the term *occupational* has roots in the word *activity*). Some of the physical care rendered by occupational therapists closely mirrors some of the rehabilitative exercises used by physical therapists. Indeed, in many rehabilitation centers, a physical therapist and an occupational therapist work together on a single patient's case.

As with many health professions, the roots of occupational therapy can be traced to postwar veteran rehabilitation; pioneers in the profession taught craft skills to soldiers with disabilities (Ambrosi & Barker-Schwartz 1995). Today, occupational therapists specialize in functional bracing and the modification of everyday items for the special needs of their patients. Functional bracing is the use of a supportive or assistive device that allows a joint to function despite anatomical or biomechanical limitations. Examples include knee braces and splints of various kinds, such as wrist and hand splints that allow people to eat with a fork. Occupational specialists also work with people to improve their concentration, motor skills, and problem-solving ability.

Occupational therapists may specialize in task-specific, work-related rehabilitation, helping clients reacquire the motor and cognitive skill needed to return to work. This type of activity may range from teaching basic skills such as coordinated movement to specific skills such as hammering, typing, or driving a car. To assist people with disabilities, an occupational therapist may be called on to evaluate the layout of schools, homes, and workplaces to suggest methods of eliminating functional barriers. The occupational therapist often employs a clinical technician called a certified occupational therapy assistant (COTA) to carry out rehabilitation plans.

Employment Settings

The occupational therapist and occupational therapy assistant (OTA) work in hospitals, rehabilitation centers, nursing homes, and orthopedic clinics and provide outpatient occupational therapy service in secondary schools and colleges. In 2008, an estimated 104,500 occupational therapists were in the U.S. work force (U.S. BLS 2009, 2011b). They instruct people with disabilities such as muscular dystrophy, cerebral palsy, and spinal cord injuries in the use of adaptive equipment like wheelchairs, walkers, and aids for eating and dressing. They may design special equipment for home or work use. They may work in cooperation with employers or supervisors to modify the work environment for the patient.

Schools often hire occupational therapists to work with a particular age group or a group with a particular disability. Occupational therapists may evaluate a child's abilities, modify equipment, and help the child participate fully in school programs and activities. Other occupational therapists work in mental health settings or in drug or alcohol rehabilitation settings. They may work in large rehabilitation centers in cooperation with physical therapists or provide home health care services.

Education and Credentials

Occupational therapists attain their education through professional 4-year undergraduate programs or in entry-level master's degree programs. Beginning in 2007, a master's degree was the minimum educational requirement. Certification or licensure as an occupational therapy assistant requires successful completion of a 2-year technical program.

Occupational therapists and occupational therapy assistants must be licensed in the states where they practice. The basic requirement for licensure is to complete an accredited program and pass the National Board for Certification in Occupational Therapy examination. Those who pass the examination earn the title of occupational therapist registered (OTR).

Professional reciprocity exists between the United States and Canada. Graduates from Canadian-accredited occupational therapy programs are eligible to sit for the U.S. examination (students in the province of Quebec must successfully complete three Tests of English as a Foreign Language [TOEFL] examinations). Graduates of accredited programs in the United States are eligible to practice in Canada.

Employment Opportunities

According to the U.S. Bureau of Labor Statistics, the employment picture for occupational therapists looks promising through 2018. As the baby boom generation moves into middle age, an age category in which the incidence of heart attacks and strokes tends to increase, demand for occupational therapist services will increase. In addition, the steady growth in the population over age 75 will ensure that demand for therapeutic services will remain high. Hospitals will continue to employ a large number of occupational therapists, including in their outpatient rehabilitation clinics.

The market for school occupational therapists also looks promising as the population of schoolchildren increases and children with disabilities continue to receive specialized attention. The median annual income of occupational therapists in 2011 was $66,780 (U.S. BLS 2011b).

Physical Therapist

▶ Physical therapists administer the patient's program and perform the required functional evaluations. Physical therapy assistants assume much of the hands-on patient care, functioning under a physical therapist's supervision.

The physical therapist (PT) is educated to provide rehabilitative care to a diverse patient population with a wide range of injuries, illnesses, and diseases. Patients include accident victims; people with low back pain, arthritis, and head injuries; and those with congenital or acquired disease states such as cancer. A physical therapist may treat individuals ranging from infants to seniors with conditions such as joint injury, burns, cardiovascular disease, and neurological deficits. Usually physical therapists are in daily contact with other health professionals such as nurses and physicians.

Therapists work approximately 40 hours per week, including some evenings and weekends. Physical therapists combine diagnostic tests with therapeutic exercise; therapeutic modalities such as ice, heat, and electrical stimulation; and manual techniques such as joint mobilization to rehabilitate or habilitate their clients. The therapy used depends on the condition being treated. Therapy for orthopedic injuries relies heavily on resistance training and proprioception activities (activities that improve knowledge of position, weight, and resistance of objects in relation to the body), whereas therapy designed to treat people with disease states and neurological conditions tends to emphasize cardiovascular aspects and neuromuscular control.

Entry-level education prepares students to enter the workforce as generalists, but once a person begins to practice professionally, the tendency is to begin to specialize in one of seven areas (see the sidebar).

Changes in the health care system, such as managed care and the limitation of reimbursement, have influenced the practice of physical therapy. More of the responsibility for patient treatment (the use of therapeutic modalities and therapeutic exercise) is moving to the physical therapy assistant (PTA), whereas the physical therapist performs patient evaluations and assumes a broader administrative role.

Physical Therapy Specialty Certifications

Specialty	Description
Cardiopulmonary	Treatment of patients with acute or chronic diseases of the cardiovascular or respiratory system
Clinical electrophysiology	Measurement of normal and abnormal electrical activity within the human body
Geriatrics	Conditions related to aging or other problems associated with older members of the population
Neurology	Treatment of patients with injuries or diseases of the brain and nervous system
Orthopedics	Treatment of patients with injuries or diseases of the muscles, bones, and joints
Pediatrics	Treatment of children in health and disease during development from birth through adolescence
Sports physical therapy	Treatment of an athletic population, normally people who have incurred injuries as the result of competition

Employment Settings

The employment settings for physical therapists and physical therapy assistants are as diverse as their patient base. Because they tend to specialize in their practice, physical therapists may work in specialized private clinics or serve specialized (e.g., pediatric or neurological) roles in hospitals, although some professionals still fill the generalist role.

About two-thirds of physical therapists work in hospitals, skilled nursing facilities, health and wellness centers, or clinics. Physical therapists may also be in private practice or provide home care services. School systems, colleges, and universities employ physical therapists to provide on-site services to students. Some physical therapists hold more than one job—one in private practice and one in a health care facility.

Education and Credentials

The wide range of potential patients and the broad scope of conditions that physical therapists treat require a vigorous and demanding academic program. The academic program also includes clinical affiliations that expose the student to several work settings and patient types. Entry-level physical therapy education is at the master's degree level, but approximately 80% of entry-level programs are clinical doctorates. The Commission on Accreditation in Physical Therapy Education (CAPTE) accredits physical therapy programs.

The American Physical Therapy Association (APTA) is the national professional organization for physical therapists. Although certification is not specifically required, the APTA offers certification for each specialty area. Licensed professionals may also seek advanced education in physical therapy at the master's and doctoral degree levels. Physical therapists who specialize in sports medicine have a strong orthopedic interest and may choose to pursue specialty certification in orthopedic or sport physical therapy. These individuals may also have education as athletic trainers or exercise physiologists.

Becoming a physical therapy assistant involves successful completion of a 2-year associate's degree program accredited by CAPTE, the same organization that accredits physical therapy programs. Many, but not all, states require licensure for the physical therapy assistant to practice. Regardless of whether they are licensed, physical therapy assistants work under

▶ Physical therapists are educated as generalists but tend to develop specialties while in the workforce.

the supervision of a physical therapist to carry out the prescribed protocol. The physical therapist must perform regular reevaluations of patients who are under the care of a physical therapy assistant.

Licensure of physical therapists and physical therapy assistants is contingent on successfully completing the appropriate program (i.e., physical therapist or physical therapy assistant) and passing an examination recognized by the state. Historically, the physical therapist's patients were required to have a referral from a physician and physician oversight. Many states now permit direct physical therapy access, thus eliminating the need for a referral. Insurance reimbursement, however, often requires a referral or physician consultation.

The regulations for physical therapists educated outside of the United States vary from state to state. In general, these individuals must complete a course of study that is substantially equivalent to that of U.S. physical therapy education and must be proficient in the English language. For more information, contact the Foreign Credentialing Commission on Physical Therapy (www.fccpt.org).

Employment Opportunities

The job outlook for physical therapists in the United States is encouraging. Employment is expected to grow much faster than the average for all occupations through 2020. As with occupational therapists, the demand for physical therapists is likely to increase as the age of the general population increases.

With advances in medical treatments, a higher percentage of accident victims are likely to survive, increasing the demand for rehabilitative care. Employers, eager to cut down on work-related injuries, are likely to employ more physical therapists to serve as worksite evaluators and as safety instructors to limit the number of injuries (U.S. BLS 2011c). Pending federal legislation imposes limits on reimbursement to therapists and may affect job growth in the short term, but the long-term employment picture is quite healthy.

In 2011 the median salary for physical therapists was $76,310 (U.S. BLS 2011c). On average, physical therapy assistants earned $49,690. Physical therapists working in home health care settings earned the highest salaries (U.S. BLS 2011d).

Therapeutic Recreation Specialist

The physiological and psychological benefits of exercise and leisure activities are described throughout this text. Therapeutic recreation specialists, or recreation therapists, use these activities plus dance, drama, arts and crafts, animals, and community reintegration outings to help people with physical, cognitive, emotional, or behavioral disabilities restore function and reduce or eliminate the effects of disability and develop independence.

Recreation therapists work cooperatively with others on the treatment team to integrate functional activities into an overall rehabilitation program. Depending on the setting, the treatment team may consist of physical therapists, occupational therapists, speech therapists, physicians, dietitians, and mental health counselors. The recreation therapist focuses on leisure activities to restore function. For example, recreation therapists can use leisure activities such as fishing, painting, or singing to restore limb dominance, to decrease anxiety, to introduce positive leisure values and alternatives, or simply to promote the individual's overall sense of well-being.

In many programs, the goals are to achieve independent function and promote social integration of the patient (Wardlaw et al. 2000). For example, a recreation therapist working in an outpatient program for people with severe and persistent mental illnesses may have a strong emphasis on working toward social skill development and leisure education. In contrast, a person working with older adults who have dementia may use leisure activities to help slow the decline in cognitive function.

Those considering a career in recreation therapy have an opportunity to work with a wide variety of people in many different settings. A strong belief in the role that recreation plays in creating a healthy lifestyle and an acceptance of people who have different ability levels

❚ Recreational therapy provides opportunities for people with special needs to participate in leisure activities that help develop physical and social skills.

serve as a personal foundation for entering this career. Recreation therapists learn diverse techniques for helping people meet their personal goals.

Employment Settings

Historically, recreation therapists have worked in inpatient and outpatient settings in psychiatry and physical rehabilitation. Increasingly, recreational therapists (as with other health care professionals) are working with older adults across their continuum of care in facilities such as subacute care settings, adult day care centers, and assisted living residences.

Many recreation therapists work in community settings, most often community recreation but also programs for people with continuing disabilities such as mental retardation or developmental disabilities, cerebral palsy, and so on. In these settings therapists often help people to continue their treatment programs after discharge from inpatient programs and introduce people with disabilities to new or modified leisure options. Recreation therapists in community settings work both in segregated programs created specifically to meet the needs of people with disabilities and as inclusion specialists, helping to include people with and without disabilities in typical recreation programming.

Recreation therapists are also employed in substance abuse rehabilitation centers, residential facilities for people with mental retardation or developmental disabilities, camps for children or adults with disabilities, and public schools. Recreation therapists usually incorporate community settings into their patients' treatment plans. Recreation therapists are often the treatment team members within inpatient settings who determine readiness for discharge by observing the patient's functioning in a community setting during an outing toward the end of a hospital stay. Approximately one-third of recreational therapy jobs are based in hospitals, and another third are in nursing care facilities. Relatively few recreational therapists are self-employed (U.S. BLS 2011e).

As with all allied health professionals, recreation therapists assess patients to determine their needs and desires, create a treatment plan in conjunction with the patient, and implement and evaluate the plan. These professionals must be familiar with limitations of medical conditions and make accurate documentation on medical records. Good interpersonal skills are necessary.

Education and Credentials

To enter the therapeutic exercise profession, a bachelor's degree in therapeutic recreation or a degree in recreation with an emphasis in therapeutic recreation is required. A limited number of institutions offer stand-alone therapeutic recreation programs; most commonly, therapeutic recreation is an emphasis area in a general recreation program. Entry level for therapeutic recreation professionals requires a bachelor's degree in therapeutic recreation, a term (quarter- or semester-long) internship under a certified recreation specialist, and successful completion of the national certification examination. A certified therapeutic recreation specialist (CTRS) is credentialed by the National Council for Therapeutic Recreation Certification.

Employment Opportunities

According to the U.S. Bureau of Labor Statistics there were 23,300 working recreation therapists in 2008. The median salary for recreation therapists is $38,370 (U.S. BLS 2009). Hospitals, which employ one-third of recreation therapists, are expected to reduce their staffs because of cost containment pressure. Some of these cuts will undoubtedly be recreation therapists. The most promising opportunities for recreation therapists are likely to be at nursing facilities, assisted living residences, and adult day care centers because of the increasing population of older adults (U.S. BLS 2011e).

Strength and Conditioning Specialists

People need proper strength and conditioning to obtain maximum physical performance, reduce the frequency of injury, and decrease the possibility of cardiovascular disease. Strength coaches design weight training programs and cardiovascular conditioning programs based on the demands inherent to specific sports and the specific needs of individual athletes. Strength and conditioning specialists may develop individualized programs in conjunction with athletic trainers, physical therapists, or physicians for athletes who have specific deficits or who have completed their rehabilitation programs.

Employment Settings

University athletic departments, professional sport teams, health clubs, and corporate fitness centers employ strength and conditioning specialists. Recognizing the contribution of training programs to athletic success, universities with large Division I-A athletic teams have invested heavily in strength and conditioning programs for athletes. These programs are commonly housed in facilities of more than 15,000 square feet (about 1,400 square meters) and may have a staff of four or more conditioning coaches, each of whom is responsible for one or more sports.

 ACTIVITY 14.5

What's My Profession?
Guess which profession is being described in this game in your web study guide.

Given the large number of athletes participating in a variety of sports, these facilities tend to be in use year-round. Likewise, an increasing number of high schools are recognizing the value of having strength and conditioning specialists on staff. Individuals who have other certifications, licenses, or areas of specialization can often implement strength and conditioning principles into their workplaces.

Education and Credentials

Those certified as strength and conditioning specialists often have other professional credentials in areas such as athletic training, exercise physiology, physical therapy, and general medicine. Credentialing of strength and conditioning specialists is not associated with a degree program, but a candidate must have a bachelor's degree and current cardiopulmonary resuscitation (CPR) certification to take the examination sponsored by the National Strength and Conditioning Association. A degree in kinesiology is preferred. To receive certification, candidates must successfully complete a two-part examination.

> A teaching degree is useful for gaining employment as a high school strength coach; college athletic departments often look for candidates with a degree in exercise physiology or a related field.

 ACTIVITY 14.6

Day in the Life

What does it take to become a strength and conditioning coach, and what does a typical work day look like? Go to your web study guide to discover how closely your ideas about what this career entails match the reality as reported by one practicing professional.

Passing both sections of the examination designates the person as a certified strength and conditioning specialist (CSCS). Credentialed professionals who become certified as strength and conditioning specialists can compete more strongly in the job market and increase their salaries. Most National Strength and Conditioning Association members hold a degree beyond the level of a bachelor's degree.

Individuals interested in becoming a strength coach for a collegiate or professional team should also possess a bachelor's degree in a related field such as exercise physiology or kinesiology. Although no uniform course of study is offered at the undergraduate level, an emphasis in fitness leadership with courses in performance assessment, program organization, and exercise leadership, in addition to the scientific core of courses normally offered in a kinesiology undergraduate program, provides excellent preparation. Obviously, an internship in a department of athletics conditioning facility is useful, as are appropriate volunteer experiences. Certification as a teacher will enhance the possibility of being hired as a high school strength coach. A graduate assistant position or internship position as a strength and conditioning coach also improves a person's marketability.

Are You Clinically or Technically Oriented?

Do you prefer to devise solutions on your own, or would you rather follow a structured sequence of steps? Clinicians devise and develop therapeutic exercise plans and are therefore called on to solve problems and make decisions. Technicians, in contrast, are experts at performing specific sets of skills.

Technical professions include physical therapy assistants and occupational therapy assistants. Clinical professions include athletic training, physical therapy, occupational therapy, and higher levels of cardiac rehabilitation. Working with student-athletes as an athletic trainer is an option for those interested in more clinically oriented careers.

Employment Opportunities

No specific data on strength and conditioning specialists are available, but this trend follows the same pattern as that for fitness workers in general, whose job numbers are expected to increase faster than average for all occupations. More people are spending more time and money on fitness, and health clubs are providing more personal services to retain their members. The average salary for strength and conditioning coaches in 2011 was $30,600 (http://strengthandconditioningcoachsalary.com/).

Trends and Opportunities in Therapeutic Exercise

As we enter the second decade of the 21st century, people are living longer than in any point in history. Despite this, obesity is prevalent in the United States. Largely as the result of inactivity, obesity increases the risk of other conditions, primarily diabetes. There are other consequences to a sedentary lifestyle, such as cardiovascular disease. To combat the physical toll and financial costs of unhealthy lifestyles, increased emphasis is being placed on the habilitative aspects of therapeutic exercise.

All the professions described in this chapter will place increased emphasis on identifying physical barriers to activity (e.g., strength, range of motion, cardiovascular limitations) and biomechanical factors that may increase the risk of injury.

The aging of the population will require education regarding the unique needs of older individuals. Courses in geriatric health care are being added to undergraduate and post baccalaureate curricula, and this topic is often emphasized in postprofessional continuing education courses.

Perhaps the biggest trend in future health care will be the increased emphasis on remote home health care. This Internet-based approach allows clinicians to prescribe, monitor, and evaluate therapeutic exercise programs using video and voice, web-based forms, and electronic monitoring devices. Perhaps the most intriguing aspect of this approach is the use of video gaming systems, such as the Xbox, that will entertain the patient or client during exercise sessions yet provide valuable medical feedback.

Advice for Therapeutic Exercise Students

You should understand by now the importance of having keen knowledge of human anatomy, human physiology, and kinesiology as well as the effects of exercise on healthy and unhealthy individuals when working in the sphere of therapeutic exercise. These areas, however, represent only a partial list of the skills and knowledge required to practice in this sphere.

Exercise is therapeutic only when it is performed in a safe and appropriate manner. Developing and delivering appropriate therapeutic exercise routines require the ability to solve problems, collect information regarding the patient's or client's condition, identify the therapeutic goals, and determine the appropriate course of action. To solve problems effectively, therapeutic exercise professionals must be able to access, manage, and interpret various forms of information and apply them to the current case. This task often requires the use of computers and an understanding of how to access Internet sites and perform literature searches.

Health care professionals must be able to communicate effectively across demographic, sociocultural, and professional boundaries. Certain settings or roles require additional areas of knowledge. For instance, professionals who work in fee-for-service environments need to become familiar with record keeping and insurance billing systems. A business background would be useful for practitioners in private practice, and a person in any type of leadership position would benefit from coursework in management and administration.

Professional ethics is an overarching principle between these professions. Therapeutic exercise professionals must adhere to basic ethical and professional principles that are defined by state laws, professional membership organizations, and certification and licensure

Skills Analysis

If you are interested in entering a therapeutic exercise profession, you will want to work toward a degree in kinesiology as well as become certified in a particular area. What other skills do you need to learn? Do you have good problem-solving and communication skills? Will you enjoy the business sides to these professions—billing, record keeping, and so on? In what ways can you test the waters of these professions to see whether they are for you?

agencies. Although the particulars vary from profession to profession and from state to state, some themes are common across the professions.

Practitioners must adhere to the profession's scope of practice—the legal parameters that define the profession. Another tenet is that the services provided must be in the patient's or client's best interest, including use of the best scientific evidence in the patient's care. Practitioners must maintain the confidentiality of medical records. When services are being billed to a third party such as an insurance company, the services billed must accurately represent the care provided. Although your instructors and supervisors will identify these issues for you, it is ultimately your responsibility to seek out this information and ensure that you remain in compliance. Failure to do so could result in criminal liability.

How does one choose an area in which to work? Although reading can provide you with an overview of these professions, the only way to discover what a profession is really like is to see professionals at work. Many professionals allow students to shadow them throughout the course of a day. Perhaps a professor or other acquaintance can help you make this connection. You may also have an opportunity to do an internship or apprenticeship or find a part-time or summer job in one of these areas.

Case Study in Ethics

You are an athletic trainer or physical therapist for Costello College. A family member refers a friend to you for examination and care. It seems that the friend, Pat, sprained an ankle while playing a pickup basketball game. Immediately following the injury Pat went to emergency department for care. Several hours later the physician ruled that there was no fracture. The laws in your state that define the scope of practice for an athletic trainer allow you to treat "anyone who is competing in, or preparing for, competition in organized athletics." The physical therapy laws in your state allow treatment only by physician referral.* You explain to Pat that you legally cannot provide care. Pat says he understands the law but views this as helping out a family friend.

Your Decision

Although friendship is important, the laws governing both physical therapy and athletic training preclude you from treating your family friend. The law specifies that the athlete must be injured during organized athletic participation for an athletic trainer to render services. However, the law also states that a physician must specifically prescribe that a patient be treated by a physical therapist. Furthermore, since you would be providing care to Pat at Costello College, that would place the college at legal risk and potentially jeopardize your employment. Would you provide care to Pat?

Resolution

Your team physician at Costello College has a local private practice. To keep your care legal, you help Pat set up an appointment with your team physician. She then can refer Pat to you for legal treatment, keeping everything within the scope of both the state laws that regulate the practice of athletic trainer and physical therapist.

*These laws are rare but do exist.

Wrap-Up

Professions arise to fulfill a societal void. The therapeutic exercise professions fill a void by either assisting people in obtaining their desired level of physical fitness (habilitation) or helping injured individuals regain lost function (rehabilitation). Part of the attractiveness of this sphere of professional practice is the wide range of the population served. Highly honed athletes, newborn infants, and the geriatric population may all rely on therapeutic exercise specialists.

Almost as diverse as the population being treated are the settings in which therapeutic exercise professionals work. From the structured environments of hospitals and laboratories to the hectic atmosphere of athletic training rooms, these work settings appeal to a wide range of interests. This chapter covers some of the more prominent therapeutic exercise professions such as athletic training, cardiac rehabilitation, occupational therapy, and physical therapy. If you are interested in any of these professions, we encourage you to seek more information from your professors and explore this area in greater depth.

More Information on . . . Careers in Therapeutic Exercise

Organizations

American Association of Cardiovascular and Pulmonary Rehabilitation

American College of Sports Medicine

American Occupational Therapy Association

American Physical Therapy Association

American Therapeutic Recreation Association

Canadian Association of Occupational Therapists

Canadian Athletic Therapists Association

National Athletic Trainers' Association

National Strength and Conditioning Association

Journals

American Journal of Occupational Therapy

Annual in Therapeutic Recreation

Athletic Therapy Today

Canadian Journal of Occupational Therapy

Journal of Athletic Training

Journal of Cardiopulmonary Rehabilitation and Prevention

Journal of Clinical Exercise Physiology

Journal of Orthopaedic & Sports Physical Therapy

Journal of Sport Rehabilitation

Journal of Strength & Conditioning Research

Physical Therapy

Strength and Conditioning Journal

Therapeutic Recreation Journal

 ACTIVITY 14.7

Use the key points review in your web study guide as a study aid for this chapter.

Review Questions

1. Describe how therapeutic exercise promotes healthy lifestyles.
2. Discuss the similarities and differences between habilitation and rehabilitation.
3. Identify the common overlaps among the therapeutic exercise professions described in this chapter.
4. What skills or attributes are unique to the professions presented in this chapter?
5. People obtain multiple credentials or specialization to make themselves more marketable. Describe some possibilities for dual credentials and cross-training based on the descriptions of professions in this chapter. What benefit would dual credentials and cross-training provide? What professions not described in this chapter would also lend themselves to dual credentials?

chapter fifteen

Careers in Teaching Physical Education

Kim C. Graber and Thomas J. Templin

chapter objectives

In this chapter we will

- describe what a physical education teacher does,

- provide insights into exciting research,

- explain how to remain current in the field,

- describe settings in which physical educators teach,

- describe highly effective teachers, and

- provide information about career options.

It was the last day of school and Mrs. Kersh's senior students in physical education would soon be graduating. In many ways, she was joining them in their journey into a new stage of life. Tomorrow, her students would march across the stage in their caps and gowns to receive their diploma. During the ceremony, Mrs. Kersh would also be recognized for her 30 years of service to the school as a physical education teacher, department head, and coach.

As she looked through her office window onto the athletic fields, a shiver ran down her spine as she watched the students in Mr. Peter's class engaged in a vigorous game of soccer. When she initially began her teaching career, Mrs. Kersh was the first African American teacher in the school, and her students were primarily Caucasian. Although she enjoyed her position, particularly because she was able to convey her enthusiasm about physical activity and sport to her students, she often thought her students were disadvantaged because they didn't have the opportunity to learn and grow through interactions with other students who represented different cultures, races, and ethnicities.

As she continued to watch Mr. Peter's class, Mrs. Kersh was struck by how the composition of students had changed over 30 years, and she smiled as she observed students of many races and ethnicities laughing and playing together. She also felt pride at watching the degree of student engagement in activity. In 1980, girls were much less active in physical education and had significantly less engagement in physical activity during class. She remembered when physical education class was characterized by students who stood in long lines while they waited for their turn to participate. Intuitively, Mrs. Kersh always knew that this was an ineffective technique, but it wasn't until research on teacher effectiveness was published that she and her colleagues realized the importance of engaging students in activity for maximal amounts of time. Fortunately, her colleagues had consistently agreed to accompany her to annual conferences to learn about effective teaching practices. They had been open to acquiring new ideas, and this had made all the difference. Tomorrow Mrs. Kersh would graduate with her students, albeit into a life of retirement. Today, however, she would proudly reflect on her career and the accomplishments of her colleagues and students.

As a participant in a physical education class, you probably were concerned primarily with improving your performance and fitness levels, playing well during competition, and enjoying the experience. You may have felt proud of your ability, hard work, and dedication when you had an especially successful performance. The truth is that if your performance improved, and if you enjoyed the experience, your teacher was largely responsible. That is, the teacher planned effective class sessions, executed appropriate instruction, kept participants actively engaged in suitable learning tasks, used proper technology, maintained a positive learning environment, and provided useful feedback.

> Good teachers work hard to create an environment in which all students can learn.

Effective teachers understand their subject and know how to convey it appropriately. Expert teachers in river rafting, for example, know which techniques should be used to propel a raft when the river is smooth as opposed to when it is at its most dangerous. They understand when to expect a change in conditions of the river and how to maneuver safely and effectively from one type of water to another. They also know which techniques are most appropriate for novices and which are best for experts. They convey instructions in a manner that is understandable and provides the best practice opportunities for learners to accomplish new skills.

As the scenario at the beginning of the chapter suggests, to teach others successfully requires a background in and knowledge about the curriculum and **pedagogy** of physical education. Without a background in pedagogy, physical activity instructors tend to "wing it," an approach that can sometimes work out in the short run but often leads to wasted time and effort—and stress for the teacher. In the opening scenario, the teachers were hired because of their expertise as instructors in different dimensions of physical activity. Not only do they have good knowledge about curriculum development, but they are effective teachers with exceptional knowledge about pedagogy. They know what it takes to get students to learn.

- If you are a teacher, how will you get and keep the students' attention?
- How much time should you spend talking, and how much time should students spend practicing?
- Will your methods work?
- How will you structure class to ensure that students have adequate time during class to improve their fitness levels?
- How will you motivate students to engage in physical activity outside of the school setting and make appropriate nutritional selections?

A person with a background in pedagogy knows the answers to these questions. The knowledgeable teacher knows that the best way to get and keep students' attention is to establish rules and procedures for good behavior in the gymnasium and on the playing field. An effective teacher with a background in pedagogy knows that practicing a skill correctly and in gamelike situations is more important than listening to a 20-minute lecture on basketball techniques. An effective teacher also understands the critical importance of physical activity to children's health. The teacher knows that approximately 33% of children are considered

How would your answers to the bulleted points on page 419 change if you were working with this group of students?

overweight (Ogden et al. 2008) and many do not receive adequate amounts of physical education or physical activity (National Association for Sport and Physical Education [NASPE] and American Heart Association [AHA] 2010) but believes that physical education can help reverse these negative trends.

✔ **ACTIVITY 15.1**

Teacher- or Student-Focused?

Use this web study guide activity to distinguish between the two types of instruction.

Webster's defines *pedagogy* as the art, science, or profession of teaching. For our purposes, we'll modify that definition a little and say that pedagogy focuses on *teaching behaviors* and *producing learning in students*. Although the distinction between teaching and producing learning might not be immediately clear, what it comes down to is focusing on the student instead of the teacher. Rather than approach each class with the question, "What am I going to teach today?," effective teachers instead ask themselves, "What are my students going to learn today?" This approach can improve interaction between the learner and the teacher and thus lead to a better education for students.

Ideally, then, those who pursue pedagogy as a career choice are concerned with producing some type of learning in students. Whether you want to become a physical education teacher, physical activity director, or high school coach, you need to learn the skills associated with effective pedagogy. Typically, prospective teachers gain pedagogical knowledge through study in an accredited agency or program, such as an education department at a university; other times they might acquire the required knowledge through working with a mentor, such as a cooperating teacher in a public school, or by attending a state teachers' conference. Ideally, though, all would-be teachers will supplement what they have learned about teaching with further information from the knowledge base that concerns itself with the study of the theories and methods of teaching physical education.

How Would You Rate Your Experiences in Physical Education Class?

Reflect on your own experience by answering the following questions:

1. Who stands out in your memory as the best physical education teacher you've ever had?
2. What qualities did this teacher have that made him or her effective?
3. Who stands out in your memory as the worst physical education teacher you've ever had?
4. What weaknesses did this teacher possess, or what qualities did he or she lack, that made him or her ineffective?
5. What do you think are the most important traits that a teacher should possess?

What Is the Teaching of Physical Education?

This chapter will introduce you to the teaching of physical education (see figure 15.1), sometimes referred to as **sport pedagogy**. As you read this chapter, think of your educational experiences thus far, asking yourself whether you received effective instruction and why or why not. Consider how you might incorporate some of the findings from research into your teaching, particularly if these findings contradict your experiences as a student. Because most of the research on pedagogy focuses on teaching in public schools, much of this chapter is about public school physical education. We will, however, try to make connections to other types of physical activity instruction, for example the instruction delivered by community recreation directors, coaches, or exercise instructors for older adults.

Similar to those in other specializations described in this book, experts in physical education teaching have many career options. Generally, most elect to obtain certification and become teachers and coaches in public schools. On the basis of state certification requirements, individuals elect a grade level in which they intend to teach and enroll for courses that specifically address how to instruct children of that age. Although many choose to work at the elementary level because they like interacting with young children and appreciate the enjoyment that youngsters experience while engaged in physical activity, others decide to work in middle or high schools because they also wish to coach. As a means of supplementing their salary in the summer months, many teachers also work as camp counselors, swim instructors, and recreation leaders.

Some people who enter departments of kinesiology for purposes of becoming physical activity instructors do not intend to teach in the public schools. Instead, they decide that

FIGURE 15.1 Physical education is one of the spheres of professional practice centered in physical activity.

they are better suited to teaching in private schools or community recreation centers. Some may elect to become specialists, such as wilderness experts or golf professionals. Others will choose to work at a college or university as an instructor or professor. Most, however, elect to become instructors of physical activity because they are interested in improving the health and wellness levels of the population, engaging people in activities that they can continue throughout their life span, and working with children in schools.

What Do You Think You Know About Teaching?

Experiences and beliefs acquired during pretraining will influence all instructors, but the best instructors base their programs and teaching methods on sound pedagogical precepts.

Research has long demonstrated that undergraduate students who are interested in teaching often believe that they already know all there is to know about teaching and have little more to learn (Lanier & Little 1986). This belief results from your experiences with and observations of professional practice while you participated in physical education or worked as a coach, camp counselor, or supervisor in a recreation center. Consider that you have spent approximately 13,000 hours as a student. Add to these the hours that you spent in other learning environments, such as summer camps, recreational sport programs through school or the park district, or private instruction, and you'll find that you've accumulated many strong beliefs about teaching (Lortie 1975; Schempp 1989).

Despite the logic of emulating someone whose qualities you admire, relying only on experience has some drawbacks. Because teachers influence different students in different ways, you can't assume that everyone shared your positive experience with a particular instructor. Nor can you assume that a teacher is effective because everyone liked him or her. For example, say you were a starter on your high school's basketball team. Because of your status, your physical education teacher may have allowed you to skip class on game days, singled you out to demonstrate new basketball skills to classmates, and given you more opportunities than others to play basketball during class. Such personal experiences might lead you to believe that the pedagogy was sound. You may believe that athletes *should* be released from class before a game, that better players *should* demonstrate their skills for the benefit of their classmates, and that higher-skilled students *should* receive more opportunities for activity than lower-skilled students receive. Despite your personal experience, however, all these beliefs may be mistaken, because they are based on unsound, unfair, or unproven teaching principles. But the scenario might have been different. Your teacher may have realized that not all students in class were as highly skilled as you were. In response, this teacher may have spent many hours planning to ensure that all students had an opportunity to improve their skill level and to experience success. This teacher's practices were grounded in effective pedagogy.

Although it is probably human nature to accept and imitate the behavior of people you respect and reject the behavior of those you do not respect when forming your own pedagogy, a better approach is to analyze the available research about teaching and learning and base your behavior and values on this analysis. To some degree you will always refer to personal experience when making decisions about instruction and designing curriculum, but this experience should never be your only criterion.

Certification and Continuing Education

Although the majority of students who are admitted into a department of kinesiology or physical education certification program intend to teach in the K through 12 schools, others may wish to become recreation specialists, fitness trainers, coaches, or outdoor education experts. Regardless of their career path, students must meet many program requirements before formal admission into a teacher education program and before certification. These requirements typically involve the following:

- Grades or grade point average: In most teacher education programs, students must maintain a minimum grade point average. In fact, most programs require a minimum grade point average just for admission into the program. Often, students must maintain a grade of B or higher in their professional education courses to continue in the program and graduate with an emphasis in teacher education.

- Praxis I: Before admission into certain courses (e.g., methods courses) or a particular phase in a program, students are required to complete a preprofessional skills test (PPST or Praxis I), entailing a series of tests developed to assess basic proficiency in reading, writing, and math. Students must achieve minimum scores in each area to progress in the program.

- Praxis II: Before admission into certain courses or student teaching, students must pass Praxis II. This exam evaluates specialty content knowledge and pedagogical knowledge (methods and curricular knowledge). The Educational Testing Service administers the test.

Although the requirements for certification vary from state to state and country to country, they usually involve a series of exams. In addition, the program in which you are enrolled is responsible for ensuring that you have developed adequate knowledge and have certain competencies. State officials periodically evaluate the curriculum in these programs. If a teacher education program fails to meet established state standards, the program is placed on probation and may lose its ability to offer teacher certification programs to students.

After a student meets all program and state testing requirements, he or she may apply for a teaching license. In many cases, reciprocity between states enables a graduate from a program in one state to teach in one or more other states throughout the country. Some states require teachers to complete continuing education courses periodically to maintain a teaching license. This requirement helps ensure that teachers remain current and knowledgeable about the most effective teaching practices. For example, students may be required to take 12 credits of continuing education courses every four years in order to remain certified. They can usually accomplish this by taking courses at a college or university, enrolling in a master's degree program, or attending approved professional conferences. Requirements, however, vary considerably from state to state.

With regard to other physical activity positions like coaching, some schools mandate that coaches be certified teachers, others require only an undergraduate degree, and the remainder expect coaches to have specialized training or an endorsement from a group such as the American Sport Education Program. Similarly, fitness professionals are often expected to become certified as a personal trainer. Some students enter colleges and universities for their training, whereas others obtain certification through online courses.

> ## @ ACTIVITY 15.2
> ### Resources for Professionals
> In this activity in your web study guide, you'll explore the NASPE website.

A large number of journals are geared specifically to teachers. They are designed to provide information about recent advances in the field, innovative teaching strategies, and results from research that teachers can easily interpret. To remain current, teachers should subscribe to and read as many journals as time permits. Good journals to begin with are the *Journal of Physical Education, Recreation and Dance; Strategies: A Journal for Physical and Sport Educators;* and the Canadian *Physical Education and Health Journal.*

In the United States, the easiest way to remain current and informed is to join the National Association for Sport and Physical Education (NASPE). This association is one of five health-related associations that reside within the American Alliance for Health, Physical Education, Recreation and Dance (AAHPERD). By connecting to the AAHPERD website (www.aahperd.org), you can access additional information about the associations and about NASPE specifically (www.aahperd.org/naspe/). For your annual dues, you will receive a monthly subscription to the *Journal of Physical Education, Recreation and Dance* and will be able to attend the national convention at a reduced rate. In addition, you will be able to log on to a members-only website where you can access the most recent and important updates within the field.

In Canada you can remain informed by joining the Canadian Association for Health, Physical Education, Recreation and Dance (CAHPERD). This association has representation in all provinces and is committed to advancing the field of physical education. The association offers *Physical Education and Health Journal,* which contains much useful information for teachers. AAHPERD and CAHPERD members often attend the same national conferences. For more information you can log on to the CAHPERD's website (www.cahperd.ca/eng/index.cfm).

Considering the cost of your undergraduate education, the annual fee for joining these associations is a modest means of remaining current. Becoming a member is a professional responsibility that informed professionals take seriously. Other professional associations and journals are also listed in this chapter.

Examining Research to See What Effective Teachers Do

We can learn much about what teachers do by examining the wealth of research that has been conducted in the area. Research on teaching pertains to the teaching and learning process that occurs in educational settings such as public or private schools. Researchers in a typical study might try to determine whether students learn better through one method of instruction than another. These researchers might look at how many times an instructor provides feedback to individual students during class and whether that feedback is specific or general. Or, researchers might investigate whether students who receive more practice opportunities during class learn more than those who receive fewer opportunities. Many of the findings about effective teaching are transferable from one setting, like a school, to another setting, like a recreation center.

Research on teaching is concerned with the scientific study of the processes of teaching and learning and is deeply rooted in what happens in education generally. Over the past two decades, scholars have undertaken many different types of investigations and have examined many aspects of teaching.

Over the last several decades, researchers have studied the characteristics of effective teachers and their classrooms. Although the majority of research examining the effectiveness of physical activity instruction has occurred within the realm of physical education, many of the findings are easily transferable to other educational settings. As such, research on effective teaching in physical education can inform other types of physical activity instructors.

Teaching Expertise

We asked you earlier to consider how experience may or may not relate to effective teaching. Although experience can enhance teacher effectiveness, it doesn't guarantee effectiveness

(Graber 2001). Unfortunately, some teachers become bored or burned out with teaching and lose interest in learning about ways in which they can improve their teaching or influence student leaning. Although we would like to believe that this group represents a small minority of teachers, boredom and burnout are problems in many schools, particularly at the high school level. These teachers no longer learn from experience, nor are they interested in reading professional journals, taking graduate classes, or attending professional conferences. On the bright side, thousands of effective teachers continue to learn from their experiences. Some of these teachers also have a high degree of expertise. Experts, and those who continue to learn about teaching by reading and by attending professional conferences, are probably less likely to experience burnout.

> Teaching experience alone does not guarantee expertise.

Berliner (1988) proposed that teachers progress through a series of stages as they gain expertise. Some scholars believe that expertise relates to being able to perform well in a particular domain. For example, Peyton Manning, Bubba Watson, Candace Parker, and Lindsey Vonn can be characterized as having high levels of performance skill in their respective sports of football, golf, basketball, and skiing. Other scholars, however, argue that although high skill level is important, it does not ensure that a physical education teacher has teaching expertise (Dodds 1994).

Appropriate Practice Experiences for Students

Your experiences in physical education, recreation, or camp settings were probably similar to those that many others had as children. If so, your teachers and counselors probably had you play dodgeball, kickball, "steal the bacon," and other such games. As you progressed into the middle and high school grades, you may have received a few days of instruction before playing softball, basketball, football, volleyball, soccer, and the like. If you are now interested in pursuing a career in the physical activity field, you probably had above-average skills in these games and sports—you might have enjoyed eliminating other students in dodgeball or scoring the winning goal in soccer. Such memories probably remain satisfying for you and likely influenced your interest in a career in physical education.

Unfortunately, probably not all your classmates shared your positive experience when engaged in physical activity. For those students who were less skilled—the ones you quickly eliminated from games such as dodgeball—physical activity time was, at best, supervised recess and, at worst, an opportunity for embarrassment. For most of those students, no instruction was available and thus they had little chance of improving their skills. Think about it: Did you acquire your athletic expertise by participating in physical education class? Or did you acquire skill through extracurricular activities organized by coaches (or parents) who were determined to help you improve your playing ability?

Defending physical education class and nonstructured recreation programs as opportunities to improve skill level is sometimes difficult. For example, if a student spends one or two days practicing a volleyball bump pass before being expected to perform the pass correctly during a game, how much chance does the student have to succeed? Consider that serious volleyball players spend hundreds of hours developing their skills. For students to succeed at any skill, they must be exposed to *appropriate practice*. When designing activities, instructors need to consider whether drills simulate game situations. For example, the most appropriate method of practicing a volleyball set is to practice setting the ball to a target area from a position in which the volleyball is bump passed to the setter, not tossed. This drill closely resembles a game situation. (Remember the principle of quality and the principle of quantity discussed in chapter 3.)

Active Learning Time

Do you recall the many times as a student you had to wait in line for a turn? If you're like many students, you sometimes had to wait through half the class for an opportunity, say, to bounce once or twice on a trampoline. Or maybe you played volleyball with 11 others on your side of the net, waiting minutes at a time for a chance to touch the ball. If this sounds familiar, you're not alone in your experience. For various reasons, thousands of students each year receive inadequate opportunities to practice and learn new skills.

Research conducted in the classroom (Brophy & Good 1986) has clearly linked time on task to student achievement. Time on task, also called **engaged time**, is defined as the time students spend actually doing physical activity or sport. Engaged time in physical activity settings is traditionally low. In fact, research from several decades ago indicates that students in physical education classes spent only about 30% of class time engaged in physical activity (Anderson & Barrette 1978; Metzler 1979, 1989). Historically, students have spent 50% to 70% of time waiting, transitioning, and being managed by the teacher. Fortunately, today's teachers are being trained to engage students in activity for at least 50% of the time, which is what experts recommend (Siedentop & Tannehill 2000). There also is increased emphasis on making sure that students are *appropriately* engaged (performing correctly with frequent success). Such time is often called **academic learning time** or **functional learning time**.

Considering that physical education class is the only opportunity throughout the entire school day that some students have to be engaged in physical activity, teachers must keep students active (and successful). Given the obesity epidemic and the high rate of physical inactivity that characterizes people throughout the world, physical education teachers must also be committed to keeping students active in appropriate learning activities for the majority of the class period.

Are You Suited for a Career in Teaching Physical Education?

Do you enjoy helping others learn? Do you like working with children? How do you feel about long but fulfilling work days? Would you like to work in an environment where you can share your enjoyment of physical activity with others?

If you responded positively to the questions above, you might be well suited for a career as a physical education or physical activity instructor.

Effective Class Management and Discipline

When teaching rules and routines, instructors can best assist students by having high expectations, being firm but warm, developing clear rules, and describing how rules will be enforced.

Probably the most common concern among new teachers is class management and discipline. *Class management* involves organizing students in such a way that learning is most likely to occur, whereas *discipline* involves teaching rules, enforcing them when they are broken, and rewarding exceptional behavior. Regardless of their experience level, teachers can never prepare for all the situations that they might encounter. How do you handle a student who refuses to take a time-out? How do you deal with a student who ridicules other students?

The best that instructors can do is to take proactive measures that tend to decrease potential difficulties. George Graham (2001) has suggested ways to create a positive learning environment. In general, he suggests using the first few classes of a new academic year or unit to establish and implement rules and procedures for good behavior. For example, in a physical education class, students need to learn the new teacher's signals, such as start and stop signals, and other signs to look or listen for during class that tell them to behave in a certain way or that a particular behavior choice is good or poor. If students learn such lessons early, the entire school year will run more smoothly for the class.

In a study of seven effective elementary specialists, instructors were observed teaching students the stopping and starting signals for class by using the signals a total of 346 different times during the first few days of instruction (Fink & Siedentop 1989). After the first few classes, most students were behaving appropriately; teachers promptly reprimanded those who did not follow the signals. By implementing this routine during the first few days of every unit, these teachers quickly constructed an environment in which learning could take place.

Assessment

Effective teachers understand the importance of assessing student achievement. They use formative assessment to determine whether students are learning the content of a unit as it progresses, and these teachers use summative assessment at the end of a unit to determine whether students have acquired the skills and knowledge of the unit. Without formal assessment, it is impossible to determine whether children have learned. Unfortunately, some teachers do not assess students to determine whether they have learned as a result of class. These teachers base students' grades primarily on dressing out and attending class. Although you might believe that it isn't fair to grade students in physical education because some are naturally less skillful than others, do you believe that students in math or science should not be graded because the curriculum is less intuitive for some than for others?

Many states have state learning standards that specify what students should know and be able to do as a result of physical education. Many of these standards are based on the NASPE National Standards for Physical Education (2004). As schools are increasingly being held accountable for standards-based student learning, a greater number of teachers are beginning to use formal assessments in physical education. NASPE recently published a series of assessments that match the national standards and are designed to inform teachers,

Case Study in Assessment

After several offers to teach, Sarah accepted her first position as a physical education teacher at Whitefish Middle School. She did so because of the unconditional support of Mr. Carson, the principal. Before and while in college, Sarah observed that many physical education teachers did not care about assessment, but to her it was extremely important. Sarah decided she would grade her students on their effort, skill improvement, attitude, and knowledge. Each of those four topics would have different values pertaining to their importance. The effort portion of a student's grade would include dressing appropriately for activity and participating during class. Skills would be tested during each unit and would be based on improvement only. Students would take skill pretests before each unit and a posttest at the end of the unit. Student attitude was another priority for Sarah, and she believed that cognitive exams should be administered. Finally, Sarah would use the Fitnessgram to assess her students' fitness levels. Although this measurement tool would not affect their grade, she would see how they progressed over time in various dimensions of health-related fitness.

Sarah met her colleague, Mr. Jones, a veteran of 20 years at Whitefish, and explained her approach to assessment. After a few minutes, Sarah could tell that the interaction wasn't going well as Mr. Jones' face turned bright red. He was furious and stated, "No way are you going to do that here. If the kids show up and dress, they get an A. That's the way it's going to be." Sarah was shocked and slowly retreated as she said, "Let's see what the principal, Mr. Carson, has to say about this!"

Your Decision

Do you think Sarah should have made such a fuss over the methods used in teaching her class? Based on your experience in high school physical education classes, how likely is it that other veteran physical educators might have the same reaction as Mr. Jones? Do you think Sarah should have threatened to go to the principal? How would you have handled the situation?

Resolution

Sarah, Mr. Jones, and Mr. Carson met to discuss the assessment challenge. Even though he had been a colleague of Mr. Jones for more than ten years, Mr. Carson was emphatic in his support for Sarah. He told Mr. Jones that he supported her and that perhaps it was time for the "old dog to learn some new tricks" from this beginning teacher. He explained that because of the new state standards related to teacher evaluation; it was time for all of the teaching staff at Whitefish to implement state of the art techniques, including student assessment. With some hesitation, Mr. Jones consented and Sarah felt relieved yet certainly cautious about future interactions with her new colleague.

students, parents, and others whether students are meeting the national standards for their age level. These assessments have undergone vigorous testing to ensure their validity and reliability (NASPE 2010, 2011). Fitnessgram is another measurement that teachers can use to accurately assess students for health-related fitness.

> ✔ **ACTIVITY 15.3**
>
> **Is This Teacher Effective?**
> In this activity in your web study guide, you'll observe and evaluate a teacher in action and you'll rate the effectiveness of a former Teacher of the Year.

► Students manipulate the learning environment when they engage in off-task behaviors or become **competent bystanders**. By ignoring off-task behavior, instructors encourage further manipulation.

Students learn accountability through clearly stated and consistently enforced expectations. One instructor characteristic that promotes student accountability is called "with-it-ness." Teachers demonstrate their **with-it-ness** by knowing what's happening in the learning environment and by displaying this awareness through oral or other communication with students. By holding students accountable for their actions, these teachers reduce the likelihood that students will try to manipulate the learning environment.

Specific Feedback

Besides providing clear instructions, teachers must also provide clear, specific, and immediate feedback (Rink 2006). Unfortunately, teachers often provide too many instructions, and students either become bored and stop listening or cannot remember all that the teacher said. In contrast, during practice, teachers seldom provide students an adequate amount of correct, prompt, and specific feedback.

The research and our own experiences indicate that teachers make the following mistakes when providing feedback.

► Teachers can increase the probability that instruction will be effective by providing appropriate learning activities; maximal active learning time; and correct, prompt, and specific feedback.

- Feedback is often incorrect. For example, a teacher may not see the fundamental error committed by a student and may focus instead on a secondary feature of performance.
- Teachers sometimes focus on an aspect of performance that does not require feedback while neglecting an area that does require feedback. For example, Stroot and Oslin (1993) found that teachers neglected to provide feedback on a component of the overhead throw because they were unable to diagnose deficits, choosing instead to provide feedback on aspects that students had already mastered.
- Some teachers do not time their feedback so that the learner receives prompt help when practice trials are defective in form or when they produce unsuccessful results. Practicing incorrect performance for an extended period has predictably negative results!
- Teachers provide less feedback during game play. Feedback provided during game play can be valuable; it gives all students information about ways to improve.

Alternative Curriculums

You and most of your classmates were likely exposed to a traditional curriculum in which instructors designed activities with little student input and used a teacher-dominated approach. Although a traditional approach can be effective if used by a creative teacher, alternative curriculums offer strong promise and can be implemented in many different environments.

Physical education today is much more exciting than ever. Although the obesity epidemic is a serious health concern, it can be credited with motivating teachers to improve their physical education programs. You may have heard the term *new physical education*. Teachers who incorporate novel and exciting learning ideas in their classes, particularly those who emphasize fitness and nontraditional activities like cycling, rock climbing, and in-line skating, are said to be teachers of this new movement of physical education. They are sometimes profiled in the national media as having changed the face of physical education.

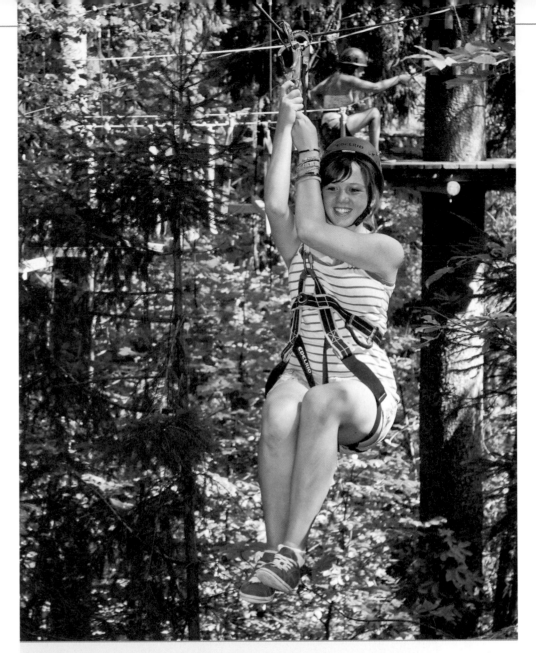

▌ Alternative curriculums offer exciting possibilities to students who may not be interested in traditional activities.

- The *elective* curriculum allows learners to choose one activity from a wide selection of activities. Learners are encouraged to develop skills that will transfer into a lifetime interest in that activity. Many large-school programs offer the elective curriculum to students.

- The *fitness curriculum* emphasizes cardiorespiratory efficiency, muscular strength and endurance, flexibility, and body composition, particularly at the secondary level. Students participate in health-related fitness activities such as running, swimming, circuit training, cycling, in-line skating, and aerobics (Corbin et al. 2001). This type of curriculum is becoming increasingly popular because it is perceived as an opportunity to address such health concerns as the obesity epidemic during physical education. If implemented appropriately, a fitness curriculum can provide students with an opportunity to enjoyably engage in moderate to vigorous physical activity. Physical educators must not forget, however, that it is equally important to teach skills to students so they can successfully engage in multiple activities throughout their life span.

- The *sport education* model appears to hold promise, particularly as a means of improving secondary physical education (Siedentop 2004). In this model, teachers treat students as athletes. Students are formed into teams and not only learn how to play but also assume some responsibility for roles as managers, coaches, trainers, officials, statisticians, and tournament administrators.

Effective teachers are concerned with implementing curricular models that are interesting to students and produce the greatest opportunity for student learning.

- The *wilderness and adventure education* curriculum introduces learners to such activities as canoeing, backpacking, camping, white-water rafting, skiing, first aid, ropes courses, climbing walls, and new and cooperative games. Evidence suggests that wilderness and adventure education can be highly appealing to students (Dyson 1995).

- The *social development model* is concerned with teaching students self-control and responsibility and draws heavily on the pioneering efforts of Hellison (1995). This model is particularly effective in environments in which students have not had opportunities to develop personal responsibility and positive social skills.

- The *teaching games for understanding* model has received increasing attention from both teachers and researchers. Whereas the traditional way of teaching games is first to ensure that students have acquired the motor skills necessary for game play, the teaching games for understanding model emphasizes teaching tactical awareness, game strategies, and game appreciation while placing less emphasis on initial motor skill development (Griffin et al. 1997).

Equity Issues and Student Needs

The learning environment in which physical education occurs is often an inequitable environment in which some students receive less (or more) attention from teachers because of race, gender, physical ability, physical appearance, or socioeconomic status. Despite measures such as Title IX and Public Law 94-142, students continue to be disadvantaged for reasons beyond their control.

Teachers must ensure that low-skilled or obese students will not be ridiculed or embarrassed.

In physical education settings, one of the most common forms of discrimination is a bias against students with less ability. Elimination games, for example, are a common activity. Students with the lowest levels of skill are usually the first eliminated, although they are the ones most in need of practice. Furthermore, classmates often ridicule or bully these students. If you were not highly skilled, how would you feel when it was your turn to roll on a scooter during a relay race, particularly if your team was in the lead?

Another common means of discrimination is to single out obese and out-of-shape students in negative ways. Interestingly, the obese student who gets lapped while participating in the mile run may actually be working harder than the student who crosses the finish line first. For this reason, an increasing number of teachers are beginning to use technology such as heart rate monitors as standard practice to more fairly assess student performance.

All learners carry personal problems into the learning environment at one time or another. Instructors need to be aware of this and be ready to adjust their teaching approach as necessary.

Similarly to low-skilled students, females are often treated in negative ways. In 1984, Patricia Griffin published the results of a classic study in which she described six patterns of female involvement during coeducational physical education classes. Whereas some females were highly skilled and participated vigorously, others had much greater difficulty when participating in game situations. Some barely participated in physical activity during class and instead assumed the role of "cheerleader," "femme fatale," "lost soul," or "system beater." If you are female, do you see yourself on this list? Although Griffin (1984) conducted the study over 20 years ago, her descriptions of female behaviors during physical activity are far from obsolete. It remains true that far too few instructors make efforts to ensure that girls have an equitable, enjoyable, and successful experience in physical activity.

Teachers must also consider other factors that might affect the students they teach. Students come to school with a host of problems that interfere with their ability to learn. Studies have shown, for example, that students' grades typically decline when their parents undergo conflict or divorce. In some areas such as inner cities, learners face a multitude of serious problems (gang involvement, poverty, lack of adequate nutrition) that get in the way of or even prevent learning (Tyson 1996). As a result, instructors often confront a need to address the emotional and material needs of students and have far less time for concern about physical activity (Ennis 1994).

Teacher–Coach Role Conflict

How well you are able to fulfill multiple roles while teaching physical education may determine your ability to perform the teaching role competently. For example, a teacher may have a second job outside the school setting, may have family responsibilities, or may volunteer in a variety of civic organizations. Inside the school the teacher may assist in a variety of duties such as cafeteria monitor, recess supervisor, bus supervisor, or student organization sponsor.

Another primary role that most teachers of physical education assume is that of coach for one or more sport teams at the school. In fact, many people enter the teaching ranks, and more specifically physical education teaching, because of a strong desire to coach. Many physical education teachers competed on interscholastic or intercollegiate teams before becoming a teacher. Researchers have studied the dual roles of teaching and coaching over the last decade and have found that for some teacher-coaches, if not many, a conflict in performance of both roles may arise (Locke & Massengale 1978). Specifically, the research has focused on how one can be an effective teacher of physical education while also striving for success as a coach.

> Both teaching and coaching are satisfying career choices, but people must be careful to fulfill the obligations of each role if they elect to engage in both simultaneously.

ACTIVITY 15.4

Conflicting Roles
Decide whether four different teacher-coaches show signs of role conflict in this activity in your web study guide.

Role conflict is defined as two or more incompatible roles that are difficult to perform simultaneously. For the teacher-coach, this definition suggests that an individual finds the concurrent roles and the expectations linked to the two roles difficult, if not impossible, to navigate. If one examines the many characteristics linked to teaching versus coaching, both compatibility and incompatibility between the two roles become clearer.

By recalling your observations of former teachers and coaches and reflecting on each role, you should recognize that the successful performance of both roles may be difficult for some people. The research suggests that because of this difficulty, many individuals withdraw (called *role withdrawal* or *retreatism*) to one role over the other. Those who prioritize coaching over teaching will withdraw or retreat to coaching; those who prioritize teaching will concentrate on teaching. When this happens, one role suffers, and the clients (either students or athletes) will likely suffer the consequences of a less than satisfactory experience.

You will occupy many roles in your life, both personal and professional, but as you prepare for them and occupy them, think about your ability to perform them concurrently and the implications for your own success and that of others.

Key Pedagogical Principles Based on Research

1. Begin to develop expertise by acquiring experience and new knowledge.
2. Provide appropriate practice.
3. Provide a high amount of academic learning time.
4. Always be concerned about class management and discipline.
5. Assess for student learning.
6. Provide clear, specific feedback.
7. Develop knowledge about alternative curricular models.
8. Ensure an equitable learning environment that addresses the individual needs of all learners.
9. Be mindful of teacher–coach role conflict.

What Instructors Need to Know About Effective Pedagogy

Although professional practice knowledge is a powerful source of information, such knowledge is not a substitute for thoughtful consideration of the available research literature on effective teaching.

Although research has the potential to inform teachers how to teach most effectively, most teachers do not subscribe to research journals. Instead, in many cases, teachers derive the knowledge that informs teaching practice primarily through their experiences as students in the public schools and other physical activity settings and through certification programs, informal discussions with other instructors, and in-service workshops and conferences focusing on the practitioner rather than the researcher.

Besides what they learn from experiences outside actual instructing time, teachers also acquire much of their pedagogical expertise in the form of professional practice knowledge. You learned in chapter 12 that this type of knowledge underlies the practice of all professions; teaching physical education is no different. Over hours, weeks, and years of instructional experience, teachers learn (often the hard way!) which techniques work well with students and which do not. Instructors and coaches often refer to their professional practice knowledge when explaining their instructional decisions to others.

Nevertheless, as valuable as personal experience can be, it is a serious mistake for teachers or coaches to look to personal experience as their only guideline for how to teach or coach. Enough scholarly research has now been done that any teacher, no matter how experienced, can learn from research.

Although the available research sometimes seems inaccessible to instructors, because it can be difficult to understand or does not apply directly to their situations, scholarly study has significantly influenced the way in which instructors teach physical education. Instructors now have a vocabulary for discussion of such issues as management time, reducing off-task student behaviors, and promoting quick lesson transitions. This all indicates that teachers are gradually becoming aware of the research on effective teaching and making attempts to incorporate the results into their own practice.

Settings for Teaching Physical Education

We all have been exposed to excellent physical education settings. Typically, these settings have good facilities and outstanding teachers. They are a joy to experience and significantly contribute to our quality of life. Learning is fun, and we come away with an expanded set of physical skills. Teachers in this type of setting are highly motivated and thrive in a workplace that promotes their effectiveness. In essence, factors promoting a high quality of work life are present: respect from students, teacher participation in decision making, collegial support and stimulation, a high sense of efficacy among teachers, resources to enable effective performance, and a common sense of vision within the school (Stroot 1996). Regrettably, the quality of life for teachers in other settings may be just the opposite. Settings that are underfunded, poorly maintained, and underequipped and that have no mission often produce unmotivated teachers.

The quality of the instructional environment in a school often can determine the success and satisfaction of both teachers and students. Understanding the influence of the workplace or the context in which one is employed is extremely important to the teacher of physical education. Like other workplace settings, schools contain a host of contextual factors that one must understand and cope with to be successful. Social, psychological, political, economic, and other organizational factors may emerge within the individual work setting or class. For example, having access to a power base along with having the moral and financial support of colleagues and administrators within a school setting is critical for a teacher.

Larger influences outside the workplace may affect what transpires within the workplace as well. Issues related to family, population demographics, civil rights, health care, drugs and violence, educational reform and policy, and advances in information technology may influence what occurs in the school setting (Tyson 1996). How instructors cope with contextual factors influences their success as instructors and, ultimately, the success of learners.

Many of the research findings are interesting to all instructors of physical activity, regardless of the setting in which they work. Some findings are positive, whereas others offer unflattering portraits of the school environment.

Life and Work in the Physical Education Setting

Teacher effectiveness and satisfaction are often linked to the characteristics of the work setting. Understanding the specific nature of any workplace is important as one contemplates working within a given activity arena. An important early research study (Locke 1975) yielded three important results:

1. Diverse activity and isolation from other adults characterize the work life of some physical education teachers.
2. Differences among students influence teacher effectiveness.
3. The curriculum of physical education is distinctly different from that of other subjects relative to space, activities, and relationships.

> In contrast to the exciting and profitable learning settings observed in most elementary schools, secondary physical education settings have been described as often being nothing more than a waste of time, in which embarrassment, humiliation, discomfort, and student inactivity appear to be the norm (O'Sullivan 1994).

Other studies have illustrated the positive and negative aspects of school physical education—what life in the gym is like for teachers and students. These studies have shown that although elementary physical education is in a relatively healthy state, secondary physical education suffers from a variety of difficult social and political problems—some of which appear now to be reaching a point of crisis.

Some scholars have vividly depicted the malaise that existed within American society relative to the role of secondary physical education in the mid-1980s. Cases of struggle appeared to be more commonplace than profiles of excellence. Griffin (1986) summarized the various obstacles to excellence at that time:

- Lack of teacher or program evaluation
- Lack of formal incentive or reward for good teaching
- Lack of professional support for teacher and program development activities
- Inadequate facilities, equipment, and scheduling
- Failure to include teachers in decision making
- Higher value placed on compliance and smooth operations than on teaching competence
- Acceptance of mediocrity
- Isolation

In discussing these obstacles, Griffin stated, "In most cases, the schools did not intentionally limit what could be accomplished. . . . Instead, they practiced benign neglect. The tendency of everyone to ignore physical education presented what seems to have been the most formidable obstacle to excellence" (p. 58). In essence, physical education as a subject area at the secondary level seems to be perceived as superfluous to student learning, and the physical education teacher holds an occupational status that has been described as marginal.

> Research on school physical education has revealed both positive and negative dimensions of the school environment.

Although much remains the same since these descriptions were written more than two decades ago, there is some basis for optimism. Legislative acts such as Public Law 108-265, which requires every school in the United States that participates in the National School Lunch Act to implement a school-wide plan addressing physical activity and nutrition, give new hope for significantly improving the environment for physical education teachers. In fact, teachers have never been better positioned to assume a major role in leading the school initiative to improve the health and fitness levels of students (Graber & Woods 2007; Woods & Graber 2007). Physical education teachers, who were once marginalized, are now encouraged to assume roles such as school physical activity director (Castelli & Beighle 2007).

Workplace Conditions

Among the factors that can influence the teaching and learning process are political, organizational, and personal and social factors. For example, the economic conditions of a school often influence what facilities and equipment are at the disposal of teachers and students. As a result, some schools are able to offer a wide variety of curricular opportunities for students, whereas other schools are more limited. Figure 15.2 illustrates the factors most likely to influence workplace conditions.

FIGURE 15.2 Workplace conditions influencing the teacher.

Within the school context, teachers' success and satisfaction, or lack thereof, are linked to many variables. Class size, for example, has the potential to influence what material a teacher is able to cover, the amount of feedback he or she can provide to individual students, and the degree to which students have sufficient opportunities to practice. School policy may dictate how often students receive instruction or how they will be graded. Tyson (1996) suggests that the following as conditions negatively influence physical education teachers' work.

• *Curriculum content.* Teachers at both the elementary and secondary levels often teach too many physical activities and too many classes in a day to be effective. Teachers often believe that they have no authority to determine the curriculum in terms of time allocation and number of activities.

• *Class size.* Class size is generally quite large in physical education—well beyond the size of other classes.

• *Staffing.* Physical education teachers are often isolated and left alone to fulfill instructional demands.

• *Policy.* Local and state policies often require the minimum in physically educating students at the elementary and secondary levels. Grading policies dictate a teacher's instructional approach.

• *Students.* Physical education teachers who are preoccupied with the control of disruptive students feel less effective. Student gender, ability, and other characteristics greatly influence a teacher's effectiveness.

• *Status and rewards.* Physical education teachers are often perceived as marginal faculty relative to the overall goals of a school. Administrators, parents, and students perceive physical education as less important than other school subjects. Physical education teachers who receive strong support from administrators, colleagues, students, and parents feel more successful in their work. They often believe, however, that others appreciate them more for their work in coaching than for their work in teaching. Teachers do feel a sense of accomplishment based on student progress.

- ***Facilities.*** Teachers often express the need for better facilities and equipment to enhance their instruction.

> → Workplace conditions have the potential to facilitate or constrain the instructor of physical activity.

Roles for Physical Education Professionals

In contrast to some of the negative characterizations of school physical education presented earlier, there are positive examples of excellence in the schools. The wellness legislation mentioned earlier, Public Law 108-265, has great potential for improving the quality of elementary and secondary school physical education. As physical educators become acknowledged as school wellness experts and as greater emphasis is placed on physical activity across the curriculum, it is likely that physical education teachers will become less marginalized and instead will be perceived as school leaders in promoting physical activity and health through quality school programs (Graber & Woods 2007; Woods & Graber 2007). Of course, program improvement is not easy, and it must be led by dedicated individuals who have the ability to negotiate a variety of obstacles such as low budgets and sometimes less than ideal support from colleagues. Teachers who are successful are likely to do the following:

- Use innovative instructional strategies
- Develop novel curriculums
- Integrate physical education with other subject matters
- Develop unique ways to promote learning progressions
- Use exemplary classroom management strategies
- Become involved in professional development activities
- Model athletic skill and fitness
- Implement after-school programs for students and adults

In addition, successful teachers are likely to have support such as the following:

- High-profile public relations programs
- Supportive colleagues and administrators
- Adequate funding
- Administrators who promote equitable learning settings

These teachers demonstrate a sincere interest in and enthusiasm for teaching, a genuine concern for students, and a continued desire to grow and develop as teachers. The settings in which they work are not unlike other physical activity settings (e.g., fitness clubs, private golf and tennis clubs, recreation centers) in which the welfare of participants or customer service is a high priority.

ACTIVITY 15.5

Day in the Life
What does it take to become a physical education teacher at the high school level, and what does a typical work day look like? In this activity in your web study guide, you'll discover how closely your ideas about what this career entails match the reality.

Typical Teaching Responsibilities

Before becoming a certified teacher, most people will have spent many hours in the schools as part of practicum and student teaching experiences, during which teachers in training observe and assist experienced teachers. Although no experience can completely prepare new teachers for all that they will encounter once they enter the instructional setting as certified teachers, practicum and student teaching experiences provide teachers in training with good exposure to students that will ease the transition into the teaching role.

Teachers can expect to meet with students for approximately 180 days each school year. Although most schools operate on a nine-month calendar, some school districts convene throughout the year (year-round schooling). In the latter case, students attend school for the same number of days as those attending for nine months, but students in year-round schooling have longer breaks more frequently throughout the year.

Physical education teachers help students learn movement skills, sport skills, and lifespan physical activities.

Teachers at the elementary level generally teach 8 to 10 classes per day for 30- to 45-minute periods. These teachers meet with students from one to five times per week and sometimes split their time between two schools. Those at the secondary level teach fewer classes (five to seven) for 45- to 60-minute periods. These teachers usually meet with students three to five times per week at the same school. Whereas physical education teachers at the elementary level may be the only specialists at their schools, those in secondary settings often have several colleagues in physical education.

Although U.S. national recommendations state that elementary school students should receive at least 150 minutes of weekly physical education and secondary students at least 225 minutes of physical education per week for the entire school year (NASPE and AHA 2010), students in most states receive far less school-based physical education instruction. For example, the percentage of students who receive physical education declined from 42% to 28% between 1991 and 2003. Fortunately, however, because of increased concern about children's health, 44 states have introduced legislative acts since 2005 to improve or increase the amount of physical education that students receive (Cawley et al. 2006).

With regard to content, teachers at the elementary level tend to teach short units of instruction that emphasize basic skill development and introduce children to concepts such as fast, slow, force, and direction. Teachers at the middle school level tend to teach longer units that introduce students to a variety of sports, games, and fitness and exploratory activities. Those at the high school level typically teach the longest units as a means of helping students acquire adequate skill to participate proficiently in a few activities throughout the life span. Teachers at all levels also may be required to teach classes specifically designed for children with special needs. As a result, most teacher education programs offer courses that provide future teachers with knowledge concerning the needs of special students and appropriate ways of structuring the curriculum.

Although teacher salaries are modest, the rewards of teaching can be great. Witnessing the satisfaction of a child who accomplishes a new skill or seeing a previously inactive student become an active participant can offer satisfaction that is difficult to find elsewhere.

Rewards for Outstanding Teaching

Often teachers labor for most of their careers without receiving adequate reward for their efforts, but local and national groups regularly recognize hardworking, competent teachers. Teachers of the year are those who have received state, regional, or national recognition for being among the best in their profession by displaying special characteristics linked to excellent teaching, outstanding curriculums, and the promotion of student learning. You may have wondered who these people are. Where do they teach? What and how do they teach? What led to their development as great teachers? We have all known great teachers throughout our lives. Can you identify the teachers whom you considered great and who had an influence on your life?

For every subject area, teachers across the United States have received recognition on an annual basis for excellence in teaching. These teachers are selected through various award programs. For example, the Council of Chief State School Officers (CCSSO) has sponsored a program since 1952. Through a nomination process (see www.ccsso.org), a nominee for and ultimate winner of this award should

- inspire students of all backgrounds and abilities to learn;
- have the respect and admiration of students, parents, and colleagues;
- play an active, useful role in the community as well as in the school; and
- be poised, be articulate, and possess the energy to withstand a taxing schedule.

USA Today recognizes an All-USA Teacher team. A number of organizations sponsor this program:

- National Association of Secondary School Principals
- National Middle School Association

- National Association of Elementary School Principals
- National Education Association
- American Association of Colleges for Teacher Education

This program honors 60 teachers, with 20 teachers honored on first, second, and third teams. The criteria for selection center on how a teacher addresses student needs and the influence of the teacher on students, the school, and community. Since 1997 the program has honored a few teachers of physical education. In 2002 Phil Lawler, a middle school teacher from Naperville, Illinois, was recognized as a member of the first team.

ACTIVITY 15.6

Excellence

In this activity in your web study guide, you'll identify behaviors that characterize excellent teachers.

More closely related to our discipline, the National Association for Sport and Physical Education, with funding from Sportime, sponsors a Teacher of the Year (TOY) program. The first NASPE Teacher of the Year was Regina McGill from Bettendorf, Iowa. The NASPE TOY selection process begins at the state level. State winners are then considered for district honors. From there, a special panel of judges consider district winners for the national awards at the elementary, middle, and high school levels. TOY nominees on all three levels are expected to meet the following criteria (www.aahperd.org/naspe/template.cfm?template=naspeAwards/criteria/toy.html):

- Conduct a quality physical education program that reflects the NASPE standards and guidelines for K through 12 physical education programs
- Use various teaching methods and plans for innovative learning experiences to meet the needs of all students
- Serve as a positive role model for personal health and fitness, enjoyment of activity, fair play, and sensitivity to the needs of students
- Participate in professional development opportunities
- Provide service to the profession through leadership, presentations, or writing

In a study of four Teachers of the Year from the state of Georgia, DeMarco (1999) discovered that they were given unique nicknames like Betty the Movement Educator Extraordinaire, Cara the Passionate Professional, Tommy the Gentle Giant, and Angela the Athletic Social Worker. In addition, they all met the criteria just listed. These teachers maximized student engagement in motor tasks and managed the physical education class with a high level of effectiveness (i.e., students were not disruptive). DeMarco found that all four teachers had a strong belief in the value of their work and their ability to do that work, and they loved students and teaching. Furthermore, students, fellow teachers, administrators, and parents greatly admired these TOYs. Finally, all four teachers were active in professional organizations (in local teaching organizations and at the state and national levels of AAHPERD).

Individuals enter the teaching profession for a variety of reasons. Some elect to teach because they enjoy working with children or like working in a school setting. Others decide to teach because it provides them with an opportunity to coach, supervise intramurals, or also serve in an administrative capacity. Although we can't all be teachers of the year, activity 15.7 lets us learn about the career selection and daily work life of an award-winning teacher.

Although most students who major in physical education will ultimately teach in the public schools, a variety of options are available (Graber et al. 2011). Most teachers, particularly at the high school level, will also choose to coach. Some people will pursue graduate education

ACTIVITY 15.7

Interview with a Practicing Professional
In this interview in your web study guide, award-winning teacher Tammy Brant talks about her motivation for becoming a teacher, what her typical day is like, and the advice she'd give to students considering the teaching profession.

and obtain a master's degree, which will enable them to be instructors at a college or university, and some will obtain a doctorate degree so that they can teach and conduct research in a higher education setting. Some may elect to work in a fitness facility or community recreation center. Others may simply decide to volunteer their time as a youth sport coach.

Physical Education Teacher in the Schools

Most people who become certified teachers elect to work in a public or private school at the elementary, middle, or high school level. Their university or college course work will have consisted of courses designed to teach them about the process of teaching (instructional strategies, teaching styles) and about the product of teaching (curriculum). Future teachers in most states will have to pass one or more state certification exams designed to ensure that they have basic skills in subjects such as reading and math, have acquired knowledge about the subject matter in which they elect to teach (physical education), and have the competence to teach effectively. As a means of helping university professors design an undergraduate curriculum that adequately prepares future teachers, NASPE (2008) has developed a set of national standards that most teacher education programs throughout the United States strive to achieve (see the sidebar listing the NASPE national standards). These standards relate to the knowledge and skills that you will be expected to apply as a teacher of physical education.

The 6 NASPE National Standards for Initial Physical Education Teacher

- **Standard 1: Scientific and Theoretical Knowledge.** Physical education teacher candidates know and apply discipline-specific scientific and theoretical concepts critical to the development of physically educated individuals.
- **Standard 2: Skill-Based and Fitness-Based Competence.** Physical education teacher candidates are physically educated individuals with the knowledge and skills necessary to demonstrate competent movement performance and health-enhancing fitness as delineated in NASPE's K-12 Standards.
- **Standard 3: Planning and Implementation.** Physical education teacher candidates plan and implement developmentally appropriate learning experiences aligned with local, state, and national standards to address the diverse needs of all students.
- **Standard 4: Instructional Delivery and Management.** Physical education teacher candidates use effective communication and pedagogical skills and strategies to enhance student engagement and learning.
- **Standard 5: Impact on Student Learning.** Physical education teacher candidates use assessments and reflection to foster student learning and to inform instructional decisions.
- **Standard 6: Professionalism.** Physical education teacher candidates demonstrate dispositions essential to becoming effective professionals.

Reprinted from *Moving into the future*, with permission from the National Association for Sport and Physical Education (NASPE), 1900 Association Drive, Reston, VA 20191, USA.

Teaching in Higher Education Settings

If you decide to become a professor of physical education, in which capacity you will likely prepare people to become teachers, you should acquire public school teaching experience and maintain ties to the public schools. This arrangement will provide you with greater credibility in the eyes of both your students and the teachers in the public schools who will supervise your students during their practicum teaching experiences.

A career option that some individuals elect to pursue is to teach at a college or university. By obtaining a master's degree, you would be qualified to work as an instructor who might be responsible for teaching skills courses (e.g., basketball, aquatics, racket sports) or supervising student teachers. Although some people elect to remain employed as instructors throughout their life span, others believe that this career option is limited. Because these individuals do not have a doctorate, they cannot serve on important college or university committees and cannot teach or advise graduate students.

By obtaining a doctorate, which is the precursor to the professorial ranks, people gain additional career responsibilities that are often considered desirable. Most professors of pedagogy have worked at one point or another as teachers in the public schools. Many elect to work at the college or university level because they want to improve teacher education. An increasing number of professors also want to engage in research and mentor graduate students. They design studies intended to help us better understand the work of teaching and how to promote effective instruction that leads to student learning. In addition, professors present papers at professional conferences; collaborate with teachers in the public schools; remain informed about state certification requirements; and serve on departmental, college, and university committees.

Regardless of whether you elect to pursue graduate studies, you should have some public school teaching experience. Would an instructor or professor in a university teacher education program have greater credibility in your eyes if he or she had several years of experience as a teacher? In most cases, your answer will be yes. Public school experience can open the door to other interesting career possibilities. In many cases, it is a requirement if you wish to be hired at the college or university level.

Adapted Physical Education

Adapted physical education is a program of skill instruction and exercise that has been modified so that all students can participate. In the United States, federal Public Laws 94-142, 101-476, and 105-17 stipulate that students with disabilities will be provided with a program of physical education that includes motor skills; fundamental skills such as running, catching, and throwing; and skills in games, sports, and aquatics. The range of disabilities that fall under these laws is large, including Down syndrome, hearing disorders, spina bifida, muscular dystrophy, and a host of others. If a child has unusual restrictions, the instruction must be provided through an individualized program developed by the child's teacher, parents, and other interested parties. Where the disability permits, children must be mainstreamed into the regular classroom or gymnasium as a way of meeting Public Law 94-142, which means that the physical education teacher must be prepared to teach a curriculum to children with a diverse range of abilities.

In larger school districts, students with similar disabilities may be grouped for physical education instruction. Obviously, one cannot presume to teach students with serious disabilities without understanding something about disabilities and how they affect the child's performance. Some colleges and universities offer certification or endorsement in physical education special education; others have full-blown certification programs. Your state probably requires at least one course in adapted physical education as part of your teacher certification program, but you may want to specialize in adapted physical education and focus exclusively on this exciting area.

Beyond the public schools, specialized schools for children with mental disabilities, those with severe orthopedic disorders, and those with limited or no capacity for seeing or hearing usually hire a staff of adapted physical education teachers. If working with children and adults who have been challenged by disease and injury is a career possibility that excites you, give some thought to obtaining a master's degree in adapted physical education, a background that will prepare you for the challenges of such a career.

Case Study in Adapted Instruction

Town Middle School had a great adapted physical education program that helped students with disabilities stay active. The teacher responsible for the wonderful program was Ms. Kelly. The students with disabilities were uniformly placed in Ms. Kelly's adapted physical education class. "Coach" Stevens, the other physical education teacher, never had students with disabilities in his class—that is, until one day when Coach Stevens began taking attendance. As he called the name "Joe Henninger," nobody responded. Coach Stevens expected that Joe was absent until some students explained to him that Joe was hearing impaired. At first, Coach Stevens believed it was student prank, but he soon discovered that the students were telling the truth. Coach Stevens watched a student walk up to him who was speaking and signing at the same time. With a worried look on his face, Coach Stevens thanked Joe for introducing himself. During the class activity, knockout basketball, Joe played the game well and even won a few times. Coach Stevens was surprised that Joe looked athletic and had great skills. Unfortunately, Joe's athleticism did not change Coach Stevens' mind about asking the principal, Mr. Hicks, to place Joe in Ms. Kelly's care in the adapted physical education class.

After a meeting with Mr. Hicks, Coach Stevens, and Ms. Kelly, Mr. Hicks agreed to move Joe back to the adapted class. Joe was extremely disappointed and appealed to his parents to change the situation. He felt that his rights were being violated.

Your Decision

Do you think that Coach Stevens' reaction to Joe was fair? Based on your experience with disabled students, how likely is it that other experienced teachers would have the same response to Joe? Do you think Ms. Kelly could have been more forceful in advocating for Joe? What role does federal law have in the local school district's decision on Joe's placement?

Resolution

In a meeting with Joe's parents, their attorney, Mr. Hicks, Coach Stevens, Ms. Kelly, and the school district's attorney, it was concluded that under Federal Public Law 94-142, Joe was entitled to be placed in Coach Stevens' class and that Coach Stevens should make every effort to accommodate Joe in the promotion of a meaningful learning environment.

Coaching

Many people who enter teaching do so because they desire a career in coaching. (Chapter 16 covers careers in coaching in depth.) That is, they are less interested in teaching and more interested in coaching. Although a career in teaching physical education offers entry into coaching, a teacher-coach can become less committed to teaching than to coaching (see previous discussion of teacher–coach role conflict). In such cases, students in the classes of these teacher-coaches often suffer.

Of course, many teacher-coaches do an outstanding job of balancing the workload of teaching and coaching. These people believe that physical education is critical because it offers an opportunity for students to acquire skills regardless of ability. They carefully plan lessons that promote skill development in all students. These teacher-coaches also believe that athletics are important because they offer students an opportunity for an advanced level of competition. They carefully manage their time to ensure success in both teaching and coaching. Coaching, however, is not an easy job, and coaches may confront situations that are challenging and unexpected.

Those who elect to coach spend many hours at their schools each day. They often arrive at school earlier than other teachers to plan classes and practices. Although some coaches receive an extra release period during the day to plan, many spend their evenings and weekends planning practices and scheduling competitive events. Immediately before or after school they spend several hours practicing with students, engaging them in drills and fitness activities that will enable them to compete successfully against other students. Some coaches coach only one sport during the year, whereas others elect to coach year-round and even during the summer months. Some become so engaged in the coaching role that they offer summer clinics or camps for students.

Although most physical education teachers elect to coach in the public schools, some prefer coaching at a more competitive level. These people choose to coach at colleges or universities. Most begin their careers as graduate teaching assistants enrolled in master's degree programs. As they gain experience and acquire name recognition, a few of these people obtain an assistant or head coaching position in higher education. At the most competitive level, Division I, these individuals only coach. Their career advancement is contingent on winning. As a result, they face enormous pressure to recruit and produce winning teams. For more information on coaching at a competitive level, see chapter 16.

Case Study in Coaching

Before the season started, rumors started filling the hallways of Eden High School. Students were talking about "hazing." Bonnie wondered whether the buzz was about an incident that had happened at a school nearby that had been in the news recently or was just gossip. She didn't really think much of it in her role as a physical education teacher-coach, because hazing never seemed to be a problem within the school. She definitely didn't think this was a problem within her classes or with her girls' basketball team. Then she learned the problem was, in fact, at Eden. To her dismay, Bonnie learned that hazing was an annual ritual of the boys' basketball team—one known by the head coach of the team. The seniors on the basketball team made the freshmen (some in her physical education class) in the program sneak out of their homes at midnight and meet in a secluded area behind the local mall in subfreezing weather. Forced to strip down to only their underwear, these boys were subjected to an hour of strenuous physical exercises. The freshmen were sworn to secrecy about the "private meeting" and were told to go home. As it turned out, one of them "blew the whistle" after becoming very ill that very evening and visiting the emergency room of the hospital.

Your Decision

What is your interpretation of the responsibility of the head coach of the boys' team? How about the role of the school principal or athletic director? Should rituals be accepted in interscholastic sports? If so, what type of rituals? What do you think should happen to the seniors involved as well as the head coach?

Resolution

Separate meetings were called with the seniors and the freshmen involved as well as their parents. Also, in attendance were school administrators, the athletic director, the basketball coach, and representatives from the school faculty. The situation was presented by the school principal, testimony was obtained from all parties, and the school hazing policy was reviewed. Subsequently, the evidence was also reviewed by a school panel that was established to rule on the case. The panel ruled that the seniors should be suspended from school for a week and dismissed from the team. The coach, who knew of this ritual, was permanently dismissed from the coaching role and placed on probation for the remainder of the year as a teacher.

ACTIVITY 15.8

Day in the Life

What does it take to become a K-8 physical education teacher, and what does a typical work day look like? In this activity in your web study guide, you'll discover how closely your ideas of what this career entails match the reality.

Trends and Opportunities in Teaching Physical Education

If you intend to become a teacher of physical education, your prospects for locating employment are high. In many states, schools have a desperate need for physical education teachers. In Illinois, for example, students at all grade levels are required to take physical education daily. As a result, severe teacher shortages exist in many areas of the state. In some states, the need for teachers is so severe that undergraduate students can apply for a loan that will be forgiven if they agree to teach in that state for a certain length of time after they become certified. Rapidly growing states and counties also have a tremendous need for quality physical education teachers. If you are willing to move to a part of the country that needs physical education teachers, you are likely to find employment quickly.

You can improve your employability by ensuring that you acquire as many skills as possible while enrolled in a teacher education program as an undergraduate. Although you are likely aware that physical education teachers who also are quality coaches are always in demand, you may be unaware that physical education teachers often locate employment because they also possess the following:

- Certification as an athletic trainer
- Certification to teach health education
- Ability to speak a second language fluently
- Endorsement to teach in additional subject areas
- Certification to teach driver's education

Selecting a Career

List the advantages and disadvantages of each of the following career options.

- Teaching physical education in the public schools
- Teaching in higher education settings
- Teaching adapted physical education
- Coaching in the public schools
- Coaching in higher education such as a Division I school

Schools that currently do not offer physical education will likely add it as a required curricular area in the future. The obesity epidemic has created greater awareness of the need for students to engage in physical activity daily. For some students, school is the only place where they have an opportunity to be active. As more schools add physical education or increase the time devoted to the subject, the need for teachers will increase. Never has there been a better time to enter the profession.

For individuals who wish to teach or coach outside of the K through 12 setting, the education they receive while enrolled in a teacher certification program will place them in good standing for whatever teaching-related position they may eventually assume. For example, have you ever observed community soccer coaches who allow students to stand in long lines waiting for a turn while one student at a time kicks the soccer ball? People who matriculated through a teacher certification program, regardless of whether they will ever teach in a K through 12 school, would not allow students to wait in lines. Recreation supervisors know that positive and specific feedback is most useful, and outdoor recreation guides make a special effort to equitably include both males and females in activities.

Advice for Physical Education Students

When effective teachers are asked to describe when they ultimately learned how to teach, their response is that they have never stopped learning to teach. That is, they believe that learning to teach is a lifelong process that continues well after one becomes certified and leaves the influence of a teacher education program. Our goal in writing this chapter was not only to acquaint you with the teaching profession but also to emphasize that learning to teach is an ongoing process. In providing advice, we encourage you to engage in professional discourse with colleagues, attend professional conferences, and become familiar with knowledge generated from research investigations. Just as you would expect your physician to remain current with the most recent advances in the field, parents and students will expect you to remain current.

Wrap-Up

The objective of this chapter was to introduce the knowledge base in the pedagogy of physical education. Exciting developments have occurred over the past two decades, and new knowledge will continue to emerge as researchers ask more sophisticated questions and develop new methods for collecting data. The information in this chapter is but a small sample of the types of knowledge currently available to those in pedagogy.

We hope that you will aspire to be like one of the Teachers of the Year—someone who is committed to student learning, engages in effective teaching practices, works toward developing subject matter expertise, and is professionally involved and current. We also hope that you will receive the same satisfaction in your career as have the thousands of teachers who enter the workplace each day—with the belief that they can make a difference in the lives of children.

If you are considering embarking on a teaching career, you will need to know much more about pedagogy than we are able to discuss here. You will have greater success if you consistently and objectively reevaluate your current beliefs in light of new research-based knowledge about effective teaching. As a member of the next generation of teachers, you can make a difference.

More Information on . . . Careers in Physical Education

Organizations

American Alliance for Health, Physical Education, Recreation and Dance (AAHPERD)

American Educational Research Association (AERA)

Association Internationale des Ecoles Supérieures d'Education Physique (AIESEP)

Curriculum and Instruction Academy (a division of NASPE)

National Association for Kinesiology and Physical Education in Higher Education (NAK-PEHE)

National Association for Sport and Physical Education (NASPE)

Special Interest Group (SIG; a division of AERA that focuses on research on learning and instruction in physical education)

Journals

Action in Teacher Education

American Educational Research Journal

College Student Journal

Educational Technology

Elementary School Journal

High School Journal

International Journal of Physical Education

Journal of Educational Research

Journal of Physical Education, Recreation and Dance

Journal of Research and Development in Education

Journal of Teacher Education

Journal of Teaching in Physical Education

Kinesiology Review

Physical Educator

Quest

Sport, Education and Society

Strategies: A Journal for Physical and Sport Educators

Teaching and Teacher Education

 ACTIVITY 15.9

Use the key points review in your web study guide as a study aid for this chapter.

Review Questions

1. How do the experiences that you acquired as a student of physical activity connect to the examples of effective and ineffective teaching discussed in the chapter?

2. How can you develop expertise in one or more of the content areas of physical education?

3. Describe three ways in which a teacher you have observed could increase the amount of learning time available to students.

4. To reduce management time, what routines are most important for students to learn? Which routines would you emphasize? Describe how you would implement those routines.

5. Which curricular framework would you choose to implement as an instructor? Why?

6. How would you interact with a student who is experiencing learned helplessness?

7. How can you avoid the pitfalls of teacher–coach role conflict?

8. Explain how you would work to improve a high school physical education program.

chapter sixteen

Careers in Coaching and Sport Instruction

Joseph A. Luxbacher and Shirl J. Hoffman

chapter objectives

In this chapter we will

- acquaint you with the wide range of professional opportunities available in the sphere of coaching and sport instruction;

- familiarize you with the nature of the work and qualifications for coaching and professional sport instruction;

- inform you about the educational requirements and life experiences necessary to become a qualified and successful coach or sport instructor; and

- help you identify whether one of these professions matches your skills, aptitudes, and professional desires.

Tomas McKenzie, a former collegiate basketball player at the Division I level, earned a graduate degree in physiology and kinesiology and teaches at a local community college. He also coaches an elite AAU basketball team composed entirely of high school juniors who plan to play at the college level. AAU basketball is generally considered to be one of the best ways of being seen by college coaches and obtaining a scholarship. The team had advanced easily through regional competition and was traveling to compete in a national tournament that would attract a number of college coaches from top Division I programs as well as coaches from elite Division II schools. Obviously this was a very important opportunity for players to display their talents to a captive audience. Parents of the players were well aware of this fact and were hoping that the money they had invested in their sons' AAU play would be returned through the award of college scholarships. Participating on an elite AAU team required more than mere talent—it required a substantial outlay of money. The costs of travel to and from games and tournaments, hotels, meals, equipment, entry fees, and the like totaled thousands of dollars over the course of a year. Most considered it money well spent, as AAU basketball provides the best possible avenue to obtain a college scholarship.

Everything went as planned. The team played well, winning first and second round games with little difficulty, and advanced to the championship game. On the evening prior to the championship match, the team's leading scorer and a definite Division I prospect was caught out of his room after curfew vandalizing hotel property. Coach McKenzie had a dilemma—what to do? The biggest game of the season was just ahead, and his star player had broken the law and violated team rules as well. Should the coach overlook the mishap, perhaps imposing a penalty after the tournament was completed, or take immediate action? Although winning the tournament had been the ultimate goal, Coach McKenzie felt strongly that it was more important to demonstrate that no player is above the law or considered to be more important than the team. He decided to suspend his star player and not play him in the final game. Despite the player's absence, the team played very well. The game wasn't decided until the last shot—and that shot clinched a victory for their opponents. For Coach McKenzie and the team the outcome was disappointing, but he believed that players had learned an important lesson about accountability to one's self and to the team. He had stood by his principles in suspending the star player, and the team had responded by battling to the end and almost winning the championship. As he was leaving the postgame locker room Coach McKenzie was verbally accosted by the father of the suspended player, who threatened to sue him. He said that he and his wife had invested thousands of dollars in their son's basketball career, and that McKenzie's decision not to play him in the final had virtually guaranteed that all of that money had been wasted. Not only would he not get a scholarship, but college coaches would do a background check and discover the reason for the suspension. The conversation abruptly ended as the father told Coach McKenzie that he would be hearing from the parents' lawyer.

This scenario combines several real-life situations. It points to the fact that coaching can be a very exciting, satisfying, and enjoyable profession, but it also can pose hazards for coaches and players alike. By its inherent nature coaching brings into close contact individuals with different personalities who may have different expectations of the sport experience, whose short- and long-term goals conflict, and who perceive the world of competitive athletics in different ways. As a consequence, coaching can make for an interesting yet challenging way to make a living.

In the previous chapter, we focused on careers in teaching physical education. Although teaching is a major professional track in kinesiology, some students set their sights on careers as coaches or as sport instructors in nonschool settings. Of course, physical education teachers in schools often take on coaching positions as additional duties, but an increasing number of opportunities to coach in nonschool settings are available as well. In addition, as swimming, tennis, and golf facilities open, sport camps and instructional sport academies continue to expand, and public interest in community nonprofit sport initiatives increases, opportunities for employment in these non–school-based careers will also increase. This chapter focuses on careers in these areas. They are part of the coaching and sport instruction sphere of the physical activity professions shown in figure 16.1.

All the professions in this sphere are concerned with developing either modest or high levels of motor skill among those being served. The participants served by coaches and instructors vary tremendously, including young children, university students and athletes, and professional sport figures. The settings in which coaches and professional sport instructors work vary greatly; they include public and alternative schools, community agencies and organizations, colleges and universities, sport clubs and academies, community agencies, and professional organizations operated as businesses. Although institutions and organizations employ most coaches and instructors, many are freelancers—self-employed instructors (in gymnastics, golf, baseball, tennis, or swimming, for example) who operate their own small companies.

What Is Coaching and Sport Instruction?

Coaching and sport instruction have a long and storied tradition, extending back as far as the ancient Greeks and Romans. We know, for example, that coaches hired by early Greek municipalities and supervised by public officials called *gymnasiarchs* taught young people how to compete in various forms of athletics. Although held in high esteem, coaches could

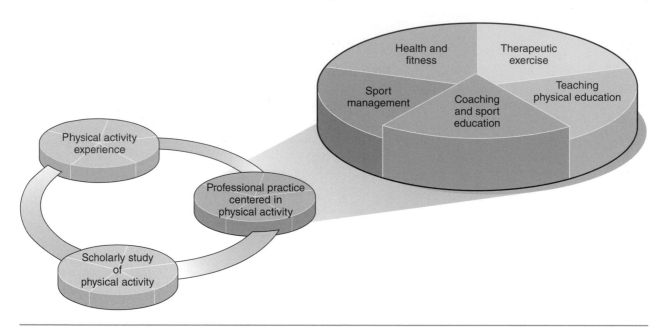

FIGURE 16.1 Coaching and sport instruction compose one of the spheres of professional practice centered in physical activity.

be demanding taskmasters. One notable Greek vase depicts a trainer (or coach) flogging two pankratiasts (athletes who performed a vicious form of wrestling popular in the early Olympic Games) for gouging each other in the face while wrestling.

Now, many centuries later, coaches continue to be popular and well regarded, and many continue to have reputations as demanding taskmasters (although none flog their athletes as the early Greek trainer did!). Some ply their trade in the anonymous quarters of community gymnasiums and school athletic fields; others work in the glaring spotlight of the collegiate and professional ranks. Some coaches earn very little; others earn millions of dollars for their efforts. Regardless of fame or fortune, coaching and serving as a sport instructor can be a fascinating and rewarding career. This chapter will help you determine whether it might be a career for you to consider.

This chapter includes both coaching and sport instruction, which are similar in many ways. Both are performed by professionals who have mastered knowledge about physical activities and the skills and techniques required to efficiently and effectively transmit that knowledge to other people. Both require expertise in designing practice experiences that stimulate learning and conditioning experiences that enhance performance. You will recall that chapter 3 addressed these two types of physical activity experiences.

But coaching and instruction also differ in significant ways. An instructor's efforts are usually—but not always—directed toward novices or those lacking a high level of proficiency, such as young children in a community youth sport development program, middle-aged women enrolled in a beginning swimming class at the local community pool, or people of all ages who want to improve their golf swings by consulting a teaching professional at the local driving range. Coaches at the high school, college, and professional levels, in contrast, typically direct their planning and efforts toward a relatively elite population. Students with above-average physical proficiency for their age and gender commonly compose the bulk of school sport teams; college teams tend to be populated by an even more select group culled from the best of high school teams. As the level of competition increases, the pool of eligible performers becomes smaller and typically the pressure to win becomes greater. Even youth sport programs, whether inadvertently or by design, tend to weed out those who are less gifted in their age groups, leaving a population of relatively skilled youngsters compared with others of similar age. Sad to say, at the youth level this phenomenon appears to have become the rule rather than the exception. The less physically mature kids are typically eliminated early in the weaning process and as a consequence do not have an equal opportunity to continue with the sport. And, of course, professional teams comprise an even more highly skilled group.

Still, teaching and coaching share many functions. Both involve instruction and dealing with people on a personal basis; in fact, the dictionary definition for coaching is close to that of teaching. Both the gymnastics teacher and the gymnastics coach try to impart knowledge and skill in certain routines, to instill in gymnasts a love of the activity, and to improve their performances. Improvement comes through practice and conditioning and through explanations, instructions, and verbal feedback as well as visual feedback in video replays and demonstrations.

So how do we determine what distinguishes instruction or teaching from coaching? A good way to address the question is to distinguish between the acts of teaching and coaching and between the professions of teaching and coaching. The acts of teaching and coaching both may be designed as attempts to alter the thinking, feelings, or behavior of a particular clientele by systematically exposing them to physical activity experiences (covered in chapter 3), along with appropriate verbal and visual experiences, to bring about predetermined outcomes.

Figure 16.2 shows the overlap in the acts of instruction and coaching as well as their uniqueness. The areas that do not overlap tend to be those directly related to the clients served.

The act of sport instruction — Specific instructional activities for promoting learning in novices

Activities common to acts of sport instruction and coaching — Instructional activities designed to alter attitudes, knowledge, and behaviors with respect to physical activity

The act of coaching — Specific instructional activities designed to improve performance in elite athletes

FIGURE 16.2 The acts of sport coaching and sport instruction. Note the considerable overlap of the two.

Instruction focuses primarily on novices, coaching on more advanced performers. As a consequence, instruction typically focuses on skill development, whereas coaching encompasses skill refinement coupled with tactical (strategy) learning. Coaching is usually directed toward select or elite populations who to some extent have already acquired the skills, knowledge, and attitudes essential for performance. The overriding emphasis for coaches, particularly at higher levels of competition, must be on individual and team performance, as success is typically measured by wins and losses. Because of this, coaches, especially those in the college and professional ranks, tend to direct much of their effort toward improving skills already learned, rather than teaching new or basic skills, and then meshing improved technical ability with sport-specific tactics and a high level of muscular fitness, aerobic fitness, or both.

Thus, whereas coaches may spend a disproportionate amount of time motivating and conditioning athletes and refining and retaining acquired skills, sport instructors are likely to focus their efforts on helping their clients acquire new skills and learn how to apply them in real-life settings.

But making too fine a distinction between the acts of teaching and coaching is probably not useful because the roles associated with these positions can switch back and forth between each other. For example, it is safe to assume that most coaches consider themselves first and foremost as teachers. Conversely, sport instructors, particularly at the youth level, commonly envision themselves as coaches. Youth sport coaches, for example, spend much of their time

Case Study in Determining Practice Schedule

Mohammed was the newly appointed coach of the high school track team. He had plenty of experience as a runner, having competed at the international level, but very little experience performing or coaching field events. He was concerned about his lack of knowledge about the weight events, particularly the shot put. He had read books on the skill but wanted to be sure he took a scientific approach that incorporated all he had learned in kinesiology classes. He reflected on what he had learned in classes in motor learning and performance, biomechanics, and exercise physiology to come up with a basic practice plan. Relying on an analysis described in chapter 3, he recognized that the shot put was a closed skill that required improvement in both physical performance capacity (strength mainly) and skill development (given that a coordinated series of perfectly executed movements was essential to attaining the skill goal). He reviewed his notes from his classes and believed he had a very good handle on the mechanics of the event and understood the best way to develop athletes' strength. How he should arrange practices, however, was still an open question.

Your Decision

How would you design practices for this event? Would you vary the practice conditions by changing the distances and directions the athletes projected the shot? Would you have the athletes perform the movements without the shot during early trials? Would you have them practice the skill repetitively, giving them at least 25 successive practice trials each session?

Resolution

Guided by what he had learned in his motor learning and performance course about open and closed skills, Mohammed recognized that the shot put event was a closed skill, and he decided that he would have his athletes focus on developing a stereotyped, effective movement pattern rather than varying the practice conditions from trial to trial. He realized that expert performance of this skill ultimately depended on executing a power-producing series of movements and that the weight of the ball might cause his athletes to develop mechanically incorrect technique in early trials. Therefore, he decided to move gradually by having athletes work on technique and then introducing the ball. Finally, he recognized that "massed practice trials" would lead quickly to fatigue and faulty performance patterns, so he adopted a "distributed" schedule in which his athletes performed only 15 trials each practice interspersed by time taken to walk to retrieve the shot and time taken to rehearse the movements without the ball.

teaching children how to perform basic skills. The emphasis is not, and should not be, on who wins the games but rather on maximizing each player's level of expertise. High school varsity or college coaching, however, usually involves more of the unique aspects of coaching identified in figure 16.2. Sometimes professional sport instructors serve in a role that we might define as coach-instructor, as when they supply instruction on a one-to-one basis to professional golf and tennis players. Some of the more successful golfers regularly visit these instructors (also called teaching professionals or swing doctors) to correct flaws that have caused their performance to deteriorate. Many professional tennis players have staff in their entourage who perform the same function. Regardless of whether one calls them teaching professionals, golf or tennis instructors, or coaches, such people play fundamentally different roles and work in quite different settings than do coaches of scholastic athletics or college-level athletics or those who supply sport instruction to large groups in municipal recreation settings.

Although considerable overlap exists between the acts of coaching and instruction, the same is not true between the professions of coaching and sport instruction. By "profession," we refer to the entire range of duties of a coach or instructor beyond the direct acts of coaching or instruction. Instructors and coaches have different professional responsibilities that they must carry out in substantially different occupational subcultures. Figure 16.3 shows the differences and similarities in the instruction and coaching professions. As you can see, similarities between the two professions are considerably fewer than similarities between the acts of instruction and coaching shown in figure 16.2. There are two major differences:

1. Instructors tend to spend relatively more of their time with on-task duties, by which we mean direct involvement in the dissemination of knowledge and molding of behavior of students. Coaches spend proportionately less of their time on this and more time on the off-task duties of recruiting, scouting, reviewing films, compliance (eligibility) issues, scheduling, budgeting, and fundraising.

2. The nature of the off-task duties of the two jobs differs. Whereas the instructor spends time maintaining records, repairing and maintaining equipment, advertising classes, and (if teaching in an institution) attending to institutional demands, coaches are more likely to become absorbed with the off-task duties described in figure 16.3.

> The *acts* of sport coaching and sport instruction are more similar than distinct; the *professions* of sport coaching and sport instruction are more distinct than similar.

Responsibilities of sport instructors

- Repairing and maintaining equipment
- Organizing records of student progress
- Advertising classes
- Meeting with parents
- Monitoring security of facility
- Keeping financial records
- Participating in professional organizations
- Continuing education or certification

Responsibilities of coaches

- Monitoring security and preparing playing facilities and locker rooms
- Scheduling and planning trips
- Recruiting players
- Meeting with parents
- Scouting other teams
- Reviewing game film
- Talking to media
- Hiring personnel
- Evaluating officials
- Motivating and counseling athletes

FIGURE 16.3 Differences and similarities in the coaching and sport instruction professions.
Based on the study of DeMarco 1999.

Clearly, then, in the most fundamental analysis, the act of instruction is essentially the same as the act of coaching. To some extent all teachers are coaches and all coaches are teachers. But when we consider the professional responsibilities of the two roles, we see that the jobs are really quite different. Even in cases in which people hold both roles in the same job, as when school or college physical education teachers also coach, the demands of the two roles are distinct. So, although we recognize similarities in many of the professional responsibilities of coaches and teachers, we also need to keep in mind that each operates in a distinctly different occupational subculture and each requires a unique set of professional knowledge and skills. In addition, the ultimate goals to be achieved are not always identical. Let's look more closely at the settings in which coaches and sport instructors work.

Settings for Coaching and Sport Instruction

The setting in which one works is important because it usually defines the job expectations and duties of those who work there. Coaches and sport instructors work in many different settings, and their duties and responsibilities vary accordingly. Most settings fall into one of the following categories: community based, institutional, or commercial. Let's consider each of these in detail.

Community Settings

Community organizations offer a variety of settings for coaching and sport instruction. Because most of the organizations offering sport programs in these settings are nonprofit, volunteers fill many of the coaching positions, although some youth and adult sport leagues hire part-time coaches. Parks and recreation departments, for example, usually sponsor athletic leagues for both youths and adults. Other organizations such as the YMCA and YWCA, Young Men's Hebrew Association (YMHA) and Young Women's Hebrew Association (YWHA), American Youth Soccer Association, U.S. Volleyball Association, and U.S. Fencing Association provide similar opportunities to children and adults. Municipal recreation authorities often oversee community sites for coaching and sport instruction. Although recreation has developed into a specialized profession in its own right, the career aspirations of recreation graduates and kinesiology graduates overlap. If a career in municipal recreation interests you, explore the possibility of obtaining an undergraduate minor in recreation or taking some courses in recreation leadership and management.

Community organizations are civic minded; they have as their first goal improvement of the social, physical, and moral development of the local community. (Think back to chapter 12 in which we talked about the two value systems that influence professional conduct. Coaches and instructors working in community programs are usually guided by a social trustee, civic professionalism orientation.) Usually these organizations rely on municipal funding, user fees, dues, and private and public contributions from agencies such as United Way or on business or corporate sponsorships of athletic teams. Some are sponsored directly by tax money; most are funded through donations and corporate sponsors. The Boys & Girls Clubs of America, for example, serve 4.5 million boys and girls at 4,300 locations in 50 states with a trained staff of 49,000 only through the funding (approximately $230 million) provided by major corporations in the country. The Amateur Athletic Union sponsors a number of sport programs and holds local as well as national competitions. The Joy of Sports Foundation in Washington, DC, is one of the largest community programs to serve at-risk young people through sport. The foundation has both paid and volunteer staff.

Instructors in community settings may work in pool areas teaching swimming classes, in gymnasiums or weight rooms facilitating adult group exercise programs, in testing rooms conducting stress tests and fitness evaluations, or in conference rooms engaging in health and nutrition consulting.

Profile of Boys & Girls Clubs of America

- Overview: Boys & Girls Clubs of America provide opportunity for safe and fun places for youth during out-of-school hours. The BGCA provides varied and diverse programming for participants.
- Scope: national (approximately 4,300 locations in 50 states)
- Locations: urban, suburban, rural
- Setting: community-based organization, public schools, recreation centers
- Participants: kindergarten through high school students
- Funding: public grants, corporations, foundations, investment income

Typical worksites for coaches of community-based sport teams are gymnasiums; skating rinks; soccer, football, softball, and baseball fields; and tracks. Generally, games are played on the organization's property or on that owned by municipalities or schools.

Settings for Coaching

Community settings	Institutional settings	Commercial settings
YMCA or YWCA	Middle schools	National and Olympic facilities
Boys & Girls Clubs	Senior high schools	Private clubs, camps, sport academies
Municipal recreation facilities	Private secondary schools	Professional sport organizations
National youth sport programs	Community colleges	Freelance, entrepreneurial
Various nonprofit organizations	Colleges and universities	

Institutional Settings

Most of the sport instruction in institutional settings takes place in physical education class (see chapter 15). As an arm of education, physical education is part of the formal curriculum in public and private schools, colleges, and universities. But sport also is an important part of the extracurricular program in educational institutions. As a high school student, you probably gained an appreciation of the importance that people in your school and community placed on interscholastic sports. As a college student, you are undoubtedly aware of how important sport programs have become in higher education. Many universities are identified more by their football and basketball programs than by their academic excellence.

> For middle and high school coaches, directing the team is merely one of many responsibilities they have at the school. For most, the primary responsibility is teaching classes.

ACTIVITY 16.1

Day in the Life
What does it take to become a director of recreation, and what does a typical day look like? In this activity in your web study guide, you'll discover how closely your ideas about what this career entails match the reality as reported by one practicing professional.

High School and Middle School Settings

In public and private schools, coaches usually serve on the teaching faculty or in an administrative capacity as well, although some school districts hire part-time coaches as fewer full-time teachers opt to take on these responsibilities. Sometimes these part-time coaches, although they may have expertise in their chosen sport, have little training in kinesiology or education. Fortunately, school boards and superintendents increasingly are recognizing the need to hire only fully qualified coaches to work with student athletic teams. In most cases, coaching is an add-on responsibility to the primary responsibility as a teacher of physical education or other subject. Scholastic coaches, particularly in football and basketball, are often in the public eye. Because athletic teams are considered representatives of the school to the broader community, coaches must pay careful attention to the conduct of their athletes and, of course, their own conduct. Usually, practices and games take place on school property, although travel to other schools is necessary for away games. Coaches usually work during after-school hours, and in some sports the season can last more than 5 months. Coaches usually are among the most popular and admired teachers in the school. In many private schools, all students are required to be a member of at least one interscholastic team, and coaches must accept all willing and committed students as part of the team regardless of their athletic talent. Consequently, the range of talent among the sport team members in private schools may vary greatly. In these situations the coach must be able to deal effectively with both highly skilled and unskilled players.

College and University Settings

Work settings at the college level vary according to the type of institution and level of athletic competition. Coaches at community colleges usually coach one sport and teach physical education classes, although this can vary from National Junior College Athletic Association (NJCAA) Division I through Division III schools. Teaching and coaching take place in the same facility, which generally consists of an office complex, gymnasiums, weight rooms, swimming pools, and an athletic training room. Because the programs at many community colleges and technical colleges cater to part-time adult students, athletic programs sometimes do not receive strong emphasis. At these schools, athletic programs and facilities may be modest, although that is not always the case. Not surprisingly, the larger the school and the more emphasis the school places on sport, the larger and better equipped the facility will be. Athletic programs at larger community colleges usually participate as members of the NJCAA, which sponsors national championship competitions each year in a variety of sports. Currently more than 500 institutions are members of the NJCAA, which offers sports in Divisions I through III to more than 45,000 athletes.

Salaries Surge Upward for Division I Football and Basketball Coaches

Success (i.e., winning) translates into big bucks in Division I athletics. In January 2012, *USA Today* (Berkowitz & Upton 2012) reported that new head football coaches at major college programs will be paid an average of 35% more the following season than was paid in 2011. The average basic compensation at the schools making changes will increase to more than 1.5 million dollars next season. This increase in pay will in turn spur a corresponding increase in pay for all head coaches in the NCAA's Football Bowl Subdivision. Top payouts for newly-appointed coaches in 2012 include Urban Meyer (Ohio State, $4 million), Charlie Weis (Kansas, $2.5 million), Bill O'Brien (Penn State, $2.3 million), and Mike Leach (Washington State, $2.25 million). Some of the top-earning men's basketball coaches include Florida's Billy Donovan and North Carolina's Roy Williams, who make approximately $3.5 and $1.6 million per year, respectively. At least five NCAA Division I women's college basketball coaches are making more than $1 million, led by UConn's Geno Auriemma at just over $1.6 million, according to a *USA Today* survey. All of these salaries are substantially more than the presidents of those respective universities earn, as well as the President of the United States.

The worksites at some small 4-year colleges are similar to those of community colleges. Coaches at small colleges are likely to hold faculty posts in the physical education or kinesiology department in addition to their coaching responsibilities. Sometimes coaches in small institutions gain tenure as faculty members.

Where coaches also serve as instructors in the physical activity academic program, they must juggle the responsibilities assigned by the head of the kinesiology department with those assigned by the athletic director. Coaches at smaller colleges often become very involved in the social and academic life of the school. Usually, the pressure to win and the outside obligations at small (Division III) colleges are not as burdensome as those that coaches at the highest levels experience, although self-imposed pressure to win is a potential cause of stress at all levels of competition. Sometimes coaches in small institutions gain tenure as faculty members.

Among larger American colleges and universities, coaches are commonly employed full-time to coach a single sport. At universities whose teams rank among the elite, a head coach often operates more as a CEO than as an instructor, overseeing a vast operation involving high-pressure recruiting and intense publicity. This is the rule rather than the exception with the sports of Division I football and Division I basketball (men and women). Coaches at large public universities are now often the highest-paid public employees in their states, earning several times more than the university president or the governor (see sidebar). Along with the large salary comes heightened expectations—if head coaches are not highly successful, quite likely they will be replaced. When the head coach is fired, most of his or her assistants are fired as well.

Coaches in 4-year colleges and universities work in an athletic facility consisting of office space, gymnasiums, weight and conditioning rooms, swimming pools, and an athletic training room. Schools offering football commonly have a separate stadium and office complex for the football staff. Coaches and athletic administrators (e.g., athletic directors and their assistants) are not the only ones who have offices in the athletic facility. Compliance officers, academic tutors and counselors, sport marketers and promotion personnel, media relations officers, facility directors, and event directors typically have offices there as well. Although the structure of university athletic departments varies, all athletic faculty report to an athletic director, who in turn reports directly to a vice president or to the university president.

Other Institutional Settings

Opportunities for coaching are available in many other institutional settings, including residential and nonresidential schools for children with disabilities (e.g., schools for the deaf or blind), residential juvenile institutions for troubled or disadvantaged youth, detention facilities, and military settings. Increased opportunities are now available for coaches, instructors, and administrators in some large churches that have excellent facilities in which to conduct sport and exercise programs. In all these institutional settings, the general philosophy and special mission of the institution will guide coaches and coaching practices. Generally, sport in such settings is intended to accomplish an educational or recreational mission rather than provide entertainment for spectators. Those considering a coaching job in such settings should make sure that they will feel comfortable working in the environment of the specialized institution, because coaching is usually highly integrated into the mission that the institution has been established to serve.

Commercial Sport Settings

Commercial sport settings are those in which sport instruction or coaching is offered as a product in a business venture, usually by self-employed professionals. These settings range from the huge arenas, stadiums, and practice facilities of professional teams to tennis, swimming, or squash facilities at swanky country clubs and health spas. An increasing number of human performance clubs are focusing attention on development of sport skills and improved athletic performance, in addition to exercise. Coaches and teaching professionals who work with elite athletes often have the most diverse worksites of all teaching and coaching professionals. Some instructors and coaches work at a fixed facility, whereas others may teach and coach at a variety of locations. Some itinerant golf or tennis teachers work out of their homes or the trunks of their cars. Other sport instructors own and operate their own martial arts studios, gymnastics schools, or swimming clubs. Coaches of professional sport teams in football, baseball, soccer, hockey, and basketball work in fixed settings as employees of sport organizations. Professional coaches work at permanent facilities similar to those in colleges and universities; such facilities contain office space, practice and game areas, weight rooms, and athletic training rooms. In this type of organization, coaches usually report to a general manager and team owner.

ACTIVITY 16.2

Settings

Consider the challenges of working in different settings in this activity in your web study guide.

▶ In the elite sport setting, some teachers and coaches have a stable worksite, whereas others teach and coach at a variety of facilities. Still others are self-employed and may work out of their homes.

Elite athletes hire a select number of highly qualified coach-instructors. These professionals tend to freelance. Such coaches are often on the road, traveling to their clients' hometowns. For example, figure skating coaches work with high-caliber competitive figure skaters at the skaters' home rinks. Professional coaches who are employed by sports academies have the luxury of remaining at a fixed location.

A few sport instructors who work with elite athletes have the luxury of a stable worksite. These sites are usually schools or academies with residences on or near the campus where budding teenage Olympic or professional athletes can stay for several months of the year. For example, affluent parents who believe that their children have a promising career in basketball, baseball, ice hockey, golf, soccer, or tennis might consider enrolling their children in the IMG Academy. Aspiring young golfers may enroll full-time at the International Junior Golf Academy in Hilton Head, South Carolina, where they spend much of the day practicing and learning golf from highly skilled instructors and attending campus-based preparatory schools.

Sport Instruction on a Large Scale: IMG Academy

IMG Academy is the largest, most successful multisport training and education institution in the world. More than 12,000 junior, collegiate, adult, and professional athletes from more than 80 countries attend each year. Although IMG may be best known for its famous alumni, including Andre Agassi (tennis), Vince Carter (basketball), and Adam Dunn (baseball), you don't have to be a serious competitor to take advantage of the programs. IMG Academy is a gated, 400-acre campus in Bradenton, Florida, consisting of golf driving ranges and practice facilities, tennis courts, baseball fields, turf training fields, indoor basketball facilities, sport performance facilities, gyms, clubhouses, and dormitories. It has two cafeterias (one for adults and professionals and one for students) and operates an on-campus restaurant for students in the evening. IMG Academy's divisions include these:

- The IMG Bollettieri Tennis Academy
- The IMG Leadbetter Golf Academy
- IMG Soccer Academy
- IMG Baseball Academy
- IMG Basketball Academy
- IMG Lacrosse Academy
- IMG Madden Football Academy
- IMG Performance Institute (IPI)
- IMG Pendleton School
- The IMG University and Language Center
- Sports Therapy Center
- The Wellness Spa
- Evert Tennis Academy (in Boca Raton, FL)

Young athletes stay in campus dormitories and have access to a highly competent teaching and coaching faculty. For this, their parents pay more than $30,000 per year. Tuition for the on-campus prep school and fees for private lessons and other amenities can cost an additional $35,000.

When we think of coaching and sport instruction settings, we usually think of the popular sports we see every day on television. However, coaches work in a variety of settings. Rowing, kayaking, and canoeing instructors work on rivers. With competitive rock climbing becoming more popular, we find instructors at indoor climbing walls, in the great outdoors scaling mountains, and even in towns and cities climbing stone or brick buildings. As interest in eco-tourism and outdoor adventure experiences continues to expand, the need for instructors and coaches in outdoor recreation pursuits is expected to increase accordingly. Each summer, directors of camps, some with large budgets and impressive facilities and equipment, hire expert sport instructors to oversee instructional operations. Positions as sport instructors also are available at large resorts that sponsor extensive sport programs.

Roles for Coaching and Sport Instruction Professionals

In the previous chapter we saw that physical education teaching is an exciting and rewarding career, although it often poses problems for the teacher. The challenge of teaching a large number of different skills to large numbers of students who possess a wide range of ability levels (some of whom would prefer not to be there), in less than optimal environments and often with inadequate time to accomplish instructional objectives, is somewhat daunting. By contrast, coaching and professional sport instruction are free of most of these disadvantages. Professional sport instructors usually work with relatively small numbers of highly motivated learners, athletes, or clients under optimal conditions in which attention can be focused exclusively on one skill or activity. Let's examine more closely the duties and responsibilities of sport instructors and coaches.

Community Physical Activity Program Instructors

Community-based programs of physical activity classes are those offered by municipalities through departments of recreation and other nonprofit agencies. For the most part, nonprofit organizations are on the front lines of strengthening communities by providing needed services to individuals and families. Physical activity program leaders who work in nonprofit organizations can help meet this challenge. They also teach older people in physical activity programs offered by community centers, recreation facilities, retirement centers, and other nonprofit agencies. Adult programs may include soccer, softball, tennis, swimming, golf, adult exercise, yoga, dance, and martial arts. Working in nonprofit agencies such as these can be rewarding for professionals interested in serving an older clientele. As with employment in all nonprofit settings, financial compensation tends to be less than that available in commercial or other institutional settings, but knowing that your services are valued can be a form of compensation in its own right.

Community Youth Sport Development Leaders

A particularly exciting area of community work focuses on underserved youth. Over the past 2 decades, many communities have experienced damaging social, economic, and political influences. These pressures have had a direct effect on the way in which human services are made available today.

Throughout the United States, communities are looking for a solution to the crises facing America's youth. The typical approach has been to focus on what is wrong with young people and then develop strategies and programs that will solve the problem. This traditional approach to helping youth, especially underserved youth, is well intended, but it focuses on kids' deficiencies and problems rather than their capacities, skills, and assets. The new approach assumes that young people have knowledge and positive qualities that can be enhanced. Increasingly, community-based organizations are using this asset-based approach in their physical activity settings for youth.

Physical activity is a particularly effective way to focus on kids' capacities, skills, and assets. The unique energy, human interactions, and creativity that are harnessed by physical activity allow community-based physical activity program leaders to see kids' potentials and personal qualities that might otherwise go unnoticed. Furthermore, youth development philosophy holds that young people who are currently having difficulty in school or at home, or even

▶ Nonprofit organizations and park and recreation centers offer the majority of physical activity programs in community settings.

those involved with the juvenile justice system, can, when given the proper opportunities, beat the odds and become contributing members of society.

Because of the assorted needs of community-based organizations, physical activity professionals at this level perform a range of tasks beyond instruction. Typically they assume responsibility for directing the physical activity programs, including marketing the program, budgeting, and overseeing a staff. In addition, they may be involved with a variety of managerial duties as well as tutoring, mentoring, literacy programs, and a host of other national community services. If you are a kinesiology major who wants to work with youth but not in a public school, becoming a community physical activity instructor may be an alternative. Salaries for some of these community youth development positions tend to be lower than those of public school teachers. But, as with all nonprofit ventures, the attraction is having the opportunity to make a difference in kids' lives and knowing that you are contributing to your community.

Professional Sport Instructors

▶ The professional sport instructor is one who offers services (private or group lessons) solely in an activity of his or her expertise. More than any other teacher, the teaching professional is similar to a coach, although he or she is not saddled with all of the administrative duties commonly performed by the head coach of a team or group.

Teaching professionals and *professional sport instructors* offer services (private or group lessons) in a specialized physical activity in a professional setting. Swim, golf, and tennis professionals who work either as freelance entrepreneurs or at clubs may also hold positions as members of the physical activity instruction faculty at colleges and universities. Sometimes they teach beginners but may also coach highly skilled individuals who are competing in leagues and tournaments. Outside the university, the teaching professional can be found at private, semiprivate, or public teaching facilities instructing beginning, intermediate, advanced, or elite players. Golf professionals, for example, may work at a municipal golf course giving lessons to the public and conducting golf camps or managing the pro shop or the clubhouse. And, of course, some serve as coaches or instructors to players on the Professional Golf Association and Ladies Professional Golf Association golf circuits.

Professional sport instructors typically are self-employed and are hired as independent contractors. Employment security for certified coaches and instructors depends on the satisfaction of the clients and the owners or managers of the instructional facility. Thus, certified professional sport instructors, like professionals in other fields, have a high degree of control over their careers. Their continued employment depends on their ability to provide services that their clients find valuable and are willing to pay for. Unlike coaches at institutions, professional instructors usually are not covered by group insurance policies and retirement plans. More employment opportunities are available at the beginner or intermediate levels than at the elite or professional levels. Although many people want to learn to swim, golf, rock climb, and sky dive, only a select few (in relative terms) want to work hard and pay an instructor to become proficient enough to perform at an elite competitive level. Incomes for private instructors can range from only a few thousand dollars a year (the golf instructor whose office is the trunk of her car, for example) to more than $100,000 a year for the instructor-entrepreneur (owner, operator, and instructor of a successful golf instruction school).

Certification and Education of Community Physical Activity Program Instructors and Professional Sport Instructors

Do you recall our discussion in chapter 12 about professionalism and the importance of certification procedures to ensure that only the most highly qualified people can serve in professional capacities? Community physical activity and sport programs require various types of credentials for the instructors who lead them.

Certification for Community Physical Activity Instructors

No certification programs exist specifically for youth sport development work focused on underserved children. A bachelor's degree is required for most positions, preferably from a program that offers a minor specialization in youth sport development. Increasingly, colleges

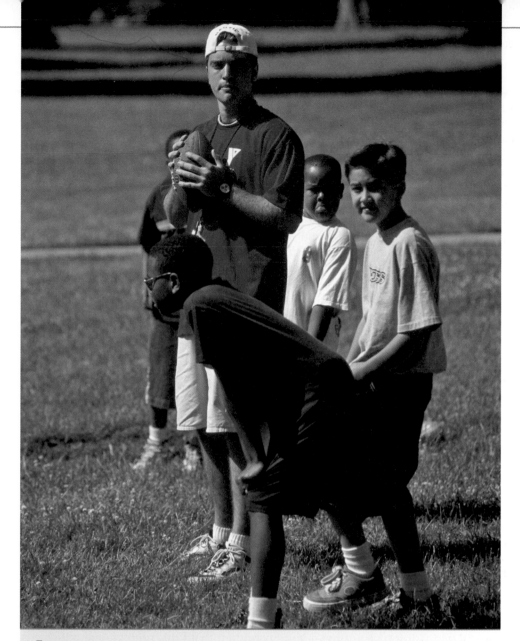

Coaches who work with underserved children not only help them develop athletic skill, they also teach them to be personally and socially responsible young people.

and universities around the country are beginning to offer bachelor's and master's degrees in kinesiology with a concentration in what is termed either *community youth sport development* or *community leadership* (Hellison et al. 2000). These programs teach students how to develop, implement, and evaluate community-based physical activity programs for young people. Such programs also include courses on youth agency administration, principles of community organization, grant writing and fund-raising, social work, and community development. Elective courses in social work, counseling, and management may be helpful. Internships that involve working with children daily are an important part of these programs.

Certification for Professional Sport Instructors

Although certification in the form of college degrees is usually sufficient for coaches of the most popular sports (basketball, football, baseball) in the United States, certification also can be important in cases in which the coach or instructor works in a noninstitutional setting such as a sport club. Most organized sports offer some sort of instructor or coaching certification. For example, most coaches of age-group teams in large soccer clubs are certified (licensed) by the United States Soccer Federation (commonly referred to as U.S. Soccer), the National

Soccer Coaches Association of America (NSCAA), or both. Likewise, to teach swimming or to work as a lifeguard will in most cases require you to secure American Red Cross certification as a water safety instructor. This certificate will enable you to teach swimming lessons and coach the swim or synchronized swim teams at a private club, YMCA, or community center. If you are interested in teaching scuba diving, you will need to acquire the proper certifications from PADI (Professional Association of Diving Instructors), which start at the assistant instructor level and move to the open-water scuba instructor level.

The United States Fencing Coaches Association (USFCA) has a rigorous certification program offered through Coaches College, in which would-be instructors move through five levels. At levels I and II they master skills in use of the three weapons (épée, saber, foil), and at levels III through V they specialize in one of the weapons. To pass the first part of the certification program they must write a paper, propose a 1-year training cycle, and serve as an intern at Coaches College. At the highest level, they must develop a fencer under their supervision. Only when the fencer places in the USFA ranking list or in the top 12 of NCAA fencers will the USFA grant certification.

The men's Professional Golf Association (PGA) and the Ladies Professional Golf Association (LPGA) also offer instructor certification programs. One becomes a golf professional under either PGA or LPGA auspices only by passing an arduous education and testing program that can take up to 8 years to complete. The PGA certification program requires study, training, experience, and testing in every aspect of golf—not only playing aptitude and teaching technique but also club repair, analysis and correction of swing flaws, supervision of tournaments, and the business side of golf including management of golf shop operations. Certification programs also exist for many other sports (see pp. 468-471). If you plan a career as an instructor or coach in one of these sports, you should begin working on the certification for that sport as soon as possible.

Coaching

As with becoming a teacher, becoming a coach means building relationships that can change the life of another for the better.

We all have known people who have touched our lives as no one else could. Often the emotions associated with our memories of these people—the feelings of respect, awe, deference, and admiration—seem as vibrant and real as when we first experienced them. For many current and former athletes, a coach is in the forefront of their minds and hearts.

Many opportunities are available to those who want to coach. Individuals coach a variety of people in many different settings—young children to adults, recreational athletes to elite athletes. Some college and professional coaches are paid hundreds of thousands or even millions of dollars, whereas others coach simply for the joy it brings. Coaches can be found in sport programs in communities, secondary schools, colleges and universities, and professional sports. These programs have different coaching requirements, philosophies, and expectations, and the coaches in these programs often differ in their educational background and coaching experience.

Youth Sport Coaches

Municipal recreation departments usually offer a wide range of competitive youth and adult sports, for which the director or supervisor of recreation is responsible. The range of activities offered depends on the ingenuity of the director and the available budget and facilities; activities can include age-group swimming programs conducted in public swimming pools as well as highly organized competitive programs in football, baseball, softball, soccer, basketball, tennis, and other popular sports. Generally, but not always, these programs tend to deemphasize winning and emphasize inclusiveness and fair play. Volunteers recruited and trained by the director of the program usually staff community programs. The director also is responsible for hiring and training officials to oversee the contests. Games usually take place during evening hours, on weekends, or during the summer.

Organizations that rely on volunteer coaches often face the difficult task of finding qualified, certified coaches. Because many states do not have specific standards or certification requirements for volunteer coaches, anyone interested in coaching is often deemed qualified. Generally, community organizations are short of volunteers, so anyone who wants to coach at this level can usually find a coaching position. Often a parent of one of the team members assumes the role of coach. Coaches who have had formal education in kinesiology may serve as administrators of large community-based youth sport leagues or as instructors of coaching clinics that are offered to laypersons who coach in youth sport league. If you are interested in pursuing a career in coaching, volunteering to coach a local youth sport team while still an undergraduate student will provide you with excellent ground-floor preparation. In addition, the experience will allow you to determine whether coaching is the right profession for you. Consider that the essentials of coaching at the youth level differ significantly from those required at the high school or college level of competition.

A relatively new development in youth sports is private sport programs in which for a fee, children join "travel teams" that, as the name implies, travel around a county or state to experience competitive interleague play. Proceeds from fees are used to pay the coach. Some civic programs offer competition on a regional or national scale. For example, age-group swimming programs compete at local, regional, and national levels, as do youth soccer programs. In theory the more gifted competitors are selected for the traveling teams, which offer the players opportunities to compete with others of the same ability. In practice, however, there are inherent problems with player selection in many of the travel programs, and the added competitive pressures imposed on young athletes are not always in their best interest. Little League sponsors a national baseball tournament that leads eventually to the Little League World Series in Williamsport, Pennsylvania, which is televised nationally.

Some of these programs hire professional coaches. Pop Warner (football) and Amateur Athletic Union (basketball), which are private organizations, are two of several national programs in the United States for children that are more competitive than community-based programs. Coaches at this national competitive level usually spend more time performing the nonteaching duties associated with coaches, such as scouting and recruiting new talent, viewing game films, and organizing team travel. Large soccer clubs field teams in several age groups (U-10 up through U-18 ages, on both the boys and girls side), and the pressure to win, even at the young ages, is intensified. Winning clubs attract more players, which in turn generates more money for the club. Coaches usually are experienced players or are

When Commercialism Threatens Youth Sports

A growing problem confronting those committed to the educational benefits of youth sport programs is the tendency for organizations, media, and corporations to exploit them for commercial gain. Some may question, for example, whether the Little League World Series, televised around the world, is appropriate for young children. Others question the increasing trend for professional sport organizations to sign athletically promising young children to professional contracts and commercial endorsements. In recent years, sponsors have signed young children to lucrative multiyear contracts in the hope that they will mature into professional standouts. For example, several years ago Nike signed Freddy Adu, an exceptionally skilled young soccer player, to a $1 million endorsement at age 13; although Adu now plays professionally, he has not achieved the level of prominence in the soccer world that was expected. Other top professional teams have adopted Nike's philosophy. Bayern Munich, Germany's richest and historically most successful professional soccer team, recently sold Mehmet Ekici, who had been signed with the club at 8 years of age, to Werner Bremen for 5 million euros. In 2011, Real Madrid, a top Spanish club, signed Argentine prodigy Leonel Angel Coira to its youth academy after seeing him in tryouts. Leonel is 7 years old! The signing reflects the philosophy of top teams that lock in fresh talent as early as possible to avoid paying huge transfer fees when their potential blossoms. Obviously, with respect to these young athletes, soccer is no longer a "youth sport," and the pressure imposed on young athletes has intensified.

licensed by U.S. Soccer or NSCAA. The coaching staff is typically hired and coordinated by the club's director of coaching, who oversees the entire program. As a consequence, coaching directors of large soccer clubs often earn a salary commensurate with most college soccer coaches and are held to the same high expectations. If their teams are successful (i.e., win), then these coaches are likely to keep their positions; if not, they may experience the same fate as college and professional coaches who do not win.

Coaching in Institutional Settings

Community coaches volunteer in community-based sport programs that range from educational, inclusive, and relatively noncompetitive programs to those that emphasize competition, performance, and winning.

Institutional coaches are those who work in educational institutions such as secondary schools (middle, junior high, or high schools), colleges, or universities. As mentioned earlier, institutional coaches may also serve in specialized residential institutions for young people who are disabled or socially disruptive. Some coaching opportunities are also available in the armed services. In each case, the coach strives to evoke each athlete's potential and guide the athlete in overcoming personal challenges and achieving personal success within his or her limitations. Coaching duties include teaching physical skills; keeping individual and team statistics; scheduling practices, games, and tournaments; and managing and maintaining equipment. Working at an institution requires physical activity professionals to relinquish some autonomy in the carrying out of their duties. Generally, coaches working in institutions do not enjoy the freedom of the freelancer, although they do have the benefit of a steady and reliable income. Complying with institutional expectations and conducting oneself in accordance with established codes of conduct are essential. Ultimately, the institutional coach serves at the behest of the institution's administrator or its board.

Coaching in Middle and Secondary Schools

Secondary-level school coaches typically spend their days teaching and receive an additional stipend for coaching.

According to the National Federation of State High School Associations (2011), more than 7.5 million high school students participated in athletic programs in 2010-2011. An enormous number of people are required to coach these millions of students. For those who love sport and love being around young people and serving as a positive role model, coaching can be an exciting and rewarding career. At the same time, you should factor a certain dose of realism into your decision about whether to pursue coaching as a career. The most important point may be the fact that coaches work many hours for relatively low compensation. Even high school coaches of major sports such as football often earn only a small stipend on top of their regular teaching salaries. The committed coaches who love what they do would tell you that the money is merely icing on the cake; the rewards of getting close to the kids on their teams, of playing a substantial role in their development, and of having the respect of the community at large are more than adequate compensation. You should weigh these considerations before you make your decision.

Middle, junior high, and high school coaches usually are certified public school teachers who teach a full schedule and receive a stipend for taking on the extracurricular activity of coaching. Certified, part-time coaches who aren't members of the school faculty may be hired to coach specific teams. Although the qualifications for coaches in the public schools vary by state, a teaching or coaching certification and background check are typically required. The coach is one of the most visible representatives of the school. Unlike teachers who work in the confined spaces of the classroom, with only their students, parents, and principal assessing their performance, the coach performs in front of all the preceding as well as the community at large. The daily newspapers cover his or her successes and failures. The coach's performance is often the center of conversation in local diners and coffee shops. Everybody is allowed to hold an opinion (and to voice it) about the effectiveness of the coach.

The number of contact hours between coaches and athletes varies depending on the governing body regulating public secondary school athletics. Typically, secondary school coaches spend 8 to 10 hours a week in after-school practices. When competitive play begins, the number of practice hours may decrease to 8 and game hours may increase to between 4 and 6. Some coaches have early morning practice schedules; others have late afternoon or early evening schedules.

The Parent Problem

Parents who desperately want their children to succeed at sports can become a serious problem at the site of competition for youth sport coaches and organizations. Within the past few years, we have read about a father who was sent to jail for beating and berating a coach because the coach took his 11-year-old son out of a game; a father (a dentist) who sharpened the nose guard of his son's football helmet so that the son could slash opponents; and a police officer who gave a Little League pitcher $2 to hit a 10-year-old player with a fastball during a game. In Massachusetts, a hockey player's father who was supervising a practice session was beaten to death by another father because of a disagreement concerning the rough play allowed during practice. Such incidents have led to the drafting and promulgation of sportsmanship codes for parents. In some states parents must sign the code before their children will be permitted to play. Perhaps this approach will help correct many of the problems, but the coaches, directors, and referees of such programs must remain vigilant, not hesitating to exclude parents who threaten to turn the fun of sport into the tragedy of violence. Organizations such as the Institute for the Study of Youth Sports (ISYS) at Michigan State University and the National Alliance for Youth Sports (NAYS) have been instrumental in educating parents and coaches in this regard. The "Sport Parent Code of Conduct," by NAYS, and the "Bill of Rights for Young Athletes," written by Dr. Vern Seefeldt and Dr. Rainer Martens and distributed by the ISYS, have been widely adopted by youth sport programs across the United States.

Besides spending time in practices and games, coaches have many responsibilities behind the scenes (see figure 16.3). To varying degrees, all coaches are involved in organizing and scheduling team meetings and practices, ordering equipment, maintaining and overseeing inventory, checking athletes' eligibility, arranging transportation to and from events, talking to the media, preparing for home events, fund-raising, and organizing end-of-season banquets. The extent of coach involvement in these activities depends in part on the size of the coaching staff and the number of support personnel. For example, some schools have a full-time athletic director (AD), who is responsible for organizing home contests. Other schools may have a part-time AD, who must delegate that responsibility to another coach. Likewise, some sports (especially football) require many assistant coaches to whom head coaches can delegate specific duties.

The differences in coaching at the various educational levels tend to be a function of the athletes' developmental differences. Some middle school athletes are "diamonds in the rough" with the potential to be excellent athletes but are physically undeveloped (i.e., prepubescent and small) compared with others on the team. Others may have just gone through a growth spurt and need to relearn a given skill in their new-sized body.

Middle or junior high school coaches have the unique responsibility of introducing young athletes to the world of interscholastic sport, which requires athletes to balance the attention they give to their academics and their sport. The middle school coach's primary tasks are

- to help students interested in athletics develop a positive view of themselves as athletes and
- to teach them important, complex physical skills and the basic strategies needed to become competitive student-athletes.

These coaches also must introduce their athletes to a concept of competition in which winning takes on greater importance than it often does at the community level.

The high school coach is primarily involved with students who are interested in athletics and want to invest the time needed to become good athletes. The high school coach has the opportunity to build on the foundation set by the middle or junior high school coach by teaching more complex tasks and game strategies. Competition and the importance of winning take on new meaning at the high school level. The increased emphasis on winning necessitates that coaches choose their teams wisely. Coaches spend more time reviewing game films, developing strategies specific to a given opponent, and scouting and recruiting potential team members.

Coaching in Colleges and Universities   Winning is clearly extremely important to all intercollegiate coaches; the pressure to win at this level, especially at large universities, is much greater than that at the high school level. The enormous financial stakes of big-time college athletics can reduce it from an enjoyable, educational experience for everyone to a brutal exercise in survival. As anyone who reads the sport pages knows, the coach who doesn't win his or her share of games is soon fired.

The quest of college coaches is to maximize the athletic potential of already skilled athletes, build on the foundation established in high school, and fine-tune each athlete's ability to move that athlete toward becoming a member of the elite. Most college coaches have the advantage of working with a relatively small number of highly skilled athletes in well-equipped facilities. Given the popularity of basketball and football in American society, these two sports tend to receive most of the resources of the athletic department. Coaches of "minor" sports such as cross country, volleyball, softball, soccer, tennis, and golf usually have modest budgets and resources at their disposal.

Like secondary-level coaches, college coaches have many behind-the-scenes responsibilities. To varying degrees all coaches are involved in team, facility, and equipment management and in budgeting. At the college or university level, talent scouting, recruiting, fund-raising, public relations, and athletic eligibility take on greater significance. The degree of significance often depends on the type of college in which the coach works.

> College coaches aim to maximize the athletic potential of already skilled athletes but also have many other responsibilities, including team and facility management, budgeting, monitoring academic progress, and recruiting.

ACTIVITY 16.3

Day in the Life
What does it take to become a head baseball coach at the college level, and what does a typical work day look like? In this activity in your web study guide, you'll discover how closely your ideas about what this career entails match the reality as reported by one practicing professional.

Employment opportunities at the college or university level are not as plentiful as they are at the secondary level. Because there are no national or state qualifications or requirements for coaching at the college or university level, most colleges and universities searching for a coach look at applicants with an established track record in the sport. Typically, establishing a name requires college playing experience, coaching experience, a history of successful seasons, and being known by many other coaches and administrators. Many Division I coaches began their careers at the secondary level and moved through the lower divisions of college athletics. Others managed to secure an internship during their undergraduate years or sign on as team manager and, with a great deal of effort, work their way up to an assistant coaching position.

Coaching in smaller colleges offers many opportunities for coaches to get close to their players and to meet and interact with them in many different settings. Although pay generally is not high, the rich quality of life and reduced pressure to win offer some compensation. Coaching in larger institutions where coaches' jobs, and their assistants' jobs, depend on the team's win–loss record usually requires an enormous commitment in time and energy, sacrifice of one's family and social life, and the fortitude to withstand the anxiety in years when the team fails to win many games. If you happen to be one of the few perennially successful coaches, however, coaching can be a lucrative and exciting career.

The Long Shot: Probability of Competing in Athletics Beyond High School

A National Collegiate Athletic Association (NCAA) survey on the estimated probability of competing in athletics beyond high school provides sobering numbers to those who envision playing at the college and professional levels. Although many high school players dream of playing at the next level, data published by the NCAA show the low probability of these dreams being fulfilled:

Athletes	Women's basketball	Men's basketball	Baseball	Men's ice hockey	Football	Men's soccer
High school athletes, n	538,933	545,844	471,025	36,912	1,108,441	398,351
High school senior athletes, n	125,409	155,955	134,579	10,546	316,697	113,815
NCAA athletes, n	15,708	17,500	31,264	3,944	67,887	22,573
High school athletes who move to NCAA, %	3.6%	3.2%	6.6%	10.7%	6.1%	5.7%
NCAA athletes who become professional, %	0.9%	1.2%	11.6%	3.7%	1.3%	1.0%
High school athletes who become professional, %	0.03%	0.03%	0.60%	0.10%	0.08%	0.04%

These percentages are based on estimated data only and should be considered approximations only of the actual percentages.

Adapted, by permission, from NSCA, 2012, *Estimated probability of competing in athletics beyond the high school interscholastic level* (Indianapolis, IN: National Collegiate Athletic Association). Available: www.ncaa.org/wps/wcm/connect/public/ncaa/pdfs/2011/2011+probability+of+going+pro

The NCAA has estimated that an athlete's chance of competing in his or her chosen sport at the college level is not great (NCAA 2011). For example, only about 3% of high school senior basketball players will play NCAA-sponsored basketball. These figures do not take into account the opportunities that are available to compete in the lower divisions of the NCAA, National Association of Intercollegiate Athletes (NAIA), and National Junior College Athletic Association (NJCAA).

Coaching in Professional Sport

Earlier in this chapter we discussed the teaching professional. When one thinks of teaching pros, names such as Tim Gullikson (tennis), Jim Suttie (golf), and Jim Counsilman (swimming) may come to mind. Who comes to mind when you think of professional coaches? Such a question may evoke images of Jim Leyland and Joe Girardi (baseball), George Karl and Phil Jackson (basketball), and Bill Belichick and Tom Coughlin (football). Although the number of professional sport coaches is small compared with the number of secondary school and college coaches, professional sport coaches typically have national prominence.

As with other professional coaching positions, facilitating athletes' performances is central to the job of the professional sport coach. The position also includes elements of administration, athlete recruitment, and public relations events and media appearances. Generally, the professional sport coach has more nonteaching duties than does the teaching professional discussed earlier. For example, some sports, such as football and hockey, require coaches to spend a significant amount of time participating in coaching meetings and reviewing game

Generally, the professional sport coach has many nonteaching and noncoaching duties, including administration, recruitment, and media appearances. Continued employment depends heavily on producing a winning team.

films; continued employment depends heavily on the performance of their teams. The win–loss record in such sports is more important than the win–loss record in sports such as golf and tennis. Additionally, the professional sport coach deals solely with elite athletes, whereas the teaching pro is often involved with beginning and intermediate athletes.

So You Want to Be a Professional Sport Coach? Good Luck!

Securing a head coaching position at the professional level may pose the most difficult job search imaginable. Consider the following: The five major professional sports in the United States have a total of 140 head coaching positions (National Football League, 32; Major League Baseball, 30; National Hockey League, 30; National Basketball Association, 30; Major League Soccer, 18). Although it is true that the tenure of a professional sport coach may be brief and that openings do occur on a regular basis, it is also true that coaches already within the professional coaching community are often first in line to fill the most recent vacancy. This is evidenced by the fact that several coaches in each of the professional sports have coached two or three different teams within the league. They get fired or leave one job only to fill a similar position with another club.

Employment opportunities for the professional sport coach are few and far between (see sidebar). The number of professional sport teams is limited, and many individuals are qualified to coach at the professional level. At the professional level, as in the college and university setting, no national or state qualifications are needed for coaching. Prospective professional sport coaches need an established name in the sport; most acquire one by successfully coaching at the college or university level.

Job security for the professional sport coach is tenuous at best. When you coach a team sport at the professional level, you have entered an industry where success relies in large part on public support. In professional sports, public support waxes and wanes relative to a team's win–loss record. Just as the organization's success depends on producing a winning team, so does retaining your job. Many professional sport coaches thrive on this aspect of the job. They bask in the challenge of creating or maintaining a winning tradition, and accomplishing that goal brings great rewards and personal satisfaction.

Certification and Continuing Education of Coaches

Coaching has been described as both an art and a science. Whatever the actual balance between the two, coaching is a highly complex process of applying knowledge about human performance, human development, motor learning, and exercise. Coaching also involves high-order cognitive skills such as analysis and diagnosis and basic counseling skills that enable the coach to communicate with the athlete. All successful coaches must have excellent interpersonal skills and the ability to teach the techniques, tactics, and strategies of a given sport based on motor learning principles. Coaches must understand the fundamentals of exercise as it relates to training and conditioning. Managing any sport enterprise entails planning, organization, appropriate staffing, leading others to team goals, monitoring progress toward team goals, and understanding the legal liability and risk factors associated with making decisions. In the last 2 decades, first aid and drug education have become significant concerns for coaches. As the sports medicine and athletic training fields have become more established, the coach's role has expanded to include injury prevention and emergency care facilitation. When injuries do occur, the coach should be knowledgeable enough to help trained medical personnel.

Certification of Youth Sport Coaches

Certification is becoming increasingly important as the problems afflicting youth sport have become more serious. Never before has the need for highly qualified coaches who operate on the basis of a sound coaching philosophy been more obvious. Coaches for most youth sport teams are parents and others who volunteer to serve. The kinesiology graduate is most likely to be the organizer and administrator of the program, not a coach. As such, you should recognize the value of certification for your coaches. A number of certification programs are available for youth sport coaches. The National Youth Sports Coaches Association (NYSCA) and the American Sport Education Program (ASEP) offer excellent educational programs for coaches. The National Coaching Certification in Youth Sports, offered by the United States Sports Academy, is available online, on-site, or through correspondence. All these programs require a modest tuition fee. Various sport-specific certification agencies (e.g., U.S. Soccer) are also available for equipping youth sport coaches with the skills and knowledge required for successful coaching.

U.S. Soccer National Coaching Certification Program

Millions of boys and girls under the age of 19 play organized soccer in the United States, creating a huge demand for coaches at all levels. In an effort to educate coaches and provide a standard for coaching young soccer players, U.S. Soccer has instituted a national coaching certification program that offers both state and national coaching licenses for various levels of expertise and experience. Certification begins at the state level, which focuses on elementary principles of coaching and serves to prepare coaches for the 36-hour State D License course. National A, B, and C courses consist of 7 days of instruction and 2 days of extensive oral, written, and practical examinations.

U.S. Soccer also offers the National Youth License. This coaching license is designed to provide youth coaches, club coaching directors, physical education teachers, and soccer administrators with the knowledge to successfully structure soccer environments for children aged 4 to 12. The National Youth License course explores the role of the coach as a facilitator; the physical, mental, and emotional needs and capabilities of 4- to 12-year-old players; lessons from developmental psychology; and the art of teaching. Candidates are videotaped for analysis during live training sessions.

Education and Certification of Middle and High School Coaches

Your department probably offers a curriculum (usually as a minor rather than a major) to prepare you for coaching. Most regional school districts and local coaching societies sponsor continuing education programs and annual clinics for middle school, high school, and college coaches. In some states these clinics attract thousands of coaches and feature presentations by some of the most popular college and professional coaches. Some coaches, eager to learn more about the scientific side of coaching, enroll in master's degree programs that offer a concentration in coaching and sports medicine.

Coaching education experiences are available through various national coaching associations. A good example is the certification program offered by the National Soccer Coaches Association of America (NSCAA), which sponsors a coaching academy that offers residential and nonresidential courses. The residential academy offers course work toward National, Advanced, and Premier Diplomas. The National Diploma is a 50-hour course that includes fundamental coaching tactics. Graduates must pass oral, written, and practical tests. The Advanced Diploma is a 50-hour course for the experienced coach. A minimum of 10 years of coaching experience is required. The Premier Diploma is a 50-hour course that examines different systems of play, nutrition, sportsmanship, ethics, and personal and professional development. To enroll for the Premier Diploma, a candidate must have passed the Advanced Diploma with a grade of distinguished pass. The courses are very informative but also quite costly (the National Diploma course costs $1,050 for nonmembers residing at the academy) (NSCAA 2012).

➤ The American Sport Education Program and the Coaching Association of Canada offer general coaching certification programs particularly for community and scholastic coaches. Certification requires knowledge of sport psychology, sport physiology, coaching philosophy, sport pedagogy, sport management, sport first aid, and drug education.

National Soccer Coaches Association of America Coaching Program

The NSCAA Coaching Academy offers a comprehensive coaching certification program designed to enhance the knowledge of soccer coaches. The following is a brief summary of the NSCAA offerings.

Nonresidential Courses

Nonresidential courses vary in length and are intended primarily for coaches in a particular community and region, so lodging and meals are not part of the course packages. Because the NSCAA appreciates the desire of coaches to take courses germane to the level of player they are coaching, only a few of the courses have prerequisites (noted in the following table). Residential courses offered at selected sites are also available (NSCAA 2012).

Course	Coaching level	Time commitment
Youth Level I Diploma	U-5 to U-8	2.5 hours
Youth Level II Diploma	U-5 to U-10	5 hours
Youth Level III Diploma	U-6 to U-10	13 hours
Youth Level IV Diploma	U9 to U12	21 hours
Junior Level V Diploma	U11 to U18	13 hours
Junior Level VI Diploma*	U11 to U18	21 hours
Junior Level VI Supplement*	U11 to U18	8 hours

*Prerequisites required.

Adapted, by permission, from NSCAA, 2012, *Coaching courses* (Kansas City, KS: National Soccer Coaches Association of America). Available: http://www.nscaa.com/education/courses

Master Coach Diploma

The Master Coach Diploma is offered by the NSCAA in response to a growing trend in coaching education—that of continuous self-improvement. The program offers an opportunity for coaches to develop a mastery of the profession of coaching (NSCAA 2012).

The Master Coach Diploma builds on the 150 hours of field and classroom learning contained in the NSCAA National, Advanced National, and Premier Diploma courses. The Master Coach Diploma program includes divisions for youth, high school, college, club, and professional coaches and involves each coach in 20 projects designed to be completed at the coach's own pace and to enhance her or his growth as a well-rounded professional. Toward that end, all coaches in the program have the opportunity to learn and grow in fields such as sport psychology, leadership, team and organizational management, use of technology, and nutrition.

The Master Coach Diploma program is designed to be completed over a period of 18 months, with all members of a given Master Coach class beginning and completing the course on the same timetable. Although a significant portion of the course involves independent study, candidates also work with a mentor drawn from current NSCAA national staff coaches and will benefit from learning opportunities generated by interaction with other Master Coach Diploma candidates.

Education and Certification of College and Professional Coaches

Interestingly, although coaching at the high school level may require preparation in kinesiology, positions in coaching at the highest levels (college and university or professional teams) may require very little in the way of formal academic preparation. Mike Krzyzewski, the famous Duke University basketball coach, for example, holds a BA from the U.S. Military Academy; University of Maryland coach Gary Williams earned a BS in business; and Bill Belichick, coach of the highly successful New England Patriots NFL team, holds a bachelor's degree in economics. None of these coaches availed themselves of a formal education in kinesiology. (Often the academic credentials of some of our most illustrious college coaches remain buried in university personnel files, unavailable for public inspection—something very curious for institutions that supposedly take academic qualifications very seriously.) If your goal is to coach at a very high level in football or basketball in the United States, getting your foot in the door to the coaching establishment at a fairly young age and showing you can produce winning teams are likely to benefit you much more than highly polished academic credentials. Experience as a head coach in a major college sport can also set the stage for a career in athletic administration once the fire to continue coaching has been extinguished. A number of prominent football and basketball coaches have moved to administration later in their careers. The experiences gained as the CEO of a major college sport can be beneficial in the administrative role as well.

 ACTIVITY 16.4

Interview with a Practicing Professional
In this interview in your web study guide, Kirk Bruce, assistant athletic director at the University of Pittsburgh, talks about how the jobs of coaches and administrators differ and what makes for success in a career in athletics.

Ethics and Coaching

Recently a prominent coach at a major university was fired for allegedly having knowledge that players violated NCAA rules and not reporting the violations to his superiors. In another celebrated case, a college basketball coach was terminated for allegedly lying to NCAA investigators during a probe of recruiting violations. Such ethical issues are not limited to the highest levels of amateur sports. Every year, middle school and high school coaches are fired for inappropriate conduct, ranging from unethical relationships with students to psychological and physical abuse of the athletes for whom they are responsible. Issues related to ethical conduct on the part of the coach have become an important consideration at all levels of coaching. Whether it takes the form of encouraging players to cheat, pressuring other faculty to give passing grades to a star player, or performing antics during a game that incite players and spectators against officials, unethical behavior should have no part in the professional's life. The National Federation of State High School Associations has promulgated a code of ethics for coaches to ensure that athletics are integrated with the total educational program and that coaches act with the athletes' welfare in mind at all times. The code of ethics is shown on page 472. If you plan to be a coach in a middle school or high school, memorize this code.

 ACTIVITY 16.5

Professions in Coaching and Sport Instruction
Evaluate the qualifications and philosophies of five coaches in this activity in your web study guide.

In the final analysis, ethical conduct flows from a sound personal philosophy. Chapter 5 emphasized how important it is for professionals to have a solid base of values on which to make ethical decisions that affect the lives of those for whom they work. Without having developed this base and applied it to coaching, you stand a good chance of making decisions by the seat of your pants rather than from the larger perspective of the values that you hold sacred. Have you developed a personal philosophy that can guide you in making decisions as a coach?

National Federation of State High School Associations Coaches Code of Ethics

1. The coach shall be aware that he or she has a tremendous influence, for either good or ill, on the education of the student-athlete and, thus, shall never place the value of winning above the value of instilling the highest ideals of character.

2. The coach shall uphold the honor and dignity of the profession. In all personal contact with student-athletes, officials, athletic directors, school administrators, the state high school athletic association, the media, and the public, the coach shall strive to set an example of the highest ethical and moral conduct.

3. The coach shall take an active role in the prevention of drug, alcohol, and tobacco abuse.

4. The coach shall avoid the use of alcohol and tobacco products when in contact with players.

5. The coach shall promote the entire interscholastic program of the school and direct his or her program in harmony with the total school program.

6. The coach shall master the contest rules and shall teach them to his or her team members. The coach shall not seek an advantage by circumvention of the spirit or letter of the rules.

7. The coach shall exert his or her influence to enhance sportsmanship by spectators, both directly and by working closely with cheerleaders, pep club sponsors, booster clubs, and administrators.

8. The coach shall respect and support contest officials. The coach shall not indulge in conduct which would incite players or spectators against the officials. Public criticism of officials or players is unethical.

9. Before and after contests, coaches for the competing teams should meet and exchange cordial greetings to set the correct tone for the event.

10. A coach shall not exert pressure on faculty members to give student-athletes special consideration.

11. A coach shall not scout opponents by any means other than those adopted by the league and/or state high school athletic association.

Reprinted, by permission, from National Federation of State High School Associations (NFHS). *The coaches code of ethics* (Indianapolis, IN: NFHS). Available: www.osaa.org/osaainfo/coachescodeofethics.pdf.

Trends and Opportunities in Coaching and Sport Instruction

The outlook for coaching jobs is promising. According to the U.S. Department of Labor (2012), job availability in this field will increase 29% between 2010 and 2020. Coaching was listed 19th in the 2012 ranking of best jobs by *US News* (*US News* 2012). However, competition for positions at the college and professional levels is extraordinarily strong. Salaries remain fairly low. According to *US News*, the best paid 10% of coaches (at all levels) earned $63,720 in 2010. Median salary was $28,340; the bottom 10% earned $16,380. Future job opportunities as sport instructors or coaches at various commercial sport academies are less clear. One might logically expect, however, that as long as public interest in sport remains high, job opportunities in these areas will remain high.

Advice for Coaching and Sport Instruction Students

The first piece of advice that any coach or sport instructor will give to students who want to coach is to make sure that they are suited for a career in coaching. Take a moment to review the questions listed in the sidebar, Are You Suited for a Career in Sport Instruction or Coaching? If, after going through this exercise, you believe that you are suited for a career in this area, consider this second piece of advice: Carefully lay out a plan for developing a career in coaching. This process will entail at least four steps. The first step is to take coaching courses in your department or at neighboring colleges and universities. If your primary program of studies will accommodate a coaching minor, arrange to do this minor.

Are You Suited for a Career in Sport Instruction or Coaching?

If you are considering a career in coaching or sport instruction, ask yourself which setting most appeals to you. Will you enjoy working in a community, commercial, or institutional setting? Are you attracted to working with underserved youth in community youth development programs or with skilled or budding athletes in specialized sport programs? Do you enjoy spending many of your weekends "on the job," when most games are played? Do you have an entrepreneurial spirit that will support a venture as an independent sport instructor or teaching professional? Are you comfortable marketing yourself and your expertise? Consider whether you like dealing with people as much as you like playing your favorite sports. Are you an effective communicator? Do you have what it takes to be a leader? Do you like to plan and organize activities? Do you like to engage in strategy? Are you organized? How committed are you to the sport you want to coach and to the young people who play it? Do you think that you have good judgment? Will you be happy with having both your successes and failures in the public eye? Will it bother you to have others evaluate your performance, second-guess you, and even criticize you? Do you have what it takes to look beyond the win–loss column and focus on the changes and experiences of the athletes who play for you? Are you willing to experience the emotional ups and downs that typically characterize a career in coaching? Answer these questions truthfully and you will get a better idea of whether you are emotionally and intellectually suited for a career in sport instruction, coaching, or both.

The second step is to gain coaching experience. This is a very important element in the overall process of preparing for a career in coaching. On-the-job learning, particularly at the collegiate and professional levels, is often viewed as more important than the completion of coaching courses or certification licenses. Opportunities for coaching "internships" abound in most communities. Explore opportunities to become a volunteer coach in your local youth sport organization. Sport associations and recreation departments are always looking for volunteer coaches. You probably will start out as an assistant coach, so be patient; other opportunities will develop. If you have special interest in swimming, tennis, or martial arts, inquire about volunteer opportunities at local clubs or schools. If golf is your passion, make inquiries at local country clubs or golf courses for opportunities to volunteer. Opportunities to gain experience may also be available in athletic programs in local school districts. Many

Successful teaching and coaching require skills that go beyond scientific knowledge. Most successful teachers and coaches have developed communication, leadership, and organizational skills that enable them to develop rapport with their participants and foster an environment that leads to a sense of community among participants.

coaches welcome offers by coaches-in-training to help with coaching duties. Be sure to check with private schools in your area; these schools often have openings for coaching positions that they fill with college students preparing for coaching careers.

The third step toward developing a career in coaching is to stay up-to-date on the latest trends and tactics of your chosen sport. Attend clinics, purchase coaching DVDs, watch webinars, and become a voracious reader of materials related to the sport you wish to coach. As a coach, you are regarded as an expert in your sport. Read books, articles, and newsletters on coaching. Many of these will be available from national or local coaching associations for your sport of interest. Attend conferences, clinics, and other opportunities for coaching education in your sport. Teaching and coaching resources that you might find helpful are listed on pages 475-476, along with organizations and associations that can help you develop as a coach. With some diligent website navigation (beginning with the sport that interests you), you will find many more organizations.

The fourth step is to begin, while you are still an undergraduate student, building a network of active coaches. You will find no quicker way of joining the fraternity of the coaching ranks than cultivating contacts who can serve as mentors. The importance of being connected in the coaching community should not be underestimated.

Take stock of your personal abilities and develop those areas in which you detect weaknesses. Coaching success depends a great deal on communication. A course in communication skills will help, but it cannot take the place of experience. Teachers and coaches can be highly knowledgeable about the principles of communication but not be effective communicators. Or they may know the principles of leadership but not be effective leaders. Being a successful teacher or coach requires that you have not only knowledge but also the skill to share that knowledge in such a way that people listen and act accordingly. This view is reflected in the fact that the greatest athletes on the field do not necessarily make the greatest coaches. Although it is difficult to pinpoint why this is true, part of the reason may be that the sport came easily to them and as a consequence they never developed the communication skills needed to explain the reasons for their success to less talented individuals. Conversely, athletes endowed with less natural ability may have had to study the sport in greater detail and depth to achieve competitive levels of performance. They have experienced adversity, can understand the challenges confronting the average athlete, and can therefore relate and communicate more effectively with the group as a whole.

ACTIVITY 16.6

Compare Organizations
Compare two sport organizations in this activity in your web study guide.

Successful coaching places a premium on organizational skills. Detailed planning of in-season practice sessions, off-season training and conditioning programs, and game preparation is an essential component of these jobs. Remember, for most middle and high school coaches, and for some college coaches, coaching is simply one part of their total job responsibilities. Fitting it all together requires a person to be organized. Do you possess good leadership skills? Coaching effectiveness hinges largely on your ability to inspire athletes to higher levels of performance. This undertaking requires leadership. Hundreds of books on leadership, many written by business executives and successful coaches, are on the market. Begin a course of reading that will help you develop effective leadership skills. Examining your own style of communication, organization, and leadership will enable you to identify the weak spots and develop a plan for remediation.

Simply put, the more you do, the more you learn. Taking advantage of these opportunities will enhance your knowledge and skills, thus making you a more competent and effective teacher or coach when you enter the profession.

Wrap-Up

In this chapter you learned some basic information about the coaching and sport instruction professions. You also received an overview of possible careers within teaching and coaching and the work settings in which these careers take place. Both professions include instruction and offer opportunities to serve a wide range of people across many ages and ability levels. In fact, few professions offer such a wide array of populations from which to choose.

You have also met professionals in teaching and coaching and have been privy to an inside view of their impressions of their career choices. As you can see from the stories provided in the interviews in this chapter, a great range and number of opportunities are available to those who decide to enter a teaching or coaching career, not only to capitalize on personal interest but also to promote physical activity programs and serve society. Ask yourself the following question: Do you want to make a positive difference in someone's life? If the answer to that question is yes, then a career in sport instruction and coaching may be just right for you.

More Information on . . . Careers in Coaching and Sport Instruction

Organizations

Amateur Athletic Union

American Baseball Coaches Association

American Swimming Coaches Association

Black Coaches & Administrators

Boys & Girls Clubs of America

Coaching Association of Canada

Canadian Swimming Coaches & Teachers Association

Institute for the Study of Youth Sports

Joy of Sports Foundation

National Association for the Education of Young Children

National Association of Intercollegiate Athletics

National Association for Sport and Physical Education

National Collegiate Athletic Association

National Federation of State High School Associations

National Junior College Athletic Association

National Soccer Coaches Association of America

United States Olympic Committee

Women's Sports Foundation

YMCA of the USA

YWCA of the USA

Journals

American Football Monthly

Annual Review of Golf Coaching

Canadian Journal for Women in Coaching

Coach and Athletic Director

Coaching Volleyball

Inside Gymnastics
International Journal of Sports Science & Coaching
Journal of Asian Martial Arts
Lacrosse Magazine
Scholastic Coach and Athletic Director
Soccer Coaching International
Softball Magazine
The Sport Journal
Swimming World
Track & Field News
Winning Hoops
Wrestling USA Magazine

 ACTIVITY 16.7

Use the key points review in your web study guide as a study aid for this chapter.

Review Questions

1. What distinguishes sport instruction from sport coaching?
2. Describe the differences and similarities between the act of instruction and the act of coaching.
3. Contrast the professions and subcultures in coaching and sport instruction.
4. Outline the variety of work settings for coaches and sport instructors.
5. List the three professions in coaching and sport instruction that are most appealing to you. Examine and discuss the educational requirements and qualifications for these careers.
6. Consider two professions from this chapter—one in sport instruction and one in coaching. Describe the duties and responsibilities for each, including their primary purposes.

chapter seventeen

Careers in Sport Management

G. Clayton Stoldt and Mark C. Vermillion

chapter objectives

In this chapter we will

- acquaint you with the wide range of professional opportunities in sport management;

- familiarize you with the purposes and types of work done by professionals in sport management;

- inform you about the educational requirements and experiences necessary to become an active, competent professional in the field; and

- help you identify whether one of these professions fits your skills, aptitudes, and professional desires.

Consider the following scenarios representing decisions that sport managers routinely confront.

- The race director for a local marathon conducts a meeting with nearly 100 volunteers from the community to discuss assignments and race logistics. The volunteer staff is a critical aspect of administering the event, and it is important that volunteers receive both adequate training and appreciation for their contribution to the event.

- Staff members and interns in a college athletics department's communications office activate numerous social media channels in support of a student-athlete's candidacy for a prestigious national award. By activating the department's fan base to vote in an online poll, the student-athlete stands a strong chance of being selected.

- An account manager for a professional golf tournament meets with a corporate partner. The discussion centers on a report prepared by the tournament staff that details the various benefits the partner received as a result of its sponsorship last year.

- The sports director for a community recreation commission develops a schedule for the upcoming boys and girls soccer season. With 600 children registered, 500 games to be played, and six fields available, scheduling poses a significant challenge.

In each of these scenarios, sport managers and employees of sport organizations make decisions that serve the consumer while enabling the sport organization to meet its financial goals.

Students with an interest in sport and physical activity may find a career in sport management appealing. Whether the job involves the use of marketing, program management, financial, or other administrative skills, managerial positions in sport usually prove to be both challenging and rewarding. Although competition for jobs in the sport management industry is often fierce, numerous career options are available, and the field is growing.

According to Milano and Chelladurai's (2011) moderate estimates, sport is a $189 billion industry in the United States, which is larger than the oil and gas industry or computer and electronics manufacturing. Based on data from 2005, the $189 billion figure included more than $79 billion spent on sport equipment, apparel, and footwear; almost $49 billion spent by sport spectators; more than $14 billion spent on sponsorships, television, Internet, and magazine advertisements; and more than $1 billion spent individually on both fantasy sports entry fees and sport video games. Because of the many job opportunities in sport management, it makes up one of the spheres of professional practice centered in physical activity (see figure 17.1).

> ▶ The thriving sport management industry offers numerous viable options for students looking for careers in sport.

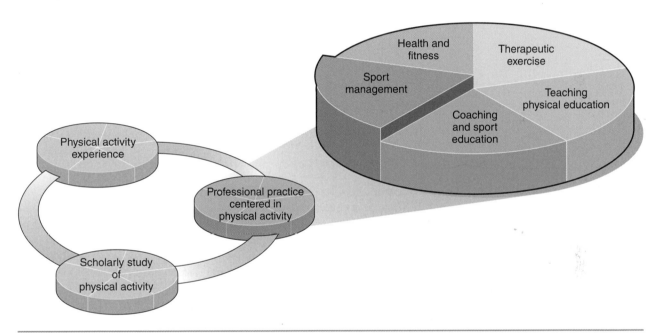

FIGURE 17.1 Sport management is one of the spheres of professional practice in physical activity.

The purpose of this chapter is to examine professions within the sport management industry. The career opportunities within this industry are many, varied, and expanding. Job opportunities exist in professional sport, amateur sport, sport participation, sporting goods, and support services. Graduates with a well-rounded degree can choose from a variety of roles—from event management to human resource management to marketing management. We hope that this chapter helps you decide whether a career in sport management is right for you.

What Is Sport Management?

The **sport industry** is composed of several distinct sectors, but it is clearly big business. One moderate estimate showed sport as the 15th largest industry in the United States (Milano & Chelladurai 2011). Sport management scholars have defined the term *sport industry* in a number of ways (Li et al. 2001; Meek 1997; Milano & Chelladurai 2011; Pitts et al. 1994). In this chapter we use the following definition: "The market in which the products offered to its buyers are fitness, sport, recreation, and leisure related. These products include goods (e.g., baseball bats), services (e.g., sport marketing, health clubs), people (e.g., professional players), places (e.g., golf courses), and ideas" (Pitts et al. p. 18). On the basis of this definition, the sport management sphere of professional practice centered on physical activity can intersect with career opportunities in three of the other spheres (health and fitness, coaching and sport education, and therapeutic exercise).

As discussed in chapter 3, physical activity professionals are expert manipulators of physical activity experiences. In other words, their expertise lies in manipulating physical activity experiences to improve performance and health. They may serve as personal trainers, teachers, athletic trainers, or coaches. Many sport management professions relate indirectly to the manipulation of physical activity itself (e.g., fitness center administration), and others focus on the manipulation of the elements supporting spectatorship (e.g., promotions, marketing, public relations) to maximize customer satisfaction. Generally, these professionals provide support services, facilities, and other amenities to make physical activity and spectatorship possible. For example, a sport manager may be employed as one of the following:

- Marketing coordinator for a local speedway (see the section titled Sport Entertainment)
- Program coordinator for a YMCA (see Sport Participation)
- General manager for a golf course (see Sport Participation)
- Athletic director in a high school (see Sport Entertainment)
- Development officer for a college or university athletic department (see Sport Entertainment)
- Event manager for a triathlon (see Sport Services)
- Manager of a sports medicine retail store (see Sporting Goods)

Given the overlap between sport management and other spheres of professional practice, students interested in building careers as sport managers sometimes find themselves competing for jobs with students who have studied in related areas such as recreation, leisure, or sport tourism. Although similarities exist among these areas, so do distinctions.

As noted in other parts of this text, recreation or leisure involves physical activity that results in personal enjoyment, excitement, and fulfillment. Recreation commonly focuses on the consumer as an active participant (Gullion 1998). Students in recreation programs often prepare for careers in corporate fitness, nonprofit recreation, military recreation, parks and recreation, campus intramurals, and other forms of leisure services. Sport management students may have similar career interests, as the subsequent section in this chapter titled Sport Participation details. Graduates of sport management programs sometimes land jobs with parks and recreation departments, nonprofit fitness providers, and campus intramural programs.

Sport tourism involves people whose primary motivation in traveling outside their communities is to participate in physical activity, watch sport events, or visit sport attractions, for example, a hall of fame (Gibson & Fairley 2011). Students in sport tourism programs often prepare for careers as special event managers, travel coordinators, and sport commission administrators. Again, sport management students may pursue similar career paths and land jobs in sport tourism.

What distinctions exist among the areas of study? Given the diversity in the way that sport management, recreation and leisure programs, and sport tourism programs are structured, it is difficult to define fixed differences. Still, we can say that recreation programs commonly offer greater depth in terms of specific forms of recreational programming and related considerations. Sport tourism programs address how to develop and manage sport-related travel opportunities and determine what the effect of such initiatives may be. Sport management programs, in contrast, tend to focus on the sport-specific application of business principles such as management and marketing in varied for-profit and nonprofit settings (see the sidebar).

Sport Management and Other Areas of Study

The following comparison of course descriptions may illustrate distinctions among recreation, sport tourism, and sport management. The first set of courses is a part of undergraduate curriculum for the recreation, park, and leisure studies program at the University of Minnesota. The second set is from the concentration in tourism and hospitality management that is a part of the bachelor's program in recreation, parks, and tourism at the University of Florida. The third set is from the sport management program at Wichita State University.

Selected Courses in Recreation, Park, and Leisure Studies (University of Minnesota 2011)

- Orientation to leisure and recreation: Introduction to the history and development of the parks and recreation movement; sociological, economical, psychological, and political considerations of leisure and recreation in contemporary society; interrelationship between professional and service organizations; orientation to the professional field.

- Recreation programming: Various methods, skills, and materials needed for planning, developing, implementing, and evaluating professional recreation programs for diverse populations in various settings.

- Leisure and human development: Exploration of relevant issues concerning many roles of leisure in human development from influence on healthy fetal development to viability until death. Examination of diverse, multicultural perspectives on leisure, its centrality throughout history, and its influence on how civilizations define themselves.

Selected Courses in Tourism, Recreation, and Sport Management (University of Florida 2011)

- Ethical issues in sport: Review and study of ethical issues that affect sport.

- History of sport and exercise sciences: The development of sport and exercise sciences in the United States from colonial America to the present. Interpretations of sport and exercise sciences and their relation to society.

- Sport facility design and management: Analysis of the various elements involved in planning and management of sport facilities.

Selected Courses in Sport Management (Wichita State University 2011)

- Organization and administration of sport: Discusses the fundamental aspects of management within any sport-related entity. Addresses management, marketing, facility management, human resources, legal issues, budgeting and finance, purchasing, and communication.

- Legal aspects of sport (I): Provides students with the knowledge, understanding, and application of how the following legal issues influence the sport industry. Specific content addressed includes the legal system, statutory law, risk management, tort law (negligence and intentional torts), contracts, and employment-related issues within the sport industry. A primary objective of this course is to enhance the decision-making and problem-solving ability of each individual student regarding legal issues in sport and physical activity.

- Marketing sport and physical activity: Introduces tools and concepts used to market sport and physical activity. Emphasizes marketing strategies that are applicable to the sport administrator, teacher or coach, and exercise professional.

University of Minnesota 2011, University of Florida 2011, Wichita State University 2011.

481

The sport industry has grown significantly in both variety and complexity since the 1800s. Technological developments, greater discretionary money and time for individuals, flexible work assignments, changing demographics and psychographics, evolving consumer needs and desires, and other societal influences have caused the sport industry to blossom.

Sport managers are more involved in the activities and job responsibilities surrounding an event than they are in the sport or activity itself.

Although the academic study of sport management is relatively new, management of the sport industry has a long history. In the early 1800s, people planned how to promote boxing and billiards, and by the mid-1800s, those responsible for the promotion of baseball games scrutinized the placement of baseball fields outside urban areas. The 1880s and subsequent years brought about distribution issues that required the input of sport managers. During the 1890s, sporting goods manufacturers commonly hired professional advertising agencies (Fielding & Miller 1996). By 1920, the structure (e.g., management, marketing, legal issues and related regulations, financial structure, distribution) of organizations in the sport industry was well established.

Settings for Sport Management

The sport industry can be divided into three segments, and all three segments share functional roles within the industry (see figure 17.2). For instance, professional sport franchises and nonprofit sport organizations require people working in management, marketing, and other functions. Likewise, both nonprofit organizations such as the YMCA and for-profit organizations such as ice-skating arenas need people with expertise in financial and legal services. A great deal of interdependence exists among the various sport industry segments, a point that is illustrated later in the chapter. Furthermore, the tasks listed under the different functional areas (e.g., event management, financial management) are not exhaustive. Human resource management, for example, could also include motivating, compensating, retaining, and evaluating.

The number of in-house tasks versus the number of tasks that are outsourced varies within the industry and within individual sport organizations. Similarly, industry segments differ in how each function is allocated among existing employees. Larger sport organizations often assign specialized responsibilities to individuals based on areas of expertise. People in smaller organizations are likely to be responsible for performing a variety of the job functions indicated.

Examining Your Interactions With Sport

To consider the prevalence and popularity of sport, you need only examine your interactions with sport and leisure during a typical day. Sport is an integral part of the lives of many people, and that is good news for sport managers because consumer demand for sport and leisure activities results in job opportunities for sport management students.

Think about your average week. How much time do you spend watching sport programming on television? How much time do you spend reading about sports online, discussing sports on social media platforms, or playing fantasy sports? How much time do you spend engaging in sport-related activities such as working out, golfing, or cycling? How much money do you spend on apparel featuring the logos of sport teams or the brand marks of sporting goods manufacturers? How much time do you spend in sport-related conversation?

Sport Entertainment

A proliferation of **sport entertainment** opportunities has created an abundance of jobs. At the professional level, a multitude of major and minor league teams as well as motor sports, golf, tennis, and mixed martial arts events entertain sport fans. Expansion teams, new professional leagues, and the professionalization of alternative sports (e.g., action sports) continue to fuel professional sport entertainment options and related career opportunities. Each entertainment production (i.e., the game or event) represents a tremendous undertak-

FIGURE 17.2 Sport management industry settings and some common functional roles in each segment.
Based on Meek 1997.

ing, combining the skills, competencies, and knowledge of many people working in separate functional areas. As indicated in figure 17.2, each professional sport entertainment organization provides career opportunities in the following areas:

- Event management
- Financial management
- Human resource management
- Legal management
- Marketing management
- Public relations management
- Venue management

The amateur sport entertainment sector mirrors professional sport in many ways. As with professional sport, the entertainment alternatives are vast and the career opportunities abundant in settings such as Olympics-related events and college athletics. Career opportunities in this industry segment often include the following:

- Development
- Public relations
- Marketing
- Operations
- Ticket sales
- Compliance
- Student services
- Facilities

Sport Participation

Whereas managers in the entertainment sector enable consumers to enjoy the sport performances of others, those in the participation sector provide opportunities for customers to engage in sport activities themselves. The sport participation sector includes both for-profit and nonprofit organizations, and the following sections address both.

For-Profit Participation Segment

For-profit sport participation organizations include fitness and health clubs, bowling alleys, roller-skating rinks, miniature golf courses, golf courses, ice-skating rinks, sport parks, and more. As indicated in earlier chapters, the health and fitness industry has many facets. An individual pursuing a career as a personal trainer, for example, would have different professional responsibilities and educational training than a person employed in a sport management capacity. According to the International Health, Racquet and Sportsclub Association (IHRSA 2011), by the beginning of 2011 there were 29,890 health and fitness clubs operating in the United States. The growth of almost 1,000 health clubs in 6 years, even during the economic recession of 2008-2011, helps to illustrate the increasing presence of for-profit sport participation organizations.

Of particular interest to potential sport managers is the breadth of interest in sport participation (see chapter 3). The sport industry is continuing to diversify. According to the Sporting Goods Manufacturing Association (2011), sport participants continue to embrace traditional activities such as bowling and tennis and are reengaging in activities like bicycling that they may have enjoyed as children. Sport participation is ever evolving and includes new activities such as bicycle polo and cardio tennis (The Consilience Group 2007).

The for-profit sport participation segment has also recognized growth within existing product mixes, supplementing core products such as resistance training with newer offerings such as spin classes. More and more businesses are catering to specific target markets such as unfit women and time-pressed professionals. Employment opportunities vary, typically according to the size of the organization, but job opportunities are available in a variety of functional areas:

- Management
- Sales
- Marketing
- Program planning
- Human resources
- Risk management

For-profit sport organizations are in an enviable position as their target markets and product lines expand.

Considering Sport Entertainment Management

Think about the last time you attended or watched on television a professional or amateur sport event. Identify as many of the peripheral activities or elements (e.g., concessions) associated with the main event (i.e., the game) as you can. Are you interested in managing events or venues? If so, begin developing organizational, time management, leadership, and communication skills. All of these are necessary for a successful career in sport entertainment management.

Nonprofit Participation Segment

As discussed in chapter 16, the nonprofit sport participation segment is replete with jobs for qualified sport management professionals. Nonprofit organizations providing participation opportunities in sport include YMCAs, YWCAs, Boys & Girls Clubs, and hospital-affiliated fitness centers. As in other industry segments, individual jobs in these organizations may be specialized in one of the functional areas identified in figure 17.2 or may be more general. For example, a YMCA located in a major metropolitan area may offer job opportunities in specialized functional areas, whereas a Boys & Girls Club may employ one person to be responsible for all recreational-related areas.

 ACTIVITY 17.1

Careers in Sport Participation Management
Research for-profit and nonprofit organizations in this activity.

Like many other areas in sport, the nonprofit sport participation sector offers significant opportunities. IHRSA (2007) reports that nearly 20% of all U.S. health club consumers are members of YMCAs or YWCAs. In 2010, the YMCA had 2,600 branches with almost 20,000 full time employees in more than 10,000 communities in the United States. The Y also has a global presence, with 45 million people worldwide in 124 countries being affected by programs, services, or facilities run by the Y (The Y 2011).

Other nonprofit organizations are common in the field. Combined facilities based in hospitals, universities, municipalities or towns, and churches or faiths hold a 15% market share in the United States (IHRSA 2007).

Support Services

The support services segment includes an even wider array of sport organizations than do the previous two segments, sport entertainment and sport participation. Included in the support services segment are both the sporting goods industry and a variety of service providers. The following sections explore each of these vast categories.

Sporting Goods

The sporting goods industry represents a significant portion of the overall sport industry. Milano and Chelladurai (2011) estimated that almost $24 billion was spent on sporting goods in the United States in 2005. Couple that with another $6 billion spent on goods licensed to display the names or logos of sport properties, and it becomes clear that the industry sector is sizable. Sport managers play a vital role in ascertaining what sport products people need and want. Employment opportunities in the sporting goods industry include the following:

Sport management encompasses many different roles and settings.

- Research analysts
- Retail store managers
- Manufacturing sales representatives
- Accounting managers
- Licensing administrators

Sport Services

As a result of the enormous growth in the sport industry, sport-related services have become more specialized. Today, sport managers can specialize in sport insurance, sport event management, or sport marketing. This segment of the industry is known as **sport services**. The following are examples of organizations that offer sport services:

- Sports & Fitness Insurance Corporation: provides insurance for health clubs, fitness centers, martial arts studios, personal trainers, and dance schools
- *Team Marketing Report:* provides information services on sport consumers, ways to increase ticket and sponsorship revenues, individuals making sponsorship decisions, and more
- Collegiate Licensing Company: manages the licensing programs of universities and colleges
- Exclusive Sports Marketing: promotes and manages events such as triathlons and other recreational sporting events
- Anthony Travel: service provider specializing in working with university athletic departments, professional sport organizations, individual athletes, and fans

The sport service segment is likely to realize continued growth as organizations outsource job functions to reduce personnel costs while at the same time receiving specialized services for seasonal sport product offerings.

Roles for Sport Management Professionals

People perform many similar professional roles regardless of the sport industry segment in which they are employed. Rather than list separate professions in the sphere of sport management, as in previous chapters we describe the sport industry in terms of functional roles and related tasks that may be assigned to one or more employees within various sport organizations.

Event Management

In all physical activity professions, the staging of an event is a tremendous undertaking. Many organizations hire people with specific expertise in the area of **venue** and **event management**. These employees are responsible for the many facets associated with hosting an event, including risk management, security, venue setup, concessions, scheduling, game presentation, customer service, and many more. Although a single person may cover all of those responsibilities for a small-scale event, large staffs are commonly employed for a major event. For instance, the 2011 Super Bowl had a host committee staff of 10 people (Nishimura 2011), and 10,154 volunteers worked in roles ranging from airport greeters to support personnel at the NFL Experience ("Super Bowl" 2011). Ultimately, the event attracted 103,219 game attendees (Chase 2011).

> ► Sport event management is a complex function because it requires skills specific both to securing, organizing, and executing the event and to creating partnerships with other organizations that may have a stake in the event.

Crowd Control and Security

At the beginning of 2012, an incident of soccer fan violence in Egypt left more than 70 people dead and hundreds of others injured (Manfred 2012). The riot was the latest event highlighting the need for crowd control and security at sporting events around the world. Crowd control measures are quite varied and have been an issue of concern for many decades. Muret (2011) noted the crowd control strategies used in mega-events in the United States, such as the NFL's Super Bowl, and the unique problems of controlling a crowd of more than 93,000 people. LaVetter and Choi (2010) discussed not only the risk to participants of toppling goal posts in college football but also the school's liability and risk associated with such events. Adequate crowd control strategies and security should be identified prior to any sporting event, regardless of size or anticipated attendance.

Who is responsible for crowd control and security at a sporting event? What factors play into safely controlling a crowd?

Clearly, venue and event management requires sport managers to consider a multitude of factors: traffic control, crowd control, resource availability, management and coordination of paid personnel and volunteers, and more. Opportunities in this area are widespread and vary from the Super Bowl to an evening of Little League baseball at a local park.

Sport commissions—local entities that attempt to attract sporting events and franchises to their cities to boost the local economy—have seen an enormous increase in the United States in recent years. The National Association of Sports Commissions (NASC) has grown from 15 members in 1992 to more than 550 members (NASC n.d.). Similarly, nearly 100 Canadian municipalities have joined the Canadian Sport Tourism Alliance (CSTA) in an effort to cultivate travel to their towns and cities for sporting events (CSTA n.d.). Clearly, employment opportunities in this sector of the field are abundant.

Once an event is scheduled, sport commission personnel collaborate and cooperate with others (e.g., team owners, employees, and volunteers) in the sport industry to ensure that the event is a success. A sport commission that secures a bid to bring a bowling championship to its city, for example, would work closely with practitioners in the local bowling community to ensure that adequate facilities are available, customer service is exemplary, scheduling is organized, and volunteers are gathered and committed.

> ## Considering Career Options in the Sporting Goods Industry
>
> Do you have a special interest in sport equipment and apparel? Perhaps you like to explore the newest merchandise on the market or you imagine designing your own. Maybe you like mining data and researching trends to predict the newest fad in athletics. You might be great at getting people excited about new products or services. All of those skills and interests could contribute to a career in sporting goods.

Financial Management

► High demand for sport-related products creates a positive environment for the financing of a variety of sport facilities including health clubs, swimming pools, arenas, and stadiums.

Financial management is critical to the operation of most sport-related organizations. Financial managers may administer budgets, oversee income allocation, pursue development opportunities, and handle investments. Managing financial situations for sport organizations requires knowledge and understanding of financial tools (e.g., financial statements, budgeting procedures) as well as a solid understanding of the sport product. Without having an understanding of the sport product, a person is not equipped to capitalize on potential revenue opportunities.

Two areas of sport financial management that provide a significant number of opportunities are accounting and facility finance. All sport organizations, and almost all employees in the sport industry, are involved in some manner of accounting and budgeting. Sport marketers, for example, who are responsible for promotions, advertising, and sponsorships, must ensure that expenses don't exceed revenues. Similarly, sport event managers must balance expenses, such as those associated with facility rental, ushers, concession purchases, and parking attendants, with anticipated revenues generated from sources like tickets and sponsorships.

Financing facilities has become an important issue in the sport industry. More than $9 billion was spent on facility construction in 2005, according to Milano and Chelladurai (2011). Sport organizations and their public constituents have used a variety of sources to fund their projects, including private sector money (e.g., corporate sponsorships), public support (e.g., subsidies, tax abatements), and organizational revenue (e.g., personal seat licenses, luxury seating) (Howard & Crompton 2005). The growth in facility construction has provided interested individuals with a number of finance-related jobs in the sport industry.

Human Resource Management

► Employees arguably represent the greatest asset of any sport organization; therefore, the human resource function is critical in selecting, training, and retaining high-caliber people.

Human resource management, also known as personnel management, may be one of the most important job functions in the sport industry (Chelladurai 2006; Miller 1997). Employees of sport organizations are arguably their organization's greatest asset. Happy and content employees are more loyal to the organization, resulting in reduced costs associated with turnover and in employees who are more experienced. Employees are responsible for the entire sport product delivery process including customer service, operations, and other crucial functions. A gap in any job functional area dilutes the quality of the sport product. Low-quality products eventually fall below the financial break-even point, causing sport organizations to go out of business.

Human resource managers work to combine employees' talents and desires with the needs of the sport organization to create a pleasant and legally defensible work environment. These managers have traditionally dealt with the hiring, training, maintenance, and dismissal of employees. Recent trends, however, reveal a far broader scope associated with human resource management. Today, human resource managers may organize and manage such benefits as day care for employees' children, employee stock ownership plans, professional development opportunities, fitness facilities, and counseling services, all designed to attract and keep high-quality employees.

Human resource managers in sport organizations with unions have added responsibilities. Unions negotiate agreements with management regarding compensation, working conditions, and other issues. As evident from previous professional sport work stoppages, union and management relations are critical to the profitability of sport organizations. Sport managers employed outside the professional sport domain also encounter union issues when dealing with concession and merchandising vendors, security personnel, and officials and umpires.

Risk Management

As society has grown increasingly litigious, sport administrators have become more concerned about **risk management**. Many issues facing sport managers today are complex and warrant advice from attorneys and other legal experts, but all sport managers should have knowledge necessary to minimize risks within their organizations.

Risk management evolved from a predominant concern about lawsuits to a more encompassing concern that includes risks of all kinds (Sharp et al. 2010). Today, the sport industry views risk management as including, but not limited to, losses resulting from inadequate or improper security, food service distribution, warnings, supervision, instruction, and facility and equipment design. Many sport organizations have designated risk management personnel to protect against losses of all kinds. Risk management groups assist sport entities in identifying potential hazards, suggesting and implementing ways to manage potential losses, evaluating existing risk management plans, and continually improving existing risk management efforts.

Risk management is especially important in the amateur and professional sport entertainment industry segments. Besides being attentive to the concerns already listed, the entertainment industry segment must comply with complex rules and regulations established by governing organizations. Fearing losses to their athletic organizations and programs, some colleges have responded by using outside legal assistance in their compliance efforts. Many others have designated particular administrators as compliance officers.

> ► Risk management involves taking steps to reduce losses of all kinds. Risk management requires the support of all employees, from top management down to the front line.

ACTIVITY 17.2

What's in Your Closet?
Gauge your personal interest in sports apparel and equipment in this activity in your web study guide.

The mass media has chronicled the consequences of compliance-related failures, and such situations are unfortunately common. From fall 2007 to late fall 2011, 79 National Collegiate Athletic Association (NCAA) institutions reported 86 separate violations in several sport programs (NCAA 2011). Major violations occur frequently. In the last 10 years, according to the NCAA, 175 major infractions have been reported to NCAA officials (NCAA 2011). A university compliance officer strives to keep the athletic department in compliance with the rules and regulations detailed in the lengthy NCAA compliance manual; in university policies and procedures; and in local, state, and federal legislation and ordinances.

Considering Your Experiences as an Employee

Think of an organization where you work or have worked in the past. Did that organization create a friendly work environment that was conducive to productivity and efficiency? What did the organization do well? What could it have done to improve the work environment? Does tackling such issues interest you? Can you see yourself being an instrument of change in an organization, recruiting others, establishing workplace policies, or ensuring company compliance with employment guidelines? If so, human resource management might be the profession for you.

Marketing Management

Marketing management provides a burgeoning career opportunity for qualified students. Some even suggest that effective marketing is the key to success for any sport product, whether it is a new running shoe, a sport clinic, or a sporting event. A great sport product, offered at an affordable price and available in a favorable location, generates little revenue unless people are aware of it. Sport marketers may spend time working in a variety of marketing management areas that include, but are not limited to, the following:

- Research and development can range from sophisticated analyses of consumer opinions and economic impact studies to focus groups or brainstorming sessions on how to package and present products more effectively.

- Sport promotion involves the activities that sport marketers undertake to heighten interest in products with the purpose of increasing consumption (e.g., sales, viewership). **Promotions** are tactics that marketers use to communicate with and attract buyers. Promotions come in a wide range of forms, including product sampling opportunities, premium giveaways, price reductions, player appearances, camps, and clinics (Irwin et al. 2008; Mullin et al. 2007).

- Sport sponsorship involves an exchange between a company and a sport organization. For example, a sponsoring company may provide either cash or in-kind resources (e.g., products, services) in exchange for title sponsorship of an event such as the AT&T USA Track and Field Championships. Sponsorships represent powerful promotional platforms that build brand awareness for sponsoring companies. They provide an important revenue stream for sport organizations, but sport managers must be mindful of which brands make for the best and most appropriate partners for their sport organizations. The case study ahead illustrates this issue.

- Advertising is "any paid, nonpersonal (not directed to individuals), clearly sponsored message conveyed through the media" (Mullin et al. 2007, p. 237). Many sport-related organizations purchase **advertising** to promote their events or sell their services or products.

- Merchandising efforts extend far beyond apparel. Customers who visit sport specialty stores may choose from an array of licensed products, including apparel, toys, automobile accessories, and even pasta in the shape of team mascots. Sport trading cards continue to be big business, with some of the rare or high-demand cards commanding thousands of dollars, and the market for sport video games is thriving. Some sport organizations conduct their own **merchandising** programs, whereas others sell licenses to other businesses to place their names and logos on various products.

- Distribution connects the sport product or service with the ultimate consumer. For sporting goods manufacturers, **distribution** includes building relationships with distributors and retailers, securing optimal shelf space, and maintaining low-cost product transportation systems. For sport managers in the entertainment or participation sector, distribution concerns may include facility design, accessibility, flexibility, and attractiveness because the sport product (e.g., game) is produced and consumed in the same location (e.g., stadium). Broadcasting arrangements may be considered a distribution-related issue because they focus on getting the sport product to consumers (Mullin et al. 2007). The National Football League reaps nearly $5 billion in television rights fees each year (Deitsch 2011), illustrating the potential magnitude of such agreements in the sport entertainment industry.

Public Relations Management

Public relations management in sport seeks to achieve positive relationships between the sport organization and its most important constituents (Stoldt et al. 2006, 2011). This involves diverse activities ranging from media relations management to community relations to investor relations.

A number of people within sport organizations work in media relations. Almost all professional and college sport organizations assign at least one person to media relations. These

Case Study in Assessment

Randy adamantly believes in the positive impact that sport can have on a variety of groups or persons. As a result, Randy was very excited about taking a position as director of marketing and sponsorships for the athletic department at a public (state-supported) Division II university in the midwestern United States. Randy sees his organization as playing an important role in not only entertaining the community but also educating and informing the community of worthwhile causes in the region while developing student athletes into productive men and women in society.

Decision

After taking the position, Randy was presented with a sponsorship opportunity from a well-known beer distributor and brewery. The sponsorship would bring much-needed revenue to an athletic department struggling with decreased donor contributions brought about by an economic recession. However, Randy is aware that alcohol abuse is prevalent on many campuses, and he wonders what kind of message is sent when a college athletic department promotes beer brands. He has many factors to weigh before making a decision regarding the sponsorship opportunity. Specifically, Randy needs to check university policy and current sponsorship trends within college athletics while simultaneously weighing the relevant pros and cons.

Resolution

To financially benefit the athletic department, careful consideration of the beer distributor's proposal should be taken. In accordance with the athletic department's vision (and Randy's too), however, any signs posted within athletic facilities could be used as an awareness opportunity. For example, the posted signs, although associated with the beer distributor's name and logo, could educate community members about the dangers of excessive alcohol consumption and the importance of responsible drinking behavior. This resolution allows the distributor to be associated with the athletic department; brings a much-needed financial boost to the athletic department; and educates both college students and community members on the importance of responsible drinking behavior.

employees produce organizational media, manage organizational websites and social media platforms, track statistics, arrange interviews, supply story ideas, and generally service media requests. These individuals provide an integral link between the organization and its stakeholders.

Although the media may be more integrally involved in sport than in any other industry in the United States, **public relations management** involves much more than just media relations. Sport entertainment organizations often employ people specifically to work in fan relations, and sporting goods businesses generally realize the advantages of having customer relations employees who provide a variety of services ranging from handling complaints to giving special instructions regarding the assembly of products.

Finally, some sport managers work in community relations. For example, these employees may seek opportunities for organizational representatives to make public appearances or speeches. These public relations professionals may coordinate donations to good causes, promote the organization's environmental initiatives, and execute other tasks pertaining to social responsibility.

Program Management

Program management is common in all sport industry segments. Health clubs, parks and recreation departments, athletic programs, and sport services all are involved with the provision and delivery of sport programs. A parks and recreation department, for example, may offer a

---

Recent Research: Corporate Social Responsibility in Sport

The November 2009 issue of *Journal of Sport Management* focused entirely on corporate social responsibility (CSR) in sport. According to Bradish and Cronin (2009), CSR "can be broadly understood as the responsibility of organizations to be ethical and accountable to the needs of their society as well as to their stakeholders" (p. 692). The journal addressed a variety of views and ideas focusing on CSR, including community development via sporting events (Misener & Mason 2009) and whether sport fans find CSR initiatives important in their consumption processes (Walker & Kent 2009). After a comprehensive summary of CSR (Godfrey 2009), Babiak and Wolfe (2009) identified both internal and external motivations for sport organizations' participation in CSR activities.

---

youth sport program, an aquatics program, and a fitness program. Each program, in turn, is staffed with its own employees who deliver desired activities (e.g., lessons, clinics, tournaments).

Program management activities vary across functional areas and sport industry segments. Two of the more common task areas are scheduling and maintenance of facilities and equipment.

Scheduling

Scheduling is one of the primary concerns for program managers. Many a Little League coordinator has dealt with scheduling complications stemming from too many teams with access to too few facilities. Sport entertainment administrators must also address scheduling and related considerations. Scheduling games, events, and even practices can be complicated when other events in the community affect public demand or when one facility must accommodate multiple demands. Scheduling travel can be complex, and some athletic programs employ administrators whose primary function is to coordinate travel.

Facility and Equipment Maintenance

Maintenance of facilities and equipment is a crucial task for program managers. Those working in sport participation organizations must maintain and update facilities to meet the needs and demands of consumers. Included is the responsibility to keep facilities and equipment clean, safe, and operative. In addition, sport managers must monitor the external environment (e.g., competitors, new product developments, and changing consumer demographics and psychographics) for new products and other developments that may enhance consumer offerings and satisfaction.

Considering Career Options in Sport Program Management

Think of a sport organization with which you are familiar. What programs does the organization offer? Do you have the skills and aptitudes necessary to run such programs? If you are interested in program management, what skills would you need to develop? After you've identified those skills, think about how you can start to develop them. What opportunities are available to you now? Which will require some research and planning to get involved in? Make a plan for your skill development.

ACTIVITY 17.3

Day in the Life
What does it take to become an athletic director at the college level, and what does a typical work day look like? In this activity in your web study guide, you'll discover how closely your ideas about what this career entails match the reality as reported by one practicing professional.

Prospects and Opportunities for Careers in Sport Management

The sport management industry offers a plethora of job opportunities in a variety of industry segments and functional areas. Objectively determining the most popular industry segments is difficult, because they all appear to be hiring qualified students of sport management. A review of various websites and trade publications provides insight regarding available job opportunities. For example, TeamWork Online advertises administrative jobs available in most of the major professional sport leagues as well as in large sport entertainment companies.

Data from the U.S. Department of Labor also illustrate the vast array of job opportunities available in sport management. Table 17.1 offers information regarding the number of jobs and mean salaries for selected positions and sectors of the industry.

> Many exciting employment opportunities exist for qualified individuals interested in sport management careers. The proliferation of professional sport teams, society's continuing interest in health and fitness, progressive technology, and eclectic consumers with varying needs and demands explain, in part, the promising growth.

TABLE 17.1 Mean Salaries for Selected Jobs in Sport

Industry sector	Job type	Annual mean salary
Spectator sport	Chief executive	$201,640
	Financial manager	$114,520
	Sales manager	$91,100
	Market research and marketing specialist	$54,130
Fitness and recreational sport centers	Chief executive	$161,310
	Sales manager	$79,890
	Supervisor, personal services	$38,580
	Recreation worker	$22,320

From U.S. Department of Labor, Bureau of Labor Statistics 2010.

Education and Qualifications

The North American Society for Sport Management (NASSM) website lists hundreds of colleges and universities with sport management programs at the bachelor's, master's, and doctoral levels (NASSM n.d.). More than 200 of those programs are located in the United States, but the site also lists programs in Canada, Europe, Australia, New Zealand, India, and Africa.

If you were to review the program information for even a handful of the institutions listed on the NASSM website, you would likely find considerable variation among the programs in terms of focus, breadth and depth, housing (e.g., business, education), and experiential learning requirements. High-quality programs may be structured in a variety of ways, and no single program is the best fit for all prospective students.

The Commission on Sport Management Accreditation (COSMA) is an accrediting body whose purpose is "to recognize excellence in sport management education" (COSMA n.d.). Among the many factors COSMA considers when reviewing programs are curricula that "reflect the mission of the institution and its academic unit, and are consistent with current, acceptable practices and principles of professionals in the academic and sport marketplace communities" (COSMA 2010, p. 9). Programs seeking accreditation at the undergraduate level are expected to feature curricula covering seven broad areas (COSMA 2010):

1. The social, psychological, and international foundations of sport
2. The management of sport, which includes not only sport management principles but sport leadership, sport operations management, event and venue management, and sport governance

3. Ethics in sport management

4. Sport marketing and communication

5. Sport finance, accounting, and economics

6. Legal aspects of sport

7. An integrative experience such as the development of a strategic management plan or policy manual, an internship, or some other capstone experience (e.g., thesis, comprehensive examination)

▶ Sport management students should capitalize on available internship opportunities with credible sport management organizations that provide learning opportunities.

Although all the content areas listed are critical to the attainment of a quality sport management education, the internship and other integrative experiences (e.g., practicum) deserve special emphasis. Sport organizations look for people with practical, real-world sport experience. Because the sport management industry is extremely competitive, the references, knowledge, and work experience that a student can gain from internship site supervisors are invaluable. Consequently, the internship or other form of integrative experience is essential to the student's ability to secure a job.

Students may obtain internships in any of the sport industry segments discussed earlier. Responsibilities associated with internships tend to be situation specific. For example, an internship at a Division I institution may be highly specialized. In other words, the student intern may work specifically in one area, such as compliance, marketing, facilities, operations, or tickets. In contrast, an internship at a Division III institution may combine roles. In this situation, the student intern may work in a variety of areas, including marketing, facilities, operations, and tickets.

▶ Multifaceted individuals can make decisions that are in the best interest of the sport organization as a whole rather than decisions that benefit only particular business functions of the organization.

Figure 17.2 either directly or indirectly identifies all the competency areas recognized by COSMA. As you can see, each functional area is somewhat dependent on, and linked to, the others. Moreover, many benefits can be gained from an understanding of each area. For example, if you work in a youth sport program, you will be better able to provide successful program offerings if you understand budgetary limitations, managerial issues, consumer behaviors, risk management principles, and societal implications. Similarly, if you work with a professional hockey franchise, you will have a better understanding of the intricacies involved and communication required of individuals working in different functional areas of a large event. You would learn, for example, that executing promotions requires the close cooperation of the people working in the area of facilities. Sound effects (e.g., music, clapping hands), lighting, fixtures, and so on must be coordinated so that the fans are entertained, the sponsor's promotion is flawlessly executed, and promotion participants are not injured.

When you understand how various issues and functional areas influence your product, be it a sport event, clinic, or apparel item, you are better able to communicate with colleagues, employees, and customers. Your conceptual understanding makes you an invaluable asset to your sport organization. In addition, by being knowledgeable in a variety of competency areas, you enhance your marketability in the profession.

Students studying for sport management careers follow different curricula than do students studying for careers as teachers, coaches, personal trainers, athletic trainers, and so on. As you have seen, sport management students are often not required to take courses in motor learning, exercise physiology, or anatomy and physiology. Course work in theoretical kinesiology, however, can significantly enhance a person's marketability in the sport participation sector. For example, students working in the health club industry can improve their marketability by combining their sport management degree with course work in appropriate areas of theoretical kinesiology. This well-rounded education will enable a sport manager to make better hiring decisions and provide better supervision.

Trends and Opportunities in Sport Management

In recent years, *Street & Smith's SportsBusiness Journal*, a weekly professional publication, has featured a series of interviews on a wide range of topics with executives in or in partnership with the sport industry. One of the questions posed to the executives was what they liked about the future of the sports business. Here is a list of the responses offered by some of those interviewees ("What I Like," 2010, 2011a, 2011b, 2011c, 2011d, 2011e, 2011f):

- Sarah Hirshland, senior vice president, Wasserman Media Group: "Consumers continue to prove how important sports are in their lives. They want more, they want it live and they want to share it with their growing base of friends."

- Tony Schiller, executive vice president and partner, Paragon Marketing Group: "Real partnerships between ownership and players. Dollars are becoming more scarce, so more creativity in game presentation and revenue distribution are increasingly essential."

- Conrad Smith, COO, Minnesota Lynx: "Combining the in-person experience with the comforts of watching a game on television at home."

- David Butler, CEO, Paciolan: "Fans continue to want more access, more information and a more personal involvement, and that passion is a great sign for our industry."

- Andrew Walker, chief marketing officer, WTA: "Convergence with entertainment."

- Mark King, CEO and president, TaylorMade-Adidas Golf: "Incredible growth fueled by new media."

- Charlie Besser, president & CEO, Intersport: "The globalization of sports . . . the world is shrinking fast!"

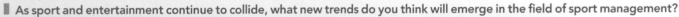

❚ As sport and entertainment continue to collide, what new trends do you think will emerge in the field of sport management?

These comments have an array of implications; however, three seem particularly pertinent. First, as the lines blur between sport and entertainment, and as new technologies make higher levels of direct engagement possible, the future of the sport business is bright. Sport management professionals who are able to deliver more compelling experiences to sport spectators and participants can expect to enjoy successful careers. Engagement is no longer limited to the sport facility itself or to the delivery of sport programming via the mass media. New and social media now make it possible for consumers to interact anytime and anyplace.

A second implication is that as sport management becomes more competitive and complex, sport managers who are able to forge successful partnerships will be in high demand. The nature of those partnerships will vary from setting to setting. But whether the task is to negotiate mutually acceptable agreements between player personnel and management or to secure corporate support for a community running event, managers with strong collaborative skills and creative minds will enjoy advantages in the competitive marketplace.

A third implication is that the sport industry is growing more global by the day. Global sport properties such as the World Cup are now commanding extravagant rights fees from media companies wanting to televise the matches on American television. NBA stars now enjoy celebrity status in countries as far away as China. And diverse populations are now being recognized for their achievement in global events such as the Paralympics. Sport managers who embrace the opportunities and challenges associated with globalization are well positioned for career growth.

Advice for Sport Management Students

As you know from reading about other physical activity professions in previous chapters, simply meeting educational requirements will not ensure your success as a sport management professional. The field of sport management is hypercompetitive, so students need to enhance their prospects for establishing a career by gaining as much practical experience as possible, seeking a mentor, and building a professional network.

ACTIVITY 17.4

What's My Profession?
Test your understanding of the sport management professions by playing this game in your web study guide.

Besides completing the required courses and obtaining your degree, you will need practical experience in your area of interest. Such experience is invaluable in helping you improve your communication skills and enhancing your problem-solving skills. Do not underestimate the value of a mentor, someone in the field who can give you practical insights into your new career. Keep up with the profession by contacting sport management organizations and reading professional journals; resources to get you started are listed on page 498. The following sections offer details on strategies that you can pursue to prepare for a career in sport management.

ACTIVITY 17.5

Day in the Life
What does it take to become a public relations director in the professional sports world, and what does a typical work day look like? In this activity in your web study guide, you'll discover how closely your ideas about what this career entails match the reality as reported by one practicing professional.

Practical Experience

You should always be at work building your resume. Practical experience is a necessity in the competitive job market of the sport industry. Almost all programs have internship requirements, and we have already described the benefits of internships. The best internships, however, often go to students with the most job experience. Try to find part-time work in the sport industry and volunteer for special events that may arise (e.g., fun runs, triathlons, Special Olympics). Such experiences will allow you to apply what you learned in the classroom to the job setting while simultaneously gaining valuable resume material.

Mentors

Mentors are people working in a student's chosen profession who provide students with guidance, learning opportunities, and contacts in the field. Students with mentors gain insights that they cannot obtain in the classroom, and they have an advocate working on their behalf when they seek future jobs. Some mentor relationships occur spontaneously when someone with education and experience takes an interest in a student. You can also arrange for a mentor directly by asking someone to be your mentor. A variety of people can serve as mentors, including job supervisors, university professors, and family friends. You can even have more than one mentor. Having multiple mentors increases your exposure to knowledge and opportunities.

ACTIVITY 17.6

Interview with a Practicing Professional
In this interview in your web study guide, Brian Hargrove talks about his career in event planning and marketing, what trends he sees in the field, and how students interested in pursuing a career in this field can best prepare themselves.

Communication Skills

Sport managers working in virtually every functional area discussed in this chapter need strong communication skills. Most undergraduate curriculums require students to take one or more classes to develop these skills. Because most jobs in the sport industry are service oriented, however, we encourage sport management students to take more communications-related courses than the minimum number required. The ability to communicate effectively with others, write well, and use new communication technologies is essential.

Problem-Solving Skills

Sport managers making hiring decisions want qualified employees with an educational and employment background that will allow them to be successful on the job. A characteristic that often distinguishes successful practitioners is their ability to analyze a problem effectively and generate creative solutions. We encourage you to enroll in courses that will help you develop such skills and to seek work experiences that require such skills. For example, by volunteering to work in a fitness center's membership campaign, you will gain experience that is more valuable than what you would gain by simply checking in members at the front desk.

Generalization

Sport-related jobs are plentiful but in high demand. As previously mentioned, by working in a variety of functional areas, you gain two advantages. First, by experiencing different facets

Assessing Your Sport Management Aptitude

Do you have the skills and aptitudes needed to pursue a career in sport management? Do you communicate clearly and confidently with people? Are you adept at problem solving? What types of experiences would be useful to pursue during your undergraduate years? Would you like to have a mentor? Have you considered seeking volunteer opportunities or part-time work within sport organizations to gain practical experience?

of the field, you will be better positioned to identify which areas of work most interest you. Second, you may expand the number of jobs for which you are qualified when you begin your job search. Once you have established a career path or advanced to graduate school, you should work to identify a specialization.

Wrap-Up

The sport management professions offer a broad range of exciting career opportunities for qualified individuals. In areas from sporting goods retail to commercial bowling, sport managers in managerial and administrative positions are looking for individuals with a sport management education. Education, work experience, professionalism, volunteerism, networking, a good work ethic, and a positive attitude are characteristics that sport management employers seek. Individuals in sport management careers often hold jobs that are interesting because of the variety they offer, exciting because of the activities they involve, and challenging because of the ever-changing nature of the field. Not surprisingly, many sport administrators report high levels of job satisfaction. If you are looking for a fulfilling job and have a keen interest in sport, the sport management professions may be for you.

More Information on . . . Careers in Sport Management

Organizations

North American Society for Sport Management (NASSM)

Sport Marketing Association

Journals

Athletic Business

Athletic Management

International Journal of Sport Communication

International Journal of Sport Management

Journal of Legal Aspects of Sport

Journal of Sport Administration & Supervision

Journal of Sport Management

Sport Marketing Quarterly

Street & Smith's SportsBusiness Journal

ACTIVITY 17.7

Use the key points review in your web study guide as a study aid for this chapter.

Review Questions

1. Discuss the breadth of the sport industry and the related career opportunities it provides with specific reference to the industry segments discussed throughout this chapter.

2. Explain how the following would help an individual succeed in the sport industry: (a) good communication skills, (b) problem-solving skills, (c) practical experience, (d) a mentor.

3. What type of course work and related academic content does a quality sport management program provide?

4. Identify three functional areas within the sport industry and elaborate on the types of jobs that might be specific to those professional roles.

5. Identify two existing sport organizations for each segment of the sport industry.

6. Elaborate on the type of sport management–related job you would be most attracted to and why.

abilities—Genetically endowed perceptual, cognitive, motor, metabolic, and personality traits that are susceptible to little or no modification by practice or training.

academic learning time—Time in which students are actively learning.

activities of daily living (ADLs)—Physical activities in the sphere of self-sufficiency that involve personal grooming, dressing, eating, walking, and using the toilet.

activity experience—Training in, observation of, practice of, or participation in physical activity.

advertising—Information placed in the media by an identified sponsor that pays for the time or space.

aerobic—In the presence of oxygen.

aesthetic experience—A subjective experience in which sensations appeal to our senses of beauty, grace, and artistic appreciation.

aggression—Behavior intended to inflict harm or injury on another person.

anaerobic—In the absence of oxygen.

anemia—A hemoglobin concentration below 12 grams per deciliter in women and 13 grams per deciliter in men.

anxiety—An unpleasant, high-intensity feeling that typically results from a demand or threat.

arousal—A state of bodily energy or physical and mental readiness.

ascetic experiences—Physical activity experiences that involve either discomfort, pain, or suffering.

athletic training clinic—A health care facility, usually associated with high school, collegiate, or professional athletic teams, that specializes in the health care needs of athletes.

attitudes—Relatively stable mind-sets toward physical activity.

autotelic attitude—An attitude characteristic of experiences whose value lies in the doing; the activity is an end in itself. Intrinsically motivated individuals have an autotelic attitude.

axiology—The branch of philosophy that examines values, the purpose of human existence, and the nature of the good life.

biochemistry—The chemistry of living organisms.

biomechanics—Subdiscipline of kinesiology defined as the application of the principles of mechanics to animate motion.

breadth of capacity—A low to moderate level of competence in a wide range of physical activities.

burnout—A state of extreme mental, emotional, and physical exhaustion that occurs because of chronic stress.

cardiac output—The amount of blood pumped out of one side of the heart per minute.

cardiopulmonary rehabilitation—Exercise programs designed to improve the function of the heart, lungs, and vascular system.

carpal tunnel syndrome—A type of cumulative trauma disorder to the hand and wrist, usually suffered by carpenters, typists, packers, assembly line workers, and others who repeat the same movements for several hours each day.

choking—The sudden or progressive deterioration of performance below the typical and expected level of expertise for a person performing under pressure, whereby the person is seemingly incapable of regaining control over her performance.

clinical settings—Health and fitness facilities that operate within hospital or medical facilities.

closed skill—A motor skill in which performers must coordinate their movements with a predictable, usually stationary, environment.

code of ethical principles and standards—Guidelines that are usually published and disseminated by professional associations to their members to ensure that the welfare of clients always receives top priority from practitioners.

cognitive skills—Human acts that require complex modes of thought, including rational analysis and problem solving, to achieve a predetermined goal. Most professionals rely on highly developed cognitive skills.

cohesion—The tendency for groups to stick together and remain united in pursuing goals.

community settings—Health and fitness facilities that operate within community centers, churches, or other nonprofit organizations such as YMCAs or Jewish Community Centers.

competent bystanders—Well-behaved students who avoid participation without attracting notice.

competition—A principle or framework for organizing physical activity in which participants compare their performances with each other's or with a standard for the purpose of increasing enjoyment.

conditioning—The temporary end state of training reflected in the performer's possession of an adequate level of strength, endurance, and flexibility to carry out desired tasks.

cumulative trauma disorders—Injuries to muscle, tendons, nerves, and ligaments brought about through repetitive motion of a body part.

custodial care facility—Long-term care facility that caters to the medical, rehabilitative, and specialized needs of a patient, including assistance with the activities of daily living.

deductive reasoning—A method of reflection that starts with one or more broad premises and moves toward specific conclusions.

depth of capacity—A high level of competence in a narrow range of physical activities.

descriptive reasoning—A method of reflection that involves looking at one example of an event and describing its essential qualities.

developmental skills—Skills performed in nonsport settings where rules and competition are irrelevant.

discipline—Organized body of knowledge considered worthy of study, usually studied in a college or university curriculum.

disinterested spectating—A form of watching sport contests in which the observer is nonpartisan in his or her feelings about the outcome.

dispositions—Short-term, highly variable psychological states that may affect our enjoyment of physical activity.

distribution—The process of connecting the sport product or service with the ultimate consumer.

dualism—When applied to the nature of persons, a doctrine that emphasizes the radical distinctiveness and independence of mind and body.

electromyography (EMG)—A system for monitoring and recording the electrical activity in muscles.

emblems—Body movements such as hand signals that can be translated easily into explicit messages.

employee assistance programs (EAPs)—Programs supported by business and industries to help employees with personal problems that may affect their work performance.

endurance—The ability to sustain physical activity for prolonged periods.

engaged time—Time that students spend engaged in activity.

epistemology—The branch of philosophy that is concerned with how we know things and the foundations on which such knowledge rests.

ergonomists—Engineers who seek to improve the safety and efficiency of work through analysis of workers' movements and conditions in the workplace.

ethics—The branch of philosophy that analyzes right behavior and the kinds of responsibilities that individuals have toward others and themselves.

event management—The planning, organizing, execution, and evaluation of sport-related events (e.g., Super Bowl, Little League baseball tournaments).

exercise—Physical activity intended to improve one's physical performance in an activity, to improve one's health, or to regain performance that has been reduced as a result of injury or disease.

exercise physiology—The physiology of physical activity; also called *exercise science*.

exercise science—See *exercise physiology*.

experiential knowledge—Self-knowledge and knowledge about physical activity derived from performing or watching physical activity.

expressive movements—Movements used in a physical activity as a way of expressing one's own emotion or personality; differentiated from instrumental movements.

extrinsic approach to physical activity—Approach whereby physical activity is valued because of the benefits that come from participating.

extrinsic motivation—Motivation derived from pursuing and obtaining rewards outside the activity itself, such as money and status.

face-to-face contact activities—Contests such as football, wrestling, basketball, and soccer in which individuals interact with opponents' attempts to achieve the goal by physically manipulating their movements.

face-to-face noncontact activities—Contests such as volleyball, tennis, and baseball in which individuals interact with opponents to maximize their own chances of winning but do not physically manipulate their opponents.

feedback (intrinsic and extrinsic)—Information about the movement provided to the learner during and after a movement; this information may come from external sources (e.g., instructor, videotape) or internal sources (muscles, joints, through the nervous system).

financial management—Tasks involving financially related activities such as the prudent investment and use of assets, monetary development opportunities, and short-term and long-term budgeting.

focus group—A group interview for people's shared understandings. It involves an organized discussion with selected individuals on their views and experiences. Focus groups are used for sociological research and marketing purposes.

free time—Personal time that has not been encumbered with obligations; also called *discretionary time*.

functional learning time—Time in which students are appropriately engaged.

game spectator knowledge—Knowledge about the game one is watching, including the players, strategies, and competitive tactics.

gender—Social position based on a set of norms or expectations about how we should behave that are linked to societal understandings of sexuality and procreation.

gestures—Movements used to communicate our intentions to others; gestures may be classified as illustrators, emblems, or regulators.

habilitation—The processes and treatments leading to the acquisition of skills and functions that are normal and expected for an individual of a particular age and status.

habilitative therapeutic exercise—Processes and treatments leading to the acquisition of skills and functions that are considered normal and expected for an individual of a certain age, status, and occupation.

health-related exercise—Exercise undertaken to develop or maintain a sound working body.

health-related fitness—Developed through physical activity experience, capacities and traits associated with low risk of hypokinetic diseases.

helping professions—Professions primarily committed to providing services.

hemoglobin—An iron-containing protein found inside red blood cells that carries most of the oxygen in the blood.

holism—When applied to the nature of persons, a position that underscores the interdependence and interrelatedness of thought and physicality.

home maintenance activities—Activities in the sphere of self-sufficiency that are intended to improve or repair conditions of living in one's apartment or house.

human agency—A theory suggesting that people are actively involved in developing or "constructing" their own sports.

human factors engineers—See *ergonomists*.

human movement biomechanics—Study of the structure and function of human beings using the principles and methods of mechanics of physics and engineering (see reviews in Atwater 1980; Hatze 1974; Winter 1985).

human resource management—Tasks involving, for example, the recruitment, orientation, retention, and evaluation of employees as well as benefit program planning and design and implementation of interval grievance procedures.

hyperplasia—An increase in muscle mass because of the splitting of muscle fibers.

hypertrophy—An increase in muscle mass because of the enlargement of the muscle fiber.

hypokinetic diseases—Diseases such as heart disease, obesity, and high blood pressure that are directly associated with low levels of daily physical activity.

hypothermia—A condition in which the body temperature falls below 95 °F (35 °C), causing heart rate and metabolism to slow; can be life-threatening.

illustrators—Gestures used to demonstrate or complement what is being said.

imagery—Use of all the senses to create or re-create an experience in the mind.

impersonal competition—Physical activities such as mountain climbing and longdistance swimming in which an individual attempts to better an established record in an activity that does not involve opponents.

inductive reasoning—A method of reflection that starts with specific cases or examples and moves toward broad, general conclusions.

inpatient facility—Rehabilitation hospital, nursing home, or other institution where patients spend extended periods of time for purposes of receiving medical or other health-related treatment.

instrumental activities of daily living (IADLs)—Activities in the sphere of self-sufficiency such as shopping, telephoning, cooking, or doing laundry that are less personal than activities of daily living.

instrumental movements—Movements used in a physical activity to accomplish the goal of the action; differentiated from expressive movements.

internalization—The process by which an activity gradually comes to be valued for its intrinsic qualities.

International Classification of Functioning, Disability and Health—Developed by the World Health Organization, a diagnostic approach that focuses on the patient's needs and considers environmental factors to maximize function.

internship—Culminating educational experience for kinesiology majors planning professional careers. It involves extended work at one or two professional sites where students work under the supervision of a veteran professional.

interpersonal skills—Skills that enable the physical activity professional to listen to and communicate with clients, students, or patients.

interventions—A collective term for therapy provided to a patient or client intended to produce a positive outcome.

intrinsic motivation—Motivation derived from the rewards inherent or within the actual activity, such as enjoyment and feelings of accomplishment.

intuitive knowledge—Knowledge gained through physical activity that doesn't depend on rational or conscious processes.

isokinetic—An exercise in which the muscle changes length at a constant rate of velocity.

isometric—An exercise in which tension is produced by the muscle without any change in length.

isotonic—An exercise in which the muscle changes length while maintaining a constant tension.

kinematics—A description of a movement, based on mechanical physics, in which movement characteristics like position, velocity, and acceleration are recorded.

kinesiology—The discipline or body of knowledge that studies physical activity through performance, scholarly analysis, and professional practice.

kinesiology theory—Theoretical knowledge about physical activity as embodied in the subdisciplines of kinesiology.

learning—A permanent alteration in the functioning of the nervous system that enables performers to achieve a predetermined goal consistently.

leisure—A state of being in which humans find deep satisfaction and contentment, often accompanied by feelings of wonder, celebration, excitement, and creativity.

leisure activities—Physical activities that nourish or maintain the disposition of leisure.

leisure studies or recreation—An area of study or department in a college or university that focuses on preparing individuals for careers in the leisure industry.

liberal studies—Also known as the liberal arts and sciences, an area that is regarded as the core of a higher education and is responsible for inducing students to love the truth and to develop independence in thought and action.

marketing management—The production, pricing, promotion, and distribution of a sport-related product or service in a way that inspires transactions between the sport organization and its consumers.

materialism—A position on human nature that describes the person as made of only one thing—namely, atoms.

maximal oxygen uptake or $\dot{V}O_2max$—The highest rate of oxygen uptake during heavy dynamic exercise.

mechanical, market-driven professionalism—Mode of practice whereby technique, methodology, profit, and prestige assume priority over clients' wants and needs.

memory drum theory—Developed by Franklin Henry (1960), a theory proposing that rapid and well-learned movements are not consciously controlled but are run off automatically (as an older computer uses a memory drum to store and retrieve data).

mentors—Professional people who can provide less-experienced professionals or preprofessionals with learning opportunities, responsibility, knowledge, and contacts who can contribute to the attainment of a successful career or continued improvement within an existing career.

merchandising—The sale of goods and services related to the sport organization (e.g., various team apparel or other products with team logos such as coffee mugs and note pads). Merchandising provides sport managers opportunities to enhance brand image, brand loyalty, and customer satisfaction and to generate revenue.

metabolic rate—The rate at which the body uses energy.

metaphysics—The branch of philosophy that is concerned with the nature of reality, that is, with what exists.

mixed methods—A research approach that involves collecting numerical data (e.g., via instruments) and gathering textual data (e.g., via interviews) so that both quantitative and qualitative methods are used. This approach provides a more complete picture of the phenomenon being studied.

modernization theory—A theory emphasizing that the rise of modern sport occurred during the Industrial Revolution as American society shifted away from an agricultural and local orientation and developed city-based industries rooted in science and technology.

motivation—A complex set of internal and external forces that influence individuals to behave in certain ways.

motor performance fitness—A capacity developed through physical activity experience that enables people to perform daily activities with vigor. This type of fitness often incorporates an element of skill.

motor skill taxonomy—A classification system that categorizes motor skills according to their common critical components.

motor skills—Human acts that require efficient, coordinated movements to achieve a predetermined goal. Some professionals such as dentists and surgeons rely on motor skills as well as cognitive skills.

movement—Any change in the position of body parts relative to each other.

muscular endurance—Ability of a muscle to repeatedly exert force over a prolonged period.

myofibril—The part of the muscle fiber that contains the contractile elements.

mystical knowledge—Knowledge about another dimension of reality apprehended through participation in sport and exercise.

nanobiomechanics (nanoscale biomechanics)—The investigation of biological phenomena of individual living cells. Forces measured are on the order of trillionths of a newton (1 N ≈ 0.225 pounds [0.102 kilograms]).

neuromuscular injuries—Injuries to nerves or muscles that cause pain and limit or prevent the performance of physical activities.

novel learning tasks—A movement task with which the subject does not have prior experience; usually a simple movement like linear positioning or tracking.

open skill—A motor skill in which performers must coordinate their movements with an unpredictable, usually moving, environment.

outpatient facility—A short-term care facility where patients do not stay overnight.

peak experience—Special type of mystical experience that runners and others engaged in strenuous sports and exercises can experience.

pedagogy—The art, science, and profession of teaching.

perceptions—Meaningful constructs or messages based on the interpretations of sensations from past subjective experiences.

perceptual skills—Skills that enable you to identify and recognize problems that have occurred or are likely to occur in professional settings.

personality—The unique blend of psychological characteristics and behavioral tendencies that makes individuals different from and similar to each other.

phenomenology—A method of reflection that examines the content of consciousness and gives credence to differences encountered in normal subjective life.

physical activity—Movement that is voluntary, intentional, and directed toward achieving an identifiable goal.

physical fitness—A capacity developed through exercise that enables one to perform the essential activities of daily living, engage in an active leisure lifestyle, and have sufficient energy remaining to meet the demands of unexpected events.

physical performance capacity—Qualities of physical activity such as flexibility, muscular endurance, cardiovascular endurance, and strength that are developed through training rather than learning.

plasma volume—The volume of extracellular fluid found in the blood.

power—(a) Ability to do what you want without being stopped by others; (b) the rate at which work is done (work ÷ time or force × velocity).

practice—(a) A type of physical activity experience that involves cognitive processing and leads to improvement in skill (learning); (b) repetition of a task, often with an instructor's guidance or feedback, to promote learning.

practice theory—Knowledge concerning the client, the method, and the outcomes that guides practitioners in performing their duties as professionals.

practitioners—Those who use knowledge to bring about predetermined objectives. Professionals are practitioners.

prehabilitation—Therapy that occurs following an injury but before surgery; involves developing as much strength and range of motion as possible before surgery.

preprofessional—Professionals-in-training whose goal in pursuing undergraduate studies is to become highly competent practitioners.

principle of quality—Principle stating that experiences that engage us in the most critical aspects of an activity are most likely to improve our capacity to perform that activity.

principle of quantity—Principle stating that increasing the frequency of experiences that engage us in the critical components of a physical activity will increase our capacity to perform that activity.

private practice—An entrepreneurial venture in which a professional establishes his or her own workplace with its own client pool.

professional practice centered on physical activity—Knowledge and skills required to successfully serve as a professional who systematically manipulates physical activity experiences to achieve predetermined goals.

professional practice knowledge—Knowledge derived from integration of performance, scholarly study, and practical experience about appropriate ways to deliver professional services.

program management—Activities such as scheduling and equipment maintenance that sport organizations must address to ensure operational effectiveness.

promotions—Activities such as personal selling, sales promotions, advertising, and publicity that are designed to attract consumers to sport-related products and services.

psychoanalytic self-knowledge—Knowledge about one's deep-seated desires, motivation, and behavior gained through participation in sport and exercise.

psychological inventory—Standardized or objective measure of a specific sample of behavior, typically in the form of a questionnaire.

public relations management—Tasks in areas such as media relations and community relations that are designed to achieve positive relationships between the sport organization and its most important constituents.

qualitative analysis—Evaluation of aspects of a movement without using a measuring instrument (e.g., evaluating through visual observation).

quantitative analysis—Evaluation of movement based on numerical measures obtained using an instrument (e.g., using a stopwatch).

rational knowledge—Knowledge about facts, concepts, and theories gained through reason, logic, and analysis.

reaction time (simple and choice)—The speed of response to a light or sound. Pressing a button after seeing a signal involves simple reaction time; choosing from among more than one button to press, depending on which signal one sees, involves choice reaction time.

receptor—A specialized nerve ending found at the end of sensory neurons that detects changes in the environment.

regulators—Hand and body movements used to guide the flow of conversations, such as in greetings or when parting company.

rehabilitation—Physical treatment, exercise, and educational or counseling sessions that lead a person to attain maximum function and a personally acceptable level of independence.

rehabilitative therapeutic exercise—Processes and treatments designed to restore skills or functions that were previously acquired but have been lost to disease, injury, or behavioral traits.

retention-based facilities—Commercial (for-profit) health and fitness facilities that make a large percentage of profits from ongoing monthly membership dues. These facilities engage in activities directed at helping members meet their goals and have a positive experience in the facility so that they will continue to pay for their membership on a month-by-month basis.

risk management—The prevention of loss associated with, for example, inadequate financial planning, employee management, facility and equipment maintenance, customer service, or short- or long-term planning efforts.

sales-based facilities—Commercial (for-profit) health and fitness facilities that make a large percentage of profits from membership initiation fees and prepaid dues. In following this business model, sales-based facilities do not have significant incentive to satisfy current members. This emphasis on sales can result in overcrowding and lower-quality programs and services.

self-reflection—A process whereby one experiences the subjective experiences of an activity performed in the past.

sensation-seeking activities—Physical activities that involve high speed, danger, or disorientation of the body in space.

sensations—Raw, uninterpreted information collected through sensory organs.

service—Human acts intended to improve the quality of life for others.

side-by-side competitive activities—Contests such as golf and swim racing in which individuals do not directly interact in striving to accomplish the goal.

skill—The quality of physical activity experience that underlies the performance of motor skills.

skilled movement—Movement that requires accuracy of direction, force, and rhythm or timing to accomplish predetermined goals.

social loafing—A decrease in individual performance within groups.

social trustee, civic professionals—Adherence to the creed "Healthy people and a good society first. Myself and my profession second in service of this greater good" (Lawson, 1998a).

socialization—The process by which someone becomes accustomed to the norms and expectations of a subculture. Preprofessionals are judged to have been socialized into the profession when they understand the roles and responsibilities associated with the specific subculture of that profession.

Socratic self-knowledge—Knowledge about our capacities and limitations that enables us to perform physical activity safely within the range of our abilities.

specificity of practice—The principle that only practice conditions similar to game performance will benefit future game performance.

spheres of physical activity experience—Various dimensions of everyday life in which physical activity plays an important and distinctive role.

spheres of professional practice—Categories of physical activity professions that are similar with respect to general objectives, methods, educational requirements, working environments, and other factors.

spheres of professional practice in physical activity—Groupings of different career paths in the physical activity professions according to similarities in the types of preparation required, clients served, and contexts in which the work is performed.

spheres of scholarly study of physical activity—Categories of knowledge about physical activity that are studied, taught, and researched by kinesiologists.

sport—Physical activity in which movement is performed to achieve a specific goal in a manner specified by established rules and in competitive contexts.

sport entertainment—Sport industry sector consisting of organizations that seek to attract consumers to watch athletes compete (e.g., NASCAR, NFL, NCAA institutions).

sport industry—"The market in which the products offered to its buyers are fitness, sport, recreation, and leisure related. These products include goods (e.g., baseball bats), services (e.g., sport marketing, health clubs), people (e.g., professional players), places (e.g., golf courses), and ideas" (Pitts et al. 1994, p. 18).

sport pedagogy—The study of the teaching of physical education.

sport services—Sport industry sector consisting of businesses that provide needed services to other sport organizations (e.g., event management, representation services).

sport spectacles—Staged competitions designed and promoted for audiences and intended to evoke an entire range of human emotions by virtue of their grandeur, scale, and drama.

sports medicine—A field of medicine and therapeutic exercise that specializes in the treatment, prevention, and rehabilitation of athletes and others who are involved in sports and other forms of strenuous exercise. Sports medicine also involves the investigation of training methods and practices.

strength—The maximal amount of force exerted by a muscle group.

stress—A process in which individuals perceive an imbalance between their response capabilities and the demands of the situation.

stroke volume—The amount of blood pumped out of one side of the heart per beat.

subdisciplines—Divisions that represent extensions of established disciplines such as psychology, physiology, and history. They are a way of dividing the scholarly study of physical activity to facilitate teaching and research.

subjective experience—Individual reactions, feelings, and thoughts about events.

subjective physical activity experience—Individual reactions, feelings, and thoughts that accompany either the performance or observation of physical activity.

task analysis—The systematic examination of a particular physical activity to determine its critical components.

technical definitions—Specialized meanings of terms used to convey information to others within a technical field. The technical definition of *physical activity*, for example, differs from how people define and use the term in everyday language.

theoretical knowledge—Knowledge of concepts and principles and the research strategies used to discover them. Theoretical knowledge in kinesiology is knowledge about physical activity, embedded in the subdisciplines, that is acquired by formal study of the subdisciplines.

therapeutic exercise—Exercise and movement applied in a systematic and scientific way to develop or restore muscular strength, endurance, and flexibility; neuromuscular coordination; cardiovascular efficiency; and other health performance factors.

tidal volume—The amount of air inhaled or exhaled per breath.

training—Physical activity carried out for the express purpose of improving athletic, military, work-related, or recreation-related performance.

underwater weighing—The procedure in which a person's body weight is measured while he or she is completely submerged to determine the person's body volume.

ventilation—The process in which gases are exchanged between the atmosphere and the alveoli of the lungs.

ventilatory threshold—The point during a graded exercise test at which ventilation begins increasing at a faster rate than $\dot{V}O_2$ does.

venue—A facility in which sport competitions or activities take place; includes arenas, stadia, tracks, and so on.

vertigo—The sensation that comes from disorientation of the body in space, often experienced in conjunction with dangerous activities.

with-it-ness—Teacher awareness of all events transpiring in a learning environment.

workplace knowledge—Practical, mundane knowledge that is not grounded in theory and that people use to perform everyday tasks in the workplace. Knowing where items are stored or how to clean, repair, or calibrate equipment are examples of workplace knowledge.

references

Chapter 1

Newell, K.M. (1990a). Physical activity, knowledge types, and degree programs. *Quest, 42,* 243–268.

Newell, K.M. (1990b). Kinesiology: The label for the study of physical activity in higher education. *Quest, 42,* 269–278.

Sherrington, C.S. (1940). *Man on his nature.* Cambridge: Cambridge University Press.

U.S. Centers for Disease Control and Prevention. (2011, March 1). Physical activity for everyone. www.cdc.gov/physicalactivity/everyone/glossary/index.html.

Part I

Epstein, J. (1976). Obsessed with sport. *Harper's Magazine,* pp. 253–255.

Sheehan, G.A. (1978). *Running and being: The total experience.* New York: Warner Books.

Chapter 2

Ahlskog, J.E., Geda, Y.E., Graff-Radford, N.R., & Peterson, R. (2011). Physical exercise as a preventive or disease-modifying treatment of dementia and brain aging. *Mayo Clinic Proceedings, 86*(9), 876–884.

Ainsworth, B.E., Haskell, W.L., Whitt, M.C., Irwin, M.L., Swartz, A.M., Strath, S.J., O'Brien, W.L., Bassett, D.R., Schmidtz, K.H., Emplaincourt, P.O., Jacobs, D.R., & Leon, A.S. (2000). Compendium of physical activities: An update of activity codes and MET intensities. *Medicine and Science in Sports and Exercise, 32,* S498–S516.

American Association of Neurological Surgeons. (2011). Sports-related head injury. www.aans.org/Patient%20Information/Conditions%20and%20Treatments/Sports-Related%20Head%20Injury.aspx.

Bauder, D. (2012). Super Bowl ratings record. Huff Post Media, February 6. www.huffingtonpost.com/2012/02/06/super-bowl-ratings-record-tv-giants-patriots_n_1258107.html.

Bernstein, L. (2011). Corporate fitness programs survive hard times. *Washington Post* online. www.washingtonpost.com/lifestyle/wellness/corporate-fitness-programs-survive-hard-times/2011/06/21/gHQA37OHyH_story.html.

Boden, B.P., Tacchetti, R.L., Cantu, R.C., Knowles, S.B., & Mueller, F.O. (2007). Catastrophic head injuries in high school and college football players. *American Journal of Sports Medicine, 35,* 1075–1081.

California School Boards Association. (2009). California Project Lean. www.CaliforniaProjectLEAN.org.

Carr, J.H., & Shepherd, R.B. (1987). *A motor relearning programme for stroke.* Rockville, MD: Aspen.

Cawley, J., Meyerhoefer, C., & Newhouse, D. (2007). The impact of state physical education requirements on youth physical activity and overweight. Wiley Interscience, February 27. www3.interscience.wiley.com/cgi-bin/abstract/114129506/ABSTRACT?CRETRY=1&SRETRY=0.

Csikszentmihalyi, M. (1990a). *Flow: The psychology of optimal experience.* New York: Harper & Row.

Csikszentmihalyi, M. (1990b, January). *What good are sports?* Paper presented at the Commonwealth and International Conference of Physical Education, Sport, Health, Dance, Recreation, and Leisure, Auckland, New Zealand.

Czaja, S.J. (1997). Using technologies to aid the performance of home tasks. In A.D. Fisk & W.A. Rogers (Eds.), *Human factors and the older adult* (pp. 311–334). New York: Academic Press.

De Grazia, S. (1962). *Of time, work and leisure.* Garden City, NY: Doubleday.

Dong, N.L., Block, G., & Mandel, S. (2004, February 12). Activities contributing to total energy expenditure in the United States: Results from the NHAPS study. *International Journal of Behavioral Nutrition and Physical Activity, 1*(4). www.ijbnpa.org/content/1/1/4.

Focus. (2011). Productivity wars: Europe vs. the USA. www.focus.com/fyi/productivity-wars-europe-vs-usa/.

Human Resources and Skills Development in Canada. (2011). Canadians in context: Aging population. www4.hrsdc.gc.ca/.3ndic.1t.4r@-eng.jsp?iid=33.

Human Resources and Social Development Canada (HRSDC). (2005). Advancing the inclusion of people with disabilities: Executive summary. www.hrsdc.gc.ca/eng/disability_issues/reports/fdr/2009/page00.shtml.

Hutsmyre, C., & Dixit, J. (2009, January/February). Marked for mayhem. *Psychology Today, 42*(1), 80–83.

Jackson, A.S., Sui X., Hebert, J.R., Church, T.S., & Blair, S. (2009). Role of lifestyle and aging on the longitudinal change in cardiorespiratory fitness. *Archives of Internal Medicine, 169*(19), 1781–1787.

Johnson, L.D., Delva, J., & O'Malley, P.M. (2007, October). Sports participation and physical education in American secondary schools: Current levels and racial/ethnic and socioeconomic disparities. *American Journal of Preventive Medicine, 33*(4, Suppl. 1), S195–S208.

Katz, S., Ford, A.B., Moskowitz, R.W., Jackson, B.A., & Jaffee, M.W. (1963). Studies of illness in the aged: The index of ADL: A standardized measure of biological and psychological function. *Journal of the American Medical Association, 185,* 914–919.

Kohn, A. (1992). *No contest: The case against competition.* New York: Houghton Mifflin.

Kranz, L. (2002). *Jobs rated almanac* (6th ed.). Fort Lee, NJ: Barricade Books.

Kroemer, K., Kroemer, H., & Kroemer-Elbert, K. (1994). *Ergonomics: How to design for ease and efficiency.* Englewood Cliffs, NJ: Prentice Hall.

Locke, L.F. (1996). Dr. Lewin's little liver patties: A parable about encouraging healthy lifestyles. *Quest, 48*(3), 422–431.

Manton, K.G. (2008). Recent declines in chronic disability in the elderly US population: Risk factors and future dynamics. *Annual Review of Public Health,* 29: 91–113.

Manton, K.G., & Gu, X. (2001). Changes in the prevalence of chronic disability in the United States black and nonblack population above age 65 from 1982 to 1999. *Proceedings* of the National Academy of Sciences of the United States of America, 98(11), 6354–6359.

Morris, D. (1994). Bodytalk: The meaning of gestures. New York: Crown Trade.

Morris, J.N., Heady, J.A., Raffle, P.A.B., Roberts, C.G., & Parks, J.W. (1953). Coronary heart disease and physical activity of work. Lancet, 2, 1111–1120.

National Association for Sport and Physical Education (NASPE). (2004). Moving into the future: National physical education standards (2nd ed.). New York: McGraw-Hill.

National Institute on Aging. (n.d.). Global health and aging. www.nia.nih.gov/sites/default/files/nia-who_report_booklet_oct-2011_a4__1-12-12_5.pdf.

National Institute for Occupational Safety and Health. (2000). NIOSH update: Working women face high risk from work stress, musculoskeletal injuries, other disorders NIOSH finds. www.cdc.gov/niosh/updates/womrisk.html.

National Sporting Goods Association (NSGA). (2011). 2011 sports participation ranked alphabetically. www.nsga.org/files/public/2011_Participation_Ranked_Alphabetically_web.pdf.

Neilson Wire. (2009). TV viewing among kids at eight-year high: http://blog.nielsen.com/nielsenwire/media_entertainment/tv-viewing-among-kids-at-an-eight-year-high/print/.

Novak, M. (1976). The joy of sports. New York: Basic Books.

Phenix, P.H. (1964). Realms of meaning. New York: McGraw-Hill.

Pieper, J. (1952). Leisure: The basis of culture. New York: Pantheon Books.

Pressman, S.D., Matthews, K.A., Cohen S., Matire, L.M., Scheier, M., Baum, A., & Schulz, R. (2009). Association of enjoyable leisure activities with psychological and physical well-being. Journal of Psychosomatic Medicine, 71, 725–732.

Rideout, V.J., Foehr, U.G., & Roberts, D.F. (2010). Generation M²: media in the lives of 8-18 year-olds. Kaiser Family Foundation. www.kff.org/entmedia/upload/8010.pdf.

Simon, R.L. (2004). Fair play: The ethics of sport (2nd ed.). Boulder, CO: Westview Press.

Stanne, M.B., Johnson, D.W., & Johnson, R.T. (2009). Does competition enhance or inhibit motor performance? A meta-analysis. Psychological Bulletin, 125(1), 133–154.

Statistics Canada. (2011). Physical activity of Canadian children and youth. www.statcan.gc.ca/pub/82-003-x/2011001/article/11397-eng.htm.

Strieber, A. (2010). Jobs rated 2010: A ranking of 200 jobs best to worst. www.careercast.com/jobs-rated/jobs-rated-2010-ranking-200-jobs-best-worst.

Tudor-Locke, C., Craig, C., Cameron, C., & Griffiths, J. (2011). Canadian children's and youth's predetermined steps/day, parent-reported TV watching time, and overweight/obesity: The Canplay Surveillance Study. International Journal of Behavior Nutrition and Physical Activity, 8(1), 66–75.

U.S. Bureau of Labor Statistics. (2011). Nonfatal occupational injuries and illnesses requiring days away from work, 2010. www.bls.gov/news.release/archives/osh2_11092011.pdf.

U.S. Bureau of Labor Statistics. (2012). American time use survey. www.bls.gov/news.release/atus.nr0.htm.

U.S. Census Bureau. (2002). Participation in various leisure activities: 1997. Statistical abstract of the United States, 2002 (p. 751). Washington, DC: Government Printing Office.

U.S. Census Bureau. (2012). Statistical abstract of the United States 2012. Washington, DC: U.S. Government Printing Office.

U.S. Centers for Disease Control and Prevention. (2006). Physical education curriculum analysis tool. www.cdc.gov/healthyyouth/PECAT/pdf/PECAT.pdf.

U.S. Centers for Disease Control and Prevention and National Center for Health Statistics. (2009-2010). National Health and Nutrition Examination Survey. www.cdc.gov/nchs/data/nhanes/survey_content_99_10.pdf.

U.S. Centers for Disease Control and Prevention. (2010a). Summary health statistics for the US population: National health interview study. www.cdc.gov/nchs/data/series/sr_10/sr10_251.pdf.

U.S. Centers for Disease Control and Prevention, National Center for Health Statistics. (2010b). Prevalence of overweight, obesity and extreme obesity among adults: United States, trends 1960-1962 through 2007-2008. www.cdc.gov/NCHS/data/hestat/obesity_adult_07_08/obesity_adult_07_08.pdf.

U.S. Centers for Disease Control and Prevention. (2011, December 9). State-specific prevalence of no leisure-time physical activity among adults with and without doctor-diagnosed arthritis—United States, 2009. Morbidity and Mortality Weekly Report. www.cdc.gov/mmwr/preview/mmwrhtml/mm6048a1.htm?s_cid=mm6048a1_w.

U.S. Centers for Disease Control and Prevention. (2012a). Childhood obesity facts. www.cdc.gov/healthyyouth/obesity/facts.htm.

U.S. Centers for Disease Control and Prevention. (2012b). Deaths and mortality. www.cdc.gov/nchs/fastats/deaths.htm.

U.S. Centers for Disease Control and Prevention. (2012c). Physical activity facts. www.cdc.gov/healthyyouth/physicalactivity/facts.htm.

U.S. Centers for Medicaid and Medicare Services. (2012). National health expenditure projections: 2011-2021. www.cms.gov/Research-Statistics-Data-and-Systems/Statistics-Trends-and-Reports/NationalHealthExpendData/Downloads/Proj2011PDF.pdf.

U.S. Department of Health and Human Services. (1990). Healthy People 2000. DHHS Publication No. (PHS) 91-50213. Hyattsville, MD: Public Health Service.

U.S. Department of Health and Human Services (USDHHS). (1996). Physical activity and health: A report of the surgeon general. Atlanta, GA: U.S. Department of Health and Human Services, Centers for Disease Control and Prevention, National Center for Chronic Disease Prevention and Health Promotion.

U.S. Department of Health and Human Services (USDHHS). (1997, July). Musculoskeletal disorders (MSAs) and workplace factors: A critical review of epidemiologic evidence for work-related musculoskeletal disorders of the neck, upper extremity, and low back. B.P. Bernard (Ed.). Cincinnati: National Institute for Occupational Safety and Health. www.cdc.gov/niosh/docs/97-141/.

U.S. Department of Health and Human Services (USDHHS). (2000, November). Healthy People 2010: Understanding and improving health (2nd ed.) Washington, DC: U.S. Government Printing Office.

U.S. Department of Health and Human Services (USDHHS). (2002). Healthy people 2010 (Vol. 2). McLean, VA: International Medical.

U.S. Department of Health and Human Services, Administration on Aging. (2010). Projected future growth of the older population. www.aoa.gov/aoaroot/aging_statistics/future_growth/future_growth.aspx#age.

U.S. Department of Health and Human Services, Office of Disease Prevention and Health Promotion. (2012). Healthy

People 2020. Washington, DC. www.healthypeople.gov/2020/default.aspx.

Zweigenhaft, R.L. (1977). The empirical study of signature size. *Social Behavior and Personality, 5*(1), 177–185.

Chapter 3

Anspaugh, D.J., Hamrick, M.H., & Rosato, R.D. (1991). *Wellness.* St. Louis: Mosby Year Book.

Arnot, R.B., & Gaines, C.L. (1984). *Sportselection.* New York: Viking Press.

Bernstein, N. (1967). *The coordination and regulation of movements.* London: Pergamon.

Bouchard, C., Malina, R.M., & Perusse, L. (1997). *Genetics of fitness and physical performance.* Champaign, IL: Human Kinetics.

Brill, P.A., Burkhaulter, H.E., Kohl, H.W., Blair, S.N., & Goodyear, N.N. (1989). The impact of previous athleticism on exercise habits, physical fitness, and coronary heart disease risk factors in middle-aged men. *Research Quarterly for Exercise and Sport, 60,* 209–215.

Cauley, J.A., Donfield, S.M., LaPorte, R.E., & Warhaftig, N.E. (1991). Physical activity by socioeconomic status in two population based cohorts. *Medicine and Science in Sports and Exercise, 23,* 343–352.

Dishman, R.K., & Sallis, J.F. (1994). Determinants and interventions for physical activity and exercise. In C. Bouchard, R.J. Shephard, & T. Stephens (Eds.), *Physical activity, fitness, and health: International proceedings and consensus statement* (pp. 214–238). Champaign, IL: Human Kinetics.

Ebihara, O., Ideda, M., & Myiashita, M. (1983). Birth order and children's socialization into sport. *International Review of Sport Sociology, 18,* 69–89.

Fleishman, E.A., & Hempel, W.E. (1955). The relationship between abilities and improvement with practice in a visual reaction time discrimination task. *Journal of Experimental Psychology, 49,* 301–312.

Garcia, A.W., Broda, M.A.N., Frenn, M., Coviak, C., Pender, N.J., & Ronis, D.L. (1995). Gender and developmental differences in exercise beliefs among youth and prediction of their exercise behavior. *Journal of School Health, 65,* 213–219.

Gentile, A.M. (1972). A working model of skill acquisition with application to teaching. *Quest,* Monograph XVII, 2–23.

Gill, F.B. (1989). *Ornithology.* New York: W.H. Freeman.

Heltne, P.G. (1989). Epilogue: Understanding chimpanzees and bonobos, understanding ourselves. In P. Heltne & L. Marquardt (Eds.), *Understanding chimpanzees* (pp. 380–384). Cambridge, MA: Harvard University Press.

Jago, R., Fox, K.R., Page, A.S., Brockman, R., & Thompson, J.L. (2010). Parent and child physical activity and sedentary time: Do active parents foster active children? BMC *Public Health, 10*(1), 194–202.

Kalakanis, L., Goldfield, G.S., Paluch, R.A., & Epstein, L.H. (2001). Parental activity as a determinant of activity level and patterns of activity in obese children. *Research Quarterly for Exercise and Sport, 72*(3), 202–209.

King, A.C., Blair, S.N., Bild, D., Dishman, R.K., Dubbert, P.M., Marcus, B.H., Oldridge, N.M., Paffenbarger, R.S., Powell, K.E., & Yeager, K.Y. (1992). Determinants of physical activity and interventions in adults. *Medicine and Science in Sports and Exercise, 24*(6), S221–S236.

Kirby, J., Levin, K.A. & Inchley, J. (2011). Parental and peer influences on physical activity among Scottish adolescents: A longitudinal study. *Journal of Physical Activity and Health, 8,* 785–793.

LaBarre, W. (1963). *The human animal.* Chicago: University of Chicago Press.

Lortie, G., Simoneau, J.A., Hamel, P., Boulan, M.R., Landry, F., & Bouchard, C. (1984). Responses of maximal aerobic power and capacity to aerobic training. *International Journal of Sports Medicine, 5,* 232–236.

Loy, J.W., McPherson, B.D., & Kenyon, G. (1978). *Sport and social systems.* Reading, MA: Addison-Wesley.

MacArthur, D.G., & North, K.I. (2005). Genes and human elite athletic performance. *Human Genetics, 116*(5), 331–339.

Meehan, S.K., Bubblepreet, R., Wessel, B., & Boyd, L.A. (2011). Implicit sequence-specific motor learning after subcortical stroke is associated with increased prefrontal brain activations: An fMRI study. *Human Brain Mapping, 32*(2), 290–303.

Neufeldt, V. (Ed.). (1988). *Webster's new world dictionary* (3rd college edition). New York: Simon & Schuster.

Ostrander, E.A., Huson, H.J., & Ostrander, G.K. (2009). Genetics of athletic performance. *Annual Review of Genomics and Human Genetics, 10,* 407–429.

Parker, S.T. (1984). Playing for keeps: An evolutionary perspective on human games. In P.K. Smith (Ed.), *Play in animals and humans* (pp. 271–293). New York: Blackwell Science.

Pate, R.R. (1988). The evolving definition of fitness. *Quest, 40,* 174–179.

Poulton, E.C. (1957). On prediction in skilled movements. *Psychological Bulletin, 54,* 467–479.

Public Health Agency of Canada. (2012). Healthy living. www.phac-aspc.gc.ca/pau-uap/fitness/definitions.html.

Sallis, J.F., & Hovell, M.G. (1990). Determinants of exercise behavior. *Exercise and Sport Sciences Reviews, 18,* 307–330.

Storm, C. (2005). Who's got game? State and county leaders in per capita production of professional major league athletes. Paper presented at the Annual Meeting of the Association of American Geographers. www.sportsgeography.com/projects.htm.

Torpy, J.M., Lynm, C., & Glass, R.M. (2005). JAMA patient page: fitness. *Journal of the American Medical Association, 294,* 3048. http://jama.ama-assn.org/cgi/content/full/294/23/3048#JPG1221F1#JPG1221F1.

U.S. Centers for Disease Control and Prevention. (2011). Glossary of terms. www.cdc.gov/nccdphp/dnpa/physical/everyone/glossary/index.htm.

U.S. Centers for Disease Control and Prevention. (2012). Physical activity facts. www.cdc.gov/healthyyouth/physicalactivity/facts.htm.

U.S. Department of Health and Human Services (USDHHS). (1996). *Physical activity and health: A report of the surgeon general.* Atlanta: U.S. Department of Health and Human Services, Centers for Disease Control and Prevention, National Center for Chronic Disease Prevention and Health Promotion.

U.S. Department of Commerce, Census Bureau. (2012). Participation in selected sports activities: 2009. *Statistical Abstract of the United States, 2012.* www.census.gov/compendia/statab/2012/tables/12s1249.pdf.

Warren, T.Y., Barry, V., Hooker, S.P., Xuemei, S. Church, T.S., & Blair, S.N. (2010). Sedentary behaviors increase risk of cardiovascular disease mortality in men. *Medicine & Science in Sports & Exercise, 42*(5), 879–885.

Zakarian, J.M., Hovell, M.F., Hofstettere, C.R., Sallis, J.F., & Keating, K.J. (1994). Correlates of vigorous exercise in a predominately low SES and minority high school population. *Preventive Medicine, 23,* 314–321.

Chapter 4

Atkinson, G. (2004). Effects of music on work-rate distribution during a cycling time trial. *International Journal of Sports Medicine, 25*(8), 611–615.

Berkowitz, L. (1969). *Roots of aggression: A reexamination of the frustration-aggression hypothesis.* New York: Atherton Press.

Black, S.J., & Weiss, M.R. (1992). The relationship among perceived coaching behaviors, perceptions of ability and motivation in competitive age-group swimmers. *Journal of Sport and Exercise Psychology, 14*(3), 309–325.

Clough, P., Shepherd, J., & Maughan, R. (1989). Motives for participating in recreational running. *Journal of Leisure Research, 21*(4), 297–309.

Csikszentmihalyi, M. (1990a, January). *What good are sports?* Paper presented at the Commonwealth and International Conference of Physical Education, Sport, Health, Dance, Recreation, and Leisure, Auckland, New Zealand.

Csikszentmihalyi, M. (1990b). *Flow: The psychology of optimal experience.* New York: Harper & Row.

Dishman, R.K., & Sallis, J.F. (1994). Determinants and interventions for physical activity and exercise. In C. Bouchard, R.J. Shephard, & T. Stephens (Eds.), *Physical activity, fitness, and health: International proceedings and consensus statement* (pp. 214–238). Champaign, IL: Human Kinetics.

Donnelly, P. (1977). Vertigo in America: A social comment. *Quest, 27,* 106–113.

Fox, L.D., Rejeski, W.J., & Gauvin, L. (2000). Effects of leadership style and group dynamics on enjoyment of physical activity. *American Journal of Health Promotion, 14,* 277–283.

Giamatti, A.B. (1989). *Take time for paradise: Americans and their games.* New York: Summit Books.

Goldstein, J.H., & Arms, R. (1971, March). Effects of observing athletic contests on hostility. *Sociometry, 34,* 83–90.

Herring, M., Jacob, M.L., Suveg, C., Rishman, R.K., and O'Connor, P.J. (2011). Feasibility of exercise training for short term generalized anxiety disorder: A randomized, controlled trial. *Psychotherapy and Psychosomatics, 81*(1), 21–28.

Heyman, S. (1994). The hero archetype and high-risk sports participants. In M. Stein & J. Hollwitz (Eds.), *Psyche and sports* (p. 198). Wilmette, IL: Chiron.

Hyland, D.A. (1990). *Philosophy of sport.* New York: Paragon House.

Jackson, S.A., & Csikszentmihalyi, M. (1999). *Flow in sports.* Champaign, IL: Human Kinetics.

Kenyon, G.S. (1968). A conceptual model for characterizing physical activity. *Research Quarterly, 39,* 96–104.

Kleinman, S. (1968). Toward a non-theory of sport. *Quest, 10,* 29–34.

Kouros, Y. (1990, March). A war is going on between my body and my mind. *Ultrarunning,* p. 19.

Krawthwohl, D.R., Bloom, B.S., & Masia, B.B. (1964). *Taxonomy of educational objectives: Handbook II: Affective domain.* New York: David McKay.

Kretchmar, R.S. (1985). "Distancing": An essay on abstract thinking in sport performances. In D.L. Vanderwerken & S.K. Wertz (Eds.), *Sport inside out: Readings in literature and philosophy* (pp. 87–103). Fort Worth, TX: Texas Christian University Press.

Kretchmar, R.S. (1994). *Practical philosophy of sport.* Champaign, IL: Human Kinetics.

Lipsyte, R. (1970). *Assignment: Sports.* New York: Harper and Row.

Lorenz, K. (1966). *On aggression.* New York: Harcourt, Brace & World.

Madrigal, R. (1995). Cognitive and affective determinants of fan satisfaction with sporting event attendance. *Journal of Leisure Research, 27*(3), 205–227.

Martin, T.W., & Berry, K.J. (1974). Competitive sport in post-industrial society: The case of the motocross racer. *Journal of Popular Culture, 8,* 107–120.

McAuley, E., & Courneya, K.S. (1994). The subjective exercise experiences scale: Development and preliminary validation. *Journal of Sport and Exercise Psychology, 16,* 163–177.

McAuley, E., & Jacobson, L.B. (1991). Self-efficacy and exercise participation in sedentary adult females. *American Journal of Health Promotion, 5,* 185–191.

McAuley, E., Wraith, S., & Duncan, T.E. (1991). Self-efficacy perceptions of success and intrinsic motivation for exercise. *Journal of Applied Social Psychology, 16,* 139–155.

McIntyre, N. (1992). Involvement in risk recreation: A comparison of objective measures of engagement. *Journal of Leisure Research, 24,* 64–71.

Neulinger, J., & Raps, C. (1972). Leisure attitude of an intellectual elite. *Journal of Leisure Research, 4,* 196–207.

Ogilvie, B. (1973, November). The stimulus addicts. *Physician and Sportsmedicine, 1*(4), 61–65.

Petrie, A. (1967). *Individuality in pain and suffering.* Chicago: University of Chicago Press.

Ravizza, K. (1984). Qualities of the peak experience in sport. In J.M. Silva & R.S. Weinberg (Eds.), *Psychological foundations of sport* (pp. 452–462). Champaign, IL: Human Kinetics.

Reid, H.L. (2002). *The philosophical athlete* (p. 52). Durham, NC: Carolina Academic Press.

Sachs, M.L. (1981). Running addiction. In M.H. Sacks & M.L. Sachs (Eds.), *Psychology of running* (pp. 116–125). Champaign, IL: Human Kinetics.

Sallis, J.F., Haskell, W.L., Fortnam, S.P., Vranizan, M.S., Taylor, C.B., & Solomon, D.S. (1986). Predictors of adoption and maintenance of physical activity in a community sample. *Preventive Medicine, 15,* 331–341.

Scanlan, T.K., Stein, G.L., & Ravizza, K. (1988). An in-depth study of former elite figure skaters: II. Sources of enjoyment. *Journal of Sport Psychology, 11*(1), 65–83.

Sheehan, G. (1978). *Running and being: The total experience.* New York: Warner Books.

Shi, J., & Ewing, M. (1993). Definition of fun for youth soccer players. *Journal of Sport and Exercise Psychology* (NASPSPA abstracts), *15,* S74.

Sigfusdottir, I.D., Asgeirsdottir, B.B., Sigurdsson, J.F., Gudjohnsson, G.H. (2011). Physical activity buffers the effects of family conflict on depressed mood: A study on adolescent boys and girls. *Journal of Adolescence, 34*(5), 895–902.

Smoll, F.L., & Schutz, R.W. (1980). Children's attitude toward physical activity: A longitudinal analysis. *Journal of Sport Psychology, 2,* 137–147.

Thomas, C.E. (1983). *Sport in a philosophic context.* Philadelphia: Lea & Febiger.

Wankel, L.M. (1985). Personal and situational factors affecting exercise involvement: The importance of enjoyment. *Research Quarterly, 56*(3), 275–282.

Wankel, L.M., & Krissel, P.S.J. (1985). Methodological considerations in youth sport motivation research: A comparison of

509

open-ended and paired comparison approaches. *Journal of Sport Psychology, 7,* 65–74.

Wann, D.L. (1995). Preliminary validation of the Sport Fan Motivational Scale. *Journal of Sport & Social Issues, 19,* 377–396.

Williams, J.F. (1964). *The principles of physical education* (8th ed.). Philadelphia: Lea & Febiger. (Original work published 1927)

Willis, J.D., & Campbell, L.F. (1992). *Exercise psychology.* Champaign, IL: Human Kinetics.

Zaichowsky, L.B. (1975). Attitudinal differences in two types of physical education programs. *Research Quarterly, 46,* 364–370.

Zillman, D., & Cantor, J.R. (1976). A disposition theory of humor and mirth. In T. Chapman & H. Foot (Eds.), *Humor and laughter: Theory, research, and applications.* London: Wiley.

Zillman, D., Bryant, J., & Sapolsky, B.S. (1979). The enjoyment of watching sport contests. In J.H. Goldstein (Ed.), *Sports, games and play: Social and psychological viewpoints* (pp. 297–336). Hillsdale, NJ: Lawrence Erlbaum.

Chapter 5

Boeck, S., & Staimer, M. (1996, December 6). NFL drug suspensions. *USA Today,* p. 1C.

Brown, W.M. (1980). Ethics, drugs, and sport. *Journal of the Philosophy of Sport, 7,* 15–23.

Csikszentmihalyi, M. (1990). *Flow: The psychology of optimal experience.* New York: Harper & Row.

Kretchmar, R.S. (1975). From test to contest: An analysis of two kinds of counterpoint in sport. *Journal of the Philosophy of Sport, 2,* 23–30.

Kretchmar, R.S. (1994). *Practical philosophy of sport.* Champaign, IL: Human Kinetics.

Kretchmar, R.S. (1996). Philosophic research in physical activity. In J.R. Thomas & J.K. Nelson (Eds.), *Research methods in physical activity* (pp. 277–290). Champaign, IL: Human Kinetics.

Kretchmar, R.S. (2004). Walking Barry Bonds: The ethics of the intentional walk. In E. Bronson (Ed.), *Baseball and philosophy: Thinking outside the batter's box* (pp. 261–272). Chicago: Open Court.

Kretchmar, R.S. (2005). *Practical philosophy of sport and physical activity* (2nd ed.). Champaign, IL: Human Kinetics.

Lasch, C. (1979). *The culture of narcissism: American life in an age of diminishing expectations.* New York: Warner Books.

McGinn, C. (2008). *Sport.* (The Art of Living Series). Stocksfield, UK: Acumen.

Meier, K. (1980). An affair of flutes: An appreciation of play. *Journal of the Philosophy of Sport, 7,* 24–45.

Metheny, E. (1965). *Connotations of movement in sport and dance.* Dubuque, IA: Brown.

Metheny, E. (1968). *Movement and meaning.* New York: McGraw-Hill.

Morgan, W. (1982). Play, utopia, and dystopia: Prologue to a ludic theory of the state. *Journal of the Philosophy of Sport, 9,* 30–42.

Morgan, W.J. (1994). *Leftist theories of sport: A critique and reconstruction.* Urbana, IL: University of Illinois Press.

Mosher, C., & Atkins, S. (2007). *Drugs and drug policy. The control of consciousness alteration.* Thousand Oaks, CA: Sage.

Nagel, T. (1987). *What does it all mean? A very short introduction to philosophy.* Oxford, UK: Oxford University Press.

Osterhoudt, R.G. (1991). *The philosophy of sport: An overview.* Champaign, IL: Stipes.

Perry, C. (1983). Blood doping and athletic competition. *International Journal of Applied Philosophy, 1*(3), 39–45.

Pieper, J. (1952). *Leisure: The basis of culture.* New York: Pantheon Books.

Simon, R. (2010). *Fair Play: The ethics of sport* (3rd ed.). Boulder, CO: Westview Press.

Slusher, H.S. (1967). *Man, sport and existence: A critical analysis.* Philadelphia: Lea & Febiger.

Suits, B. (1978). *The grasshopper: Games, life and utopia.* Toronto: University of Toronto Press.

Thompson, P.B. (1982). Privacy and the urinalysis testing of athletes. *Journal of the Philosophy of Sport, 9,* 60–65.

Todd, T. (1987). Anabolic steroids: The gremlins of sport. *Journal of Sport History, 14,* 87–107.

Torres, C. (2000). What counts as part of the game? A look at skills. *Journal of the Philosophy of Sport, 27,* 81–92.

Torres, C. (2002). *Play as expression: An analysis based on the philosophy of Maurice Merleau-Ponty.* Unpublished doctoral dissertation, The Pennsylvania State University, University Park.

Torres, C., & McLaughlin, D. (2003). Indigestion? An apology for ties. *Journal of the Philosophy of Sport, 30,* 144–158.

Weiss, P. (1969). *Sport: A philosophic inquiry.* Carbondale, IL: Southern Illinois University Press.

Chapter 6

Adelman, M.L. (1986). *A sporting time: New York City and the rise of modern athletics, 1820–70.* Urbana: University of Illinois Press.

Adventure Cycling Association. (2012). www.adventurecycling.org/routes/index.cfm.

American Physical Therapy Association. (2012). www.apta.org/PTEducation/Overview/http.

Baker, W.J. (1988). *Sports in the Western world.* Urbana: University of Illinois Press.

Berryman, J.W. (1973). Sport history as social history? *Quest, 20,* 65–73.

Berryman, J.W. (1989). The tradition of the "six things nonnatural": Exercise and medicine from Hippocrates through ante-bellum America. *Exercise and Sport Sciences Reviews, 17,* 515–559.

Betts, J.R. (1952). Organized sport in industrial America. *Dissertation Abstracts, 12*(1), 41. (University Microfilms No. 3322)

Blair, S.N., Mulder, R.T., & Kohl, H.W. (1987). Reaction to "Secular Trends in Adult Physical Activity: Exercise Boom or Bust?" *Research Quarterly for Exercise and Sport, 58*(2), 106–110.

Boyle, R.H. (1963). *Sport—Mirror of American life.* Boston: Little, Brown.

Carey, A.R., & Mullins, M.E. (1997, June 16). Toning up. *USA Today,* p. 1C.

Dulles, F.R. (1940). *America learns to play: A History of popular recreation.* New York: D. Appleton-Century.

Eitzen, D.S., & Sage, G.H. (1993). *Sociology of North American sport* (5th ed.). Madison, WI: Brown & Benchmark.

Eitzen, D.S., & Sage, G.H. (1997). *Sociology of North American sport* (6th ed.). Madison, WI: Brown & Benchmark.

Futterman, M. (2011, March 31). Has baseball's moment passed? *The Wall Street Journal,* p. D11.

Gerber, E.W. (1971). *Innovators and institutions in physical education.* Philadelphia: Lea & Febiger.

Gorn, E., & Goldstein, W. (1993). *A brief history of American sports.* New York: Hill & Wang.

Gorn, E.J. (1986). *The manly art: Bare-knuckle prize fighting in America.* Ithaca, NY: Cornell University Press.

Guttmann, A. (1978). *From ritual to record: The nature of modern sports.* New York: Columbia University Press.

Guttmann, A. (1984). *The games must go on: Avery Brundage and the Olympic movement.* New York: Columbia University Press.

Guttmann, A. (1991). *Women's sports: A history.* New York: Columbia University Press.

Hardy, S. (1982). *How Boston played: Sport, recreation, and community 1865–1915.* Boston: Northeastern University Press.

Hartwell, E.M. (1887). On the physiology of exercise. *Boston Medical and Surgical Journal,* 31 (March) 297–302; (April) 321–324.

Hartwell, E.M. (1899). *On physical training. Report of the commissioner of education for 1897–1898.* Vol. 1. Washington, DC: U.S. Government Printing Office.

Henry, F.M. (1964). Physical education: An academic discipline. *Journal of Health, Physical Education, and Recreation, 35*(7), 32–33, 69.

International Health, Racquet and Sportsclub Association (IHRSA). (2008). Profiles of success. www.docstoc.com/docs/120761813/profiles-sample\.

Karpovich, P.V., Morehouse, L.E., Scott, M.G., & Weiss, R.A. (Eds.). (1960). The contributions of physical activity to human well-being. *Research Quarterly, 31*(2), part II [special issue].

Lapchick, R. (2007). *The 2006 racial and gender report card: Major League Baseball.* Institute for Diversity and Ethics in Sport, University of Central Florida. www.tidesport.org/RGRC/2006/2006_RGRC_MLB.pdf.

Lapchick, R. (2012). The racial and gender report: Major League Baseball. Institute for Diversity and Ethics in Sport, University of Central Florida. www.tidesport.org/RGRC/2012/2012%20MLB%20RGRC.pdf.

Leonard, W.M. (1998). *A sociological perspective of sport.* Needham Heights, MA: Allyn & Bacon.

Levine, P. (1992). *Ellis Island to Ebbets Field: Sport and the American Jewish experience.* New York: Oxford University Press.

Lucas, J.A., & Smith, R.A. (1978). *Saga of American sport.* Philadelphia: Lea & Febiger.

Lippi, G., Banfi, G, Favaloro, E. Rittweger, J., & Maffulli, N. (2008). Updates on improvement of human athletic performance: focus on world records in athletics. *British Medical Journal, 87,* 7–15.

Massengale, J.D., & Swanson, R.A. (Eds.). (1997). *The history of exercise and sport science.* Champaign, IL: Human Kinetics.

Murphy, W. (1995). *Healing the generations: A history of physical therapy and the American Physical Therapy Association.* Lyme, CT: Greenwich.

National Athletic Trainers' Association. (2011). Profile of athletic trainers. www.nata.org/sites/default/files/Profiles-of-Athletic-Trainers_01-12.pdf.

National Center for Health Statistics. (2011). Leisure-time physical activity status. Centers for Disease Control and Prevention. www.cdc.gov/nchs/data/series/sr_10/sr10_249.pdf.

National Center for Health Statistics. (2007). *Health, United States, 2007, with chartbook on trends in the health of Americans.* Hyattsville, MD: National Center for Health Statistics, Division of Data Services.

National Federation of State High School Associations. (2012). 2010-11 athletics participation summary. www.nfhs.org/content.aspx?id=3282\.

North American Society for Sport Management. (2012). www.nassm.com/InfoAbout/SportMgmtPrograms/United_States.

Outdoor Foundation (2012). *2012 Outdoor recreation participation topline report.* Outdoor Industry Association. www.outdoorfoundation.org/research.participation.2012.topline.html\.

Park, R.J. (1980). The *Research Quarterly* and its antecedents. *Research Quarterly for Exercise and Sport, 51*(1), 1–22.

Park, R.J. (1981). The emergence of the academic discipline of physical education in the United States. In G.A. Brooks (Ed.), *Perspectives on the academic discipline of physical education* (pp. 20–45). Champaign, IL: Human Kinetics.

Park, R.J. (1987a). Physiologists, physicians, and physical educators: Nineteenth century biology and exercise, hygienic and educative. *Journal of Sport History, 14*(1), 28–60.

Park, R.J. (1987b). Sport, gender and society in a transatlantic Victorian perspective. In J.A. Mangan & R.J. Park (Eds.), *From "fair sex" to feminism: Sport and the socialization of women in the industrial and post-industrial eras* (pp. 58–93). London: Frank Cass.

Park, R.J. (1989). The second 100 years: Or, can physical education become the renaissance field of the 21st century? *Quest, 41*(1), 2–27.

Paxson, F.L. (1917). The rise of sport. *Mississippi Valley Historical Review* 4, 143–168.

Peiss, K. (1986). *Cheap amusements: Working women and leisure in turn-of-the-century New York.* Philadelphia: Temple University Press.

Pope, S.W. (1997). Introduction: American sport history—toward a new paradigm. In S.W. Pope (Ed.), *The new American sport history: Recent approaches and perspectives* (pp. 1–30). Urbana: University of Illinois Press.

Rader, B.G. (1990). *American sports: From the age of folk games to the age of televised sports.* Englewood Cliffs, NJ: Prentice Hall.

Ramlow, J., Kriska, A., & LaPorte, R. (1987). Physical activity in the population: The epidemiologic spectrum. *Research Quarterly for Exercise and Sport, 58*(2), 111–113.

Regalado, S.O. (1992). Sport and community in California's Japanese American "Yamato Colony," 1930–1945. *Journal of Sport History, 19*(2), 130–143.

Rice, E.A., Hutchinson, J.L., & Lee, M. (1969). *A brief history of physical education.* New York: Ronald.

Rudolph, F. (1962). *The American college and university: A history.* New York: Vintage.

Shafer, R.J. (Ed.). (1980). *A guide to historical method.* Homewood, IL: Dorsey.

Smith, J.K. (1979). *Athletic training: A developing profession.* Unpublished master's thesis, Brigham Young University, Provo, Utah.

Smith, R.A. (1988). *Sports and freedom: The rise of big-time college athletics.* New York: Oxford University Press.

Staley, S.C. (1937). The history of sport: A new course in the professional training curriculum. *Journal of Health, Physical Education and Recreation, 8,* 522–525, 570–572.

Stephens, T. (1987). Secular trends in adult physical activity: Exercise boom or bust? *Research Quarterly for Exercise and Sport, 58*(2), 94–105.

Struna, N.L. (1996a). Historical research in physical activity. In J.R. Thomas & J.K. Nelson (Eds.), *Research methods in physical activity* (pp. 251–275). Champaign, IL: Human Kinetics.

Struna, N.L. (1996b). *People of prowess: Sport, leisure, and labor in early Anglo-America.* Urbana: University of Illinois Press.

Struna, N.L. (1997). Sport history. In J.D. Massengale & R.A. Swanson (Eds.), *The history of exercise and sport science* (pp. 143–179). Champaign, IL: Human Kinetics.

Swanson, R.A., & Spears, B. (1995). *History of sport and physical education in the United States.* Madison, WI: Brown & Benchmark.

U.S. Bureau of Labor Statistics (USBLS). (1961). *Occupational outlook handbook* (Bulletin No. 1300). Washington, DC: U.S. Government Printing Office.

U.S. Bureau of Labor Statistics (USBLS). (1982). *Occupational outlook handbook* (Bulletin No. 2200). Washington, DC: U.S. Government Printing Office.

U.S. Bureau of Labor Statistics. (2011). *Occupational outlook handbook,* 2010-11 edition. Physical therapists. United States Department of Labor. www.bls.gov/ooh/.

U.S. Bureau of the Census (USBOC). (1966). *Statistical abstract of the United States, 1966.* Washington, DC: U.S. Government Printing Office.

U.S. Bureau of the Census (USBOC). (2012). *Statistical Abstract of the United States, 2012.* Washington, D.C.: U.S. Government Printing Office.

US Lacrosse. (2011). 2010 participation survey. www.uslacrosse.org/LinkClick.aspx?fileticket=Oh4-ZLjz9F4%3D&tabid=2800/st.

Van Dalen, D.B., & Bennett, B.L. (1971). *A world history of physical education: Cultural, philosophical, comparative.* Englewood Cliffs, NJ: Prentice Hall.

Vertinsky, P. (1990). *The eternally wounded woman: Women, doctors and exercise in the late nineteenth century.* Manchester, UK: Manchester University Press.

Weber, E. (1970). Gymnastics and sport in fin-de-siècle France: Opium of the classes? Paper delivered at the 85th Annual Meeting of the American Historical Association, Boston, MA.

Whorton, J.C. (1982). *Crusaders for fitness: The history of American health reformers.* Princeton, NJ: Princeton University Press.

Wiggins, D.K. (1980). The play of slave children in the plantation communities of the old south, 1820–1860. *Journal of Sport History, 7*(2), 21–39.

Wilson, W. (1977). Social discontent and the growth of wilderness sport in America: 1965–1974. *Quest, 27,* 54–60.

Chapter 7

Abbate, G. (2002). Gap between rich, poor widening. *Globeandmail.com.* July 7. www.globeandmail.com/serlet/ArticleNews/printarticle/gam/20020719/UGAPPCC.

Acosta, R.V., & Carpenter, L.J. (1994). The status of women in intercollegiate athletics. In S. Birrell & C.L. Cole (Eds.), *Women, sport, and culture* (pp. 111–118). Champaign, IL: Human Kinetics.

Acosta, R.V., & Carpenter, L.J. (2008). Women in intercollegiate sport: A longitudinal, national study: Thirty one year update, 1977-2008. www.acostacarpenter.org.

Andrews, D. (Ed.). (1996). Deconstructing Michael Jordan: Reconstructing postindustrial America. *Sociology of Sport Journal, 13*(4) [special issue].

Berkowitz, S., & Upton, J. (2011a, March 30). An analysis of salaries for college basketball coaches. *USA Today.* www.usatoday.com/sports/college/mensbasketball/2011-coaches-salary-database.htm.

Berkowitz, S., & Upton, J. (2011b, April 4). An analysis of salaries for women's college basketball coaches. *USA Today.* www.usatoday.com/sports/college/womensbasketball/2011-coaches-salary-database.htm.

Birrell, S., & McDonald, M.G. (2000). Reading sport, articulating power lines. In S. Birrell & M.G. McDonald (Eds.), *Reading sport: Critical essays on power and representation* (pp. 3–13). Boston: Northeastern University Press.

Brady, E., Upton, J., & Berkowitz, S. (2011, November 7). Salaries for college football coaches back on the rise. *USA Today.* www.usatoday.com/sports/college/football/story/2011-11-17/cover-college-football-coaches-salaries-rise/51242232/1.

Brown, G. (2011, November). College sports participation rates continue upward trend. *Latest News.* www.ncaa.org/wps/wcm/connect/public/NCAA/Resources/Latest+News/2011/November/College+sports+participation+rates+continue+upward+trend,

Burnsed, B., Gloeckler, G., Grover, R., Lawyue, M., Lowry, T. & Polek, D. (2008, September 30). The Power 100: The most influential people in sports, from golf to skateboarding to . . . www.businessweek.com/stories/2008-09-30/the-power-100.

Carlston, D.E. (1983). An environmental explanation for race differences in basketball performance. *Journal of Sport and Social Issues, 7*(2), 30–51.

Coakley, J. (2007). *Sports in society: Issues and controversies* (9th ed.). Boston: McGraw-Hill.

Coakley, J. (2009). *Sports in society: Issues and controversies* (10th ed.). Boston: McGraw-Hill.

Douglas, D.D., & Jamieson, K.M. (2006). A farewell to remember: Interrogating the Nancy Lopez farewell tour. *Sociology of Sport Journal, 23*(2), 117–141.

Dunning, E., Murphy, P., & Williams, J. (1988). *The roots of football hooliganism: An historical and sociological study.* London: Routledge.

Economic Policy Institute. (2011). The state of working America. Washington, DC: Economic Policy Institute. http://stateofworkingamerica.org/inequality.

Eitzen, D.S. (2006). *Fair and foul: Beyond the myths and paradoxes of sport* (3rd ed.). New York: Rowman & Littlefield.

Eitzen, D.S., & Sage, G.H. (2009). *Sociology of North American Sport* (8th ed.). Boulder, CO: Paradigm.

Fausto-Sterling, A. (2000). *Sexing the body.* New York: Basic Books.

Gibbs, A. (1997). Focus groups. *Social Research Update.* http://sru.soc.surrey.ac.uk/SRU19.html.

Gruneau, R. (1983). *Class, sports, and social development.* Amherst: University of Massachusetts Press.

Guttmann, A. (1978). *From ritual to record: The nature of modern sports.* New York: Columbia University Press.

Hargreaves, J. (2001). *Heroines of sport: The politics of difference and identity.* London: Routledge.

King, C.R. (2004). This is not an Indian: Situating claims about Indianness in sporting worlds. *Journal of Sport & Social Issues, 28,* 3–10.

Kinkema, K.M., & Harris, J.C. (1998). MediaSport studies: Key research and emerging issues. In L.A. Wenner (Ed.), *MediaSport* (pp. 27–54). London: Routledge.

Kochman, T. (1981). *Black and white styles in conflict.* Chicago: University of Chicago Press.

Lapchick, R. (2011a, March 3). *The 2010 racial and gender report card: college sport.* The Institute for Diversity and Ethics in Sport. www.tidesport.org/racialgenderreportcard.html.

Lapchick, R. (2011b, April 21). *The 2011 racial and gender report card: Major League Baseball.* The Institute for Diversity and Ethics in Sport. www.tidesport.org/racialgenderreportcard.html.

Lapchick, R. (2011c, June 16). *The 2011 Racial and gender report card: National Basketball Association.* The Institute for Diversity and Ethics in Sport. www.tidesport.org/racialgenderreportcard.html.

Lapchick, R. (2011d, September 15). *The 2011 racial and gender report card: National Football League.* The Institute for Diversity and Ethics in Sport. www.tidesport.org/racial-genderreportcard.html.

Lapchick, R. (2011e, October 5). *The 2011 racial and gender report card: Major League Soccer.* The Institute for Diversity and Ethics in Sport. www.tidesport.org/racialgenderreportcard.html.

Lorber, J. (1994). *Paradoxes of gender.* New Haven, CT: Yale University Press.

Majors, R. (1990). Cool pose: Black masculinity and sports. In M.A. Messner & D.F. Sabo (Eds.), *Sport, men, and the gender order: Critical feminist perspectives* (pp. 109–126). Champaign, IL: Human Kinetics.

McDonald, M.G. (1996). Michael Jordan's family values: Marketing, meaning, and post-Reagan America. *Sociology of Sport Journal, 13,* 344–365.

McIntosh, P.C. (1963). *Sport in society.* London: C.A. Watts Publisher.

Metheny, E. (1965). *Connotations of movement in sport and dance.* Dubuque, IA: Brown.

National Federation of State High School Associations. (2012). *Athletics participation survey totals.* www.nfhs.org/content.aspx?id=3282.

Neal, M.A. (2006). *New black man.* New York: Routledge.

Reading, C. L. & Wien, F. (2009). *Health inequalities and social determinants of Aboriginal peoples' health.* National Collaborating Centre for Aboriginal Health.

Ross, H. (1996, March 10). Waiting for the call. . . . *Greensboro News and Record,* p. C1.

Sage, G.H. (1997). Sport sociology. In J.D. Massengale & R.A. Swanson (Eds.), *History of exercise and sport science* (pp. 109–141). Champaign, IL: Human Kinetics.

Siedentop, D. (1996). Valuing the physically active life: Contemporary and future directions. *Quest, 48,* 266–274.

SportsBusiness Daily. (2010, December). NFL commissioner Roger Goodell tops SBJ's most influential list. *Street & Smith's SportsBusiness Daily.* www.sportsbusinessdaily.com/Daily/Issues/2010/12/Dec-16/People-and-Pop-Culture/Influential.aspx.

Staurowsky, E.J. (1999, November). American Indian sport imagery and the miseducation of Americans. *Quest, 51,* 382–392.

Staurowsky, E.J. (2004). Privilege at play: On the legal and social fictions that sustain American Indian sport imagery. *Journal of Sport & Social Issues, 28,* 11–29.

Stein, P.J., & Hoffman, S. (1978). Sports and male role strain. *Journal of Social Issues, 34,* 136–150.

Stephens, T., & Caspersen, C.J. (1994). The demography of physical activity. In C. Bouchard, R.J. Shephard, & T. Stephens (Eds.), *Physical activity, fitness, and health: International proceedings and consensus statement* (pp. 204–213). Champaign, IL: Human Kinetics.

Stripling, J., & Fuller, A. (2011, April 3). Presidents defend their pay as public colleges slash budgets. *The Chronicle of Higher Education.* http://chronicle.com/article/Presidents-Defend-Their-Pay-as/126971/.

U.S. Bureau of the Census. (2007). Participation in selected sport activities: 2005. (Table 1222). http://search.census.gov/search?utf8=%E2%9C%93&affiliate=census&query=Participation+in+selected+sport+activities%3A+2005&commit.x=0&commit.y=0.

User fees, insurance rate increases and access to recreation facilities. (2003, November). *Physical Training.* http://ejmas.com/pt/ptart_orfa_1103.html.

Wilson, B., & Sparks, R. (1996). "It's gotta be the shoes": Youth, race, and sneaker commercials. *Sociology of Sport Journal, 13,* 398–427.

Women's Sports Foundation. (2009, March). Women's sports and fitness facts and statistics. www.womenssportsfoundation.org.

Chapter 8

Abernethy, B., & Sparrow, W.A. (1992). The rise and fall of dominant paradigms in motor behavior research. In J.J. Summers (Ed.), *Approaches to the study of motor control and learning* (pp. 3–45). Amsterdam: Elsevier.

Abernethy, B., Thomas, K.T., & Thomas, J.R. (1993). Strategies for improving understanding of motor expertise (or mistakes we have made and things we have learned!!). In J.L. Starkes & F. Allard (Eds.), *Cognitive issues in motor learning* (pp. 317–356). Amsterdam: Elsevier.

Adams, J.A. (1987). Historical review and appraisal of research on the learning, retention, and transfer of human motor skills. *Psychological Bulletin, 101,* 41–74.

Bayley, N. (1935). The development of motor abilities during the first three years. *Monographs of the Society for the Research in Child Development,* Whole No. 1, 1–26.

Blix, M. (1892–1895). Die lange und spannung des muskels. *Skandinavische Archiv Physiologie, 3,* 295–318; *4,* 399–409; *5,* 150–206.

Chamberlin, C., & Lee, T. (1993). Arranging practice conditions and designing instruction. In R.N. Singer, M. Murphey, & L.K. Tennant (Eds.), *Handbook of research on sport psychology* (pp. 213–241). New York: Macmillan.

Christina, R.W. (1989). Whatever happened to applied research in motor learning? In J. Skinner et al. (Eds.), *Future directions for exercise science and sport research* (pp. 411–422). Champaign, IL: Human Kinetics.

Christina, R.W. (1992). The 1991 C.H. McCloy Research Lecture: Unraveling the mystery of the response complexity effect in skilled movements. *Research Quarterly for Exercise and Sport, 63,* 218–230.

Clark, J.E., & Phillips, S.J. (1991). The development of intralimb coordination in the first six months of walking. In J. Fagard & P.H. Wolff (Eds.), *The development of timing control and temporal organization in coordinated action* (pp. 245–257). New York: Elsevier Science.

Clark, J.E., & Whitall, J. (1989). What is motor development? The lessons of history. *Quest, 41,* 183–202.

Dennis, W. (1938). Infant development under conditions of restricted practice and a minimum of social stimulation: A preliminary report. *Journal of Genetic Psychology, 53,* 149–158.

Dennis, W., & Dennis, M. (1940). The effect of cradling practices on the age of walking in Hopi children. *Journal of Genetic Psychology, 56,* 77–86.

Ericsson, K.A. (2003). Development of elite performance and deliberate practice. In J.L. Starkes & K.A. Ericsson (Eds.), *Expert performance in sports* (pp. 49–84). Champaign, IL: Human Kinetics.

Espenschade, A. (1960). Motor development. In W.R. Johnson (Ed.), *Science and medicine of exercise and sport* (p. 330). New York: Harper & Row.

Fischman, M.G. (2007). Motor learning and control foundations of kinesiology: Defining the academic core. *Quest, 59,* 67–76.

Fitts, P.M., & Posner, M.I. (1967). *Human performance.* Pacific Grove, CA: Brooks/Cole.

Galton, F. (1876). The history of twins as a criterion of the relative power of nature. *Anthropological Institute Journal, 5,* 391–406.

Gentile, A.M. (1972). A working model of skill acquisition with application to teaching. *Quest,* Monograph XVII, 2–23.

Gesell, A. (1928). *Infancy and human growth.* New York: Macmillan.

Haken, H., Kelso, J.A.S., & Bunz, H. (1985). A theoretical model of phase transitions in human hand movements. *Biological Cybernetics, 51,* 347–356.

Henry, F.M., & Rogers, D.E. (1960). Increased response latency for complicated movements and a "memory drum" theory of neuromotor reaction. *Research Quarterly, 31,* 448–458.

Kelso, J.A.S. (1995). *Dynamic patterns: The self-organization of brain and behavior.* Cambridge, MA: MIT Press.

Locke, E.A., & Latham, G.P. (1985). The application of goal setting to sports. *Sport Psychology Today, 7,* 205–222.

Magill, R.A., & Hall, K.G. (1990). A review of the contextual interference effect in motor skill acquisition. *Human Movement Science, 9,* 241–289.

Malina, R.M. (1984). Physical growth and maturation. In J.R. Thomas (Ed.), *Motor development during childhood and adolescence* (pp. 2–26). Edina, MN: Burgess International.

Malina, R.M., Bouchard, C., & Bar-Or, O. (2004). *Growth, maturation, and physical activity* (2nd ed.). Champaign, IL: Human Kinetics.

McCullagh, P. (1993). Modeling: Learning, developmental, and social psychological considerations. In R.N. Singer, M. Murphey, & L.K. Tennant (Eds.), *Handbook of research on sport psychology* (pp. 106–126). New York: Macmillan.

McGraw, M.B. (1935). *Growth: A study of Johnny and Jimmy.* New York: Appleton-Century-Crofts.

McGraw, M.B. (1939). Later development of children specially trained during infancy: Johnny and Jimmy at school age. *Child Development, 10,* 1–19.

Newell, K.M. (1984). Physical constraints to the development of motor skills. In J.R. Thomas (Ed.), *Motor development during childhood and adolescence* (pp. 105–120). Minneapolis, MN: Burgess.

Otsuji, T., Abe, M., & Kinoshita, H. (2002). After-effects of using a weighted bat on subsequent swing velocity and batter's perceptions of swing velocity and heaviness. *Perceptual and Motor Skills, 94,* 119–126.

Rosenbaum, D.A. (1991). *Human motor control.* San Diego: Academic Press.

Salmoni, A.W., Schmidt, R.A., & Walter, C.B. (1984). Knowledge of results and motor learning: A review and critical reappraisal. *Psychological Bulletin, 95,* 355–386.

Schmidt, R.A. (1975). A schema theory of discrete motor skill learning. *Psychological Review, 82,* 225–260.

Schmidt, R.A. (1988). *Motor control and learning: A behavioral emphasis* (2nd ed.). Champaign, IL: Human Kinetics.

Schmidt, R.A. (1991). *Motor learning & performance: From principles to practice.* Champaign, IL: Human Kinetics.

Schmidt, R.A., & Wrisberg, C.A. (2007). *Motor learning and performance: A situation-based learning approach* (4th ed.). Champaign, IL: Human Kinetics.

Seefeldt, V., & Haubenstricker, J. (1982). Patterns, phases, or stages: An analytical model for the study of developmental movement. In J.A.S. Kelso & J.E. Clark (Eds.), *The development of movement control and co-ordination* (pp. 309–318). Chichester, UK: Wiley.

Sherrington, C.S. (1906). *The integrative action of the nervous system.* New Haven, CT: Yale University Press.

Southard, D., & Groomer, L. (2003). Warm-up with baseball bats of varying moments of inertia: effects on bat velocity and swing pattern. *Research Quarterly for Exercise and Sport, 74,* 270–276.

Spirduso, W.W., & MacRae, P.G. (1990). Motor performance and aging. In J.E. Birren & K.W. Schaie (Eds.), *The handbook of psychology of aging* (3rd ed., pp. 184–197). San Diego: Academic Press.

Stelmach, G.E., & Nahom, A. (1992). Cognitive-motor abilities of the elderly driver. *Human Factors, 34*(1), 53–65.

Thelen, E., Ulrich, D., & Jensen, J.L. (1990). The developmental origins of locomotion. In M.H. Woollacott & A. Shumway-Cook (Eds.), *Development of posture and gait: Across the lifespan.* Columbia: University of South Carolina Press.

Thomas, J.R. (1997). History of motor behavior. In J.D. Massengale & R.A. Swanson (Eds.), *History of exercise and sport sciences* (pp. 203–292). Champaign, IL: Human Kinetics.

Thomas, J.R. (2006). Motor behavior: From telegraph keys and twins to linear slides and stepping. *Quest, 58,* 112–127.

Thomas, J.R., & French, K.E. (1985). Gender differences across age in motor performance: A meta-analysis. *Psychological Bulletin, 98,* 260–282.

Thomas, J.R., & Thomas, K.T. (1989). What is motor development: Where does it belong? *Quest, 41,* 203–212.

Thomas, J.R., & Thomas, K.T. (2008). Principles of motor development for elementary school physical education. *Elementary School Journal, 108,* 181–195 (special issue edited by L. Locke and K. Graber).

Thomas, K.T., Gallagher, J.D., & Thomas, J.R. (2001). Motor development and skill acquisition during childhood and adolescence. In R.N. Singer, H.A. Hausenblas, & C.M. Janelle (Eds.), *Handbook of sport psychology* (2nd ed., pp. 20–52). New York: Wiley & Sons.

Thomas, K.T., & Thomas, J.R. (1994). Developing expertise in sport: The relation of knowledge to performance. *International Journal of Sport Psychology, 25,* 295–312.

Ulrich, B. (2007). Motor development: Core curricular concepts. *Quest, 59,* 77–91.

Ulrich, B., & Reeve, T.G. (2005). Studies in motor behavior: 75 years of research in motor development, learning, and control. *Research Quarterly for Exercise and Sport, 75,* S62–S70.

Weiss, M.R., & Klint, K.A. (1987). "Show and tell" in the gymnasium: An investigation of developmental differences in modeling and verbal rehearsal of motor skills. *Research Quarterly for Exercise and Sport, 58,* 234–241.

Chapter 9

Anastasi, A. (1988). *Psychological testing.* New York: Macmillan.

Beilock, S. (2010). *Choke.* New York: Free Press.

Bell, J.J., & Hardy, J. (2009). Effects of attentional focus on skilled performance in golf. *Journal of Applied Sport Psychology, 21,* 163–177.

Buckworth, J., & Dishman, R.K. (2002). *Exercise psychology.* Champaign, IL: Human Kinetics.

Castaneda, B., & Gray, R. (2007). Effects of focus of attention on baseball batting performance in players of differing skill levels. *Journal of Sport and Exercise Psychology, 29,* 60–77.

Coen, S.P., & Ogles, B.M. (1993). Psychological characteristics of the obligatory runner: A critical examination of the anorexia analogue hypothesis. *Journal of Sport and Exercise Psychology, 15,* 338–354.

Crawford, S., & Eklund, R.C. (1994). Social physique anxiety, reasons for exercise, and attitudes toward exercise settings. *Journal of Sport and Exercise Psychology, 16,* 70–82.

Csikszentmihalyi, M. (1990). *Flow: The psychology of optimal experience.* New York: Harper & Row.

Easterbrook, J.A. (1959). The effect of emotion on cue utilization and the organization of behavior. *Psychological Review, 66,* 183–201.

Eys, M.A., Burke, S.M., Carron, A.V., & Dennis, P.W. (2006). The sport team as an effective group. In J.M. Williams (Ed.), *Applied sport psychology: Personal growth to peak performance* (5th ed., pp. 157–173). Boston: McGraw-Hill.

Franz, S.I., & Hamilton, G.V. (1905). The effects of exercise upon the retardation in conditions of depression. *American Journal of Insanity, 62,* 239–256.

Hellison, D. (2003). *Teaching responsibility through physical activity* (2nd ed.). Champaign, IL: Human Kinetics.

Heuze, J.P., & Brunel, P. (2003). Social loafing in a competitive context. *International Journal of Sport and Exercise Psychology, 1,* 246–263.

Jones, G., Hanton, S., & Connaughton, D. (2007). A framework of mental toughness in the world's best performers. *The Sport Psychologist, 21,* 243–264.

Jones, G., Hanton, S., & Swain, A. (1994). Intensity and interpretation of anxiety symptoms in elite and non-elite sports performers. *Personality and Individual Differences, 17,* 657–663.

Krane, V., & Williams, J.M. (2010). Psychological characteristics of peak performance. In J.M. Williams (Ed.), *Applied sport psychology: Personal growth to peak performance* (6th ed., pp. 169–188). Boston: McGraw-Hill.

Landers, D.M., & Arent, S.A. (2007). Physical activity and mental health. In G. Tenenbaum & R.C. Eklund (Eds.), *Handbook of sport psychology* (3rd ed., pp. 469–491). Hoboken, NJ: Wiley.

Mullen, R., & Hardy, L. (2010). Conscious processing and the process goal paradox. *Journal of Sport & Exercise Psychology, 32,* 275–297.

Orlick, T. (1998). *Embracing your potential.* Champaign, IL: Human Kinetics.

Shields, D.L., & Bredemeier, B.J. (2007). Advances in sport morality research. In G. Tenenbaum & R.C. Eklund (Eds.), *Handbook of sport psychology* (3rd ed., pp. 662–684). Hoboken, NJ: Wiley.

Smith, R.A., & Biddle, S.J.H. (1995). Psychological factors in the promotion of physical activity. In S.J.H. Biddle (Ed.), *European perspectives on exercise and sport psychology* (pp. 85–108). Champaign, IL: Human Kinetics.

Smith, R.E., & Christensen, D.S. (1995). Psychological skills as predictors of performance and survival in professional baseball. *Journal of Sport and Exercise Psychology, 17,* 399–415.

Smith, R.E., Schutz, R.W., Smoll, F.L., & Ptacek, J.T. (1995). Development and validation of a multidimensional measure of sport-specific psychological skills: The athletic coping skills inventory-28. *Journal of Sport and Exercise Psychology, 17,* 379–398.

Vallerand, R.J., Blanchard, C., Mageau, G.A., Koestner, R., Ratelle, C., Leonard, M., Gagne, M., & Marsolais, J. (2003). Les passions de l'ame: On obsessive and harmonious passion. *Journal of Personality and Social Psychology, 85,* 756–767.

Vealey, R.S. (2005). *Coaching for the inner edge.* Morgantown, WV: Fitness Information Technology.

Vealey, R.S., & Greenleaf, C.A. (2010). Seeing is believing: Understanding and using imagery in sport. In J.M. Williams (Ed.), *Applied sport psychology: Personal growth to peak performance* (6th ed., pp. 267–304). Boston: McGraw-Hill.

Chapter 10

Amar, J. (1920). *The human motor.* New York: Dutton.

Atwater, A.E. (1980). Kinesiology/biomechanics: Perspectives and trends. *Research Quarterly for Exercise and Sport, 51,* 193–218.

Chaffin, D.B., & Andersson, G.B. (1991). *Occupational biomechanics* (2nd ed.). New York: Wiley.

Contini, R., & Drillis, R. (1966a). Applied biomechanics. In H.N. Abramson, H. Liebowitz, J.M. Crowley, & S. Juhasz (Eds.), *Applied mechanics surveys* (pp. 161–172). Washington, DC: Spartan Books.

Contini, R., & Drillis, R. (1966b). Kinematic and kinetic techniques in biomechanics. In F. Alt (Ed.), *Advances in bioengineering and instrumentation* (pp. 3–68). New York: Plenum Press.

Cureton, T.K. (1930). Mechanics and physiology of swimming (the crawl flutter kick). *Research Quarterly, 1,* 87–121.

Cureton, T.K. (1932). Physics applied to physical education. *Journal of Health and Physical Education, 1,* 23–25.

De Wit, B., De Clercq, D., & Aerts, P. (1996). Ground reaction forces and spatio-temporal variables during barefoot and shod running. In J.M.C.S. Abrantest (Ed.), *Proceedings of the XIV symposium on bioimechanics in sports* (pp. 252–255). June 25-29, 1996, Funchal, Madeira, Portugal.

Divert, C., Mornieux G., Baur H., Mayer, & Belli, A. (2005). Mechanical comparison of barefoot and shod running. *International Journal of Sports Medicine, 26,* 593–598.

Divert, C., Mornieux, G., Freychat, P., Baly, L., Mayer, F., & Belli, A. (2008). Barefoot-shod running differences: Shoe or mass effect? *International Journal of Sports Medicine, 29*(6), 512–518.

Donahue, S.W., Sharkey, N.A., Modanlou, K.A., Sequeira, L.N., & Martin, R.B. (2000). Bone strain and microcracks at stress fracture sites in human metatarsals. *Bone, 27*(6), 827–833.

Fenn, W.O. (1929). Mechanical energy expenditure in sprint running as measured by moving pictures. *American Journal of Physiology, 90,* 343–344.

Fitzgerald, M. (2006, March/April). Nanobiomechanics. *Technology Review.* www.technologyreview.com/article/405526/nanobiomechanics/.

Fung, Y.C. (1968). Biomechanics—its scope, history, and some problems of continuum mechanics in physiology. *Applied Mechanics Reviews, 21,* 1–20.

Furusawa, K., Hill, A.V., & Parkinson, J.L. (1927). The dynamics of sprint running. *Proceedings of the Royal Society of London, 102B,* 29–42.

Glassow, R.B. (1932). *Fundamentals of physical education.* Philadelphia: Lea & Febiger.

Goss, D.L., & Goss, M.T. (2012). A review of mechanics and injury trends among various running styles. *U.S. Army Medical Department Journal, Jul.-Sept.,* 62–71.

Hart, I. (1963). *The mechanical investigations of Leonardo da Vinci* (2nd ed.). Berkeley: University of California Press.

Hatze, H. (1974). The meaning of the term "biomechanics." *Journal of Biomechanics, 7,* 189–190.

Hill, A.V. (1926). Scientific study of athletes. *Scientific American, 134*, 224–225.

Hreljac, A., Marshall, R.N., & Hume, P.A. (2000). Evaluation of lower extremity overuse injury potential in runners. *Medicine and Science in Sports and Exercise, 32*(9), 1635–1641.

Hutchinson, J.R., Schwerda, D., Famini, D., Dale, R.H.I., Fischer, M. & Kram, R. (2006). The locomotor kinematics of African and Asian elephants: changes with speed and size. *Journal of Experimental Biology, 209*, 3812–3827.

Ko, M. (2010). Plantar pressure and gait in patients with diabetes. *Lower Extremity Review*, October 2010. www.lowerextremityreview.com/article/plantar-pressure-and-gait-in-patients-with-diabetes

Lieberman, D.E., Venkadesan, M., Werbel, W.A., Daoud, A.I., D'Andrea, S., Davis, I.S., et al. (2010). Foot strike patterns and collision forces in habitually barefoot versus shod runners. *Nature, 463*(7280), 531–535.

McCloy, C.H. (1960). The mechanical analysis of skills. In W.R. Johnson (Ed.), *Science and medicine in exercise and sports* (pp. 54–64). New York: Harper & Brothers.

McDougall, C. (2009). *Born to run: A hidden tribe, superathletes, and the greatest race the world has never seen.* New York: Knopf.

Meister, D. (1999). *The history of human factors and ergonomics.* Mahwah, NJ: Erlbaum.

Nelson, R.C. (1970). Biomechanics of sport: An overview. In J.M. Cooper (Ed.), *Selected topics on biomechanics: Proceedings of the C.I.C. Symposium on Biomechanics* (pp. 31–37). Chicago: Athletic Institute.

Nigg, B.M. (1983). External force measurements with sports shoes and playing surfaces. In B.M. Nigg & B.J. Kerr (Eds.), *Biomechanical aspects of sport shoes and playing surfaces* (pp. 11–23). Calgary, AB: University Printing.

Paterson, D.H., Jones, G.R., & Rice, C.L. (2007). Ageing and physical activity: Evidence to develop exercise recommendations for older adults. *Canadian Journal of Public Health, 98*(Suppl. 2), S69–S108.

Posse, N. (1890). *The special kinesiology of educational gymnastics.* Boston: Lothrop, Lee and Shepard.

Salzler, M., Bluman, E.M., Noonan, S., Chiodo, C., & de Asla, R. (2012). Injuries observed in minimalist runners. *Foot & Ankle International, 33*(4), 262–266.

Schneck, D. (2005). Editor's page. Forensic biomechanics. *American Laboratory, 37*(8), 4.

Skarstrom, W. (1909). *Gymnastic kinesiology.* Springfield, MA: Bassette.

Squadrone, R. & Gallozzi, C. (2009). Biomechanical and physiological comparison of barefoot and two shod conditions in experienced barefoot runners. *Journal of Sports Medicine & Physical Fitness, 49*(1), 6–13.

Steindler, A. (1942, November). What has biokinetics to offer to the physical educator? *Journal of Health and Physical Education, 507*–509, 555–556.

Thomas, J.T. (1969). Keynote address—The Army looks at biomechanics. In D. Bootzin & H.C. Muffley (Eds.), *Biomechanics.* New York: Plenum Press.

Wilks, D.C., Winwood, K., Gilliver, S.F., Kwiet, A., Chatfield, M., et al. (2009). Bone mass and geometry of the tibia and the radius of master sprinters, middle and long distance runners, race-walkers, and sedentary control participants: A pQCT study. *Bone, 45*, 91–97.

Winter, D.A. (1985). *Biomechanics and motor control of human gait.* Waterloo, ON: University of Waterloo Press.

Withnall, R., Eastaugh, J., & Freemantle, N. (2006). Do shock absorbing insoles in recruits undertaking high levels of physical activity reduce lower limb injury? A randomized controlled trial. *Journal of the Royal Society of Medicine, 99*(1), 32–37.

Chapter 11

American College of Sports Medicine. (2001). Position stand: Appropriate intervention strategies for weight loss and prevention of weight regain for adults. *Medicine and Science in Sports and Exercise, 33*, 2145–2156.

American College of Sports Medicine. (2010). *ACSM's guidelines for exercise testing and prescription.* Baltimore, MD: Lippincott, Williams & Wilkins.

Armstrong, L.E., Costill, D.L., & Fink, W.J. (1985). Influence of diuretic-induced dehydration on competitive running performance. *Medicine and Science in Sports and Exercise, 17*, 456–461.

Åstrand, P.-O. (1991). Influence of Scandinavian scientists in exercise physiology. *Scandinavian Journal of Medicine and Science in Sports, 1*, 3–9.

Bar-Or, O. (1983). *Pediatric sports medicine for the practitioner.* New York: Springer-Verlag.

Berryman, J.W. (1995). *Out of many, one: A history of the American College of Sports Medicine.* Champaign, IL: Human Kinetics.

Blair, S.N., Kohl, H.W., Paffenbarger, R.S., Clark, D.G., Cooper, K.H., & Gibbons, L.W. (1989). Physical fitness and all-cause mortality: A prospective study of healthy men and women. *Journal of the American Medical Association, 262*(17), 2395–2401.

Booth, F.W. (1989). Application of molecular biology in exercise physiology. In K.B. Pandolf (Ed.), *Exercise and sport sciences reviews* (Vol. 17, pp. 1–27). Baltimore: Williams & Wilkins.

Bouchard, C., Lesage, R., Lortie, G., Simoneau, J. A., Hamel, P., Boulay, M. R., Perusse, L., Theriault, G., & Leblanc, C. (1986). Aerobic performance in brothers, dizygotic and monozygotic twins. *Medicine and Science in Sports and Exercise, 18*, 639–646.

Celsing, F., Blomstrand, E., Werner, B., Pihlstedt, P., & Ekblom, B. (1986). Effects of iron deficiency on endurance and muscle enzyme activity in man. *Medicine and Science in Sports and Exercise, 18*, 156–161.

Clarkson, P.M., & Haymes, E.M. (1995). Exercise and mineral status of athletes, calcium, magnesium, phosphorus, and iron. *Medicine and Science in Sports and Exercise, 27*, 831–843.

Coggan, A., & Coyle, E. (1991). Carbohydrate ingestion during prolonged exercise: Effects on metabolism and performance. In J.O. Holloszy (Ed.), *Exercise and sport sciences reviews* (Vol. 19, pp. 1–40). Baltimore: Williams & Wilkins.

Davis, J.M., Burgess, W.A., Sientz, C.A., Bartoli, W.P., & Pate, R.R. (1988). Effects of ingesting 6% and 12% glucose/electrolyte beverages during prolonged intermittent cycling in the heat. *European Journal of Applied Physiology, 57*, 563–569.

Dill, D.B. (1967). The Harvard Fatigue Laboratory: Its development, contributions, and demise. In C.B. Chapman (Ed.), *Physiology of muscular exercise* (pp. 161–170). New York: American Heart Association.

Dill, D.B., Talbott, J.H., & Edwards, H.T. (1930). Studies in muscular activity. VI. Responses of several individuals to a fixed task. *Journal of Physiology, 69*, 267–305.

Fagard, D. (2001). Exercise characteristics and the blood pressure response to dynamic physical training. *Medicine and Science in Sports and Exercise, 33*, S484–S492.

Gollnick, P.D., Timson, B.F., Moore, R.L., & Riedy, M. (1981). Muscular enlargement and number of fibers in skeletal muscles of rats. *Journal of Applied Physiology, 50,* 936–943.

Gonyea, W.J. (1980). Role of exercise in inducing increases in skeletal muscle fiber number. *Journal of Applied Physiology, 48,* 421–426.

Hagberg, J.M. (1987). Effect of training on the decline of VO_2max with aging. *Federation Proceedings, 46,* 1830–1833.

Haymes, E.M., & Wells, C.L. (1986). *Environment and human performance.* Champaign, IL: Human Kinetics.

Hultman, E. (1967). Physiological role of muscle glycogen in man, with special reference to exercise. *Circulation Research, 21*(Suppl. 1), 99–114.

Ivy, J.L., Katz, A.L., & Cutler, C.L. (1989). Muscle glycogen resynthesis after exercise: Effect of time on carbohydrate ingestion. *Journal of Applied Physiology, 64,* 1480–1485.

Janz, K.F., Levy, S.M., Burns, T.L., Willing, M.C., & Warren, J.J. (2002). Fatness, physical activity, and television viewing in children during the adiposity rebound period: The Iowa Bone Development study. *Preventive Medicine, 35,* 563–571.

Kenney, W.L., Wilmore, J.H., & Costill, D.L. (2012). *Physiology of sport and exercise* (5th ed.). Champaign, IL: Human Kinetics.

Kroll, W.P. (1982). *Graduate study and research in physical education.* Champaign, IL: Human Kinetics.

Margaria, R., Edwards, H.T., & Dill, D.B. (1933). The possible mechanisms of contracting and paying the oxygen debt and the role of lactic acid in muscular contraction. *American Journal of Physiology, 106,* 689–715.

Morris, J.N., Heady, J.A., Raffle, P.A.B., Roberts, C.G., & Parks, J.W. (1953). Coronary heart disease and physical activity of work. *Lancet, 2,* 1111–1120.

Morrissey, M.C., Harman, E.A., & Johnson, M.J. (1995). Resistance training modes: Specificity and effectiveness. *Medicine and Science in Sports and Exercise, 27,* 648–660.

NIH Consensus Development Panel on Physical Activity and Cardiovascular Health. (1996). Physical activity and cardiovascular health. *Journal of the American Medical Association, 276,* 241–246.

Ogden, C.L., Carroll, M. D., Kit, B.K., & Flegal, K. M. (2012a). Prevalence of obesity in the United States, 2009-2010. NCHS data brief, no. 82. Hyattsville, MD: National Center for Health Statistics.

Ogden, C.L., Carroll, M. D., Kit, B.K., & Flegal, K. M. (2012b). Prevalence of obesity and trends in body mass index among US children and adolescents, 1999-2010. *Journal of the American Medical Association, 307*(5), 483–490.

Paffenbarger, R.S. (1994). 40 years of progress: Physical activity, health and fitness. In *40th anniversary lectures* (pp. 93–109). Indianapolis: American College of Sports Medicine.

Pate, R.R., Pratt, M., Blair, S.N., Haskell, W.L., Macera, C.A., Bouchard, C., Buchner, D., Ettinger, W., Heath, G.W., King, A.C., Kriska, A., Leon, A.S., Marcus, B.H., Morris, J., Paffenbarger, R.S., Patrick, K., Pollock, M.L., Rippe, J.M., & Wilmore, J.H. (1995). Physical activity and public health: A recommendation from the Centers for Disease Control and the American College of Sports Medicine. *Journal of the American Medical Association, 273,* 402–407.

Pavlou, K.N., Krey, S., & Steffee, W.P. (1989). Exercise as an adjunct to weight loss and maintenance in moderately obese subjects. *American Journal of Clinical Nutrition, 49,* 1115–1123.

Physical Activity Guidelines Advisory Committee. (2008). *Physical Activity Guidelines Advisory Committee report, 2008.* Washington, DC: U.S. Department of Health and Human Services.

Rowland, T.W. (1989). Oxygen uptake and endurance fitness in children: A developmental perspective. *Pediatric Exercise Science, 1,* 313–328.

Saltin, B. (1973). Metabolic fundamentals in exercise. *Medicine and Science in Sports, 5,* 137–146.

Sinha, R., Fisch, G., Teague, B., Tamborlane, W.V., Banyas, B., Allen, K., Savoye, M., Rieger, V., Taksali, S., Barbetta, G., Sherwin, R.S., & Caprio, S. (2002). Prevalence of impaired glucose tolerance among children and adolescents with marked obesity. *New England Journal of Medicine, 345,* 802–810.

Timson, B.F., Bowlin, B.K., Dudenhoeffer, G.A., & George, J.B. (1985). Fiber number, area, and composition of mouse soleus muscle following enlargement. *Journal of Applied Physiology, 58,* 619–624.

U.S. Centers for Disease Control and Prevention. (2007). Prevalence of regular physical activity among adults—United States, 2001 and 2005. *CDC Morbidity and Mortality Weekly Report, 56,* 1209–1212.

U.S. Centers for Disease Control and Prevention. (2008). US physical activity statistics. 1988-2008 No leisure time physical activity trend chart. www.cdc.gov/nccdphp/dnpa/physical/stats/leisure_time.htm.

U.S. Department of Health and Human Services (USDHHS). (1996). *Physical activity and health: A report of the surgeon general.* Atlanta: U.S. Department of Health and Human Services, Centers for Disease Control and Prevention, National Center for Chronic Disease Prevention and Health Promotion.

Willett, W.C., Manson, J.E., Stampler, M.J., Colditz, G.A., Rosner, B., Speizer, F.E., & Hennekena, C.H. (1995). Weight, weight change, and coronary disease in women. *Journal of the American Medical Association, 273,* 461–465.

Zwiren, L.D. (1989). Anaerobic and aerobic capacities of children. *Pediatric Exercise Science, 1,* 31–44.

Chapter 12

Albright, C.E., & Smith, K. (2006, March). Welding—trade or profession? *Techniques, 81*(3), 38.

American Kinesiology Association. (2003). Recommended core areas for coursework in a undergraduate kinesiology curriculum. www.americankinesiology.org/white-papers/white-papers/re-examining-the-undergraduate-core-in-kinesiology-in-a-time-of-change.

Feignbaum, E. (2012). Professionalism and workplace savvy. *The Houston Chronicle.* Chron.com. http://smallbusiness.chron.com/professionalism-workplace-savvy-2909.html.

Friedson, E. (2001). *Professionalism: The third logic* (p. 12). Chicago: University of Chicago Press.

Garman, A.N., Evans, R., & Krause, M.K. (2006). Professionalism. *Journal of Healthcare Management, 51*(4), 219.

Greenwood, E. (1957). Attributes of a profession. *Social Work, 2,* 45–55.

Hoffman, S.J. (2011). Examining the big picture. In American Kinesiology Association (Ed.), *Careers in sport, fitness, and exercise* (pp. 12–13). Champaign, IL: Human Kinetics.

Jackson, J.A. (1970). *Professions and professionalization.* London: Cambridge University Press.

Kuh, G.D. (2003). Learning about engagement from NSSE. *Change, 35*(2), 24–32.

Lawson, H.A. (1984). *Invitation to physical education.* Champaign, IL: Human Kinetics.

Lawson, H.A. (1998a, July). Globalization and the social responsibilities of citizen-professionals. Address to AIESEP International Conference, Adelphi University, p. 7.

Lawson, H.A. (1998b). Here today, gone tomorrow: A framework for analyzing the invention, development, transformation and disappearance of helping fields. *Quest, 50,* 225–237.

Malek, M.H., Nalbone, D.P., Berger, D.E., & Coburn, J.W. (2002), Importance of health science education for personal fitness training. *Journal of Strength & Conditioning Research, 16*(1), 19–24.

McGuigan, P.J. (2007, February). The ongoing quest for professionalism: Consumer decision processes for professional services. *Casualty and Property Insurance Underwriters Journal, 60*(2), 1–8.

Miller, K E. (2008). Energy drinks, race, and problem behaviors among college students. *Journal of Adolescent Health, 43,* 490–497.

National Strength and Conditioning Association. (2008, January 4). Code of ethics. www.nsca-lift.org/publications/CodeofEthics.pdf.

Newell, K.M. (1990). Physical activity, knowledge types, and degree programs. *Quest, 42,* 243–268.

Pettit, M.L. & Debarr, K.A. (2011). Perceived stress, energy drink consumption, and academic performance among college students. *Journal of American College Health, 59*(5), 335–341.

Satell, G. (2010). What it means to be a professional. *Digital Toronto.* www.digitaltonto.com/2010/what-it-means-to-be-a-professional/.

Schon, D.A. (1995, November/December). Knowing in action: The new scholarship requires a new epistemology. *Change,* 27–34.

The American freshman: national norms fall 2011. (2011). Cooperative Institutional Research Program and the Higher Education Research Institute. http://heri.ucla.edu/PDFs/pubs/TFS/Norms/Monographs/TheAmericanFreshman2011.pdf

Chapter 13

Anderson, D., Whitmer, R., Goetzel, R., Ozminkowski, R., Wasserman, J., & Serxner, S. (2000). Relationship between modifiable health risks and health care expenditures: A group level analysis of the HERO research database. *American Journal of Health Promotion, 15*(1), 45–52.

Bulmer, S., Howell, J., Ackerman, L., & Fedric, R. (2012). Women's perceived benefits of exercise during and after breast cancer treatment. *Women and Health, 52*(8), 771-787.

Bungum, T.J., Satterwhite, M., Jackson, A.W., & Morrow, J.R., Jr. (2003). The relationship of body mass index, medical costs, and job absenteeism. *American Journal of Health Behavior, 27,* 456–462.

Burd, S. (2009). How Safeway is cutting health-care costs. *Wall Street Journal.* http://online.wsj.com/article/SB124476804026308603.html.

DeNavas-Walt, C., Proctor, B.D., & Smith, J.C. (2012). *Income, poverty and health insurance coverage in the United States: 2011: Current population reports.* Washington, DC: U.S. Census Bureau.

Dintiman, G.B., & Greenberg, J.S. (1986). *Health through discovery.* New York: Random House.

Dychtwald, K. (1990). *Age wave.* New York: Bantam Books.

Goetzel, R.Z., Roemer, E.C., Liss-Levinson, R.C., & Samoly, D.K. (2008). *Workplace health promotion: Policy recommendations that encourage employers to support health improvement programs for their workers.* Washington, DC: Partnership for Prevention.

Goetzel, R., Anderson, D., Whitmer, R., et al. (1998). The association between 10 modifiable risk factors and health care expenditures. *Journal of Occupational and Environmental Medicine, 40*(10), 1–12.

Howden, L., & Meyer, J. (2011). Age and sex composition: 2010 census briefs. www.census.gov/prod/cen2010/briefs/c2010br-03.pdf.

Humes, K., Jones, N., & Ramirez, R. (2011). Overview of race and Hispanic origin: 2010 census briefs. www.census.gov/prod/cen2010/briefs/c2010br-02.pdf.

International Health, Racquet & Sportsclub Association. (2011). *IHRSA's profiles of success.* Boston: International Health, Racquet & Sportsclub Association.

Kaiser Family Foundation. (2012). *Health care costs: A primer. Key information on health care costs and their impact.* Menlo Park CA: Kaiser Family Foundation.

Kaiser Family Foundation and Health Research & Educational Trust. (2012). *Employer health benefits: 2012 summary of findings.* Menlo Park, CA: Kaiser Family Foundation.

Linnan, L., Bowling, M., Childress, J., Lindsay, G., Blakey, C., Pronk, S., Wieker, S., & Royall, P. (2008). Results of the 2004 national worksite health promotion survey. *American Journal of Public Health, 98*(8), 1503–1509.

O'Donnell, M. (1989). Definition of health promotion. Part III: Expanding the definition. *American Journal of Health Promotion, 3*(3), 5.

O'Donnell, M.P., & Harris, J.S. (1994). *Health promotion in the workplace.* Albany, NY: Delmar.

Prochaska, J.O., & DiClemente, C.C. (1986). Toward a comprehensive model of change. In W. Miller & N. Heather (Eds.), *Treating addictive behaviors* (pp. 3–27). New York: Plenum Press.

Tu, H.T., & Mayrell, R.C. (2010). *Research brief: Employer wellness initiative grow, but effectiveness varies widely.* Washington, DC: National Institute for Health Care Reform.

U.S. Department of Health and Human Services (USDHHS). (1996). *Physical activity and health: A report of the surgeon general.* Atlanta, GA: U.S. Department of Health and Human Services, Centers for Disease Control and Prevention, National Center for Chronic Disease Prevention and Health Promotion.

U.S. Department of Health and Human Services (USDHHS). (2008). *Physical activity guidelines for Americans.* Atlanta, GA: U.S. Department of Health and Human Services. www.health.gov/PAGuidelines/Report/A_Summary.aspx#_Toc199951149.

U.S. Department of Health and Human Services (USDHHS). (2010). *Healthy People 2020.* Atlanta, GA: U.S. Department of Health and Human Services. www.healthypeople.gov/2020/topicsobjectives2020/pdfs/PhysicalActivity.pdf.

Chapter 14

Ambrosi, E., & Barker-Schwartz, K. (1995). The profession's image, 1917–1925, Part I: Occupational therapy as represented in the media. *American Journal of Occupational Therapy, 49*(7), 715–719.

Board of Certification (BOC). (2012). *State regulatory agencies.* www.bocatc.org/index.php?option=com_content&view=article&id=92&Itemid=144.

Conn, V.S., Minor, M.A., & Burks, K.J. (2003). Sedentary older women's limited experience with exercise. *Journal of Community Health Nursing, 20*(4), 197–208.

Fernandez, W.G., Yard, E.E., & Comstock, R.D. (2007). Epidemiology of lower extremity injuries among U.S. high school athletes. *Academic Emergency Medicine, 14*(7), 641–645.

Lowe, R. (2011, November). Athletic trainers salaries on the rise according to latest survey. *NATA News,* pp. 12–14.

Olivares, P.R., Gusi, N., Prieto, J., & Hernandez-Mocholi, M.A. (2011). Fitness and health-related quality of life dimensions in community-dwelling middle aged and older adults. *Health and Quality of Life Outcomes, 9,* 117.

Prentice, W.E. (2011). *Arnheim's principles of athletic training* (14th ed.). New York: McGraw Hill Higher Education.

Sheikh, N., & Sharma, S. (2011). Overview of sudden cardiac death in young athletes. *Physician and Sportsmedicine, 39*(4), 22–36.

Sims, J., Hill, K., Davidson, S., Gunn, J., & Huang, N. (2006). Exploring the feasibility of a community-based strength training program for older people with depressive symptoms and its impact on depressive symptoms. *BMC Geriatrics, 30*(6), 18.

U.S. Bureau of Labor Statistics. (2009). Occupational outlook handbook. Athletic trainers. www.bls.gov/oco/ocos294.htm.

U.S. Bureau of Labor Statistics. (2011a). Occupational outlook handbook. Cardiovascular technologists and technicians. www.bls.gov/oco/ocos100.htm.

U.S. Bureau of Labor Statistics. (2011b). Occupational outlook handbook. Occupational therapists. www.bls.gov/oco/ocos078.htm.

U.S. Bureau of Labor Statistics. (2011c). Occupational outlook handbook. Physical therapists. www.bls.gov/oco/ocos080.htm.

U.S. Bureau of Labor Statistics. (2011d). Occupational outlook handbook. Physical therapist assistants and aides. www.bls.gov/ooh/Healthcare/Physical-therapist-assistants-and-aides.htm.

U.S. Bureau of Labor Statistics. (2011e). Occupational outlook handbook. Recreational therapists. www.bls.gov/oco/ocos082.htm.

Ussher, M.H., Taylor, A., & Faulkner, G. (2012). Exercise interventions for smoking cessation. *Cochrane Database of Systematic Reviews,* (1), CD02295.

Wardlaw, F.B., McGuire, F.A., & Overby, Z. (2000). Therapeutic recreation: Optimal health treatment for orthopaedic disability. *Orthopedic Nursing, 19,* 56–60.

World Health Organization. (2002). Towards a common language for functioning, disability, and health. ICF. Geneva, Switzerland. www.who.int/classifications/icf/training/icfbeginnersguide.pdf.

Chapter 15

Anderson, W.G., & Barrette, G.T. (1978). *What's going on in gym: Descriptive studies of physical education classes* [Monograph #1]. Newtown, CT: Motor Skills.

Berliner, D.C. (1988). *The development of expertise in pedagogy.* Washington, DC: American Association of Colleges for Teacher Education.

Brophy, J.E., & Good, T.L. (1986). Teacher behavior and student achievement. In M.C. Wittrock (Ed.), *Handbook of research on teaching* (3rd ed., pp. 328–375). New York: Macmillan.

Castelli, D.M., & Beighle, A. (2007). Rejuvenating the school environment to include physical activity. *Journal of Physical Education, Recreation, and Dance, 78*(5), 25–28.

Cawley, J., Meyerhoefer, C., & Newhouse, D. (2006, Fall). Not your father's PE: Obesity, exercise, and the role of schools. *Education Next, 4,* 61–66.

Corbin, C.B., Lindsey, R., Welk, G.J., & Corbin, W. R. (2001). *Fundamental concepts of fitness and wellness.* New York: McGraw Hill.

DeMarco, G.M. (1999). Physical education teachers of the year: Who they are, what they think, say, and do. *Teaching Elementary Physical Education, 10*(2), 11–13.

Dodds, P. (1994). Cognitive and behavioral components of expertise in teaching physical education. *Quest, 46,* 153–163.

Dyson, B.P. (1995). Students' voices in two alternative elementary physical education programs. *Journal of Teaching in Physical Education, 14,* 394–407.

Ennis, C. (1994). Urban secondary teachers' value orientations: Social goals for teaching. *Teaching and Teacher Education, 10,* 109–120.

Fink, J., & Siedentop, D. (1989). The development of routines, rules, and expectations at the start of the school year. *Journal of Teaching in Physical Education, 8,* 198–212.

Graber, K.C. (2001). Research on teaching in physical education. In V. Richardson (Ed.), *Handbook of research on teaching* (4th ed., pp. 491–519). Washington, DC: American Educational Research Association.

Graber, K.C., Templin, T.J., Haag, R., & O'Connor, J. (2011). Careers in physical education, sport instruction, coaching, sports officiating, and sport psychology. In American Kinesiology Association (Ed.), *Careers in sport, fitness, and exercise* (pp. 51–74). Champaign, IL: Human Kinetics.

Graber, K.C., & Woods, A.M. (2007). Stepping up to the plate: Physical educators as advocates for wellness policies—Part 2. *Journal of Physical Education, Recreation and Dance 78*(6), 19–28.

Graham, G. (2001). *Teaching children physical education: Becoming a master teacher.* Champaign, IL: Human Kinetics.

Griffin, L., Mitchell, S., & Oslin, J. (1997). *Teaching sports concepts and skills: A tactical games approach.* Champaign, IL: Human Kinetics.

Griffin, P.S. (1984). Girls' participation patterns in a middle school team sports unit. *Journal of Teaching in Physical Education, 4,* 30–38.

Griffin, P.S. (1986). Analysis and discussion: What have we learned? *Journal of Physical Education, Recreation and Dance, 57*(4), 57–59.

Hellison, D. (1995). *Teaching responsibility through physical activity.* Champaign, IL: Human Kinetics.

Lanier, J.E., & Little, J.W. (1986). Research on teacher education. In M.C. Wittrock (Ed.), *Handbook of research on teaching* (3rd ed., pp. 527–569). New York: Macmillan.

Locke, L.F. (1975). *The ecology of the gymnasium: What the tourists never see.* Amherst, MA: University of Massachusetts. (ERIC Document Reproduction No. ED 104-823)

Locke, L.F., & Massengale, J.D. (1978). Role conflict in teacher/coaches. *Research Quarterly for Exercise and Sport, 49,* 162–174.

Lortie, D. (1975). *Schoolteacher: A sociological study.* Chicago: University of Chicago Press.

Metzler, M. (1979). *The measurement of academic learning time in physical education.* Unpublished doctoral dissertation, The Ohio State University, Columbus.

Metzler, M. (1989). A review of research on time in sport pedagogy. *Journal of Teaching in Physical Education, 8,* 87–103.

National Association for Sport and Physical Education. (2004). *Moving into the future: National standards for physical education* (2nd ed.). Reston, VA: Author.

National Association for Sport and Physical Education. (2008). National standards for initial physical education teacher education. www.aahperd.org/naspe/grants/accreditation/upload/2008-National-Initial-Physical-Education-Teacher-Education-Standards-Edited-1-5-12.pdf.

National Association for Sport and Physical Education. (2010). *PE metrics: Assessing national standards 1-6 in elementary school.* Reston, VA: Author.

National Association for Sport and Physical Education. (2011). *PE Metrics: Assessing national standards 1-6 in secondary school.* Reston, VA: Author.

National Association for Sport and Physical Education & American Heart Association. (2010). *2010 Shape of the nation report: Status of physical education in the USA.* Reston, VA: National Association for Sport and Physical Education.

Ogden, C., Carroll, M., & Flegel, K. (2008). High body mass index for age among U.S. children and adolescents, 2003-2006. *Journal of the American Medical Association, 299*(20), 2401–2405.

O'Sullivan, M. (Ed.). (1994). High school physical education teachers: Their world of work [Monograph]. *Journal of Teaching in Physical Education, 13,* 323–441.

Rink, J.E. (2006). *Teaching physical education for learning* (5th ed.). Boston: McGraw Hill.

Schempp, P.G. (1989). Apprenticeship of observation and the development of physical education teachers. In T.J. Templin & P.G. Schempp (Eds.), *Socialization into physical education: Learning to teach* (pp. 13–38). Indianapolis: Benchmark.

Siedentop, D. (2004). *Sport education* (2nd ed.). Champaign, IL: Human Kinetics.

Siedentop, D., & Tannehill, D. (2000). *Developing teaching skills in physical education* (4th ed.). Mountain View, CA: Mayfield.

Stroot, S. (1996). Organizational socialization: Factors impacting beginning teachers. In S. Silverman & C. Ennis (Eds.), *Studying learning in physical education: Applying research to enhance instruction* (pp. 339–366). Champaign, IL: Human Kinetics.

Stroot, S.A., & Oslin, J.L. (1993). Use of instructional statements by preservice teachers for overhand throwing performance of children. *Journal of Teaching in Physical Education, 13,* 24–45.

Tyson, L. (1996). Context of schools. In S. Silverman & C. Ennis (Eds.), *Studying learning in physical education: Applying research to enhance instruction* (pp. 55–80). Champaign, IL: Human Kinetics.

Woods, A.M., & Graber, K.C. (Eds.). (2007). Stepping up to the plate: Physical educators as advocates for wellness policies—Part I [Special Feature]. *Journal of Physical Education, Recreation & Dance, 78*(5), 17–28.

Chapter 16

Berkowitz, S., & Upton, J. (2012, January 17). Salaries rising for new college football coaches. *USA Today,* 1A.

Hellison, D., Cutforth, N., Kallusky, J., Martinek, T., Parker, M., & Stiehl, J. (2000). *Serving underserved youth through physical activity: Toward a model of university community collaboration.* Champaign, IL: Human Kinetics.

National Collegiate Athletic Association. (2011). Estimated probability of competing in athletics beyond high school interscholastic level. www.NCAA.org.

National Federation of State High School Associations. (2011). Participation data. www.nfhs.org/.

National Soccer Coaches Association of America. (2012). Coaching Courses. www.nscaa.com/education/courses

U.S. Department of Labor. (2012). Occupational outlook handbook. Coaches and scouts. www.bls.gov/ooh/entertainment-and-sports/coaches-and-scouts.htm.

US News. Sports coach: Job profile and salary. (2012). http://money.usnews.com/careers/best-jobs/sports-coach.

Chapter 17

Babiak, K., & Wolfe, R. (2009). Determinants of corporate social responsibility in professional sport: Internal and external factors. *Journal of Sport Management, 23*(6), 717–742.

Bradish C.L., & Cronin, J.J. (2009). Corporate social responsibility in sport. *Journal of Sport Management, 23*(6), 691–697.

Canadian Sport Tourism Alliance (CSTA). (n.d.). About CSTA. http://canadiansporttourism.com/about-csta/about-csta.html

Chase, C. (2011, February 7). Cowboys stadium misses Super Bowl attendance record by 766. http://sports.yahoo.com/nfl/blog/shutdown_corner/post/Cowboys-Stadium-misses-Super-Bowl-attendance-rec?urn=nfl-317881.

Chelladurai, P. (2006). *Human resource management in sport and recreation* (2nd ed.). Champaign, IL: Human Kinetics.

Commission on Sport Management Accreditation (COSMA). (2010). Accreditation principles & self study preparation. http://cosmaweb.xykon-llc.com/manuals.

Commission on Sport Management Accreditation (COSMA). (n.d.). About COSMA. http://cosmaweb.xykon-llc.com/node/54.

Deitsch, R. (2011, December 15). NFL, networks win in extended rights deal. http://sportsillustrated.cnn.com/2011/writers/richard_deitsch/12/15/nfl.rights.deal/index.html.

Fielding, L.W., & Miller, L.K. (1996). Historical eras in sport marketing. In B.G. Pitts & D.K. Stotlar (Eds.), *Fundamentals of sport marketing.* Morgantown, WV: Fitness Information Technology.

Gibson, H., & Fairley, S. (2011). Sport tourism. In P.M. Pedersen, J.B. Parks, J. Quarterman, & L. Thibault (Eds.), *Contemporary sport management* (4th ed., pp. 226–246). Champaign, IL: Human Kinetics.

Godfrey, P.C. (2009). Corporate social responsibility in sport: An overview and key issues. *Journal of Sport Management, 23*(6), 698–716.

Gullion, L. (1998). Recreational sport. In L.P. Masteralexis, C.A. Barr, & M.A. Hums (Eds.), *Principles and practice of sport management* (pp. 452–475). Gaithersburg, MD: Aspen.

Howard, D.R., & Crompton, J.L. (2005). *Financing sport* (2nd ed.). Morgantown, WV: Fitness Information Technology.

International Health, Racquet and Sportsclub Association. (2007). Health club membership by club type. http://cms.ihrsa.org/index.cfm?fuseaction=page.viewPage.cfm&pageId=18890

International Health, Racquet and Sportsclub Association (IHRSA). (2011). IHRSA's global report on the state of the health club. www.ihrsa.org/industry-research/.

Irwin, R.L., Sutton, W.A., & McCarthy, L.M. (2008). *Sport promotion and sales management* (2nd ed.). Champaign, IL: Human Kinetics.

LaVetter, D., & Choi, Y. S. (2010). Implications of toppling goal posts in college football: Managing institutional risk. *Journal of Sport Administration & Supervision, 2*(1), 52–62.

Li, M., Hofacre, S., & Mahony, D. (2001). *Economics of sport.* Morgantown, WV: Fitness Information Technology.

Manfred, T. (2012, February 16). Egyptian soccer will now be played in empty stadiums after a riot killed 74 people two weeks ago. *Business Insider.* http://articles.businessinsider.

com/2012-02-16/sports/31065895_1_stadiums-egyptians-soccer.

Meek, A. (1997). An estimate of the size and supported economic activity of the sports industry in the United States. *Sport Marketing Quarterly, 6*(4), 15–22.

Milano, M., & Chelladurai, P. (2011). Gross domestic sport product: The size of the sport industry in the United States. *Journal of Sport Management, 25,* 24–35.

Miller, L.K. (1997). *Sport business management.* Gaithersburg, MD: Aspen.

Misener, L. & Mason, D. S. (2009). Fostering community development through sporting events strategies: An examination of urban regime perceptions. *Journal of Sport Management, 23*(6), 770–794.

Mullin, B.J., Hardy, S., & Sutton, W.A. (2007). *Sport marketing* (3rd ed.). Champaign, IL: Human Kinetics.

Muret, D. (2011, January 24). Super Bowl challenges include complicated crowd control. *Street & Smith's SportsBusiness Journal, 13* (38), 12.

National Association of Sports Commissions (NASC). (n.d.). About NASC. www.sportscommissions.org/About-NASC.

National Collegiate Athletic Association (NCAA). (2011). Legislative services database—LSDBi, major infraction case search. https://web1.ncaa.org/LSDBi/exec/miSearch.

Nishimura, S. (2011, November 6). North Texas Super Bowl Committee expects $7.15 million surplus. *Star-Telegram.* www.star-telegram.com/2011/11/16/3532074/north-texas-super-bowl-host-committee.html.

North American Society for Sport Management (NASSM). (n.d.). Sport management programs. www.nassm.com/InfoAbout/SportMgmtPrograms.

Pitts, B.G., Fielding, L.W., & Miller, L.K. (1994). Industry segmentation theory and the sport industry: Developing a sport industry segment model. *Sport Marketing, 3*(4), 15–28.

Sharp, L.A., Moorman, A.M., & Claussen, C.L. (2010). *Sport law: A managerial approach* (2nd ed.). Scottsdale, AZ: Holcomb Hathaway.

Sporting Goods Manufacturing Association (SGMA). (2011). Sports, fitness & recreational activities: Topline participation report 2011. www.aahperd.org/naspe/advocacy/governmentRelations/upload/2011-SGMA-Participation-Topline-Report.pdf.

Stoldt, G.C., Dittmore, S.W., & Branvold, S.E. (2006). *Sport public relations.* Champaign, IL: Human Kinetics.

Stoldt, G.C., Dittmore, S.W., & Pedersen, P.M. (2011). Communication in the sport industry. In P.M. Pedersen, J.B. Parks, J. Quarterman, & L. Thibault (Eds.), *Contemporary sport management* (4th ed., pp. 270–289). Champaign, IL: Human Kinetics.

Super Bowl XLV volunteers, by the numbers. (2011, January 26). http://blogs.star-telegram.com/cowboys_stadium/2011/01/super-bowl-xlv-volunteers-by-the-numbers.html.

The Consilience Group. (2007). *Physical activity trends: Business and policy implications.* Bethesda, MD: Author.

The Y. (2011). Facts and figures. http://ymca.net/organizational-profile/.

University of Florida. (2011). Undergraduate catalog 2011-12. https://catalog.ufl.edu/ugrad/current/courses/descriptions/tourism-recreation-and-sport-management.aspx.

University of Minnesota. (2011). Undergraduate catalog 2010-2012. www.catalogs.umn.edu/ug/.

U.S. Department of Labor, Bureau of Labor Statistics. (2010). May 2010 national industry-specific occupational employment and wage estimates (NAICS 711200—spectator sports). www.bls.gov/oes/current/naics4_711200.htm#00-0000.

Walker, M., & Kent, A. (2009). Do fans care? Assessing the influence of corporate social responsibility on consumer attitudes in the sport industry. *Journal of Sport Management, 23*(6), 743–769.

What I like: Sarah Hirshland, senior vice president, Wasserman Media Group. (2010, November 15). *Street & Smith's SportsBusiness Journal.* www.sportsbusinessdaily.com/Journal/Issues/2010/11/20101115/What-I-Like/Sarah-Hirshland-Senior-Vice-President-Wasserman-Media-Group.aspx?hl=trends%20future&sc=0.

What I like: Charlie Besser, president & CEO, Intersport. (2011a, May 2). *Street & Smith's SportsBusiness Journal.* www.sportsbusinessdaily.com/Journal/Issues/2011/05/02/People-and-Pop-Culture/What-I-Like.aspx.

What I like: Mark King, CEO and president, TaylorMade-Adidas Golf. (2011b, August 15). *Street & Smith's SportsBusiness Journal, 14*(6), 38.

What I like: Andrew Walker, WTA chief marketing officer. (2011c, August 29). *Street & Smith's SportsBusiness Journal, 14*(8), 30.

What I like: Dave Butler, CEO, Paciolan. (2011d, September 5). *Street & Smith's SportsBusiness Journal, 14*(9), 38.

What I like: Conrad Smith, COO, Minnesota Lynx. (2011e, September 12). *Street & Smith's SportsBusiness Journal, 14*(20), 30.

What I like: Tony Schiller, EVP and Partner, Paragon Marketing Group. (2011f, November 28). *Street & Smith's SportsBusiness Journal, 14*(31), 54.

Wichita State University. (2011). Undergraduate course catalog 2010-2011. www.collegesource.org/displayinfo/catalink.asp?pid={1127B424-BE20-437A-9F15-956571D0ACCB}&oig={78AC3888-D5B8-4B8D-8F51-BCB5769AFEB3}&vt=5.

index

Note: Page numbers followed by an italicized *f* or *t* refer to the figure or table on that page, respectively.

Shirl J. Hoffman, EdD, is professor emeritus of exercise and sport science at the University of North Carolina at Greensboro, where he served as department head for 10 years. He has served at all levels of education: elementary physical education teacher, college-level coach, and professor in both research university and liberal arts college settings.

Hoffman has published extensively on a variety of topics, including sport philosophy and ethics, religion in sport, qualitative analysis and diagnosis of movement, and motor learning and performance. He also has been a frequent commentator on problems in kinesiology and higher education. He is a former editor of *Quest* and was the charter executive director of the American Kinesiology Association (AKA). He is now editor of *Kinesiology Today*, an online journal published by the AKA.

Hoffman is a fellow emeritus of the National Academy of Kinesiology. He is the author of *Good Game: Christianity and the Culture of Sport* (2010) and was the project coordinator for *Careers in Sport, Fitness, and Exercise* (2011), published by the American Kinesiology Association.

Hoffman and his wife, Claude, reside in Greensboro, North Carolina and Boone, North Carolina. Photo courtesy of Shirl Hoffman.

Jennifer L. Caputo, PhD, is a professor at Middle Tennessee State University and serves as co-coordinator of the undergraduate and graduate exercise science programs. She teaches courses in exercise physiology, health and fitness assessment, and exercise prescription and directs exercise science theses and dissertations. Dr. Caputo is a certified strength and conditioning specialist through the National Strength and Conditioning Association and is a licensed medical bone densitometer operator. Dr. Caputo received her doctoral degree in exercise physiology from the University of North Carolina at Greensboro. Her master's degree is in sport and exercise psychology. Photo courtesy of Jennifer Caputo.

Margaret Carlisle Duncan, PhD, is a professor emerita of the department of kinesiology at the University of Wisconsin at Milwaukee (UWM). Dr. Duncan's ongoing research interest relates to media depictions of women's bodies and body practices. She has also extensively examined female athletes and women's sports in the media. A more recent interest is in the social construction of obesity and the discourses of overweight and obesity. Dr. Duncan is a former president of the North American Society for the Sociology of Sport (NASSS), a former president of The Association for the Study of Play (TASP), and a former editor of the journal *Play & Culture*. She has served on several editorial boards and has reviewed manuscripts for more than 25 scholarly journals. She was a fellow of the National Academy of Kinesiology (NAK), a Wisconsin Teaching Fellow, and a fellow for the interdisciplinary Center for 20th Century Studies (at UWM) on two different occasions. She has published several scholarly articles in sociology, sport sociology, media, and communication studies journals. She was also a coauthor (with Michael A. Messner) of four sport media studies commissioned by LA84, a nonprofit foundation dedicated to serving youth through sport and to increasing knowledge of sport and its impact on people's lives. Photo courtesy of Margaret Duncan.

Kim C. Graber, EdD, is an associate professor and associate head for undergraduate studies in the department of kinesiology and community health at the University of Illinois at Urbana–Champaign. She also is the director of the College of Applied Health Sciences Teaching Academy. She received her bachelor's degree from the University of Iowa, her master's from Teachers College Columbia University, and her doctorate from the University of Massachusetts at Amherst. Her research focuses on teacher education, children's wellness, and the scholarship of teaching and learning. Currently, she is a fellow in and the chair-elect of the AAHPERD Research Consortium. She also has served as president of the National Association for Sport and Physical Education and chair of the Curriculum and Instruction Academy. Dr. Graber is currently serving on the review boards of the *Journal of Teaching in Physical Education* and *Quest*. She has published numerous articles in peer-refereed journals and books and has presented her work at dozens of national and international conferences. She is a University of Illinois Distinguished Teacher/Scholar and received the 2009 Campus Award for Excellence in Undergraduate Teaching. Photo courtesy of Kim Graber.

Janet C. Harris, PhD, is an emeritus professor and the former director of the school of exercise and nutritional sciences (2006-2010) at San Diego State University. Prior to this, she was a professor and the director of the school of kinesiology and nutritional science at California State University at Los Angeles (1998-2006). Dr. Harris was located at the University of North Carolina at Greensboro for a major portion of her career, where she was a professor of exercise and sport science (1980-1998). Her research has focused on symbolic and expressive aspects of sports in American society, with work on topics such as athletic heroes of children and teenagers, boys in youth baseball, marketing slogans of collegiate football programs, and sports and the mass media. Her research is published in journals such as *Sociology of Sport Journal, Research Quarterly for Exercise and Sport, Quest,* and *Exercise and Sport Sciences Reviews.* She was coeditor of the first edition of *Introduction to Kinesiology* (2000) and also coeditor of *Play, Games and Sports in Cultural Contexts* (1983). She has an ongoing interest in the discipline of kinesiology and related physical activity professions and has published several papers on these topics. Dr. Harris has served as president of both the North American Society for the Sociology of Sport and the AAHPERD Research Consortium. She has been the editor of *Quest* and the *International Review for the Sociology of Sport* and the section editor of sociology and cultural anthropology for the *Research Quarterly for Exercise and Sport.* Her recreational interests include working out at the gym, eating out, traveling, attending musical performances, and gardening. Dr. Harris earned her BS and MS degrees at the University of California at Los Angeles and her PhD at the University of California at Berkeley. Photo courtesy of Janet C. Harris.

Jeremy Howell, PhD, is professor in the sport management graduate program and kinesiology department at the University of San Francisco. He writes and lectures on sport, culture, and business. He currently is a member of USF'S President's Commission on Health Professions Education, initiating new programming in graduate level health professions education. Jeremy has extensive community and professional experience and acts in an advisory capacity to both profit and not-for-profit organizations on business operations, product development, employee continuing education, and strategic philanthropy. Jeremy is also the university's NCAA faculty athletic representative and sits on the executive council of the West Coast Conference. From 2010 to 2011, he was part of a three-member interim athletic director transition team that oversaw the daily operations of the athletic department. Jeremy is a recipient of USF St. Ignatius Award, the highest university award given to the faculty person that best personifies the meaning of service to his community, profession, and university. Photo courtesy of Jeremy Howell.

Katherine M. Jamieson, PhD, completed her doctorate in kinesiology at Michigan State University in 1999. She is an associate professor in exercise and sport science and a contributing faculty member in the women's and gender studies program at the University of North Carolina at Greensboro. Her teaching and research interests center on critical cultural analyses of gender as it links with race, sexuality, and social class both in and beyond U.S. geographies and sociopolitical conditions. Photo courtesy of Katherine Jamieson.

Scott Kretchmar, PhD, is a professor of exercise and sport science at Penn State University. He is a founding member of the International Association for the Philosophy of Sport (IAPS) and has served as its president. He has been editor of the *Journal of the Philosophy of Sport,* is a fellow in the National Academy of Kinesiology, and has authored a popular text in the philosophy of sport. He has been named Alliance Scholar by AAHPERD, Distinguished Scholar by NAKPEHE, the Fraleigh Distinguished Scholar by IAPS (on two occasions), and the McCloy Lecturer by the Alliance Research Consortium. Photo courtesy of Scott Kretchmar.

Joseph A. Luxbacher, PhD, has more than 30 years' experience in the fields of health, fitness, and competitive athletics. He holds a PhD in health, physical, and recreation education and is employed in the department of athletics at the University of Pittsburgh. Dr. Luxbacher has authored more than a dozen books and numerous articles in the areas of sport (soccer), peak athletic performance, fitness, and weight control. He is a former professional soccer player and is the head men's soccer coach at the University of Pittsburgh. Luxbacher has twice been named Big East Athletic Conference Soccer Coach of the Year. He was honored as a Letterman of Distinction by the University of Pittsburgh in 2003 and was inducted into the Western Pennsylvania Sports Hall of Fame in 2005. Dr. Luxbacher, his wife, Gail, and their children, Eliza and Travis, live outside of Pittsburgh. Photo courtesy of Joseph Luxbacher.

Sandra Minor Bulmer, PhD, is a professor in the department of public health at Southern Connecticut State University, where she teaches and conducts community-based research. She was previously the director of fitness operations for Western Athletic Clubs in San Francisco. She has a doctorate in health education from Texas Woman's University and an MS with a specialization in exercise physiology from the University of Oregon. She is a certified health education specialist (CHES), ACSM health/fitness specialist, and ACSM health/fitness director. She is active in several professional associations and has served on the board of trustees for the Society for Public Health Education (SOPHE). Photo courtesy of Isabel Chenoweth/SCSU.

Kathy Simpson, PhD, is an associate professor in the department of kinesiology at the University of Georgia and co-director of the Biomechanics Laboratory. She received a doctorate in biomechanics from the University of Oregon. Her research focuses on determining how people adapt their movements to varying demands (e.g., impact forces on foot prostheses) and how these adaptations affect the functions and structure of the lower extremity. She has applied this research to areas such as the improvement of the sport performance of athletes with lower-extremity amputations and the performance of daily activities by individuals with hip or knee joint replacements or who use a lower-extremity prosthesis. She has served as the biomechanics section editor for *Research Quarterly for Exercise and Sport,* co-chair of the biomechanics interest group in ACSM, member of the executive board of the American Society of Biomechanics (ASB), and chair of the Biomechanics Academy of AAHPERD. Photo courtesy of Kathy Simpson.

Chad Starkey, PhD, AT, FNATA, is a professor and coordinator of the graduate athletic training program at Ohio University in Athens, Ohio. A graduate of West Virginia University, Dr. Starkey received his master's and doctoral degrees from Ohio University. Dr. Starkey has served on the board of directors for the Board of Certification, was the chair of the NATA Education Council, and is currently a commissioner for the Commission on Accreditation of Athletic Training Education. He has authored several textbooks focusing on sports medicine, orthopedic diagnosis, and therapeutic modalities. Photo courtesy of Chad Starkey.

G. Clayton (Clay) Stoldt, EdD, is a professor and chair of the department of sport management at Wichita State University. He teaches classes in sport public relations and sport marketing. Dr. Stoldt is the coauthor of *Sport Public Relations: Managing Stakeholder Communication* (2nd ed.), and his research activities have focused on sport public relations issues such as crisis communications, the roles of sport public relations professionals, and the application of advanced public relations practices in the field. As department chair, he administers two COSMA-accredited programs—a bachelor's program with more than 130 students and a master's program with more than 60 students. Dr. Stoldt received his doctorate of education from the University of Oklahoma in 1998. He has a master's degree in sport management and a bachelor's degree in journalism/mass communication. Prior to coming to Wichita State, he worked in the athletic department at Oklahoma City University, where he served as sports information director, radio play-by-play broadcaster, and development officer. He also served as an adjunct instructor at both Oklahoma City and the University of Oklahoma, teaching courses in sport management and mass communication. Photo courtesy of Clayton Stoldt.

Richard A. Swanson, PhD, is professor emeritus of kinesiology at the University of North Carolina at Greensboro, where he was dean of the School of Health and Human Performance for 12 years and a member of the faculty for 27 years. He previously served in faculty and administrative roles at The Ohio State University, Wayne State University and San Francisco State University. His publications include two books and several book chapters and research articles on the history of physical education, exercise science, and sport in the United States, as well as articles and book chapters on administration in higher education. Dr. Swanson is a charter member of the North American Society for Sport History (NASSH). He is currently pursuing research on the historical relationship between sport and religion in the United States. He served as archivist for the National Association for Kinesiology and Physical Education in Higher Education (NAKPEHE) for 30 years and has served extensively as a member of the governing board since 1973. He was the Dudley Allen Sargent Lecturer for that organization in 1997 and was awarded the Distinguished Scholar Award in 2008 and the Presidential Award in 2011. Dr. Swanson is also the recipient of the Distinguished Service Award (1992) from the College and University Administrators Council of the American Alliance of Health, Physical Education, Recreation and Dance (AAHPERD). He received his BS and MEd degrees from Wayne State University and his PhD from The Ohio State University. He resides with his wife, Lita, in Greensboro, North Carolina, and enjoys spending time with his two adult children and four grandchildren, playing golf, reading, traveling, and volunteering in the community. Photo courtesy of Richard Swanson.

Thomas J. Templin, PhD, is a professor and former head of the department of health and kinesiology at Purdue University. He also holds a courtesy appointment as a professor in the department of curriculum and instruction in the college of education. He received his education at Indiana University (BS, 1972; MS, 1975) and at the University of Michigan (PhD, 1978). He has been a member of the Purdue faculty since 1977 and served as an administrator from 1988 to 2006. Dr. Templin has received various honors through his career. Recently, he received the 2008 Curriculum and Instruction Honor Award from the C&I Academy within the National Association for Sport and Physical Education. In 2006, he was inducted as a fellow into the American Academy of Kinesiology and Physical Education. He was the Doris Drees Distinguished Scholar Lecturer at the University of Dayton and was the 100th Lansdowne Scholar for the faculty of education at the University of Victoria in the spring of 2003. He received the Outstanding Teacher Educator Award from the Indiana Association of Colleges for Teacher Education in 1986.

Dr. Templin has focused his research on the lives and careers of physical education teachers. He is well known as an author and editor and reviewer of numerous journal publications and books in physical education, including *A Reflective Approach to Teaching Physical Education* with Don Hellison and *Socialization into Physical Education: Learning to Teach* with Paul Schempp. Along with Russell Carson and Howard Weiss, Dr. Templin completed research funded by the Spencer Foundation to study emotion and affective events in teaching. This research is ongoing. He has served in leadership roles and has presented numerous national and international papers for various professional organizations including the American Alliance for Health, Physical Education, Recreation and Dance (AAHPERD) and the American Educational Research Association (AERA). Dr. Templin is a former president of the National Association for Sport and Physical Education, which represents 17,000 physical education and sport professionals. An avid golfer, he serves as a consultant with the Professional Golfers' Association of America through its university program, Play Golf America University, and as chair of the association's scholarship committee. Photo courtesy of Thomas Templin.

Jerry R. Thomas, EdD, is dean of the College of Education and professor of kinesiology, health promotion, and recreation at the University of North Texas. Previously he was professor of kinesiology and held administrative positions at Iowa State University, Arizona State University, and Louisiana State University. His research in motor development has focused on cognitive factors, expertise, and gender differences in children's motor skill. His best-selling graduate textbook, *Research Methods in Physical Activity,* is in its 6th edition, is widely used in the United States, and has been translated into seven other languages. He has authored more than 225 research and professional publications, books, and book chapters; presented more than 230 papers to conferences and universities in the United States and numerous other countries; has been visiting scholar twice at institutions in Australia and Korea, as well as in China, England, Thailand, and Hong Kong; served as the C.H. McCloy Lecturer and Alliance Scholar, as well as presenting numerous named lectures; received the AAHPERD Honor Award (1989-1990); and served as editor-in-chief of *Research Quarterly for Exercise and Sport* from 1983 to 1988 and subsequently on the editorial board for 4 years. Dr. Thomas was elected fellow in the National Academy of Kinesiology (formerly the American Academy of Kinesiology and Physical Education) in 1984 and served as president in 1991-1992. He has served as president of the North American Society for Psychology of Sport and Physical Activity and was presented with its Distinguished Scholar Award in 2003 and the President's Award in 2006. He served as president of the Research Consortium of AAHPERD and received its Distinguished Service Award in 2005. He was founding president (2007-2009) of the American Kinesiology Association. Photo courtesy of Jerry Thomas.

Katherine T. Thomas, PhD, is associate professor in the department of kinesiology, health promotion, and recreation at the University of North Texas. She studies factors influencing physical activity, particularly skill, in children. She has been funded to study health and physical activity in women of color and has received more than $1.4 million in external funding for research, service, and teaching. She has published several books, refereed journal articles, and chapters in scholarly books. She was a founding co-chair of one of the state teams for Action for Healthy Kids, co-chaired the committee that wrote the Iowa Association School Board model local school wellness policy, and served as task force co-chair that successfully proposed state legislation to examine physical activity and nutrition in schools. She has been involved on two USDA demonstration projects as social scientist and co-project director. Dr. Thomas remains focused on informing stakeholders and decision makers to influence practice in schools and communities and ensuring that those practices are evidence-based. Photo courtesy of Katherine Thomas.

Cesar R. Torres, PhD, is a professor of kinesiology, sport studies, and physical education at The College at Brockport, State University of New York. He is a past president of the International Association for the Philosophy of Sport (IAPS). He is the associate editor of the *Journal of the Philosophy of Sport* and serves on the editorial board of various academic journals. He has published several books and numerous articles in academic journals, edited collections, and newspapers, in both English and Spanish. In 2007, the International Society for the History of Physical Education and Sport awarded him the Reinhard Sprenger Junior Award. Photo courtesy of Cesar Torres.

Robin S. Vealey, PhD, is a professor in the department of kinesiology and health at Miami University in Ohio. Dr. Vealey's focus area is sport psychology, and she is particularly interested in self-confidence, mental skills training, and coaching effectiveness. She has authored two books, *Coaching for the Inner Edge* and *Competitive Anxiety in Sport.* She has served as a sport psychology consultant for the U.S. Nordic Ski Team, U.S. Field Hockey, elite golfers, and athletes and teams at Miami University. Dr. Vealey is a fellow, certified consultant, and past president of the Association for the Advancement of Applied Sport Psychology and former editor of *The Sport Psychologist.* She is also a fellow of the National Academy of Kinesiology and a member of the United States Olympic Committee Sport Psychology Registry. A former collegiate basketball player and coach, she now enjoys the mental challenge of golf. Photo courtesy of Robin Vealey.

Mark Vermillion, PhD, is an associate professor in the department of sport management at Wichita State University. He teaches courses regarding the sociology of sport, psychology of sport, and ethics in sport. Dr. Vermillion also teaches courses in the sociology department at Wichita State University, which pertain to sport, deviance, and violence. Dr. Vermillion recently coauthored a chapter regarding crime, deviance, and violence in college sports in *Sports in Higher Education: Issues and Controversies in College Athletics.* His research has addressed topics such as participation factors in college athletics, coaching interactions, disabled sport dynamics, and various manifestations of crime or deviance in sports. He received his doctorate in sociology from Oklahoma State University in 2006, where he specialized in both crime and deviance and social psychology. He has a master's degree in sociology from Wichita State University and a bachelor's degree in social science from Kansas State University. In addition to conducting academic research, Dr. Vermillion engages in applied research, such as consulting with NCAA- and NJCAA-affiliated athletic departments regarding survey methods or behavior development plans for student-athletes. He has worked with local sport organizations, including consulting and research services for the Preferred Health Systems Wichita Open annual golf tournament. Photo courtesy of Mark Vermillion.